Alfred J. Meixner, Monika Fleischer, Dieter P. Kern, Evgeniya Sheremet,
Norman McMillan (Eds.)
Optical Nanospectroscopy

Also of Interest

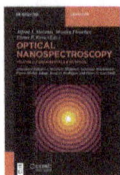

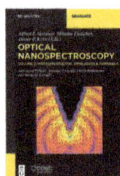

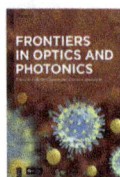

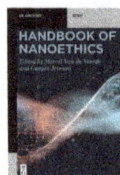

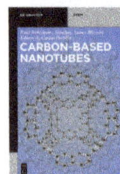

Optical Nanospectroscopy

Applications

Edited by
Alfred J. Meixner, Monika Fleischer, Dieter P. Kern,
Evgeniya Sheremet, Norman McMillan

DE GRUYTER

Editors

Prof. Dr. Alfred J. Meixner
Eberhard Karls University Tübingen
Institute of Physical and Theoretical Chemistry
Center LISA⁺
Auf der Morgenstelle 18 and 15
72076 Tübingen
Germany

Prof. Dr. Monika Fleischer
Eberhard Karls University Tübingen
Institute for Applied Physics and Center LISA⁺
Auf der Morgenstelle 10 and 15
72076 Tübingen
Germany

Prof. Dr. Dieter P. Kern
Eberhard Karls University Tübingen
Institute for Applied Physics and Center LISA⁺
Auf der Morgenstelle 10 and 15
72076 Tübingen
Germany

Prof. Dr. Evgeniya Sheremet
National Research Tomsk Polytechnic University
Research School of Chemistry and Applied
Biomedical Sciences
Lenina ave. 30
643050 Tomsk
Russian Federation

Dr. Norman McMillan
Advanced Nano Technologies Ltd
Unit 2A Nangor Road Business Park
Nangor Road, Park West
Dublin D12 Y771
Ireland

ISBN 978-3-11-044289-2
e-ISBN (PDF) 978-3-11-044290-8
e-ISBN (EPUB) 978-3-11-043498-9

Library of Congress Control Number: 2022944728

Bibliographic information published by the Deutsche Nationalbibliothek
The Deutsche Nationalbibliothek lists this publication in the Deutsche Nationalbibliografie;
detailed bibliographic data are available on the Internet at http://dnb.dnb.de.

© 2023 Walter de Gruyter GmbH, Berlin/Boston
Cover image: Diffraction limited fluorescence excitation pattern of an organic dye molecule fixed in a
thin transparent polymer film. Figure by A. J. Meixner group (2017). See Volume 1, Fig. 1.5.15 for
further details.
Typesetting: VTeX UAB, Lithuania
Printing and binding: CPI books GmbH, Leck

www.degruyter.com

Editorial for the textbook Optical Nanospectroscopy

Welcome to the three-volume textbook "Optical Nanospectroscopy". This book offers a comprehensive overview over the emerging research field of nanospectroscopy, reviewing the theoretical basis of optics and spectroscopy, relevant experimental and numerical techniques, and a snapshot of state-of-the-art applications of nanospectroscopy, as well as novel developments.

Optical spectroscopy is the study of the interaction between electromagnetic radiation and matter, to reveal its optical and electronic properties and chemical structure. Nanospectroscopy can be both the spectroscopy of very small objects down to single molecules or atoms or spectroscopy performed with very high spatial resolution from regions much smaller than the wavelength of the radiation. Nanospectroscopy is a rather young and interdisciplinary field of science that developed over the past 20 years, based on new theoretical insight into the field of light–matter interaction and stimulated by new powerful experimental capabilities leading to seminal breakthroughs in single-molecule spectroscopy, near-field optical microscopy, and nanooptics.

The motivation to advance the field originates from the need to resolve increasingly smaller structures with ever finer details on the nanometer length scale, which has been made possible by the tremendous recent advances in nanoscience and associated technologies. Imaging small complex structures with a spatial resolution of a few nanometers or even better and revealing their chemical structure or their local optical, electronic, and chemical properties is of supreme importance in microelectronics, material science, engineering, nanotechnology, chemistry, life science, and medicine.

This book is based upon work from the COST Action MP1302 Nanospectroscopy, supported by COST (European Cooperation in Science and Technology, www.cost.eu). COST is a funding agency for research and innovation networks. Its Actions help connect research initiatives across Europe and enable scientists to grow their ideas by sharing them with their peers, boosting their research, their careers, and innovation.

The COST Action Nanospectroscopy took place from 2013 to 2017. With the support of COST a network of experts in optical nanospectroscopy from all across Europe and beyond was established. More than 230 principal investigators plus their group members from 35 countries combined forces in discussing the state-of-the-art and working collectively to forge the future of optical nanospectroscopy. The interdisciplinary groups contributed complementary expertise in nanostructure synthesis or nanofabrication, microscopic and spectroscopic techniques, data analysis, and numerical simulations and theoretical modeling, as well as commercial instrumentation and end-user applications of nanospectroscopy. This network therefore offered a unique pool of expertise to draw from for the contents of this book. Many participants, both leaders in the field and early-stage researchers, contributed their specific expert knowledge to the three volumes. At the same time the editorial team took care that the first two

https://doi.org/10.1515/9783110442908-201

volumes follow the coherent outline of a textbook rather than forming a collection of articles. For this purpose, the current state of the field of optical nanospectroscopy was carefully mapped in a meta-review, and a table of contents covering the relevant information was drafted to be filled by the contributors [1].

Since nanospectroscopy is a fast-developing research field with contributions from many interdisciplinary research areas having far-reaching consequences for the future of nanoscience and technology, it is imperative that a new generation of scientists and interested individuals is trained in the topic, taking into account relevant aspects from physics and chemistry as well as its potential impact on biology, medicine, engineering, consumer products, safety and security, or even cultural heritage studies. The three volumes of this book therefore aim at readers with a basic science training, e. g., students of the natural sciences at the Bachelor or Master level, or interested readers that are entering this field and desire or need to learn more about this fascinating topic with its ever-increasing range of applications and fundamental discoveries.

In the first volume the fundamentals of optical nanospectroscopy are treated. The outline starts from the basics of light–matter interaction, continues with an overview of optical spectroscopies in general, and ends with specific techniques in nanospectroscopy that either offer nanoscale resolution or are addressed to nano-objects. In the second volume, theoretical and experimental methods essential for nanospectroscopy are covered. The necessary instrumentation is explained, insights into simulation tools for modeling nanospectroscopic processes are given, and a wide variety of nanomaterials that are typically being investigated by nanospectroscopy are introduced. Finally, in the third volume individual authors highlight current examples how optical nanospectroscopy is and may be exploited in modern photonics, sensing, e. g., in environmental monitoring, the life sciences, medicine, or diverse nuanced applications in material, chemical, and biological sciences.

The making of this book was a major milestone of the COST Action Nanospectroscopy coordinated by the undersigned and would not have been possible without the time, support, and dedication of many contributors. We would like to express our gratitude and thanks to all people involved: The international textbook team Pierre-Michel Adam, Antonio Cricenti, Volker Deckert, Johannes Gierschner, Pietro Gucciardi, Christiane Höppener, Ulrich Hohenester, Mile Ivanda, Florian Kulzer, Teresa I. Madeira, Jana Nieder, Raul Rodriguez, Ludovic Roussille, Manuela Scarselli, Evgeniya Sheremet, Dietrich Zahn, and Dai Zhang, all the authors and contributors of the book, the members of the COST Action Nanospectroscopy, and especially of the Working Group 4, for fruitful discussions, the COST Association for supporting the Action, Gabriele Thomas for editing support, Konrad Kieling for encouragement, as well as the editorial team at De Gruyter, especially Nadine Schedensack, for valuable assistance throughout the whole process of creating this book. We hope you enjoy reading about the art of optical nanospectroscopy.

Bibliography

[1] Perry S, McMillan ND, Rodriguez RD, Mackowski S, Sheremet E, Fleischer M, Kneipp K, Zahn DR, Adam P-M, Madeira TI, Meixner AJ. Novel advanced scoping meta-review methodology for defining a graduate level textbook in an emerging subject area. LIBER Quarterly. 2018;28(1):1–20. https://doi.org/10.18352/lq.10222.

Tübingen, November 2022

Alfred J. Meixner
Monika Fleischer
Dieter P. Kern
Sebastian Mackowski
Katrin Kneipp
Norman McMillan

Funded by the European Union

Editorial for Volume 3: Applications

The third volume "Applications" is the final volume of the three-part book "Optical Nanospectroscopy". While the first two volumes are clearly structured in the style of a textbook and designed to address students and interested individuals entering the field, the third volume has a different character in that it illustrates individual modern applications of nanospectroscopy in a collection of articles. The chapters represent the work and perspectives of the respective authors and are curated to cover a wide range of relevant topics.

Once nanospectroscopic methods have been thoroughly mastered, they can be profitably employed to deliver new insights into fundamental complex physical, biological, or chemical processes and material properties. Likewise they can be exploited to target a cornucopia of everyday questions of high societal relevance. Nanospectroscopy shows exciting promise and vast possibilities in such varied fields as sensing, quantum information, food safety and authentication, cultural heritage, or optoelectronic devices. With spectroscopic technologies increasingly moving from the research stage to the marketplace, impressive progress has been made in application areas from early-stage cancer detection to trace detection in liquids. Moving from conventional optical spectroscopy to nanospectroscopy has a great potential to overcome analytical problems and disadvantages by boosting limits of detection, multiplexing approaches, and spatial resolution to the single-nanometer scale. While many nanospectroscopic techniques are still mostly used by specialists and limited to laboratory use, real-world applications are growing and conquering new realms. The objective of the present volume is thus to highlight such directions and review the state-of-the-art in relevant fields.

The structure of the volume is grouped into different application areas: after a general introduction (Chapter 1), it takes a closer look at nanospectroscopy for new photonic technologies (Part 2), sensing (Part 3), life sciences (Part 4), and analytical and material science (Part 5). For each area examples are selected that show how the field can profit from nanospectroscopy and how existing knowledge, devices, or materials in these areas can be advanced by applying nanospectroscopic techniques. We thus get to travel to such diverse pastures as quantum systems, photovoltaics, radiation detectors, sensing of explosives, cancer cell detection, food science, or dating of historic art. In this fast-moving field, the given insights can only ever offer brief snapshots of the status quo. In a sense, they can be viewed as the tips of an iceberg of applications that are still going to emerge.

Each chapter was written by experts in the respective area. The authors have accepted the challenge to guide readers by not only describing the state-of-the-art, but also offering perspectives and directions for future developments for those entering the field. We are indebted to the contributors for their dedication, knowledge, and time. The contributors originate from countries all across Europe and beyond and range from early-stage researchers to senior experts. Sincere thanks go to Christophe

https://doi.org/10.1515/9783110442908-202

Couteau, Roy Aad, Suzanna Akil, Safi Jradi, Irene Izquierdo (France), Andreas Kaltzoglou, Athanassios G. Kontos, Polycarpos Falaras, Stavros Pissadakis (Greece), Rosa Maria Montereali, Francesca Bonfigli, Enrico Nichelatti, Massimo Piccinini, Maria Aurora Vincenti, Antonino Foti, Barbara Fazio, Cristiano D'Andrea, Maria Grazia Donato, Valentina Villari, Onofrio M. Maragò, Pietro G. Gucciardi, Antonio Cricenti (Italy), Gamze Yesilay, Ertug Avci, Mine Altunbek, Sevda Mert, Mustafa Culha (Turkey), Florian Laible, Anke Horneber, Janina Kneipp, Daniela Drescher (Germany), Douglas McMillan, Mark Heaton, Victor Hrymak, Simon Perry (Ireland), Enisa Omanović-Mikličanin, Amina Stambolić (Bosnia-Herzegovina), Vesna Vasić, Bojana Laban, Dragana Vasić-Anićijević, Vesna Vodnik (Serbia), Elena Shabunya-Klyachkovskaya (Belarus), Raul D. Rodriguez, Tuan-Hoang Tran, Dmitry Cheshev, Andrey Averkiev (Russia), Ramzi Maalej, Sameh Kessentini (Tunisia), and Gagik Shmavonyan (Armenia). Simon Perry (Ireland) is additionally gratefully acknowledged for support in database research.

This volume can be seen as one of many possible compilations of exciting emerging applications of optical nanospectroscopy. We are looking forward to the compilations that will be enabled by the ongoing development of the coming decades.

November 2022
Volume 3 Editors

Evgeniya Sheremet
Norman McMillan
Monika Fleischer

Contents

Part 3: **Sensing**

List of Contributing Authors

Roy Aad
Laboratory Light, nanomaterials &
nanotechnologies – L2n
University of Technology of Troyes &
CNRS EMR 7004
Troyes
France
American International University
Saad Al Abdullah
Kuwait

Suzanna Akil
Jean Barriol Institute, LCP-A2MC Laboratory
University of Lorraine
Metz
France

Mine Altunbek
Genetics and Bioengineering Department,
Faculty of Engineering
Yeditepe University
Istanbul
Turkey

Ertug Avci
Genetics and Bioengineering Department,
Faculty of Engineering
Yeditepe University
Istanbul
Turkey

Andrey Averkiev
Research School of Chemistry and Applied
Biomedical Sciences
National Research Tomsk Polytechnic University
Tomsk
Russia

Francesca Bonfigli
Department of Fusion and Technologies for
Nuclear Safety and Security,
Photonics Micro- and Nanostructures
Laboratory, FSN-TECFIS-MNF
ENEA C.R. Frascati
Rome
Italy

Dmitry Cheshev
Research School of Chemistry and Applied
Biomedical Sciences
National Research Tomsk Polytechnic University
Tomsk
Russia

Christophe Couteau
Laboratory Light, nanomaterials &
nanotechnologies – L2n
University of Technology of Troyes &
CNRS EMR 7004
Troyes
France

Mustafa Culha
Genetics and Bioengineering Department,
Faculty of Engineering
Yeditepe University
Istanbul
Turkey
Sabanci University Nanotechnology Research
and Application Center (SUNUM)
Istanbul
Turkey
Department of Internal Medicine,
Morsani College of Medicine
The University of South Florida
Tampa
USA

Cristiano D'Andrea
CNR – IPCF
Istituto per i Processi Chimico Fisici
Messina
Italy
BioPhotonics and Nanomedicine Lab (BPNLab)
CNR – Istituto di Fisica Applicata "Nello Carrara"
(IFAC)
Florence
Italy

Maria Grazia Donato
CNR – IPCF
Istituto per i Processi Chimico Fisici
Messina
Italy

Daniela Drescher
Department of Chemistry
Humboldt-Universität zu Berlin
Berlin
Germany

Polycarpos Falaras
National Centre for Scientific Research
"Demokritos"
Institute of Nanoscience and Nanotechnology
Agia Paraskevi
Greece

Barbara Fazio
CNR – IPCF
Istituto per i Processi Chimico Fisici
Messina
Italy

Monika Fleischer
Institute for Applied Physics and Center LISA+
Eberhard Karls University Tübingen
Tübingen
Germany

Antonino Foti
CNR – IPCF
Istituto per i Processi Chimico Fisici
Messina
Italy

Pietro G. Gucciardi
CNR – IPCF
Istituto per i Processi Chimico Fisici
Messina
Italy

Mark Heaton
Green Restoration Ireland (GRI)
Carlow
Ireland

Anke Horneber
Institute for Applied Physics and Center LISA+
Eberhard Karls University Tübingen
Tübingen
Germany

Victor Hrymak
Department of Environmental Health and Safety
Technological University of Dublin
Dublin
Ireland

Safi Jradi
Laboratory Light, nanomaterials &
nanotechnologies – L2n
University of Technology of Troyes &
CNRS EMR 7004
Troyes
France

Andreas Kaltzoglou
National Centre for Scientific Research
"Demokritos"
Institute of Nanoscience and Nanotechnology
Agia Paraskevi
Greece
National Hellenic Research Foundation
Theoretical and Physical Chemistry Institute
Athens
Greece

Sameh Kessentini
Département de Mathématiques,
Faculté des Sciences de Sfax
Université de Sfax
Sfax
Tunisia

Janina Kneipp
Department of Chemistry
Humboldt-Universität zu Berlin
Berlin
Germany

Athanassios G. Kontos
National Centre for Scientific Research
"Demokritos"
Institute of Nanoscience and Nanotechnology
Agia Paraskevi
Greece
Department of Physics
National Technical University of Athens
Athens
Greece

Bojana Laban
Faculty of Sciences and Mathematics
University of Priština in Kosovska Mitrovica
Kosovska Mitrovica
Serbia

Florian Laible
Institute for Applied Physics and Center LISA[+]
Eberhard Karls University Tübingen
Tübingen
Germany

Ramzi Maalej
Laboratoire Géoressources, Matériaux,
Environnement et Changements Globaux,
Faculté des Sciences de Sfax
Université de Sfax
Sfax
Tunisia

Onofrio M. Maragò
CNR – IPCF
Istituto per i Processi Chimico Fisici
Messina
Italy

Douglas McMillan
Green Restoration Ireland (GRI)
Carlow
Ireland

Norman McMillan
Advanced Nano Technologies Ltd.
Dublin
Ireland

Sevda Mert
Genetics and Bioengineering Department,
Faculty of Engineering
Yeditepe University
Istanbul
Turkey
Department of Genetics and Bioengineering,
Faculty of Engineering
Istanbul Okan University
Istanbul
Turkey

Rosa Maria Montereali
Department of Fusion and Technologies for
Nuclear Safety and Security,
Photonics Micro- and Nanostructures
Laboratory, FSN-TECFIS-MNF
ENEA C.R. Frascati
Rome
Italy

Enrico Nichelatti
Department of Fusion and Technologies for
Nuclear Safety and Security,
Photonics Micro- and Nanostructures
Laboratory, FSN-TECFIS-MNF
ENEA C.R. Casaccia
Rome
Italy

Enisa Omanović-Mikličanin
Faculty of Agriculture and Food Sciences
University of Sarajevo
Sarajevo
Bosnia and Herzegovina

Simon Perry
South East Technological University
Carlow
Ireland

Massimo Piccinini
Department of Fusion and Technologies for
Nuclear Safety and Security,
Photonics Micro- and Nanostructures
Laboratory, FSN-TECFIS-MNF
ENEA C.R. Frascati
Rome
Italy

Stavros Pissadakis
Foundation for Research and Technology
Heraklion
Greece

Raul D. Rodriguez
Research School of Chemistry and Applied
Biomedical Sciences
National Research Tomsk Polytechnic University
Tomsk
Russia

Elena Shabunya-Klyachkovskaya
National Academy of Sciences of Belarus
B. I. Stepanov Institute of Physics
Minsk
Belarus
Yanka Kupala State University of Grodno
Grodno
Belarus

Evgeniya Sheremet
Research School of Chemistry and Applied
Biomedical Sciences
National Research Tomsk Polytechnic University
Tomsk
Russia

Gagik Shmavonyan
Department of Microelectronics and Biomedical
Devices
National Polytechnic University of Armenia
Yerevan
Armenia

Amina Stambolić
Faculty of Agriculture and Food Sciences
University of Sarajevo
Sarajevo
Bosnia and Herzegovina

Tuan-Hoang Tran
Research School of Chemistry and Applied
Biomedical Sciences
National Research Tomsk Polytechnic University
Tomsk
Russia

Dragana Vasić-Aničijević
Vinča Institute of Nuclear Sciences – National
Institute of the Republic of Serbia
University of Belgrade
Belgrade
Serbia

Valentina Villari
CNR – IPCF
Istituto per i Processi Chimico Fisici
Messina
Italy

Maria Aurora Vincenti
Department of Fusion and Technologies for
Nuclear Safety and Security,
Photonics Micro- and Nanostructures
Laboratory, FSN-TECFIS-MNF
ENEA C.R. Frascati
Rome
Italy

Vesna Vodnik
Vinča Institute of Nuclear Sciences – National
Institute of the Republic of Serbia
University of Belgrade
Belgrade
Serbia

Gamze Yesilay
Genetics and Bioengineering Department,
Faculty of Engineering
Yeditepe University
Istanbul
Turkey
Molecular Biology and Genetics Department,
Hamidiye Institute of Health Sciences
University of Health Sciences-Turkey
Istanbul
Turkey

Part 1: **Introduction**

Spectroscopy is the study of the interaction between electromagnetic radiation and matter, to reveal optical and electronic properties and the chemical structure. Nanospectroscopy can be both the spectroscopy of very small objects down to single molecules or atoms or spectroscopy performed with very high spatial resolution from regions much smaller than the wavelength of the radiation. Optical nanospectroscopy uses ultraviolet (UV), visible, and near-infrared (NIR) light sources for obtaining nanoscale information via the interaction of light and matter. The impressive recent progress in multiexcitation, superresolution, and near-field optical techniques offers unique possibilities of probing nature at the nanoscale – far below the diffraction limit – with extremely high spatial, temporal, and spectral resolution and chemical sensitivity. These techniques can be applied to metallic, semiconducting, or molecular systems that can be prepared using bottom-up or top-down strategies, as well as biomolecules and living cells.

Impressive progress has been made to date in various nanospectroscopy application areas with a strong impact on society, from early-stage cancer disease diagnostics to trace detection in gases and liquids. Optical nanospectroscopy offers great prospects of overcoming analytical problems and disadvantages of other analytical techniques, and is progressively becoming a powerful alternative to conventional methods used, e. g., for the evaluation of food quality, authentication, and safety. Application of optical nanospectroscopy in the food and agricultural sector helps in detecting contaminants, for smart packaging, and in food forensics. Furthermore, nanospectroscopic tools show considerable potential for advancing the development of high-sensitivity sensors, standards for the quantitative comparison of instruments and techniques, or single-photon sources and detectors. In biomedicine they are used for cellular imaging and quantitation of viruses, amongst others. Nanospectroscopy approaches allow for extending methods that mostly exist in the laboratory to real-life applications (taking into account portability, nontrivial environments, mobile phone applications, databases of chemically specific spectra, data evaluation, ease-of-use, etc.), e. g., for point-of-care medical home applications. In the field of cultural heritage, methods of optical nanospectroscopy are used for analysis and characterization of new nanomaterials for the restoration, preservation, and conservation of artworks and for the minimally invasive study of historical artifacts for dating and determining provenance. Optical nanospectroscopic sensing approaches can be used for the detection of potentially hazardous compounds such as pollutants, toxins, drugs, and explosives with chemical specificity in gaseous, liquid, or solid phases. Nanospectroscopy tools are likewise applied in material sciences, to learn more about the synthesis of novel (nano)materials and for characterization of nanobiophotonic materials. Additionally, they promote advances in optical communication and energy conversion and storage.

The main limitation of optical nanospectroscopic techniques is that they are still mostly used by specialists and are not sufficiently known by potential end-users. The knowledge of appropriate methodology needs to be simplified and reachable for the

https://doi.org/10.1515/9783110442908-001

end-users. The instrumentation up to now is often highly specialized, expensive, sensitive to changes, and neither sufficiently standardized nor portable, while also being not specifically adapted to the respective needs, i. e., by supplying specific evaluation routines or databases for nonexpert users in the different application fields.

This volume emerged from the need to compile a collection of specific milestones and insights that optical nanospectroscopy is already providing in various areas of research. The status quo is recorded as a snapshot at a given point in time, and remaining challenges are discussed. The selection of the areas and methods is not arbitrary: It is based on the systematic analysis of a vast number of publications via their keywords in order to map out where the key advances are taking place, as outlined in the publication "Novel advanced scoping meta-review methodology for defining a graduate level textbook in an emerging subject area" [1]. The bar chart in Figure 1 illustrates that the number of publications on optical nanospectroscopy over the last decade is both significant and covering a wide area of application fields.

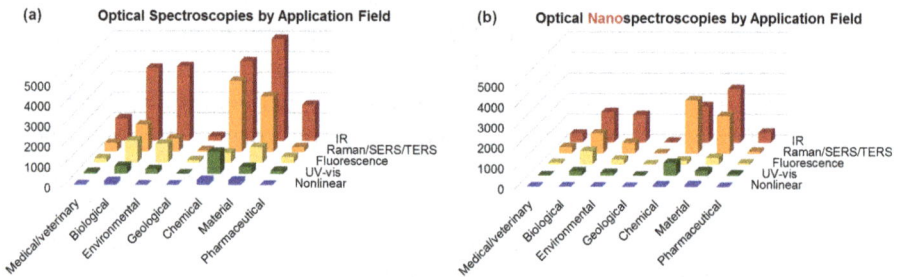

Figure 1: The number of publications from 2008 to 2021 for various optical spectroscopic modalities in various application fields based on a search using the EBSCO Essentials database (South East Technological University) [2] for (a) optical spectroscopy (overall results) and (b) the respective fractions for nano-related optical spectroscopies. For the search, the spectroscopy type was given as a subject heading, and the application area as a search term in the paper abstracts (as a conservative estimate, since the number of papers relevant to a given application is in all probability higher). For the nanosubcategory, "nano*" was added as search term in all fields.

To assist with the research for this volume, the number of publications for each spectroscopic modality has been determined for different application areas. A distinction was made between optical spectroscopic techniques in general and their nanovariants. This study, the result of which is presented in Figure 1, was carried out on the EBSCO Essentials open access database for the period January 2008 to December 2021. The bars give the number of relevant publications per spectroscopic modality for each application market. The full results are shown in Figure 1a, and the respective nanofractions in Figure 1b. The scale is the same for both sample sizes. In this rough overview, it appears that on average about 40 % of the hits refer to nanospectroscopic applications, which is a considerable number given the wide distribution of optical spectroscopic techniques. The second point worth noting is that activity in nanovolume Raman (i. e., surface-enhanced or tip-enhanced Raman spectroscopy, SERS/TERS) and infrared spectroscopy appear to be much larger than in any other modality. For the case of SERS, this trend is also reflected in the examples given in the chapters of this volume. A third result seems to be that the currently most relevant application areas for optical nanospectroscopy can be found in biological, environmental, chemical, and material sciences, with potential for growth in the medical and pharmaceutical sector. This kind of systematic literature meta-analysis can be helpful to identify new and emerging research fields and follow their development. Such an overview may be strategically insightful for young researchers looking for ways of exploring the potential of research targets they might take on with a view to shaping a research career.

This book was devised as the third volume in the trilogy "Optical Nanospectroscopy" by the same editors. It is thus based on a platform of coherently presented knowledge on the fundamentals and methods of optical nanospectroscopy discussed in Volume 1 and the instrumentation, simulation & materials treated in Volume 2. This fact is reflected in the chapter structure that takes the reader directly to the practical examples of the topic. Each chapter is preceded by a section on "Pre-knowledge" that refers to the relevant basic concepts in the first two volumes. All three volumes have been designed with early-stage researchers and interested readers with a basic science training entering the field in mind, but also to give a wide context of how optical nanospectroscopy methods can impact a variety of areas. In contrast to Volumes 1 and 2, the present volume unites independent chapters by individual authors. While the subject areas were selected following the methodology described above, the contents highlight the viewpoints and the expertise of the respective authors. The chapters also provide a broader context beyond the mere optical methodology. While focusing on optical nanospectroscopy, they likewise introduce additional approaches relevant to the specific application.

The showcased examples are roughly categorized into four application fields: photonic applications, sensing, life sciences, and analytical and material science.

With respect to **photonic technologies**, nanospectroscopy enters into such relevant areas as the investigation of quantum emitters, photovoltaic cells, and radiation imaging detectors.

Chapter 2.1 by Christoph Couteau gives an introduction into the properties of quantum emitters that exhibit an artificial atom-like behavior, in combination with photonic environments such as cavities, photonic crystals, optical fibers, quantum photonic circuits, or plasmonic nanoantennas. Nanospectroscopic methods can be employed to probe the signal from quantum emitters that are forming indispensable building blocks for quantum technologies. In Chapter 2.2 by Andreas Kaltzoglou, Athanassios Kontos, and Polycarpos Falaras, the beneficial role of nanospectroscopy in advancing different types of next-generation photovoltaics, such as multijunction, organic, dye-sensitized, quantum dot, or perovskite solar cells, is illustrated. Nano-structured interfaces are developed in order to continuously improve photovoltaic conversion efficiencies. Nanospectroscopy allows for the investigation of the opto-electronic and vibrational properties of the nanocomponents at such interfaces. In Chapter 2.3, Rosa Maria Montereali, Francesca Bonfigli, Enrico Nichelatti, Massimo Piccinini, and Maria Aurora Vincenti take a closer look at the specific challenge of measuring luminescent point defects to create radiation imaging detectors at the nanoscale. The color centers generated by ionizing radiation in lithium fluoride crys-tals and thin films are exploited for radiation detection and X-ray imaging. The optical properties of the color centers are mapped by nanoscopy, e. g., employing confocal laser scanning microscopy and scanning near-field optical microscopy or absorption and luminescence spectroscopy.

Ultrahigh-sensitivity **optical sensing** is one wide application field in which the advantages of nanostructures come fully into play. We take a closer look at minia-turized resonance shift sensors, investigate the application area of nanosensors for explosives detection, and remain in the liquid phase for lab-on-a-chip platforms and nanotensiography.

Chapter 3.1, by Florian Laible, Anke Horneber, and Monika Fleischer, illustrates the function of localized surface plasmon resonance shift sensors, where optical nanospectroscopy serves as the read-out mechanism to detect the specific binding of analytes from the liquid or gas phase to functionalized nanoantennas. With this technique, sensitivities down to single molecule binding have been demonstrated. One important application area of nanospectroscopy is highlighted in Chapter 3.2 with the detection of explosive chemicals as discussed by Roy Aad, Suzanna Akil, and Safi Jradi. In explosive trace detection, especially the portability of spectroscopic systems in combination with their chemical specificity and the high sensitivity of SERS or fluorescent sensing is of advantage. This is one characteristic example where nanospectroscopy is currently undergoing development from the laboratory to the marketplace. Chapter 3.3 gives an overview of the journey of spectroscopy towards the nanoscale by the example of forensic and agricultural science, as shown by Norman McMillan, Douglas McMillan, Raul Rodriguez, Mark Heaton, Victor Hrymak, Simon Perry, Enisa Omanović-Mikličanin, and Stavros Pissadakis. The chapter considers hardware requirements for micro- and nanovolume spectroscopy and tensiography, including a critical evaluation of the role of data analysis and correlated studies. Tech-niques both above and below the diffraction limit are evaluated, and a path towards nanoforensics is sketched.

The section on sensing already provided a first glimpse into the potential of opti-cal nanospectroscopy for learning more about biological processes at the nanoscale. This approach is further pursued in the section on nanospectroscopy in the **life sci-ences**. Especially SERS as a label-free, chemically specific technique offers new ways of biomolecular and cancer detection and can more generally be employed to probe the chemical environment in cell cultures.

Chapter 4.1 is a joint contribution by Antonino Foti, Barbara Fazio, Cristiano D'Andrea, Maria Grazia Donato, Valentina Villari, Onofrio Maragò, Ramzi Maalej, Sameh Kessentini, and Pietro Gucciardi. The authors take an in-depth look at SERS for the detection of dyes, chemicals, and biomolecules at ultralow concentrations. SERS enhancement is closely connected to the optical properties of nanoantennas, which are introduced together with optical forces between such particles. In Chapter 4.2 Gamze Yesilay, Ertug Avci, Mine Altunbek, Sevda Mert, and Mustafa Culha point out that label-free cancer detection methods can avoid possible false results found in label-based methods. SERS is a promising technique that can reduce analysis times, but further improvements of the reproducibility and reliability of the detection are required. Examples of protein, cellular, and tissue analysis with respect to cancer diagnostics are showcased. Chapter 4.3, by Janina Kneipp and Daniela Drescher, picks up on the relevance of SERS for bioapplications and its ability to probe complex biological environments, e. g., in cells and tissues. Here, SERS relies on the uptake of metal nanoparticles together with reporter molecules as SERS probes, which needs to be monitored by complementary techniques. A perspective towards in vivo subsurface molecular imaging even up to the study of larger animals is given.

Finally, in **analytical and material sciences**, optical nanospectroscopy can be applied to a multitude of complex systems to elucidate their optical, electronic, and mechanical properties at the nanoscale. Exemplarily, we illustrate its use in such varied areas as food science and agriculture, cultural heritage studies, cyanine dyes, and carbon-based materials.

Chapter 5.1, written by Enisa Omanović-Miklićanin, deals with the questions of food authentication, safety, and quality by introducing common analytical methods in general and nanospectroscopic methods and nanosensors in particular. Considering the legal frameworks and consumer expectations, optical sensors and vibrational spectroscopy methods are shown to be most promising for rapid food analysis and cost-effective monitoring. Chapter 5.2 by Elena Shabunya-Klyachkovskaya gives a detailed account of optical spectroscopic approaches for identifying dyes in cultural heritage for the purpose of pinpointing their origin, age, and authenticity. An important challenge is maintaining the integrity of the artwork, which leads to the development of specialized sample preparation techniques that are included in the chapter. The success of the approach is showcased by some impressive case studies. A more fundamental application in material science is shown in Chapter 5.3 by Vesna Vodnik, Bojana Laban, and Dragana Vasić-Aničijević. Here cyanine dyes in J-aggregates on noble metal nanoparticles are studied in a highly correlative approach that joins multiple microscopic and spectroscopic techniques. The self-organized dyes form relevant artificial light-harvesting systems that can be employed in various detection techniques and inorganic–organic hybrid systems for bioimaging, sensing, and optoelectronics. Finally, Chapter 5.4, by Gagik Shmavonyan, Tran Tuan Hoang, Dmitriy Cheshev, Andrey Averkiev, and Evgeniya Sheremet, takes us on a journey through the preparation and properties of carbon nanomaterials, graphene, and other 2D and hybrid atomic materials. Due to the strong Raman signature of carbon materials, a plethora of information beyond their mere identification can be gained by investigating such samples with high-resolution nanospectroscopic techniques, informing us about their chemical state, internal stress/strain, and configuration at the nanoscale. The many perspectives for carbon- and 2D material-based consumer products are outlined.

As demonstrated in the selected examples in this volume, optical nanospectroscopy is an emerging field of research and technology that is beginning to translate from requiring large lab equipment to showing that it may indeed take root in manifold areas throughout society. Commercialization of new nanospectroscopic techniques for sensing and analysis is being established and progressing. Instrumentation such as near-field optical microscopes, TERS platforms, surface plasmon resonance detectors, or superresolution imaging setups have been commercially available for some time now. Likewise, a broad range of custom-designed colloidal nanoparticles, nanocrystals, semiconductor quantum dots, and fluorescent markers can be purchased on the market. SERS substrates, fluorescence- and SERS-based measurement devices, and localized surface plasmon resonance sensors are being established. Still, further development and integration of operational devices will be required. Here the role of sensor miniaturization, multiplexed performance, portability, reproducibility, performance speed, and ease of use ("turnkey" instrumentation for nonexpert users) needs to be pushed further to make them attractive for additional application areas in everyday life. In addition, connecting to big data platforms will be key for more powerful chemical or medical diagnostics, not just for patients but for environmental, industrial, and consumer applications. Since these are still early days in the rapidly advancing field of optical nanospectroscopy, the current chapters can only give a snapshot of the fast-developing applications and may soon be overtaken by new developments. By showing some possible directions, we strive to give an incentive for scholars to join the effort to further advance this exciting field.

Bibliography

[1] Perry S, McMillan ND, Rodriguez RD, Mackowski S, Sheremet E, Fleischer M, Kneipp K, Zahn DR, Adam P-M, Madeira TI, Meixner AJ. Novel advanced scoping meta-review methodology for defining a graduate level textbook in an emerging subject area. LIBER Quarterly. 2018;28(1):1–20. https://doi.org/10.18352/lq.10222.

[2] https://essentials.ebsco.com/, based on the MetaReview performed by McMillan ND, Perry S, Rodriguez RD, Sheremet E.

Part 2: **New photonic technologies**

The International Year of Light and Light-Based Technologies hosted by the United Nations in 2015 highlighted the leading role of photonic technologies in the twenty-first century. Photonics is the science and technology of generating, controlling, and detecting photons, the particles of light. It was postulated that the twenty-first century will depend as much on photonics as the twentieth century depended on electronics, with huge societal and economic impact [1]. Areas such as the energy, the building and the mobility sector, communication technologies, medical diagnostics, all the way to arts and culture underlined the truly global relevance of photonics [2]. Future disruptive technologies such as quantum sensing and computing rely heavily on complex optical arrangements or photonic integrated circuits [3, 4]. The human eye as a sophisticated optical instrument enables us to navigate the world, but is limited in the size scales it can resolve. Here researchers for centuries have dedicated their efforts to developing ever more sensitive, high-resolution, and even chemically specific microscopic and spectroscopic techniques to investigate the world around us and drive further innovation. Since photonic technologies are already based on light, they are closely related to optical spectroscopic techniques. Ongoing miniaturization of on-chip devices strengthens the need to extend the spectroscopic resolution for investigation of the relevant components to the nanoscale [5]. Strain fields in advanced semiconductor structures e. g. become accessible via localized Raman spectroscopy [6, 7], facilitating sophisticated band-structure engineering. Molecular-scale fluorescence spectroscopy enables deciphering relevant processes in photosynthesis [8], etc. In the present volume, exemplary applications of optical nanospectroscopy in the emerging field of nanoemitters for quantum technologies, the continuously evolving field of photovoltaics that keeps gaining in relevance, and devices for improved photonic technologies by the example of radiation imaging detectors are highlighted.

Bibliography

[1] What is Photonics. 2015 International year of light and light-based technologies. Accessed Sep 7, 2022, at https://www.light2015.org/Home/WhyLightMatters/What-is-Photonics.html.
[2] Shaping Europe's digital future - Photonics, https://digital-strategy.ec.europa.eu/en/policies/photonics (accessed Sep 15, 2022).
[3] Kim J-H et al. Hybrid integration methods for on-chip quantum photonics. Optica. 2020;7:291–308.
[4] Bogaerts W et al. Programmable photonic circuits. Nature. 2020;586:207–16.
[5] Meireles LM et al. Synchrotron infrared nanospectroscopy on a graphene chip. Lab Chip. 2019;19:3678–84.
[6] Jain SC et al. Stresses and strains in epilayers, stripes and quantum structures of III-V compound semiconductors. Semicond Sci Technol. 1996;11:641.
[7] Tarun A et al. Tip-enhanced Raman spectroscopy for nanoscale strain characterization. Anal Bioanal Chem. 2009;394:1775–85.
[8] Gall A et al. Vibrational techniques applied to photosynthesis: resonance Raman and fluorescence line-narrowing. Biochim Biophys Acta, Bioenerg. 2015;1847(1):12–8.

https://doi.org/10.1515/9783110442908-002

Christophe Couteau

2.1 Application review of quantum emitters in nanospectroscopy

2.1.1 Key messages

- Quantum systems are governed by quantum physics, which is a discipline with special importance for nanospectroscopy. Quantum emitters can be of different types but are all modeled using "artificial atom" behavior.
- Quantum systems are unique systems with quantum mechanical properties. These are usually found at the nanoscale and as such require tools developed for nanospectroscopy.
- Quantum optics experiments and tools are necessary to characterize quantum systems and in particular to ensure that we observe single-photon sources (SPSs), indistinguishable photons, or entangled photons.
- Nanospectroscopic theory is required for characterizing quantum systems that can be used as single quantum emitters for quantum technology applications or to better understand some biophotonics or sensing effects. Quantum systems are important to understand and to extract ensemble measurements for control of photovoltaic or lighting applications.

2.1.2 Pre-knowledge

In order to facilitate the understanding of this chapter, readers may gain pre-knowledge from previous parts and chapters, in particular from Volume 1 "Optical Nanospectroscopy: Fundamentals & Methods", especially the parts elucidating atomic descriptions of light–matter interactions such as the Hertz dipole, the concept of the photon, Rabi oscillations, and Bloch equations, as the notion of optical radiating dipoles and single photons will be very present in many of the cases seen. By the same token, the theory on optical properties of solid-state band structure, excitons, and plasmons will be assumed to be known for the discussions here about plasmons and excitons. This chapter is about applications to nanospectroscopy and is consequently quite experimentally oriented, thus some basic experimental techniques are necessary and the reader is referred to Volume 1, Section 2 and the chapters on photoluminescence, single-photon spectroscopy (time correlation), and single-molecule spectroscopy (confocal). This study will mainly focus on four types of quantum emitters: molecules, quantum dots (QDs), nanocrystals (NCs) (semiconductor ones mainly, leaving aside works on perovskites [1, 2]), and color centers in diamond (although there are other defects in other materials such as silicon carbide [3] or 2D materials [4]) as they were the first systems to be used for the main observations discussed in this chapter, and as such the reader might want to refer to Volume 2, Section 6 on nanostructures used in nanospectroscopy, nanoparticles, QDs, NCs, and optical antennas (top-down).

Acknowledgement: The author would like to thank P.-M. Adam for the reading of the chapter.

https://doi.org/10.1515/9783110442908-003

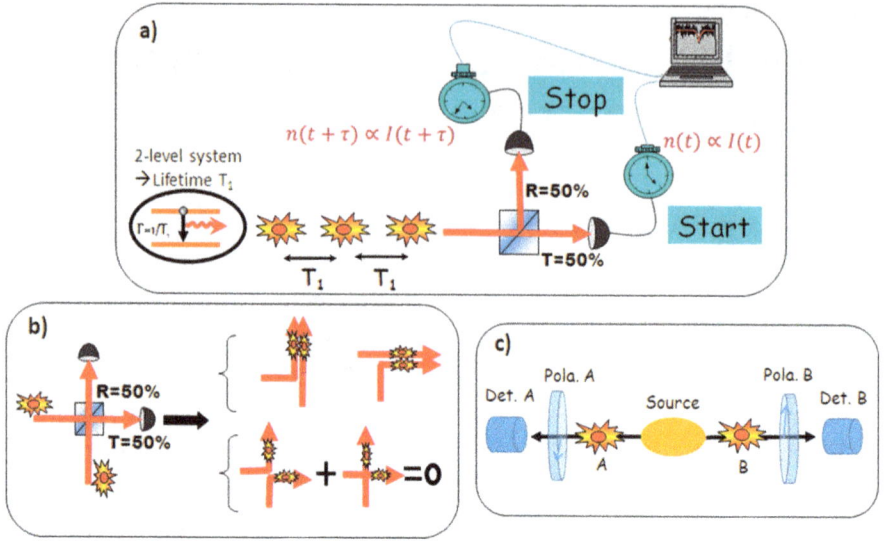

Figure 2.1.1: (a) Scheme of a typical experimental setup to measure whether an emitter emits a single photon. (b) Quantum two-photon interference effect. (c) Scheme of two entangled particles.

The discussion on quantum systems is restricted to some particular quantum states of light, namely the notion of single photons (Fig. 2.1.1a), indistinguishable photons (Fig. 2.1.1b), and entangled photons (Fig. 2.1.1c). Consequently, there will be a quick overview of these concepts. Figure 2.1.1a presents a typical experimental setup to measure whether an emitter emits a single photon. For that, we measure the so-called photon antibunching effect, which simply says that if an emitter emits one photon at a time (say like a two-level system with a certain lifetime T_1), when this single photon impinges on a 50/50 beamsplitter, then there should be no coincidental detection events from the two detectors at the same time. In Figure 2.1.1a, one detector is the "start-event" detector and the other one the "stop-event" detector. A measurement of the autocorrelation function of the signal at t with the signal itself at $t + \tau$ is then taken:

$$g^{(2)}(\tau) = \frac{\langle n(t)n(t+\tau)\rangle}{\langle n(t)\rangle^2},$$

(2.1.1)

taking the time average of the number n of photons recorded at time t and at time $t + \tau$ later. This is equivalent to recording the light intensity, as the light intensity I is directly proportional to the number of photons n according to:

$$I = \frac{\mathcal{P}}{S} = \frac{\mathcal{E}}{S \cdot t} = \frac{n \cdot h\nu}{S \cdot t},$$

(2.1.2)

with $\mathcal{P}$ being the light power, S the light beam surface, h the Planck constant, ν the frequency of light, and $\mathcal{E}$ the photon energy. For a single photon though, we can show that $g^{(2)}(0) = 0$. In other words, for zero delay between the start event and the stop event, there cannot be any double event (coincident event) in the case of a true single photon. Indeed, a single photon has to "choose" between being reflected ($R = 50\%$ at the beamsplitter) or being transmitted ($T = 50\%$ at the beamsplitter), but it cannot be both reflected *and* transmitted, and as such, a dip in the double count rate is expected to occur, reflecting the condition $g^{(2)}(0) = 0$. This type of experiment in Figure 2.1.1a is called a Hanbury Brown and Twiss experiment for historical reasons.

On the other hand, if one looks at two-photon interferences, in the case where the two photons are twins, i. e., indistinguishable, then when one photon arrives at one port of a 50/50 beamsplitter and the other photon arrives at the other port, a two-photon coalescence effect occurs. This effect was first observed in 1987 and is known as the Hong–Ou–Mandel effect [5]. It relates to the fact that photons are bosons and as such, if they are truly indistinguishable in polarization, energy, and momentum, they consequently tend to bunch together. Figure 2.1.1b illustrates this effect, where we can see that the cases where the two photons come out from different ports actually interfere destructively, and the cases where the photons come out together are left. This is a pure quantum effect that can only be understood using the quantum theory of light [6], and it is very important for quantum computation applications where quantum states with high coherence/high indistinguishability are needed.

The third quantum state envisaged for our quantum systems are the so-called Einstein–Podolsky–Rosen (EPR) pairs of particles [7], which will be photons in our case. Another name is entanglement, where two particles (say two photons A and B like in Figure 2.1.1c) are entangled in polarization where the quantum state of the two particles $|\psi\rangle_{AB}$ cannot simply be written as the product of the quantum states $|\psi\rangle_A$ and $|\psi\rangle_B$ of the two photons:

$$|\psi\rangle_{AB} \neq |\psi\rangle_A |\psi\rangle_B. \tag{2.1.3}$$

If the entanglement is polarization-based it means that when A is a vertically polarized photon $|V\rangle_A$, then B has to be horizontally polarized (for instance) $|H\rangle_B$, but the inverse is also true at the same time, and a possible entangled state could be written:

$$|\psi\rangle_{AB} = \frac{1}{\sqrt{2}} \left(|V\rangle_A |H\rangle_B + |H\rangle_A |V\rangle_B \right) \neq |\psi\rangle_A |\psi\rangle_B. \tag{2.1.4}$$

We note the use of the Dirac notation with bra and ket, standard in quantum mechanics. The EPR state is at the heart of Bell's theorem stating that there is no physical theory with local hidden variables that can reproduce all the predictions of quantum mechanics [8]. This type of entangled state, also purely quantum, is very important for quantum cryptography and communications as well as for quantum simulation.

As a broad definition, we will say that a quantum system is a physical system capable of producing one of these three aforementioned types of quantum states of light.

2.1.3 Importance of the application

For the first time in 2011, IBM implemented optical fibers on an electronic chip board (the IBM Power P775 system) demonstrating the power of a future computer using electronics AND photonics. So far, the overheads are massive by introducing optical fibers, and a new generation of integration on the chip level will be necessary for electro-optical elements. The future for photonic integrated circuits (PICs) shows great promise as light is already in use for carrying information, but it would be even more promising if it could also realize information processing and computation in a chip. These technological advances should ensure the future of a market in photonics and integrated optics for computation purposes.

In the meantime, recent developments in quantum optics and nanophotonics have resulted in better control of quantum systems [9, 10], and quantum technologies are now seen to be within reach of real applications. Harnessing light and especially "quantum light" will be the future of information and communication technologies. This chapter focuses on quantum systems, i. e., light emitters that do provide quantum states of light (mainly as SPSs) coupled to photonic devices. The focus will be primarily on near-UV to near-IR photons for optical fibers as well as satellite and free-space communications. As such, the easiest quantum emitter is any two-level system such as a single atom in an optical trap or a single electron trapped by electrodes in a 2D electron gas. To fit within the broad range of nanospectroscopy, we will focus on nanoemitters that have been recently coupled to photonic (such as photonic crystals – PhCs) and plasmonic (such as metallic nanoparticles) structures.

There are several potential applications for sensing or light harvesting but the one which perhaps holds the most hope is to go towards a future quantum computer/simulator. For that, one needs so-called nodes of stationary quantum bits (qubits) where computation is carried out (usually obtained from solid-state quantum systems) and so-called flying qubits (photons) in order to communicate between computing nodes in a complex network. Despite many improvements, efficient interfacing between the two types of qubits remains a challenging goal, and a complete control of light–matter interaction is still lacking. Photons are the natural choice for flying qubits where solid-state nanoemitters such as nitrogen-vacancy (NV) color centers in diamond [11, 12], QDs [13, 14], NCs [15], and single molecules [16] seem to be promising candidates compared to atomic systems, usually requiring complex and not always scalable setups. These are the four systems that will be considered in what follows in this chapter. Engineering a versatile platform in order to achieve efficient interfacing between quantum emitters and photons is thus an important area of research involving the nanospectroscopy of quantum systems. The emission from emitters and, in general, the matching from the nano- to the microworld is a key objective for nanophotonics.

A second key element is being able to miniaturize optical devices in the same way as electronic devices which have now attained tens of nanometers. The advantage of scaling down optical devices and in particular to go towards quantum optical circuits is to be able to add more functionalities for future quantum technologies or for the "Internet of Things" revolution that is predicted. There is, however, a major limitation with respect to downsizing optical devices that is well known, and it is called the diffraction limit roughly on the order of $\lambda/2$, thus a few hundred nanometers for our range of wavelengths. For current and future PIC devices, one needs to scale things down to what is already in place with microelectronic components, i. e., to the order of a few tens of nanometers. In this chapter, we will describe some of the strategies used to beat this limit and thus to go towards nanoscale photonic devices. This chapter is comprised mostly of two parts, the "conventional" dielectric approach, with optical cavities such as micropillars, PhCs, or optical fibers, and the metallic approach to make use of plasmonic confinement to subwavelength levels. There is however a third way, which is the hybrid metallo-dielectric photonic systems currently being envisaged.

2.1.4 State-of-the-art

2.1.4.1 Light–matter interaction

In the 1940s, people working with radars knew that it was possible to control directionality, amplitude, and range of radio waves using antennas, amplifiers, and filters for instance. As it was simply put by Edward Purcell in an article published in 1946 in *Physical Review* (actually more a long footnote than an article), "*... for a system coupled to a resonant electrical circuit, ...the spontaneous emission probability is ... increased, and the relaxation time reduced by a factor of $F_p = 3Q(\frac{\lambda}{n})^3/(4\pi^2 V)$, where V is the volume of the resonator*" and where Q is the quality factor of the resonator [17]. In our case, we are dealing with optical cavities and those are our "electrical circuits" and "resonators." This Purcell effect was first observed in the optical domain in the late 1960s by K. H. Drexhage, where thin films of controlled nanothicknesses filled with europium ions were spaced nearby a metallic mirror [18]. The lifetime of these ions was recorded as a function of the spacing; and the lifetime was modified, thus spontaneous emission was observed. This was the first demonstration of the Purcell effect in the visible range.

2.1.4.2 Cavity quantum electrodynamics and strong coupling

Following Drexhage's experiment, some tremendously significant observations were made using atoms (as is usually the case) in the early 1980s by the Haroche group in Paris and the Kleppner group at MIT [19, 20]. In both cases, the experiment consisted in measuring the lifetime of some energy transitions from atoms going through a resonant cavity and observing how it would change. In fact, in these two occurrences, one observation was enhancement [19] and the other one was inhibition of the spontaneous emission rate [20]. The birth of quantum electrodynamics (QED) came from these experiments. Not so long after, similar first results arrived from molecules in cavities [21–25]. A short while after these pioneering experiments, work was done in solid-state systems with observation of strong coupling between excitons and light in a quantum well. These new "particles" called polaritons (hybrid photon-emitter quantum states) were first observed in 1992 in an article published by Weisbuch et al., where an anticrossing in the spectrum was observed, typical of a strong coupling regime, and the characteristic photon-exciton mode splitting was determined for a 7.6 nm thick quantum well [26].

At this stage, we should specify what we mean by the strong coupling regime. When one observes the Purcell effect such as Drexhage did in 1970, one can see this effect as the modification of the spontaneous emission rate from the environment of the emitter (a mirror or a metallic nanoantenna as we shall see later on) and as such, one can talk about being in the weak coupling regime of the light–matter interaction. But if the conditions are right, i. e., if the emitter is placed in a cavity with very low losses, then the strong coupling regime can be reached where an emitter decays spontaneously, emitting its photon into the cavity. But in the case of the strong coupling regime, the emitter in the cavity has a high probability of re-absorbing the very same photon and so on and so forth. We then end up with a coherent exchange of photons between the emitter and the cavity mode, some kind of "quantum ping pong game" effect characteristic of the energy exchange between two strongly coupled resonators. The conditions for strong coupling and the quantum state of the emitter/photon are thus given by

$$g \gg \gamma, \kappa \quad |\psi\rangle_{\text{QED}} = \frac{1}{\sqrt{2}} (|e\rangle_{\text{at}}|0\rangle_{\text{ph}} \pm |g\rangle_{\text{at}}|1\rangle_{\text{ph}}), \tag{2.1.5}$$

where g is the emitter-photon coupling constant, κ is related to the incoherent leaky modes of the cavity, and γ is the incoherent spontaneous emission rate of the emitter into the noncavity modes. We assume the atom in the cavity has two levels, a ground state $|g\rangle_{\text{at}}$ and an excited state $|e\rangle_{\text{at}}$. This QED quantum state $|\psi\rangle_{\text{QED}}$ is simply the superposition state of the atom in the excited state and no photon in the cavity ($|e\rangle_{\text{at}}|0\rangle_{\text{ph}}$) and the atom in the ground state with one photon in the cavity ($|g\rangle_{\text{at}}|1\rangle_{\text{ph}}$). Figure 2.1.2 illustrates these parameters, where an emitter with a transition frequency ω_A is tuned to the energy ω_C of a cavity mode. A "standard" cavity with two mirrors is represented in Figure 2.1.2, but we will see in particular with metallic nanoantennas that this geometry is not the only one. Examples of the Purcell effect in single-molecule spectroscopy can be found, e. g., in [24, 25].

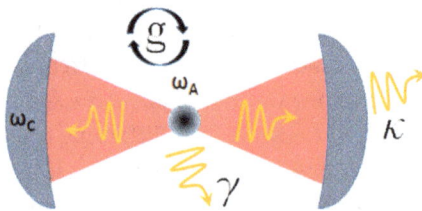

Figure 2.1.2: Two-level system with a frequency ω_A and radiative decay rate γ coupled to a cavity with frequency ω_c and loss decay rate κ, g being the coupling constant between photon and emitter.

2.1.4.3 Purcell effect and single-photon sources

Polaritons in a quantum well are a good example of a particular nanocavity as described in [26], and thus after this proof-of-concept that effects in atomic physics can be translated into solid-state systems, it was then natural to think of inserting true 0D emitters with 3D confinement such as QDs into photonic structures. These "artificial atoms" were behaving very much like a two-level system, and a configuration like in Figure 2.1.2 could thus be envisaged. This idea originated from the France-Telecom research laboratory in Bagneux in France in the mid-1990s, where a layer of QDs was grown deterministically between two distributed Bragg reflectors (DBRs) forming a Fabry–Pérot cavity. Once the DBRs and the QDs were grown, the as-grown sample surface would be etched down to form micropillars/microcavities like in Figure 2.1.3a [27, 28]. With this, the first observation of the Purcell effect by a factor of 5 was obtained followed by the first observation that a single QD could provide single photons [29] (and in fact single NCs were also proven to provide single photons by the same group in the same year [30]). At the time, there were many leaky modes (see spectrum from Figure 2.1.3a) due to the fact that many QDs were grown at once, and the cavity constructed afterwards would select/filter a certain QD that happened to be at the right place and in resonance with the cavity. Nevertheless, having these two very important conditions of being at the right place and with the right resonance simultaneously was very unlikely but was the only known way of succeeding. Indeed, if one recalls the full condition for the Purcell effect F_p to occur, this expresses the ratio of the change of decay rate between the dipole/emitter in free-space γ_0 and in the modified environment γ, and it is given by

$$F_p = \frac{\gamma}{\gamma_0} = \frac{\pi^3 c^3}{\omega_0^2} \rho_u(\boldsymbol{r}_0, \omega_0), \tag{2.1.6}$$

where ω_0 is the frequency associated with the two-level system of the electrical dipole, c is the speed of light, and $\rho_u(\boldsymbol{r}_0, \omega_0)$ is the partial local density of states (LDOS) for a dipole oscillating in a direction $\boldsymbol{u}$ related to a dipole moment $\boldsymbol{p}$. This LDOS can be obtained by the Green's formalism, it can be analytically described using Fermi's golden rule in some simple cases, or it can be obtained also using a numerical solution of Maxwell's equations, for example here using the FDTD method.

Since the first experiments in the late 1990s, technology and inventions have advanced significantly so that in situ observation of single QDs in a laser scanning microscope can allow for simultaneous selection of the right QD using a red laser beam and exposure of a photoresist to then build the cavity around it by optical lithography with a green laser beam [33]. This eventually led to the more refined structure presented as a schematic in Figure 2.1.3b where electrical injection is performed and indistinguishable photons come out of this complex structure [31]. In the late 1990s up to the mid 2000s, QD technology was very much a nondeterministic technique where

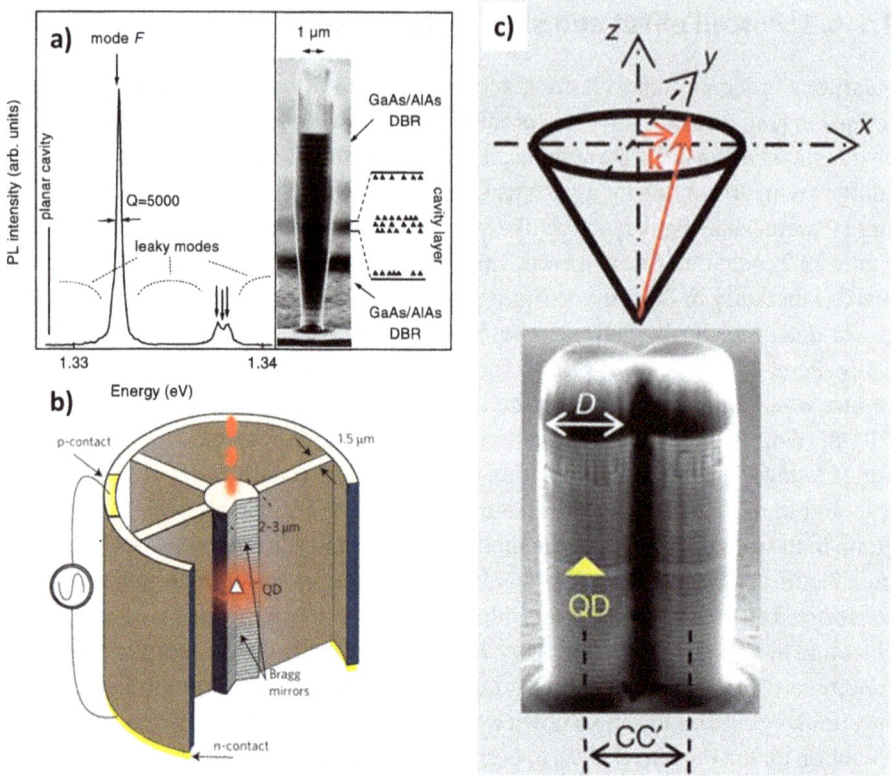

Figure 2.1.3: (a) Spectrum of III-V quantum dots in a micropillar type of Fabry–Pérot cavity (scanning electron microscopy (SEM) picture shown). Reprinted with permission from Ref. [28]. (b) Same type of structure with electrical injection of carriers. Reprinted with permission from Macmillan Publishers Ltd.: Somaschi et al., *Nature Photonics*, vol. 10, 340 (2016) [31]. Copyright 2016. (c) A coupled double cavity for producing entangled states from a single quantum dot. Reprinted with permission from Macmillan Publishers Ltd.: Dousse et al., *Nature*, 466, 217–220 (2010) [32]. Copyright 2010.

one would have to scan through many QDs before finding the right one in terms of brightness, wavelength, and quality of coherence, in order to be able to produce indistinguishable photons [34]. Nowadays, lithographic and growth techniques allow us better control over these parameters. Ideally, the excitonic emission of a QD (and an NC for that matter) should provide indistinguishable (i. e., in the same Fock state), on-demand single photons with an ideal collection efficiency, which is mostly realized by now [31]. Figure 2.1.3c presents the complex arrangement of a QD inserted in a double cavity where each cavity is coupled to one of the first two excitonic transitions of the QD in order to produce an entangled state [32].

All this work is in fact the result of years of single-QD spectroscopy, which proved to be quite complex with the creation of multiexcitons (exciton, biexciton, triexciton, and so on) or charged excitons (two electrons and one hole for instance) and how to identify them. As a promising feature, when two excitons meet in a QD forming a biexciton, the cascaded recombination (biexciton $X_2 \rightarrow$ exciton $X \rightarrow$ empty QD) should provide a "natural" source of on-demand entangled photons in polarization as predicted by the group of Yamamoto in 2000 [35]. Unfortunately, it is not that simple, as mechanical strain and stress during the growth of QDs tend to deform and elongate the dots and thus break the circular symmetry of a given QD. Figure 2.1.4a presents the ideal cascaded biexcitonic emission where an entangled state such as

$$|\psi\rangle = \frac{1}{\sqrt{2}}(|\sigma^+\rangle_{X_2}|\sigma^-\rangle_X \pm |\sigma^-\rangle_{X_2}|\sigma^+\rangle_X) \tag{2.1.7}$$

is expected to be created. But instead, the polarization states are linear (see Fig. 2.1.4b) with horizontal H and vertical V polarized emission from the X and X_2 lines, and one ends up with a mixed quantum state and not a pure quantum state like in equation (2.1.7). Nevertheless, specific techniques were used in order to tune this fine structure splitting (on the order of $\delta_0 \sim$ few GHz) by spectral filtering (see Fig. 2.1.4c for X_2 and Fig. 2.1.4d for X, in gray zones) for instance, where one selects the "right" photons, the ones with the near-zero δ_0 [36].

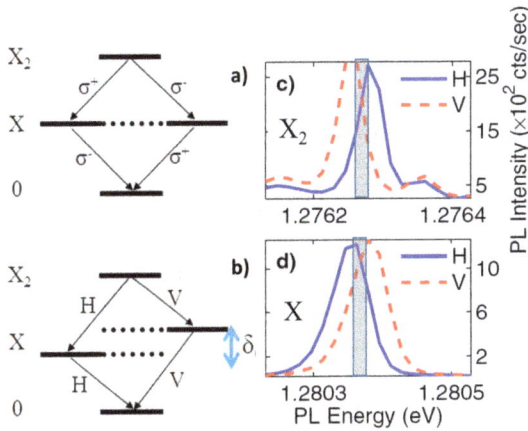

Figure 2.1.4: (a) Perfect scheme for an entangled state due to the biexciton-exciton recombination cascade. (b) A more realistic case with a splitting of the exciton state. (c) Spectral selection in gray of the "right" biexciton and (d) exciton photons for having an entangled state. Reprinted with permission from Ref. [36].

Tuning the fine structure splitting was also realized using a magnetic field [37], using an electric field [38, 39], and by mechanical stress [40]. Besides the micropillars already mentioned, other resonators have been considered such as ring resonators. In fact, the first observation of a single-photon emission with an epitaxially grown QD was realized with a QD inserted within a ring resonator and coupled to a whispering gallery mode (WGM) of a microdisk [29]. Although presenting high quality factors, these types of resonators are not very practical for SPSs in order to efficiently extract the light: micropillars and PhCs are much more appropriate for that, and ring resonators are used more for lasing devices [41]. Epitaxial QDs are very important quantum emitters considering the degree of control we have over them in terms of positioning and coupling them to photonic structures, but also to use them as efficient spin/photon interfaces [14].

2.1.4.4 Photonic crystals

As we will see, if one wants to go towards future integration of quantum systems (quantum photonics), then one needs not only to be able to extract the light efficiently but also to be able to integrate a given quantum system within a network of other quantum systems, with many types of sources and detectors being possible to fabricate. In terms of extracting as much light as possible, the idea of controlling its propagation has been around for some time, but it was only during the late 1980s that researchers thought of engineering materials such that light would bend and reflect in a specific way. The concept of PhCs was thus born out of two seminal articles published in 1987 [42, 43]. In a nutshell, the idea is that if one carefully engineers a certain material, one can create photonic bands for light, very much the same way that atoms in a periodic lattice tend to modify the propagation of electrons within a solid. For light, the atoms are for example holes that act as scatterers. In simple terms, we use the Snell–Descartes law many times and for many periodic occurrences. Then holes (say in a thin membrane) will make a periodic array of a different index ($n_{air} \sim 1$ in this case) into materials made of Si or GaAs with a high refractive index and will thus engineer a system that guides the light. This strong index contrast will modify the light propagation. Like with electrons, if the spacing between holes is chosen carefully, roughly on the order of the wavelength of the light for a PhC, then a band structure for the dispersion of light will appear, alongside with photonic band-gaps where light cannot propagate.

The first 3D PhC was realized in diamond by Yablonovich's group (one of the two discoverers of PhCs) in 1991 by drilling holes into it to be resonant in the microwave regime [44]. This is called a Yablonovite. It did not take long before people started to make these structures for optical wavelengths using e-beam lithography in order to prepare membranes. PhCs would be "drilled" into these membranes, and an emitting material in the membrane would be inserted for more complex PICs. Lasing emission from quantum wells was obtained at first [45] before Noda et al. introduced defects into PhCs in order to create high-Q cavities and to approach a system like the one presented in Figure 2.1.5 [46, 47]. In this work, they realized a silicon-based 2D photonic-crystal slab with a nanocavity with a quality factor of $Q \sim 45,000$ and a cavity mode volume of $V \sim 7.0 \times 10^{-14}$ cm^3. Figure 2.1.5a presents the resonant spectra of different cavities formed by simply adjusting a couple of holes nearby the cavity (Fig. 2.1.5b) [47].

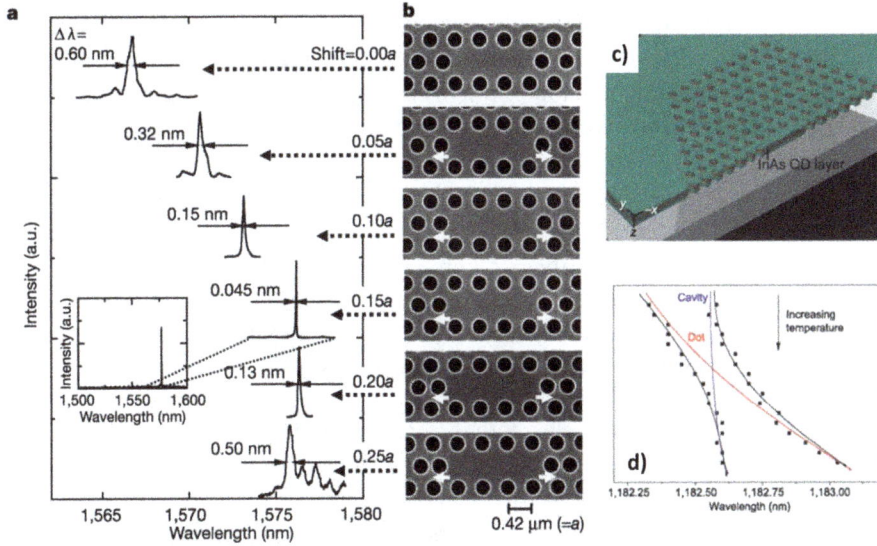

Figure 2.1.5: (a) Resonance spectra of different cavities formed by simply adjusting a couple of holes nearby the cavity. (b) SEM images of the corresponding cavities. Reprinted with permission from Macmillan Publishers Ltd.: Noda et al., *Nature*, vol. 425, 944 (2003) [47]. Copyright 2003. (c) Scheme of a layer of QDs within a photonic crystal. (d) Experimental observation of the Rabi splitting of a single quantum dot in a PhC. Reprinted with permission from Macmillan Publishers Ltd.: Yoshie et al., *Nature*, vol. 432, 200–203 (2004) [48]. Copyright 2004.

One can see a maximum of the sharpening of the spectrum for a shift of 0.15a, where a is the period of the PhC. This was done at around a telecom wavelength at 1.57 μm or so.

In order to make active quantum devices, the next step was thus to insert quantum emitters. Once again, people knew how to make QDs out of III-V semiconductors such as InAs QDs in a GaAs matrix. It was then natural to grow such a structure and then construct a PhC very much like it was done with a micropillar. Figure 2.1.5c illustrates the experiment where a layer of QDs is inserted into a PhC. The regime of strong coupling was reached with such structures as can be seen in Figure 2.1.5d, where Rabi splitting for photons and cavity modes was observed [48].

This experiment was central to many others to come, and it is still up to these days a very much studied system even in other materials such as diamond. The idea of using diamond as a platform is essential for quantum information purposes. It is well known that deep defect centers in diamond such as the nitrogen vacancy (NV) defect (where one N atom replaces one C atom next to the vacancy of another C atom) has a fine structure that is very well suited for quantum processing and quantum communications [49]. It would be thus ideal to fabricate photonic structures directly within bulk diamond already containing these NV centers. The first experiment of a PhC in a single crystal diamond was done by the Becher group in 2011 [50]. Since then, we know how to create and address single-color centers such as silicon vacancy (SiV) centers [51]. This work from Harvard University consisted in fabricating an integrated platform for scalable quantum nanophotonics where SiVs were coupled to nanoscale diamond devices. These SiV centers were placed inside diamond PhC cavities, in order to realize a quantum optical switch controlled by a single-color center. This single photon transistor is very important for quantum simulators and communications. We refer the reader to some review articles on color centers in diamond for quantum networks and quantum photonics for further reading [52].

2.1.4.5 Optical fibers and quantum systems

Ideally, in order to harness single photons from quantum systems, the simplest thing to do would be to couple photons from emitters directly into optical fibers as once the photons are in an optical fiber, they can propagate for long distances and interact with other quantum systems. This can be done, for example, using the nanomanipulation of a QD in a photonic wire that is then attached to an optical fiber as seen in Figure 2.1.6a and b. By this technique, photons emitted from a QD are directly cou-

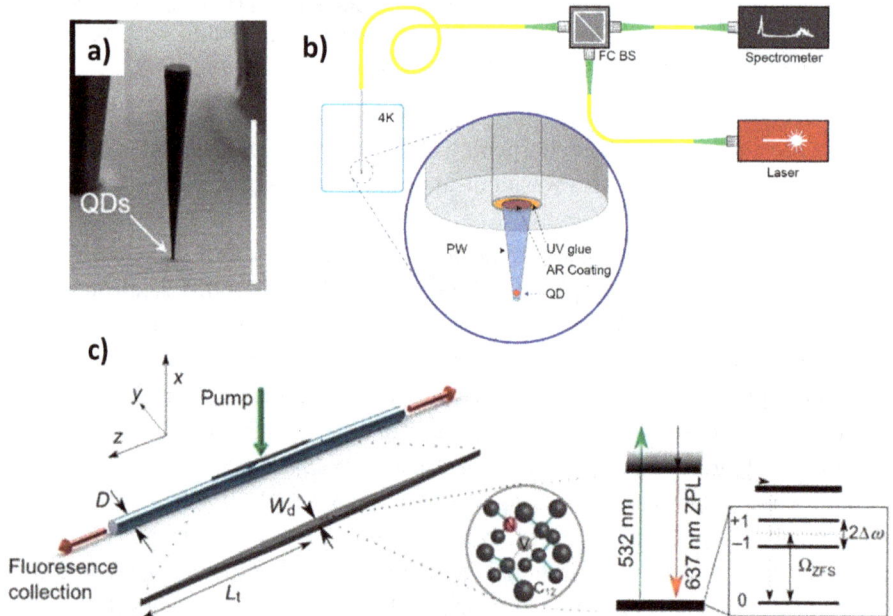

Figure 2.1.6: (a) Quantum dot in a photonic wire (b) that is then attached to an optical fiber and coupled to a Hanbury Brown and Twiss setup. Reproduced from [53], with permission from AIP Publishing. (c) NV center inside a single-mode tapered diamond waveguide that is itself integrated on a tapered silica fiber. Reprinted with permission from Macmillan Publishers Ltd.: Patel et al., *Light: Science & Applications*, vol. 5, e16032 (2016) [54]. Copyright 2016.

pled into the fiber by adiabatic mode coupling from the photonic wire to the optical fiber [53]. By the same token, a group at MIT managed to couple single photons emitted from an NV center inside a single-mode tapered diamond waveguide that was itself integrated on a tapered silica fiber as shown in Figure 2.1.6c [54]. A similar idea was demonstrated with epitaxial QDs in a nanophotonic directional coupler physically coupled to an optical fiber with a 6% coupling demonstrated [55]. It was also demonstrated that optical fibers can be used for controlling the flow of light when they are used as nanophotonic waveguides [56]. Petersen et al. showed that thanks to the strong transverse confinement of guided light, a spin–orbit interaction of light occurs, and they observed the breaking of the mirror symmetry of the scattering of light by a gold nanoparticle on the surface of a nanofiber [56]. This makes a chiral waveguide coupler in which the handedness of the incident light determines the propagation direction in the waveguide [56]. The same effect was observed using an epitaxial QD within a III-V nanobeam with the interesting perspective that spin state selection from a QD can be controlled that way [57]. This is very important for quantum information processing.

2.1.4.6 Quantum photonic circuits

Ultimately, the goal remains to go towards fully integrated quantum photonic circuits, and preferably without optical excitation, but with an electrically driven light source. A first proof-of-concept was shown by the group of Wolfram Pernice where they reported the observation of photon antibunching from an electrically driven carbon nanotube embedded within a photonic quantum circuit. The single photons were detected by waveguide-integrated superconducting single-photon detectors (SSPDs) [58]. Figure 2.1.7a presents a schematic of this experiment where several carbon nanotubes and several superconducting detectors are integrated on a chip. This is an example of an all-integrated quantum photonic circuit. This approach is a rather bottom-up one where the components are brought together.

Other approaches are studied such as the all-epitaxial approach one with III-V semiconductors [59, 57]. This is a rather top-down approach with QDs being grown and waveguides then etched down in order to guide single photons emitted from the QDs. The main advantage of using III-V semiconductors is the fact that they are high-refractive-index materials and thus guide the light very tightly when surrounded by air [59]. Figure 2.1.7d presents an SEM image of such a structure. A similar experiment was realized by the Finley group from the Walter Schottky Institute [60] with a schematic of the integrated photonic system shown in Figure 2.1.7c. In this example, InGaAs QDs emit photons waveguided along a GaAs ridge waveguide that are detected by a NbN nanowire (SSPD made of niobium nitride). This is a mix of top-down (for the light source and the waveguide) and bottom-up approaches (with the integration of the SSPD). Using this geometry, resonant fluorescence was observed, and a line width of

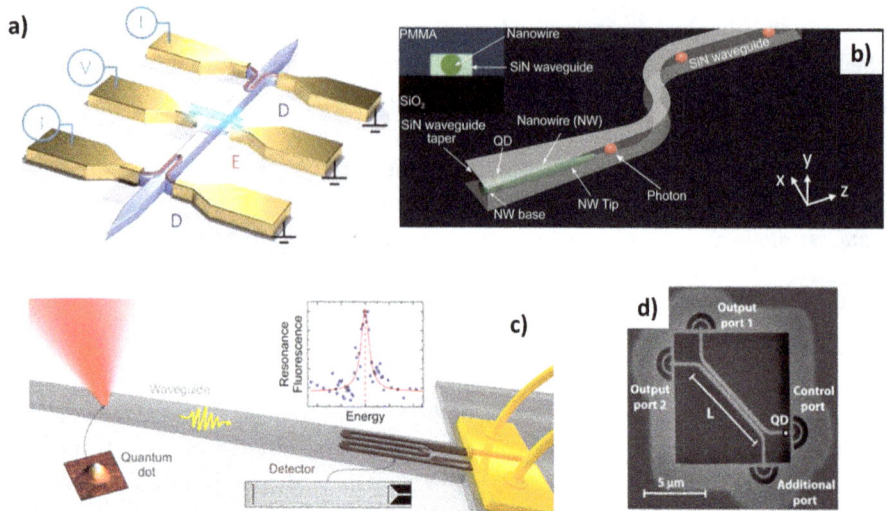

Figure 2.1.7: (a) Carbon nanotubes and superconducting detectors integrated on a chip. Reprinted with permission from Macmillan Publishers Ltd.: Khasminskaya et al., *Nature Photonics*, vol. 10, 727 (2016) [58]. Copyright 2016. (b) QD in a photonic nanowire, embedded in a silicon nitride (Si_3N_4) waveguide. Reprinted with permission from Zadeh et al., *Nano Letters*, vol. 16, 2289–2294 (2016) [62]. Copyright 2015 American Chemical Society. (c) Integration of an epitaxially grown quantum dot with a superconducting detector. Reprinted with permission from Reithmaier et al., *Nano Letters*, vol. 15, 5208 (2016) [60]. Copyright 2015 American Chemical Society. (d) Quantum dot grown in GaAs embedded in an etched-down waveguide. Reproduced from [59] with permission from AIP Publishing.

10 µeV or so was measured from a single QD, reinforcing the concept of artificial atoms with very narrow line widths [60]. We can mention the Zwiller approach using yet again a QD embedded into a photonic wire in a trumpet shape [61] in Figure 2.1.7b that was nanomanipulated and then embedded into a Si_3N_4 ridge waveguide. This material has a fairly high refractive index and is also compatible with silicon technology, which makes it particularly interesting for applications [62]. Once again, a hybrid approach bottom-up/top-down was used where grown emitters are inserted into more complex networks.

There are nowadays several platforms that can claim complete integration, and Table 2.1.1 provides a broad overview of these platforms with their different characteristics whether you work in the visible or in the telecom region (1550 nm or so). The platforms are mostly glass-based such as optical fibers, laser-written waveguides, lithium niobate, and ion-exchange waveguides, but also include high band-gap semiconductor-based platforms such as GaAs, diamond, SOI, or silicon nitride. Then quantum states of light are either created in the material, as already mentioned in the case of deep defect centers in diamond, or with the growth of QDs in III-V semiconductors. There are also hybrid solutions to these problems using different materials

Table 2.1.1: Overview of the different photonic platforms for quantum systems.

	λ range (nm)	Losses @ 600 nm (dB/cm)	Losses @ 1550 nm (dB/cm)	Index n (at 600 nm/ 1550 nm)	Compatible with fibers	Active
Silica-on-silicon (SOI)[a]	>1100	>100 dB/cm	3.6 dB/cm [67]	2.7/2.46 [68, 69]	Yes	Yes
Silicon nitride (Si$_3$N$_4$)	200–10,000	5 dB/cm	1 dB/cm	1.92/1.87 [70]	No	Yes
Laser-written in glass	300–5500	0.5–1 dB/cm	0.1 dB/cm	Δn: 0.0001–0.002 (multimode)	Yes	No
Optical fiber	300–5500	10 dB/km	0.25 dB/km	1.458/1.444	Yes	Yes
Lithium niobate	300–5500		<0.5 dB/cm [71]	2.29/2.21 [69]	No	Yes
Diamond	300–10,000	>70 % (1 mm thick)	$k = 0.009$	2.41/2.39 [69]	No	Yes
Gallium arsenide (GaAs)	>870	>100 dB/cm	<1 dB/cm	3.91/3.37 [69]	No	Yes
Ion-exchange waveguide	300–5500	0.3 dB/cm	0.1 dB/cm	Δn: 0.001–0.03 [72]	Yes	Yes

[a]We recall that silica = silicon dioxide = SiO$_2$.

and/or using plasmonic structures [63]. A roadmap on integrated quantum photonics now exists where all the different platforms are discussed in terms of current and future challenges [64], and we refer to some review articles for more on the subject of nanoscale quantum optics [65] and integrated quantum photonics [66].

2.1.4.7 Plasmonic antennas

As already seen in previous chapters, e. g., in Volume 1, Section 3, plasmonic resonances in metal can be used to control light emission from quantum emitters in various ways, using either localized surface plasmons (LSPs) or propagating surface plasmons (PSPs). Structures supporting the first type of plasmonic resonance can act as optical nanoantennas for visible light and can strongly influence parameters such as the quantum efficiency, the Purcell effect, and the fluorescence enhancement of an emitter; see Volume 2, Section 6. Figure 2.1.8a presents an example of a nanoantenna made of a bowtie geometry using gold, where molecules were inserted within the bowties [73]. Very high fluorescence enhancement factors were found in this work, more than 1000 times for some single molecules. The authors could affirm the fact that they were dealing with single molecules by looking at the blinking emission from their spectra. Although this work was one of the first ones showing such dramatic enhancement, it was a bit misleading by the fact that not everything was taken into account and the fact that the molecules chosen had low quantum efficiency to start with, of around 2.5 % only. Indeed, in order to assess the Purcell effect, one needs to look at the lifetime change. In general, the change of fluorescence intensity I for such structures is quite complex and is governed by

$$\frac{I}{I_0} = \frac{R_{exc}}{R_{exc,0}} \cdot \frac{\eta}{\eta_0} \cdot \frac{C_{coll}}{C_{coll,0}}, \tag{2.1.8}$$

where the index "0" is the case without the plasmonic structure/antenna, and no index is the case where the emitter is within the plasmonic structure. We see that three ratios are to be taken into account, where $R_{exc}/R_{exc,0}$ is the change of excitation intensity due to the electromagnetic field enhancement thanks to the structure, η/η_0 is the change of quantum efficiency, and $C_{coll}/C_{coll,0}$ is the modification of the collection of light, which is modified by the antenna effect. In the case of Kinkhabwala et al., the main modification came from the change of excitation intensity as well as the quantum efficiency. In order to make the quantum efficiency appear, we recall that the quantum efficiency can be seen as the product of the lifetime of the emitter τ times the radiative decay rate Γ_r so that we have

$$\frac{I}{I_0} = \frac{I_{exc}}{I_{exc,0}} \cdot \frac{\tau}{\tau_0} \cdot \frac{\Gamma_r}{\Gamma_{r,0}} \cdot \frac{C_{coll}}{C_{coll,0}}. \tag{2.1.9}$$

We note that we replaced R by I as the difference between the two is a factor due to dipole orientation, which in the case of many dipoles such as in this case can be assumed to be the same whether we have a structure or not, and thus can be canceled out. This is no longer true for single dipole emitters. This work also showed to some extent that the geometry does matter. As can be observed in Figure 2.1.8a, bowtie geometries were enhancing more strongly than single triangles, thus showing some kind of cavity effect of the electromagnetic field between the two triangles. We can thus see from equation (2.1.9) that in order to observe an even more dramatic fluorescence enhancement, one needs to work on all the parameters. In this case, a clear unidirectional emission was observed from a single QD placed near the antenna. In this particular case, one has to look into the interplay between LSPs and PSPs, which turns out to be quite complex [74, 75]. Various quantum emitters have been used in combination with plasmonic antennas, and NV centers in nanodiamonds are among them considering their importance for quantum technologies. Using an atomic force microscope, a single nanodiamond containing a single NV center and two gold nanospheres were assembled step-by-step [76]. It was shown that both the excitation rate and the radiative decay rate of the color center were enhanced by one order of magnitude. As has

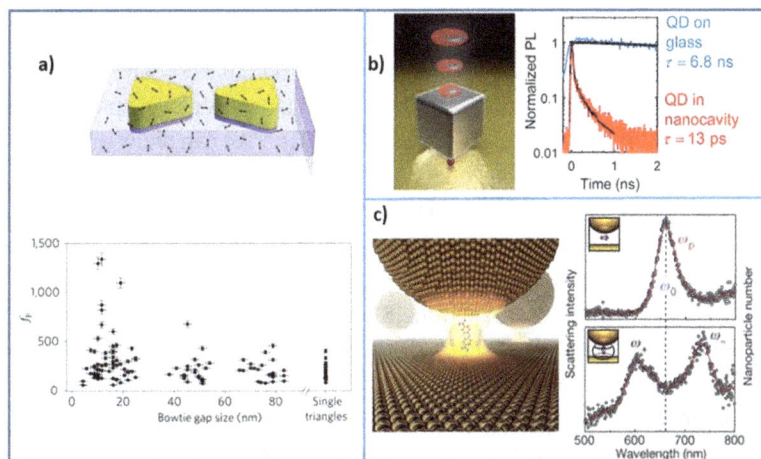

Figure 2.1.8: (a) Single molecule enhancement for different bowtie spacings and for single triangles. Reprinted with permission from Macmillan Publishers Ltd.: Kinkhabwala et al., *Nature Photonics*, vol. 3, 654 (2009) [73]. Copyright 2009. (b) Silver nanocube of 75 nm in size deposited on top of a gold film with a layer of quantum dots in between presenting a lifetime change of more than 500. Reprinted with permission from Hoang et al., *Nano Letters*, vol. 16, 270 (2015) [77]. Copyright 2015 American Chemical Society. (c) Strong coupling signature observed at the single-molecule level mediated by plasmons. Reprinted with permission from Macmillan Publishers Ltd.: Chikkaraddy et al., *Nature*, vol. 535, 127 (2016) [78]. Copyright 2016.

been known for some time for an ensemble of emitters, hotspots between diamond and gold nanoparticles provide an efficient near-field coupling [76].

At this stage, we should mention that dielectric antennas can also be very efficient as was demonstrated by the group of Sandoghdar where using a planar dielectric antenna for directional single-photon emission, they obtained a near-unity collection efficiency for a single molecule [79]. The addressing of emitters was also tackled with dielectric antennas, where a single ZnO nanowire on glass behaves like a nanowaveguide (around 10 µm long and 200 nm in diameter) that can locally excite a single-photon emitter [80].

By engineering the local excitation, one can maximize this excitation and match the absorption "impedance." The group of M. Mikkelsen from Duke University managed to play on all the parameters in order to observe an increase of the Purcell factor by more than 500 times and as a result a fluorescence enhancement of more than 1900 times while having a very strong directionality orthogonal to the substrate, which is thus very interesting for light collection [77]. The system is schematically represented in Figure 2.1.8b, where they used a nanocube of silver of about 75 nm in size, deposited on top of a gold film. A layer of low-concentration QDs was deposited in between the cubes and the film with a spacer of around 3 nm between the metals and the emitters in order to avoid direct quenching of the emission. The emitters were CdSe/ZnS NCs of about 6 nm in size. They managed to look at single QDs and thus showed that such a single-photon emitter in this geometry can show a Purcell factor of more than 500 [77]. Figure 2.1.8b presents the lifetime change from more than 10 ns to less than 13 ps when squeezed between the cube and the film, a quite tremendous decrease. As mentioned previously, this "cavity" is not a cavity strictly speaking and yet can show quite dramatic Purcell factors. In this work, the collection efficiency, the excitation efficiency, and the Purcell effect were all affected by this structure.

Since then, improvements have been made in order to place quantum emitters near plasmonic antennas with good control of the positioning. This can be realized using photopolymerization such as in the work [81] where a single NC has been placed near a gold nanocube within less than 12 nm. By the same token, a group at the University of Cambridge managed to observe the strong coupling regime at room temperature using isolated methylene-blue molecules. Figure 2.1.8c presents the schematic which consisted in sandwiching dye molecules within a gap of less than 1 nm thickness between a gold nanoparticle and a gold film [78]. This configuration has the particularity of decreasing the cavity volume of this structure to less than 40 nm^3. They reached the strong coupling regime at room temperature with dispersion curves showing a typical light–matter "mixing," with Rabi frequencies corresponding to up to 300 meV as presented in Figure 2.1.8c. It is worth mentioning that using host–guest chemistry, they could align the molecule dipole in the right direction in order to observe strong coupling and Rabi flopping. Figure 2.1.8c presents two spectra where one case presents no coupling for the wrong dipole orientation and the other one maximum coupling due to the matching dipole orientation [78].

This tour-de-force shows at last the interplay between the quality factor of a "cavity" and the mode volume of the electromagnetic field confined within this "cavity." As seen previously for the Purcell factor, we deduce that the Q/V ratio is the relevant figure of merit these days as nanofabrication and plasmonics allow researchers to move towards smaller and smaller volumes. We see in this experiment that a rather low Q can be compensated by a low mode volume V [78]. Another example is given by [82] with a single-photon emitter in a carbon nanotube that is coupled to a bowtie plasmonic cavity, with Purcell factors as high as 190.

Figure 2.1.9 is a chart summarizing some of the different structures with the mode volume in the ordinate axis and the quality factor Q in the abscissa axis. The authors in this particular work present their work in the low-Q-factor region, but with very low volumes in order to reach the strong coupling regime. For that, they define the Purcell factor P that has to be more than 10^6 at least in order to observe the Rabi oscillations. The strong coupling regime is very important in order to show strong light–matter interaction, but observing it at the nanoscale is even more important as it shows the possibility to miniaturize to a very low volume some very important effects for quantum technologies in general and sensing, chemistry, or computation in particular. We note that this is just the beginning, as many theoretical results show the potential of plasmonic approaches such as observing entanglement between two emitters mediated by LSPs [83]. This engineering of plasmonic resonance can also be coupled to

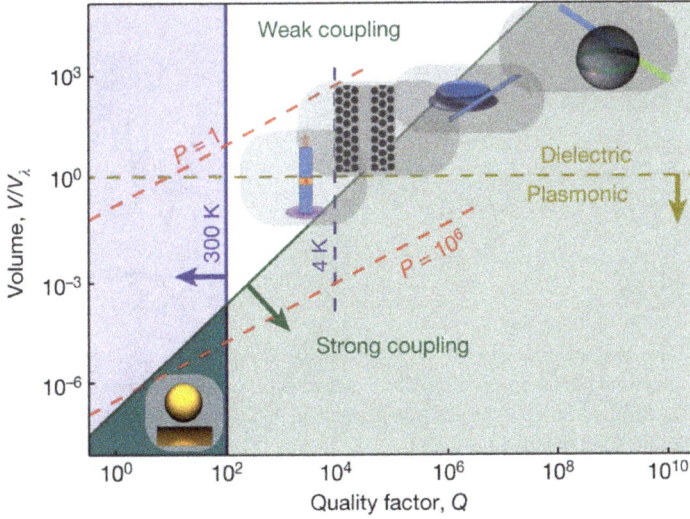

Figure 2.1.9: Chart summarizing different structures with the mode volume in the ordinate axis and the quality factor Q in the abscissa axis. Reprinted with permission from Macmillan Publishers Ltd.: Chikkaraddy et al., *Nature*, vol. 535, 127 (2016) [78]. Copyright 2016.

metamaterials in order to obtain efficient single-photon to single-plasmon coupling, such as the one realized by Roger et al., who showed the perfect absorption of single photons by a metasurface in a pioneering work [84].

2.1.4.8 Quantum plasmonics

In recent years, the field of quantum plasmonics has emerged, and with it new opportunities have appeared in quantum communications and optical coherent networks. Altewischer et al. were the first to show that plasmons at the surface of a nanohole array retained the quantum state of entangled photons [86]. For a reminder of the notion of entanglement, see Figure 2.1.1c, where a schematic of this phenomenon is given as well as equation (2.1.4) for the quantum state associated to it. Along the same lines, Fasel et al. observed energy–time (or time–bin) entanglement preservation of a plasmonic waveguide at the telecom wavelength [87]. This type of entanglement is different from the one described earlier on based on the state of polarization of photons, where in this case the "encoding" of the information is in time rather than in polarization. The reader can refer to the original proposal by Franson for an understanding of time–bin entanglement [88]. Although these were pioneering works on coupling quantum states of light with plasmons, the birth of "quantum plasmonics" can be traced back to 2007, and it came from the group of Lukin and Chang at Harvard University where the creation of single quantum plasmons was demonstrated [85]. The

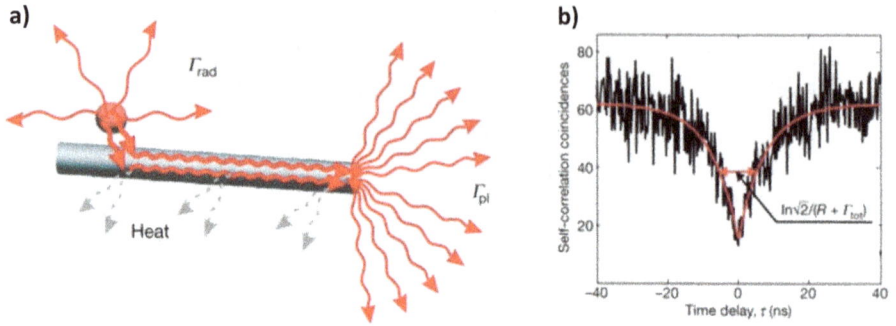

Figure 2.1.10: (a) Schematic of a single nanocrystal coupled to plasmons on a single silver nanowire and (b) photon antibunching from photons emitted by the source directly and the plasmons converted back into photons. Reprinted with permission from Macmillan Publishers Ltd.: Akimov et al., *Nature*, vol. 450, 402–406 (2007) [85]. Copyright 2007.

idea was very simple as it consisted of an SPS made of a semiconductor NC placed sufficiently close to a silver metallic nanowire. Once you excite the NC, part of its emission would couple to the plasmonic modes (PSPs in this case) of the nanowire, and single plasmons would thus be generated. Figure 2.1.10a presents a schematic of the experiment where we see that some single photons emitted by the nanosource re-radiate into free-space Γ_{rad}, some are lost by conversion into heat due to the metal, and some are converted into propagating plasmons Γ_{pl}. Figure 2.1.10b shows the standard photon antibunching between the photons collected from free-space emission and the plasmons converted back to free-space photons once the plasmons reach the end of the silver nanowire. This correlation measurement tells us that only one electromagnetic mode is emitted at a time by the NC, whether it is a single photon or a single plasmon [85]. From then on, quantum plasmonics expanded into various applications using various configurations [89, 90], notably the single-photon transistor being one of them, emerging from this group as well [91]. These demonstrations showed that plasmon-polaritons retain the quantum statistics of light quantization. We cite an experiment with nanodiamonds from 2009 [92], where it was shown that the wave–particle duality of plasmons occurs as much as it does for photons. By placing a nanodiamond, known for having a broad spectral emission, near a silver nanowire, the authors observed the single-emission behavior like it has been done by the Harvard group with NCs as well as interferences occurring at the single plasmon level by observing fringes in the broad spectrum of the nanodiamond. This single photon interference effect occurred between the plasmon transmitted directly along the nanowire and the plasmon reflected back from the opposite end of the nanowire, some kind of thin glass interference effect. In fact, it was also shown that inherent losses in plasmonics follow a Markovian process and are thus not affecting the fundamental quantum properties to be preserved by plasmons [93]. In parallel, the conservation of squeezed states of light converted into propagating plasmons was also shown, further confirming previous findings [94]. Recently, two-boson quantum interference experiments were carried out by demonstrating the Hong–Ou–Mandel effect with plasmons [95]. They showed that two waveguided plasmons retained the indistinguishability property of their respective photons. Figure 2.1.11a presents an SEM image of the structure designed for this purpose, and Figure 2.1.11b is the signature of the HOM effect described in Figure 2.1.1b. This two-plasmon interference effect definitely demonstrates that plasmons are bosons as much as photons are with the similar HOM effect. Since then, there is a plethora of research done in bringing quantum optics to nanophotonics (see for instance [96, 97] for an overview of some works in the field). In fact, in the broad range of nanophotonics, people are specifically looking into an entire quantum circuitry where the source of light, the propagation and the detection are all on the same chip and thus at the nanoscale. A first prototype experiment was realized at Delft University of Technology with a gold plasmonic directional coupler device as shown in Figure 2.1.11c [98]. At both ends of this coupler integrated superconducting SSPDs were deposited, located 20 nm below the output waveguides. The width and spacing of the superconducting meander were both 100 nm with a thickness of around 5 nm. The 150 nm thick gold waveguide was 600 nm wide and aligned for optimal absorption of the plasmonic mode in the SSPD. One can see on these optical images that for two inputs (bottom of the figures) we have only one side lighting up for the detection on the left-hand side (left Fig. 2.1.11d) or on the right-hand side (right Fig. 2.1.11d), which is a clear signature of the HOM effect.

More recently, more complex plasmonic structures have been realized in order to make metamaterials using the so-called split-ring geometry. With this device, inserted in the stationary electromagnetic wave of an interferometer, it was shown that one can realize perfect absorption (with 100 % success probability) and thus the perfect conversion of a single photon into a single plasmon [84]. Using the same geometry with polarization-entangled photon pairs, the nonlocal control of absorption of light was demonstrated with a metamaterial structure. Through the detection of one photon with a polarization-sensitive device, one could, in principle, deterministically prevent or allow absorption of a second, remotely located photon [99]. Thus, energy dissipation of specific polarization states on a heat sink is remotely controlled, promising opportunities for controlling plasmon–photon conversion and entanglement.

Finally, we should mention that ideally one would want to generate quantum states of light electrically which, for plasmons, is possible, and this alley of work still needs to be explored despite pioneering work of generating and detecting plasmons all electrically and having plasmon propagation in between [100]. In particular, the electrical generation of single plasmons remains a challenge [101].

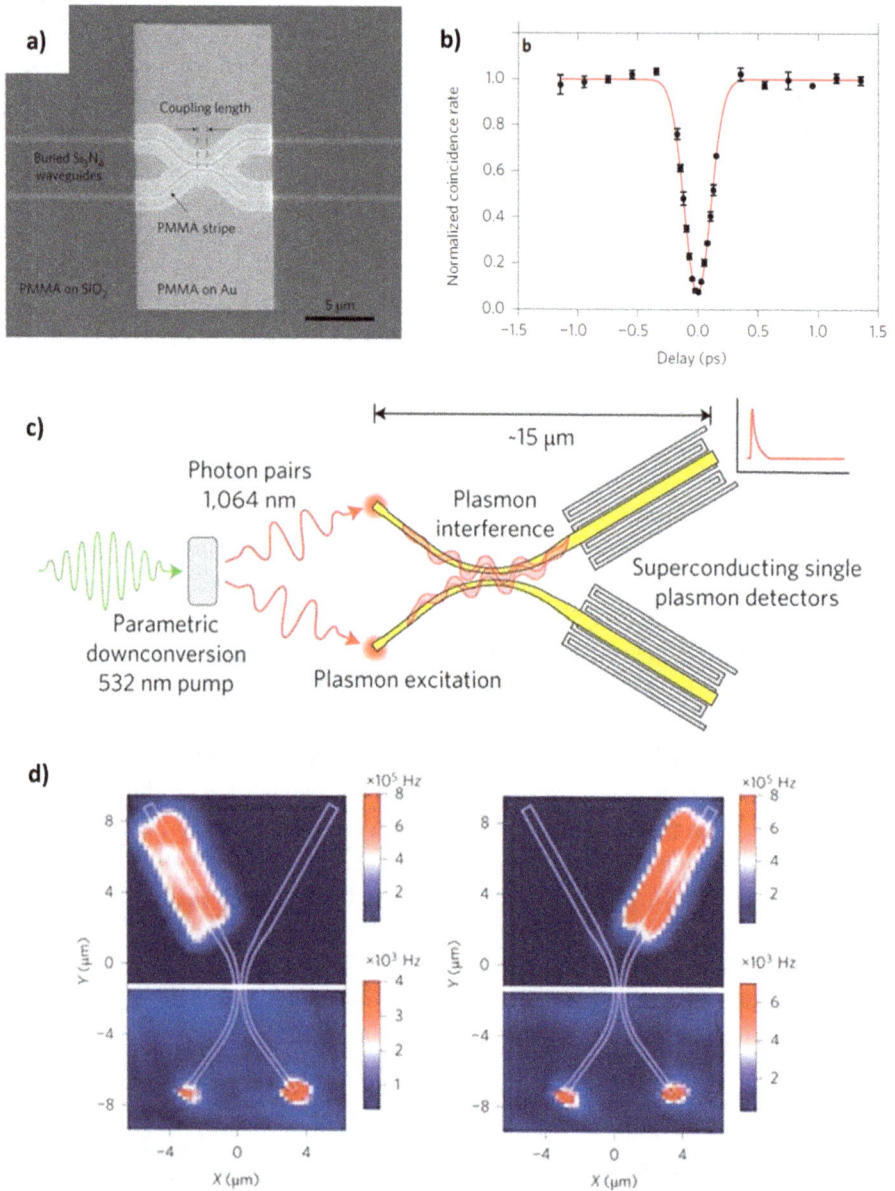

Figure 2.1.11: (a) Two-plasmon quantum interference experiments carried out with an SEM image of the structure designed for this purpose. (b) Optical signature of the HOM effect described in Figure 2.1.1b. Reprinted with permission from Macmillan Publishers Ltd.: Fakonas et al., *Nature Photonics*, vol. 8, 317 (2014) [95]. Copyright 2014. (c) Schematic of another experimental setup to integrate detectors for plasmon detection with (d) spectral images showing the experimental results associated, notably the HOM effect. Reprinted with permission from Macmillan Publishers Ltd.: Heeres et al., *Nature Nanotechnology*, vol. 8, 719–722 (2013) [98]. Copyright 2013.

2.1.5 Some challenges and solutions

Ultimately, an objective would be to realize nanophotonic circuitry where a light source and a detector are both included on an optical chip, which itself is an active photonic platform that can be manipulated optically and/or electrically. There are many challenges ahead, especially if one wants to work at the single-photon level, which would be the "perfect" control of any optical system. As seen, many quantum emitters are out there, atoms, molecules, defect centers in semiconductors, QDs, and NCs, as well as many types of detectors,[1] and it is not quite clear which approach will be the chosen one in the future. A mix of several ones might be the way to go, leading to hybrid approaches as already studied in some cases (dielectric/metallic antennas or waveguides). Will it be the top-down or the bottom-up approach that will win the race, or will it be yet again a mix of the two? Maybe the "Lego" approach will be preferable, where one constructs an optical circuit brick by brick, component by component by nanomanipulation. It is also clear that in order to either compete with or complement the current microelectronic market, optical components will have to be much smaller, and the diffraction limit will have to be overcome to produce optimal circuits at a useful scale to compete with other established technologies. In particular, one will need to have programmable and reconfigurable photonic circuits, which is far from trivial [102]. Many experimental achievements have shown so far that it is possible using plasmonics and metamaterials, but controlling and/or coupling quantum systems with plasmonic structures remains a challenge, as these structures are usually very lossy and by essence, quantum emitters do not emit a lot of light. If we refer to the current microelectronic industry and how to incorporate nano-optical components, then one has to ask the question whether we will be able to deliver quantum nanodevices that are compatible with the current mass-market technology based on silicon. It is a fair question to wonder whether for instance it does make sense to look at III-V nanodevices such as InAs QDs in a GaAs matrix whereas for the moment, Si and III-V materials are not compatible. The same goes for diamond, which is not really compatible with the silicon industry. One could try to make a nanophotonic wish list:

- Optical circuitry has to be silicon-compatible.
- Losses have to be minimal and at the single-photon level.
- Creation and detection of single photons have to be perfect: one photon in – one photon out.[2]
- Components and light propagation have to be at the nanoscale, at least on the same scale as the current microelectronics industry.
- All sorts of quantum states of light have to be able to be produced.

---

1 Not really discussed in this chapter; see, e. g., Volume 2, Section 4.
2 A kind of PhiPho device.

- Clock speed has to be at least as good as the current classical clock speed.
- Quantum states must be generated using electrical excitation.

In order to realize this wish list, most likely, hybrid approaches are the way to go, where optical adiabatic passages between one platform and another might be the key, and optical impedance matching between components might be the solution to overcome these challenges.

2.1.6 Summary and impact

It is quite clear that quantum technologies in the broad sense, whether we are talking about quantum communications, quantum sensing, quantum simulation, or quantum computing, will have a tremendous effect and importance in tomorrow's technology. It is thus also clear that quantum systems are able to provide the basic tools for such technologies and that they will work hand in hand with nanosystems. Indeed, miniaturization does not just offer small-size effects but also changes drastically the intrinsic properties of a system (Purcell effect, single-photon emission, etc.). Thus, the study and the control of quantum emitters, and notably via nanospectroscopy, is of paramount importance and will in fact be even more important as the manipulation and crucial strategic control at the nanoscale becomes more and more precise and relevant fabrication technologies improve. This is perhaps the next hurdle for nanospectroscopy in general, to be able to harness nanosystems at the quantum level, and there are certainly many opportunities emerging that offer enormous potential for research and industrial applications.

Bibliography

[1]   Chen D, Chen X. Luminescent perovskite quantum dots: synthesis, microstructures, optical properties and applications. J Mater Chem C. 2019;7:1413.
[2]   Pierini S, D'Amato M, Goyal M, Glorieux Q, Giacobino E, Lhuillier E, Couteau C, Bramati A. Highly photostable perovskite nanocubes: toward integrated single photon sources based on tapered nanofibers. ACS Photonics. 2020;7(8):2265–72.
[3]   Lukin DM, Guidry MA, Vuckovic J. Integrated quantum photonics with silicon carbide: challenges and prospects. PRX. 2020;1:020102.
[4]   Reserbat-Plantey A, Epstein I, Torre I, Costa AT, Goncalves PAD, Asger Mortensen N, Polini M, Song JCW, Peres NMR, Koppens FHL. Quantum nanophotonics in two-dimensional materials. ACS Photonics. 2021;8:85–101.
[5]   Hong CK, Ou ZY, Mandel L. Measurement of subpicosecond time intervals between two photons by interference. Phys Rev Lett. 1987;59:2044–6.
[6]   Loudon R. The Quantum Theory of Light. 3rd ed. Oxford University Press; 2000.

[7]   Einstein A, Podolsky B, Rosen N. Can quantum-mechanical description of physical reality be considered complete? Phys Rev. 1935;47:777–80.
[8]   Bell J. On the Einstein Podolsky Rosen paradox. Physics. 1964;1:195–200.
[9]   Northup TE, Blatt R. Quantum information transfer using photons. Nat Photonics. 2014;8:356.
[10]   Chang DE, Vuletic V, Lukin MD. Quantum nonlinear optics – photon by photon. Nat Photonics. 2014;8:685.
[11]   Pfaff W, Hensen BJ, Bernien H, van Dam SB, Blok MS, Taminiau TH, Tiggelman MJ, Schouten RN, Markham M, Twitchen DJ, Hanson R. Unconditional quantum teleportation between distant solid-state quantum bits. Science. 2014;345:532.
[12]   Knowles HS, Kara DM, Atatüre M. Observing bulk diamond spin coherence in high-purity nanodiamonds. Nat Mater. 2014;13:21–5.
[13]   Senellart P, Solomon G, White A. High-performance semiconductor quantum-dot single-photon sources. Nat Nanotechnol. 2017;12:1026.
[14]   Lodahl P. Quantum-dot based photonic quantum networks. Quantum Sci Technol. 2018;3:013001.
[15]   Pietryga JM, Park Y-S, Lim J, Fidler AF, Bae WK, Brovelli S, Klimov VI. Spectroscopic and device aspects of nanocrystal quantum dots. Chem Rev. 2016;116:10513–10622.
[16]   Toninelli C, Gerhardt I, Clark AS, Reserbat-Plantey A, Götzinger S, Ristanović Z, Colautti M, Lombardi P, Major KD, Deperasińska I, Pernice WH, Koppens FHL, Kozankiewicz B, Gourdon A, Sandoghdar V, Orrit M. Single organic molecules for photonic quantum technologies. Nat Mater. 2021;20:1615–28.
[17]   Purcell EM. Spontaneous emission properties at radio frequencies. Phys Rev. 1946;69:681.
[18]   Drexhage KH. Influence of a dielectric interface on fluorescence decay time. J Lumin. 1970;1–2:693.
[19]   Goy P, Raimond JM, Gross M, Haroche S. Observation of cavity-enhanced single-atom spontaneous emission. Phys Rev Lett. 1983;50:1903.
[20]   Hulet RG, Hilfer ES, Kleppner D. Inhibited spontaneous emission by a Rydberg atom. Phys Rev Lett. 1985;55:2137.
[21]   De Martini F, Innocenti G, Jacobovitz GR, Mataloni P. Anomalous spontaneous emission time in a microscopic optical cavity. Phys Rev Lett. 1987;59:2955.
[22]   Buchler BC, Kalkbrenner T, Hettich C, Sandoghdar V. Measuring the quantum efficiency of the optical emission of single radiating dipoles using a scanning mirror. Phys Rev Lett. 2005;95:063003.
[23]   Toninelli C, Delley Y, Stöferle T, Renn A, Götzinger S, Sandoghdar V. A scanning microcavity for in situ control of single-molecule emission. Appl Phys Lett. 2010;97:021107.
[24]   Chizhik AI, Chizhik AM, Khoptyar D, Bär S, Meixner AJ, Enderlein J. Probing the radiative transition of single molecules with a tunable microresonator. Nano Lett. 2011;11(4):1700–3.
[25]   Chizhik AI, Chizhik AM, Kern AM, Schmidt T, Potrick K, Huisken F, Meixner AJ. Measurement of vibrational modes in single nanoparticles using a tunable metal resonator with optical subwavelength dimensions. Phys Rev Lett. 2012;109:223902.
[26]   Weisbuch C, Nishioka M, Ishikawa A, Arakawa Y. Observation of the coupled exciton-photon mode splitting in a semiconductor quantum microcavity. Phys Rev Lett. 1992;69:3314.
[27]   Gérard J-M, Barrier D, Marzin J-Y, Kuszelewicz R, Manin L, Costard E, Thierry-Mieg V, Rivera T. Quantum boxes as active probes for photonic microstructures: the pillar microcavity case. Appl Phys Lett. 1996;69:449.
[28]   Gérard J-M, Sermage B, Gayral B, Legrand B, Costard E, Thierry-Mieg V. Enhanced spontaneous emission by quantum boxes in a monolithic optical microcavity. Phys Rev Lett. 1998;81:1110.
[29]   Michler P, Kiraz A, Becher C, Schoenfeld WV, Petroff PM, Zhang L, Hu E, Imamoglu A. A quantum dot single-photon turnstile device. Science. 2000;290:2282.

[30] Michler P, Imamoglu A, Mason MD, Carson PJ, Strouse GF, Buratto SK. Quantum correlation among photons from a single quantum dot at room temperature. Nature. 2000;406:968.

[31] Somaschi N, Giesz V, De Santis L, Loredo JC, Almeida MP, Hornecker G, Portalupi SL, Grange T, Antón C, Demory J, Gómez C, Sagnes I, Lanzillotti-Kimura ND, Lemaître A, Auffeves A, White AG, Lanco L, Senellart P. Near-optimal single-photon sources in the solid state. Nat Photonics. 2016;10:340.

[32] Dousse A, Suffczyński J, Beveratos A, Krebs O, Lemaître A, Sagnes I, Bloch J, Voisin P, Senellart P. Ultrabright source of entangled photon pairs. Nature. 2010;466:217–20.

[33] Dousse A, Lanco L, Suffczyński J, Semenova E, Miard A, Lemaître A, Sagnes I, Roblin C, Bloch J, Senellart P. Controlled light-matter coupling for a single quantum dot embedded in a pillar microcavity using far-field optical lithography. Phys Rev Lett. 2008;101:267404.

[34] Santori C, Fattal D, Vuckovic J, Solomon GS, Yamamoto Y. Indistinguishable photons from a single-photon device. Nature. 2002;419:594.

[35] Benson O, Santori C, Pelton M, Yamamoto Y. Regulated and entangled photons from a single quantum dot. Phys Rev Lett. 2000;84:2513.

[36] Akopian N, Lindner NH, Poem E, Berlatzky Y, Avron J, Gershoni D, Gerardot BD, Petroff PM. Entangled photon pairs from semiconductor quantum dots. Phys Rev Lett. 2006;96:130501.

[37] Stevenson RM, Young RJ, Atkinson P, Cooper K, Ritchie DA, Shields AJ. A semiconductor source of triggered entangled photon pairs. Nature. 2006;439:179.

[38] Muller A, Fang W, Lawall J, Solomon GS. Creating polarization-entangled photon pairs from a semiconductor quantum dot using the optical Stark effect. Phys Rev Lett. 2009;103:217402.

[39] Vogel MM, Ulrich SM, Hafenbrak R, Michler P, Wang L, Rastelli A, Schmidt OG. Influence of lateral electric fields on multiexcitonic transitions and fine structure of single quantum dots. Appl Phys Lett. 2007;91:051904.

[40] Seidl S, Kroner M, Högele A, Karrai K, Warburton RJ, Badolato A, Petroff PM. Effect of uniaxial stress on excitons in a self-assembled quantum dot. Appl Phys Lett. 2006;88:203113.

[41] Wan Y, Li Q, Liu AY, Chow WW, Gossard AC, Bowers JE, Hu EL, Lau KM. Sub-wavelength InAs quantum dot micro-disk lasers epitaxially grown on exact Si (001) substrates. Appl Phys Lett. 2016;108:221101.

[42] Yablonovitch E. Inhibited spontaneous emission in solid-state physics and electronics. Phys Rev Lett. 1987;58:2059–62.

[43] John S. Strong localization of photons in certain disordered dielectric superlattices. Phys Rev Lett. 1987;58:2486–9.

[44] Yablonovitch E, Gmitter TJ, Leung KM. Photonic band structure: the face-centered-cubic case employing nonspherical atoms. Phys Rev Lett. 1991;67:2295.

[45] Painter O, Lee RK, Scherer A, Yariv A, O'Brien JD, Dapkus PD, Kim I. Two-dimensional photonic band-gap defect mode laser. Science. 1999;284:1819–21.

[46] Noda S, Chutinan A, Imada M. Trapping and emission of photons by a single defect in a photonic bandgap structure. Nature. 2000;407:608–10.

[47] Akahane Y, Asano T, Song B-S, Noda S. High-Q photonic nanocavity in a two-dimensional photonic crystal. Nature. 2003;425:944–7.

[48] Yoshie T, Scherer A, Hendrickson J, Khitrova G, Gibbs HM, Rupper G, Ell C, Shchekin OB, Deppe DG. Vacuum Rabi splitting with a single quantum dot in a photonic crystal nanocavity. Nature. 2004;432:200–3.

[49] Nemoto K, Trupke M, Devitt SJ, Stephens AM, Scharfenberger B, Buczak K, Nöbauer T, Everitt MS, Schmiedmayer J, Munro WJ. Photonic architecture for scalable quantum information processing in diamond. Phys Rev X. 2014;4:031022.

[50] Riedrich-Möller J, Kipfstuhl L, Hepp C, Neu E, Pauly C, Mücklich F, Baur A, Wandt M, Wolff S, Fischer M, Gsell S, Schreck M, Becher C. One- and two-dimensional photonic crystal microcavities in single crystal diamond. Nat Nanotechnol. 2012;7:69–74.

[51] Sipahigil A, Evans RE, Sukachev DD, Burek MJ, Borregaard J, Bhaskar MK, Nguyen CT, Pacheco JL, Atikian HA, Meuwly C, Camacho RM, Jelezko F, Bielejec E, Park H, Loncar M, Lukin MD. Single-photon switching and entanglement of solid-state qubits in an integrated nanophotonic system. ArXiv 2016. 1608.05147v1.

[52] Ruf M, Wan NH, Choi H, Englund D, Hanson R. Quantum networks based on color centers in diamond. J Appl Phys. 2021;130:070901.

[53] Cadeddu D, Teissier J, Braakman FR, Gregersen N, Stepanov P, Gerard J-M, Claudon J, Warburton RJ, Poggio M, Munsch M. A fiber-coupled quantum-dot on a photonic tip. Appl Phys Lett. 2016;108:011112.

[54] Patel RN, Schröder T, Wan N, Li L, Mouradian SL, Chen EH, Englund DR. Efficient photon coupling from a diamond nitrogen vacancy center by integration with silica fiber. Light: Sci Appl. 2016;5:e16032.

[55] Davanço M et al. Efficient quantum dot single photon extraction into an optical fiber using a nanophotonic directional coupler. Appl Phys Lett. 2011;99:121101.

[56] Petersen J, Volz J, Rauschenbeutel A. Chiral nanophotonic waveguide interface based on spin-orbit interaction of light. Science. 2014;346:67.

[57] Coles RJ, Price DM, Dixon JE, Royall B, Clarke E, Kok P, Skolnick MS, Fox AM, Makhonin MN. Chirality of nanophotonic waveguide with embedded quantum emitter for unidirectional spin transfer. Nat Commun. 2016;7:11183.

[58] Khasminskaya S, Pyatkov F, Słowik K, Ferrari S, Kahl O, Kovalyuk V, Rath P, Vetter A, Hennrich F, Kappes MM, Gol'tsman G, Korneev A, Rockstuhl C, Krupke R, Pernice WHP. Fully integrated quantum photonic circuit with an electrically driven light source. Nat Photonics. 2016;10:727.

[59] Prtljaga N, Coles RJ, O'Hara J, Royall B, Clarke E, Fox AM, Skolnick MS. Monolithic integration of a quantum emitter with a compact on-chip beam-splitter. Appl Phys Lett. 2014;104:231107.

[60] Reithmaier G, Kaniber M, Flassig F, Lichtmannecker S, Müller K, Andrejew A, Vučković J, Gross R, Finley JJ. On-chip generation, routing, and detection of resonance fluorescence. Nano Lett. 2015;15:5208.

[61] Claudon J, Bleuse J, Malik NS, Bazin M, Jaffrennou P, Gregersen N, Sauvan C, Lalanne P, Gerard J-M. A highly efficient single-photon source based on a quantum dot in a photonic nanowire. Nat Photonics. 2010;4:174.

[62] Zadeh IE, Elshaari AW, Jöns KD, Fognini A, Dalacu D, Poole PJ, Reimer ME, Zwiller V. Deterministic integration of single photon sources in silicon based photonic circuits. Nano Lett. 2016;16:2289–94.

[63] Madrigal JB, Tellez-Limon R, Gardillou F, Barbier D, Geng W, Couteau C, Salas-Montiel R, Blaize S. Hybrid integrated optical waveguides in glass for enhanced visible photoluminescence of nanoemitters. Appl Opt. 2016;55:10263–8.

[64] Moody G et al. 2022 Roadmap on integrated quantum photonics. J Phys Photonics. 2022;4:012501.

[65] D'Amico I, Angelakis D, Bussières F, Caglayan H, Couteau C, Durt T, Kolarić B, Maletinsky P, Pfeiffer W, Rabl P, Xuereb A, Agio M. Nanoscale quantum optics. Riv Nuovo Cimento. 2019;42:153.

[66] Wang J, Sciarrino F, Laing A, Thompson MG. Integrated photonic quantum technologies. Nat Photonics. 2020;14:273.

[67] Vlasov TA, McNab SJ. Losses in single-mode silicon-on-insulator strip waveguides and bends. Opt Express. 2004;12:1622.

[68] Dwivedi S, Van Vaerenbergh T, Ruocco A, Spuesens T, Bienstman P, Dumon P, Bogaerts W. Measurements of effective refractive index of SOI waveguides using interferometers. In: Integrated Photonics Research, Silicon and Nanophotonics conference 2015. 2015.

[69] Refractive index database, https://refractiveindex.info (accessed on September 15, 2022).

[70] Subramanian AZ, Neutens P, Dhakal A, Jansen R, Claes T, Rottenberg X, Peyskens F, Selvaraja S, Helin P, Du Bois B, Leyssens K, Severi S, Deshpande P, Baets R, Van Dorpe P. Low-loss singlemode PECVD silicon nitride photonic wire waveguides for 532–900 nm wavelength window fabricated within a CMOS pilot line. IEEE Photonics J. 2013;50:2202809.

[71] Vergyris P, Meany T, Lunghi T, Downes J, Steel MJ, Withford MJ, Alibart O, Tanzilli S. On-chip generation of heralded photon-number states. Sci Rep. 2016;6:35975.

[72] Tervonen A, Honkanen SK, West BR. Ion-exchanged glass waveguide technology: a review. Opt Eng. 2011;50:071107.

[73] Kinkhabwala A, Yu Z, Fan S, Avlasevich Y, Mullen K, Moerner WE. Large single-molecule fluorescence enhancements produced by a bowtie nanoantenna. Nat Photonics. 2009;3:654.

[74] Rahbany N, Geng W, Blaize S, Salas-Montiel R, Bachelot R, Couteau C. Integrated plasmonic double bowtie / ring grating structure for enhanced electric field confinement. Nanospectroscopy. 2015;1:61–6.

[75] Rahbany N, Geng W, Bachelot R, Couteau C. Plasmon-emitter interaction using integrated ring grating-nanoantenna structures. Nanotechnology. 2017;28(18):185201.

[76] Schietinger S, Barth M, Aichele T, Benson O. Plasmon-enhanced single photon emission from a nanoassembled metal-diamond hybrid structure at room temperature. Nano Lett. 2009;9:1694–8.

[77] Hoang TB, Akselrod GM, Mikkelsen MH. Ultrafast room-temperature single photon emission from quantum dots coupled to plasmonic nanocavities. Nano Lett. 2015;16:270.

[78] Chikkaraddy R, de Nijs B, Benz F, Barrow SJ, Scherman OA, Rosta E, Demetriadou A, Fox P, Hess O, Baumberg JJ. Single-molecule strong coupling at room temperature in plasmonic nanocavities. Nature. 2016;535:127.

[79] Lee KG, Chen XW, Eghlidi H, Kukura P, Lettow R, Renn A, Sandoghdar V, Goetzinger S. A planar dielectric antenna for directional single-photon emission and near-unity collection efficiency. Nat Photonics. 2011;5:166–9.

[80] Geng W, Manceau M, Rahbany N, Sallet V, De Vittorio M, Carbone L, Glorieux Q, Bramati A, Couteau C. Localised excitation of a single photon source by a nanowaveguide. Sci Rep. 2016;6:19721.

[81] Ge D et al. Hybrid plasmonic nano-emitters with controlled single quantum emitter positioning on the local excitation field. Nat Commun. 2020;11:3414.

[82] Luo Y, Ahmadi ED, Shayan K, Ma Y, Mistry KS, Zhang C, Hone J, Blackburn JL, Stauf S. Purcell-enhanced quantum yield from carbon nanotube excitons coupled to plasmonic nanocavities. Nat Commun. 2017;8:1413.

[83] Rousseaux B, Dzsotjan D, Colas des Francs G, Jauslin HR, Couteau C, Guérin S. Adiabatic passage mediated by plasmons: a route towards a decoherence-free quantum plasmonic platform. Phys Rev B. 2016;93:045422.

[84] Roger T, Vezzoli S, Bolduc E, Valente J, Heitz JJF, Jeffers J, Soci C, Leach J, Couteau C, Zheludev N, Faccio D. Coherent perfect absorption in deeply subwavelength films in the single photon regime. Nat Commun. 2015;6:7031.

[85] Akimov AV, Mukherjee A, Yu CL, Chang DE, Zibrov AS, Hemmer PR, Park H, Lukin MD. Generation of single optical plasmons in metallic nanowires coupled to quantum dots. Nature. 2007;450:402–6.

[86] Altewischer E, van Exter MP, Woerdman JP. Plasmon-assisted transmission of entangled photons. Nature. 2002;418:304–6.

[87] Fasel S, Robin F, Moreno E, Erni D, Gisin N, Zbinden H. Energy-time entanglement preservation in plasmon-assisted light transmission. Phys Rev Lett. 2005;94:110501.

[88] Franson JD. Bell-inequality for position and time. Phys Rev Lett. 1989;62:2205.

[89] Tame MS, McEnery KR, Özdemir SR, Lee J, Maier SA, Kim MS. Quantum plasmonics. Nat Phys. 2013;9:329.

[90]   Bozhevolnyi SI, Khurgin JB. The case for quantum plasmonics. Nat Photonics. 2017;11:398.
[91]   Chang DE, Sorensen AS, Demler EA, Lukin MD. A single-photon transistor using nanoscale surface plasmons. Nat Phys. 2007;3:807.
[92]   Kolesov R, Grotz B, Balasubramanian G, Stöhr RJ, Nicolet AAL, Hemmer PR, Jelezko F, Wrachtrup J. Wave–particle duality of single surface plasmon polaritons. Nat Phys. 2009;5:470–4.
[93]   Di Martino G, Sonnefraud Y, Kena-Cohen S, Tame M, Özdemir SK, Kim MS, Maier SA. Quantum statistics of surface plasmon polaritons in metallic stripe waveguides. Nano Lett. 2012;12:2504.
[94]   Huck A, Smolka S, Lodahl P, Sørensen AS, Boltasseva A, Janousek J, Andersen UL. Demonstration of quadrature-squeezed surface plasmons in a gold waveguide. Phys Rev Lett. 2009;102:246802.
[95]   Fakonas JS, Lee H, Kelaita YA, Atwater HA. Two-plasmon quantum interference. Nat Photonics. 2014;8:317.
[96]   Xu D, Xiong X, Wu L, Ren X-F, Png CE, Guo G-C, Gong Q, Xiao Y-F. Quantum plasmonics: new opportunity in fundamental and applied photonics. Adv Opt Photonics. 2018;10:703.
[97]   Zhou Z-K, Liu J, Bao Y, Wu L, Png CE, Wang X-H, Qiu C-W. Quantum plasmonics get applied. Prog Quantum Electron. 2019;65:1–20.
[98]   Heeres RW, Kouwenhoven LP, Zwiller V. Quantum interference in plasmonic circuits. Nat Nanotechnol. 2013;8:719–22.
[99]   Altuzarra C, Vezzoli S, Valente J, Gao W, Soci C, Faccio D, Couteau C. Coherent perfect absorption in metamaterials with entangled photons. ACS Photonics. 2017;4:2124.
[100] Du W, Wang T, Chu HS, Nijhuis CA. Highly efficient on-chip direct electronic–plasmonic transducers. Nat Photonics. 2017;11:623–8.
[101] Zhang C, Hugonin J-P, Coutrot A-L, Vest B, Greffet J-J. Electrical generation of visible surface plasmon polaritons by a nanopillars antenna array. APL Photonics. 2021;6:056102.
[102] Bogaerts W, Pérez D, Capmany J, Miller DAB, Poon J, Englund D, Morichetti F, Melloni A. Programmable photonic circuits. Nature. 2020;586:207.

Andreas Kaltzoglou, Athanassios G. Kontos, and Polycarpos Falaras

2.2 Role of nanospectroscopy in the development of third-generation photovoltaics

2.2.1 Key messages

- Third-generation photovoltaic (PV) systems have reached power conversion efficiency (PCE) values of up to 47.6 %, depending largely on the relevant technology, the specific cell arrangement, and the chemical composition, structure, morphology, and size of the incorporated materials, as well as the efficient charge collection at the corresponding nanostructured interfaces.
- High-resolution, nondestructive spectroscopic techniques permit the investigation of the optoelectronic and vibrational properties of the key components in solar cells.
- The nanospectroscopy data are related to the operation mechanism and device performance and can be used for the optimization of the emerging PV technologies, namely organic, dye-sensitized, quantum dot (QD), and perovskite solar cells (PSCs).

2.2.2 Pre-knowledge

The current chapter focuses on the application of optical nanospectroscopic techniques on solar cells. It is therefore vital for the reader to be familiar with the fundamental characteristics of semiconductors (Volume 1, Chapter 1.4), electronic spectroscopy (Volume 1, Chapter 2.2), and vibrational spectroscopy (Volume 1, Chapter 2.3). Materials and characterization methods in the nanoscale are described in Volume 2, Section 6, where also detailed information on nanoparticles, QDs, and nanocrystals is given, whereas scanning-probe microscopy techniques are discussed in Volume 2, Chapter 4.5. Moreover, a background is required in electrochemical characterization methods such as electrochemical impedance spectroscopy (EIS) and cyclic voltammetry (CV) [1].

The absorption, emission, and scattering of light are widely used to determine the electronic properties of PV materials. Experimentally, the molar extinction coefficients of molecular compounds used as photoactive layers are determined by the Beer–Lambert law. The optical band-gap of semiconducting films is determined with absorption measurements, while for nontransparent powder samples and rough films with diffuse reflectance measurements this quantity is determined using the Kubelka–Munk equation. An understanding of the electronic band structure of materials as well as the n-/p-type doping in semiconductors is also required to describe the operating principles of solar cells.

Acknowledgement: The authors gratefully acknowledge funding from FP7 (Marie Curie Initial Training Network DESTINY/316494) and Horizon 2020 (Marie Curie Innovative Training Network MAESTRO/764787) EU Programs.

https://doi.org/10.1515/9783110442908-004

The vibrational modes of PV components and their interfaces are investigated with infrared (IR) and Raman spectroscopies. These methods follow different selection rules for their active modes based on the symmetry of the materials under investigation. In the case of crystalline solids, the vibrational modes are referred to as phonons, and they provide valuable information on the physical properties and chemical bonding of the thin-film materials in solar cells. Micro-Raman spectroscopy is a scattering technique that allows confocal submicrometer lateral resolution and in-depth analysis, whereas IR spectroscopy is an absorption technique with typical lateral resolution above 3 μm. A big advantage of micro-Raman spectroscopy is the in situ study of solar cell components and corresponding interfaces under electrical load. Furthermore, the Raman investigation can be associated with luminescence effects, thus combining electronic and vibrational properties in solids and elucidating phenomena that affect significantly their PV performance.

2.2.3 Importance of the application

PV systems directly convert sunlight into electrical current. They accounted in 2021 for approximately 40 % of the renewable energy production worldwide [2], with applications ranging from small portable electronic devices to large-area power generators for the utility grid. In 2021 alone, the global solar energy capacity has increased by 168 GW, reaching a total of about 1 TW [3]. This is partly due to the large reduction in the last decades of the cost/power ratio, which is estimated to be below €0.5/Watt peak[1] for commercial Si-based solar cells. Nevertheless, no significant further performance increase is expected for single-junction Si-based devices. In search of inexpensive and more efficient solar cells, scientific research has shifted towards novel PV materials and device architectures.

This chapter focuses on the use of optical and vibrational nanospectroscopy for the investigation of third-generation solar cells. Apart from the evolution of conventional single-junction solar cells into multijunction solar cells and thin-film solar cells, emerging technologies such as organic solar cells (OSCs), dye-sensitized solar cells (DSSCs), QD solar cells (QDSCs), and PSCs are studied. Emphasis is given to PSCs, which are considered as one of the 10 main breakthroughs in science for 2013 and have received tremendous attention from the research community [4]. In this context, electronic as well as IR and Raman nanospectroscopy technologies investigate in situ innovative materials and optimize the corresponding interfaces. UV-Vis spectroscopy accurately determines molar extinction coefficients of sensitizers (both transition metal complexes and organic dyes) and direct and indirect band-gaps of semiconductors (including inorganic metal oxides and hybrid organic–inorganic perovskites) as well as photoluminescence (PL) properties. Vibrational spectroscopy offers valuable information about the chemical bonding in bulk materials and on functional inter-

---

1 There have been significant price increases since the beginning of 2022, however, e. g., due to a doubling of the price of the polysilicon raw material.

faces. Macroscopic characteristics, such as the amorphous or crystalline character of the materials in use, may also be investigated by combining the above methods. Recent advances in instrumentation using Raman spectrometers with high-power and micrometer-size beams have made a large contribution to understanding, tuning and optimizing the material properties and performance of solar cells. Optical nanospectroscopy technologies also investigate the operation mechanisms and possible degradation pathways following accelerated stress tests by exposing solar cells to heat, irradiation, and humidity. These techniques can be combined with electrochemical methods (e. g., EIS), focusing on our ultimate goals to provide additional data on the device lifetime and to ensure their enhanced stability for real-world applications.

2.2.4 State-of-the-art

In general, the operating principles of solar cells involve a doped semiconductor or a photosensitizer that absorbs solar light creating electron–hole pairs. The charge carriers are then transferred to the corresponding electrodes, producing a potential barrier. The maximum voltage at zero current flow is known as open-circuit voltage (V_{OC}), whereas the maximum current is observed in short-circuit conditions (I_{SC}). In practice, the efficiency of a solar cell is determined by the J-V measurements under standard test conditions of simulated 1000 W m^{-2} solar irradiance and 25 °C cell temperature. The parameter J refers to the current density for the optically active area. The ratio of maximum electrical power (P_{max}) over the power of the incident photons is known as PCE. Another important parameter for the overall capacity of a solar cell is the fill factor (FF), as it states the deviation from the ideal behavior in the J-V curves (where FF = 1):

$$FF = \frac{P_{max}}{P_{theor}} = \frac{V_{Pmax}I_{Pmax}}{V_{OC}I_{SC}}.$$

Various effects such as blackbody radiation, radiative recombination, and spectrum losses hinder the complete conversion of incoming photons into electrical current. Theoretically, PCE is limited to 33.7 % according to the Shockley–Queisser model for a single-junction solar cell using a semiconductor with an optimum band-gap of 1.34 eV. In the 1970s, solar cells based on crystalline Si could reach PCE values of approximately 15 %. Currently, the PCE of third-generation solar cells exceeds 45 % for some laboratory-scale, multijunction devices (Fig. 2.2.1) [5, 6].

Electronic and vibrational spectroscopic techniques are generally applied to the characterization of nanostructured materials of the third-generation PV devices. Spectroscopic techniques (e. g., resonance Raman, surface-enhanced Raman spectroscopy, attenuated Fourier transform IR (FTIR) reflectance, cathodoluminescence (CL)) have advanced throughout the years in both instrumentation and theoretical

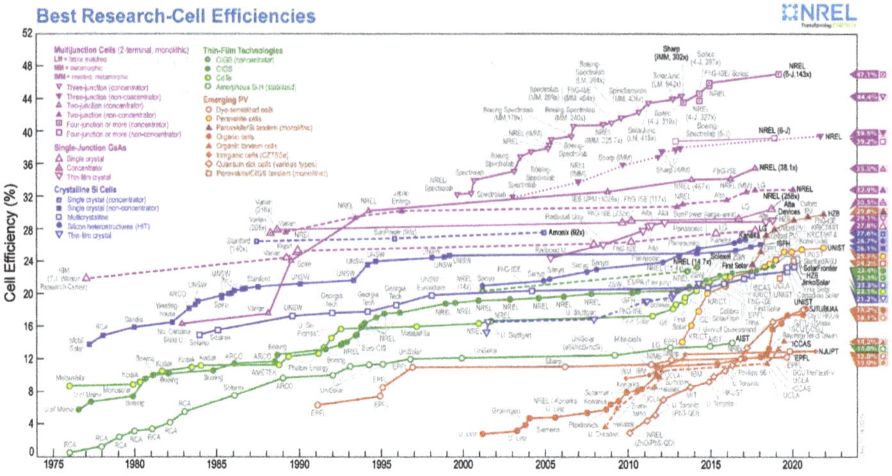

Figure 2.2.1: Efficiency chart of various types of solar cells. This plot is courtesy of the National Renewable Energy Laboratory, Golden, CO [7].

background. In this context, standard spectroscopic techniques that characterize the properties of the materials in the nanoscale can be classified as nanospectroscopic tools. In a strict terminology, however, nanospectroscopy techniques are those which improve the lateral resolution to several nm.

Fourier transform IR absorption nanospectroscopy (nano-FTIR) has reached a spatial resolution of 20 nm [8]. This is based on a scattering-type scanning near-field optical microscope (s-SNOM) equipped with a coherent-continuum IR light source. The method can determine the IR absorption spectrum of organic samples (e. g., poly(methyl methacrylate), which is used as a protective film on solar cells) with a spatial resolution of 20 nm. Electrochemical tip-enhanced Raman spectroscopy (TERS) is also a promising nanospectroscopy method, which allows for chemical imaging at the single-molecule level [9]. Moreover, electrochemical surface-enhanced Raman spectroscopy (EC-SERS) further provides the flexibility to control the applied potential in order to tune the plasmon resonance frequency as well as the interaction of molecules with their substrate.

In the following subsections, the most promising emerging technologies are classified based on the device operation principle. For completeness of the study, examples of combined use of electrochemical or spectroelectrochemical methods such as CV, EIS, and intensity-modulated photocurrent spectroscopy (IMPS) are briefly mentioned as they determine crucial parameters of solar cells, namely the diffusion coefficient, diffusion length, transport time, and recombination rate of the electrons, as well as the band-gap of the semiconductors.

2.2.4.1 Multijunction and thin-film solar cells

Multijunction (also known as "tandem") and thin-film solar cells are based on Si, binary group 13/15 compounds (e. g., GaAs), binary group 12/16 compounds (e. g., CdTe), and solid solutions of the composition $CuIn_{1-x}Ga_xS_2$ (CIGS) and $Cu(Zn,Sn)(S,Se)_2$ (CZTS). The use of multiple epitaxially grown layers of semiconducting materials with different band-gaps renders the solar spectrum exploitation more efficient while electrical resistance losses are kept to a minimum. This approach has allowed researchers to overcome the Shockley–Queisser limit, reaching a record performance of approximately 48 % for a four-junction GaInP/GaAs/GaInAsP/GaInAs system [10, 11, 7]. As in the case of traditional single-junction silicon solar cells, this category requires very high control of the materials purity and crystal quality, in order to increase electron lifetimes and avoid recombination effects. The study of the initial growth phase of a microcrystalline silicon (μc-Si:H) intrinsic absorber layer is achieved with in situ Raman measurements with high time resolution [12]. The initial layer growth is found to be influenced by the properties of the underlying seed layer. At constant deposition conditions, the initial layer growth is more amorphous than the seed layer and more amorphous than the bulk material, which is deposited afterwards (Fig. 2.2.2). The silane concentration during the initial phase can be controlled in order to decrease the amorphous volume fraction that is formed at the beginning of a μc-Si:H deposition.

With regard to CIGS and CZTS solar cells, the record PCE value is 23.4 % [7]. Polycrystalline films are made using sputtering or evaporation from the constituent elements, but due to their complex stoichiometry, many secondary phases are possi-

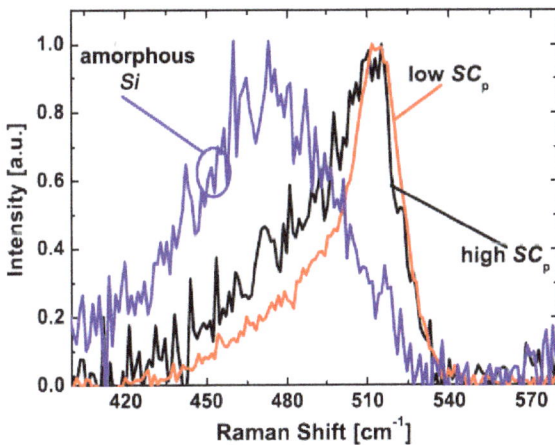

Figure 2.2.2: Stokes Raman shift of ~20 nm thick microcrystalline silicon p-layers deposited with different growth parameters after subtraction of the contribution of the underlying ZnO and normalization to their peak intensities. SC_p = applied silane concentration. Adapted with permission from [12].

ble, which potentially result in a decrease of the solar cell efficiency. Raman spectroscopy has also been used to study the crystal growth of CuInS(Se)$_2$ films for solar cell applications [13]. Raman spectra are sensitive to changes in the composition of CuIn(Se,S)$_2$ alloys [14]. By the analysis of quasi-resonant Raman scattering data obtained at a fixed excitation wavelength of 785 nm on S-rich CuIn(S,Se)$_2$ absorbers it was demonstrated that this technique is well suited for the determination of the relative content in CuIn(S,Se)$_2$ quaternary alloy layers. Resonant enhancement leads to a strong increase in the intensity of the peaks that show a strong dependence on the S content in the layers. Analysis of the broad luminescence background signal in the spectra provides an additional tool for the determination of the layer composition because of the direct linear correlation between the exciton energy and the alloy composition. CL spectroscopy is also used to analyze inhomogeneities on a submicrometer scale of active CuIn(Se,S)$_2$ absorber layers [15]. Similarly, the size of cadmium selenide nanocrystals is precisely determined by fluorescence spectroscopy [16].

2.2.4.2 Organic solar cells

OSCs use either conductive polymers or small organic molecules in order to absorb light. Although these solar devices have lower efficiency (record value 18.2 % [5]) and shorter lifetimes than Si-based systems, they are readily processable either from solution or by sublimation of the organic molecules on flexible substrates. The annealing-induced molecular ordering in bulk heterojunction (BHJ) polymer solar cells can be studied using Raman and PL spectroscopies at high resolution of ca. 10 nm [17] as well as using a combined atomic force microscopy (AFM)-IR analysis [18, 19]. Multitemperature Raman and optical microscopy of poly(3-hexylthiophene) (P3HT):[6,6]-phenyl-C$_{71}$ butyric acid methyl ester (PCBM) films were carried out to investigate the structural and morphological evolution during heating and cooling cycles of thermal annealing [20] (Fig. 2.2.3). The analysis revealed that molecular disorder in P3HT chains increases during a heating cycle and that on cooling the molecular ordering improves rapidly at ca. 100 °C due to thermally induced $\pi-\pi$ stacking. However, heating to higher temperatures leads to the formation of PCBM aggregates, which hampers the ordering of P3HT during the cooling cycle. The mechanism of morphology evolution in P3HT (donor):PCBM (acceptor) films is directly correlated with the efficiency of P3HT:PCBM solar cells.

Another report investigated in situ the molecular vibrations of two organic semiconductors in PV blends and their impact on thin-film morphology [21]. Blend films were composed of a low band-gap copolymer thieno[3,2-b]thiophene-diketopyrrolopyrrole (DPPTTT) and (6,6)-phenyl-C$_{71}$-butyric acid ester (PC$_{70}$BM). Changes in Raman spectra associated with crystallization processes of each component and their impact on thin-film morphology were studied during thermal annealing and cooling processes. Transition temperatures to crystalline phases in blends were measured at

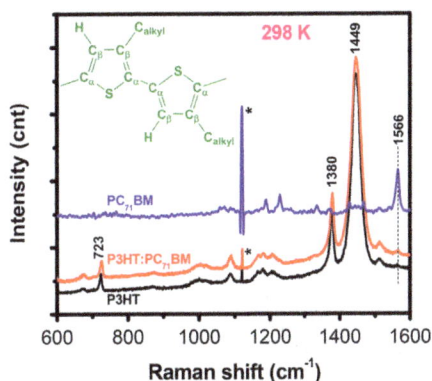

Figure 2.2.3: Raman spectra of as-cast P3HT, PCBM, and P3HT:PCBM films recorded using 514 nm excitation. The inset shows the molecular structure of P3HT. The substrate peak is represented by (*). Reprinted with permission from [20].

ca. 150 °C and 170 °C for DPPTTT and PC70BM, respectively. Such phase changes lead to modifications in local chemical composition reducing relative Raman peak intensities (I_{PC70BM}/I_{DPPTTT}) from ca. 0.4 in PC70BM-rich domains to ca. 0.15 in homogeneous areas.

Characterization with steady-state and time-resolved fluorescence spectroscopy also indicated an improved anode interlayer/BHJ interface quality in the case of hydrogen molybdenum bronze interlayers compared to stoichiometric metal oxide ones. The insertion of the s-Mo bronzes as anode interlayers universally enhanced the efficiency of organic PV cells based on BHJ mixtures of different polymeric donors and fullerene acceptors regardless of the specific combination of donor–acceptor employed. The results demonstrate that solution-processable hydrogen molybdenum bronzes can effectively be used as hole extraction layers in organic PV devices and provide a simple and versatile method to optimize polymer solar cells using simple and cost-effective materials [22].

A model BHJ of a perylene diimide (PDI) monomeric derivative has been studied (Fig. 2.2.4) [23]. In particular, this involves the process of PDI aggregation in PV layers of the model BHJ poly(indenofluorene) (PIF):PDI, in which a monomeric PDI derivative is utilized as the electron acceptor. The size of the PDI aggregates in the PIF:PDI layers was tuned by thermal annealing. Aggregate formation has a large effect on the photophysics of the PIF:PDI films and on the electrical properties of the corresponding PIF:PDI PV devices. The evolution of the PDI aggregates was monitored by fluorescence microscopy, AFM, and scanning electron microscopy.

Good performance of all-organic polymeric optocouplers on glass and poly(ethylene) substrates comprising a green-emitting polymer light-emitting diode (PLED) and a P3HT:PCBM-based polymer photodetector was observed [24]. This work demon-

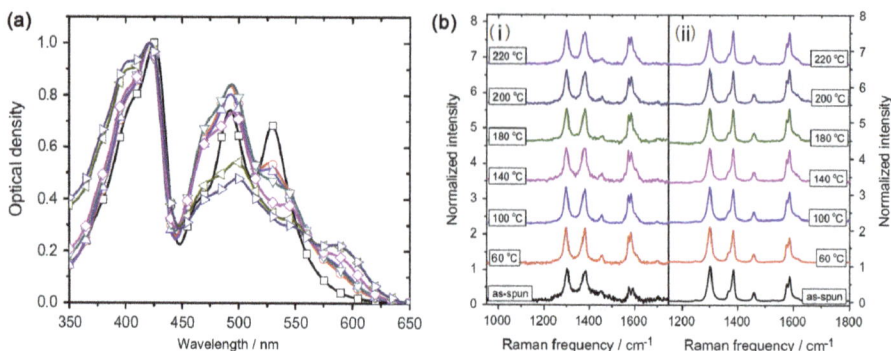

Figure 2.2.4: (a) Normalized UV-Vis spectra of PIF-Aryl:PDI 60 wt% blend films as spun (squares) and annealed at 60 °C (circles), 100 °C (up-facing triangles), 140 °C (down-facing triangles), 180 °C (diamonds), 200 °C (right-facing triangles), and 220 °C (left-facing triangles). (b) Normalized resonance Raman spectra of as-spun and thermally annealed (i) PIF-Aryl:PDI blend films and (ii) OSC devices. Reprinted with permission from [23].

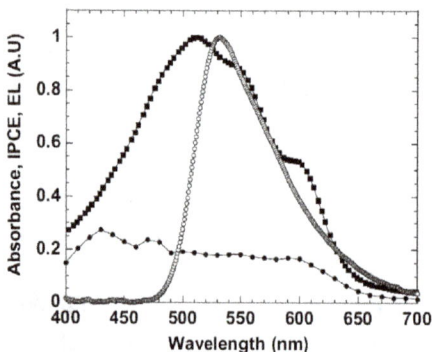

Figure 2.2.5: Absorption spectrum of P3HT:PCBM blend (full squares), electroluminescence spectrum of PLED (open circles), and incident-to-photocurrent conversion efficiency of the P3HT:PCBM photodetector (full circles). Reprinted with permission from [24].

strates the potential for all-plastic polymer optocouplers as a viable technology in future plastic optoelectronic circuits (Fig. 2.2.5).

2.2.4.3 Dye-sensitized solar cells

DSSCs have emerged as high-potential PVs for a large variety of applications, since the pioneering work of M. Grätzel in the early 1990s [25]. The innovative layered structure incorporated a large number of components where each one served a separate purpose: (a) a transparent electrode, such as indium tin oxide (ITO) or fluorine tin oxide (FTO), (b) a semiconducting mesoporous substrate (e. g., TiO_2 or ZnO), (c) a photosensitizer chemically adsorbed (in the form of a monolayer) on the semiconducting substrate that creates electron–hole pairs following excitation by sunlight, (d) an electrolyte (containing a redox couple) or hole-transporting material (HTM) (e. g., an organic HTM such as 2,2',7,7'-tetrakis(N,N-di-p-methoxyphenylamine)-9,9'-spirobifluorene (spiro-MeOTAD)), which transfers the positive charge carriers away from the photosensitizer, and (e) a counter electrode, usually made of an inert metal such as Au or Pt [26]. The use of HTMs leads to solid-state DSSCs usually endowed with robustness and long-term stability. Even though the efficiency is so far limited to 12–14 % [26, 27], there is great potential for increasing efficiency and reversibility of the photoelectrochemical processes, especially those related to the electrolyte transitions (Fig. 2.2.6).

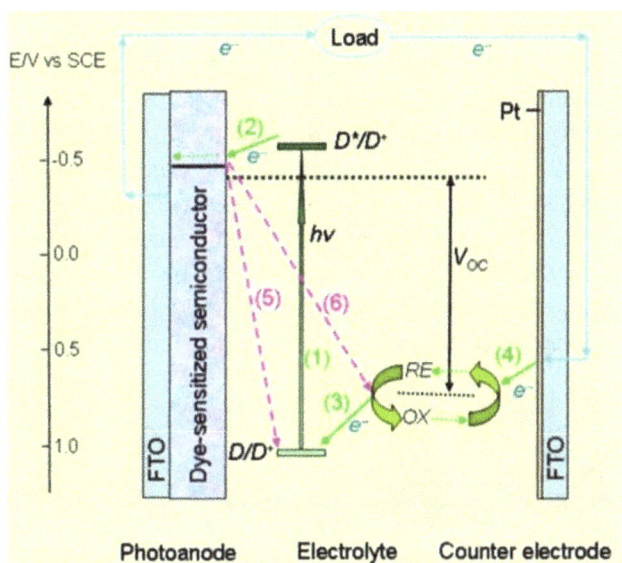

Figure 2.2.6: Energy diagram of a DSSC using a redox couple as electrolyte. The sequence of electrochemical processes is given with numbers, where green and red color denote desired and undesired processes, respectively. Reprinted with permission from [26].

Raman spectroscopy was used for the first time by the authors to investigate the dye sensitization on rough, high-surface area nanocrystalline TiO$_2$ electrodes [28]. It has been shown that it was possible to obtain and differentiate the spectra of a number of bis- or trisbipyridyl Ru(II) complexes adsorbed on the semiconductor surface in the form of a monomolecular layer. The attainment of the vibrational spectra of the chemisorbed dyes has been attributed to a high-resonance Raman effect in these chromophores [29, 30]. It has been confirmed that this effect can be observed with dyes anchored via both carboxylated bipyridyl and phosphonated terpyridyl bridging ligands.

In the last case, it has been observed that the phosphonated Ru-terpy complex and TiO$_2$ Raman vibration bands are strongly influenced by the electrical polarization of the sensitized photoelectrodes and the concentration of the electron donor-containing (I$^-$) electrolyte, thus opening new ways for the elucidation of the dye–semiconductor interaction and the optimization of the photosensitization phenomenon [31].

By combining Raman spectroscopy with EIS, the same research group characterized in situ DSSCs during their polarization, placing particular emphasis on the role of the iodide–triiodide (I$^-$/I$_3^-$) redox couple. Thus, the presence of new vibration bands in the low-wavenumber range was confirmed, assigned to intermediate species resulting from the interaction between the dye oxidized state and the iodides during the cell operation [32]. The group proved that this interaction is independent of the choice of the semiconductor and the dye and introduced the use of "resonance Raman spectrophotoelectrochemistry" in the field of DSCs as an adequate tool for device characterization and dynamic control of the corresponding interfaces [33].

In situ micro- and macro-Raman investigation of the I$^-$/I$_3^-$ redox couple in DSSCs was likewise performed by Falaras et al. [34]. Nanocrystalline solar cells using different dyes with poly-pyridyl ligands were studied. The triiodide vibrations in the 20–300 cm^{-1} range were examined by applying micro- and macro-Raman spectroscopy under variable bias of the cells (Fig. 2.2.7). Distinct regions in the current density–voltage curves correlate very well with the appearance of specific new Raman modes with variable intensity and frequency, depending on the applied potential. A strong Raman peak at 167 cm^{-1} as well as three weak peaks observed in the low-frequency range are attributed to vibrations of triiodide bound on the oxidized form of the dye via electrostatic forces.

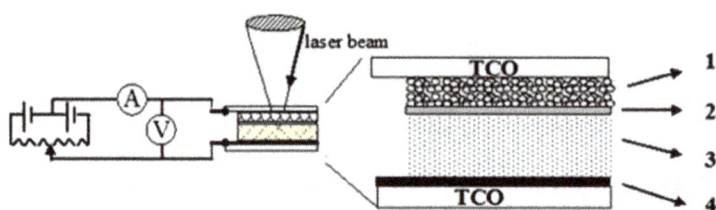

Figure 2.2.7: Schematic diagram of the Raman-spectrophotoelectrochemical setup, highlighting the high spatial resolution of the vibrational spectroscopic technique. The laser beam focuses on the photoelectrode–electrolyte interface and the *J-V* characteristics are simultaneously recorded. The DSSC structure is shown in magnification. Key: (1) porous TiO$_2$, (2) dye monolayer, (3) redox electrolyte, (4) Pt counter electrode (TCO = transparent conductive oxide). Reprinted with permission from [34].

The thermal stability of laboratory-size DSSCs based on industrially feasible materials and processes and using liquid electrolytes with two different organic solvents of high boiling point, namely methoxypropionitrile (MPN) and tetraglyme (TG), has been thoroughly investigated under prolonged thermal stress at 80 °C for 2000 h in

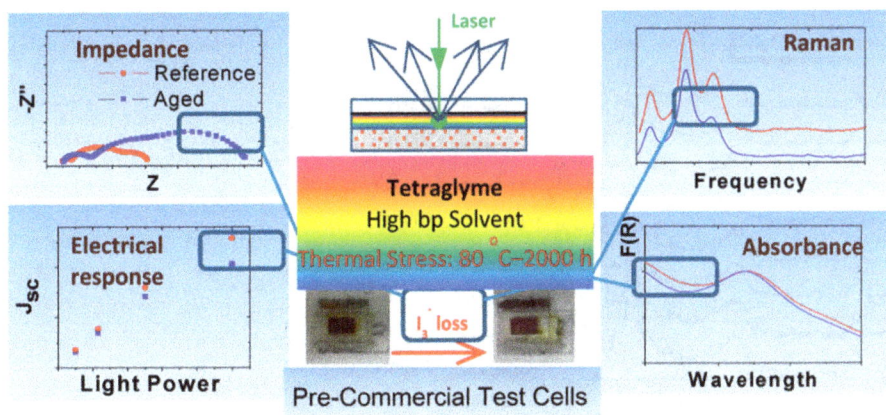

Figure 2.2.8: A combination of electrochemical, photoelectrochemical, optical, and spectroscopic techniques is applied to identify the degradation mechanisms and develop solutions toward stable DSSCs under prolonged thermal stress at high temperature. Reprinted with permission from [35].

the dark [35]. Comparative analysis of the Raman results has shown that TG presents remarkable failure resistance upon prolonged thermal stress at 80 °C for 2000 h in the dark (Fig. 2.2.8.), in perfect agreement with the evaluation of the optoelectronic properties of the devices.

The long-term stability of DSSCs following consecutive light and thermal stress has also been investigated by micro-Raman spectroscopy [36]. Changes in the Raman mode intensities have been observed for new and aged DSSCs as a function of the polarization bias, occurring both in the dye and in the anatase semiconductor (Fig. 2.2.9). This behavior along with the variation of the anatase Raman modes implies a modification of the interfacial electric field at the photoelectrode/electrolyte interface upon aging.

An electrochemical and in situ Raman spectroelectrochemical study of 1-methyl-3-propylimidazolium iodide ionic liquid (IL) electrolyte ($MPIm^+I_x^-$) in the absence ($x = 1$) and presence ($x = 3$) of iodine is presented by Jovanovski et al. [37] (Fig. 2.2.10). Cyclic voltammetric measurements in combination with a platinum disk microelectrode revealed a remarkable difference before and after addition of iodine to the IL, implying the electrochemical formation of polyiodide ions I_5^- at potentials more positive than +0.6 V in the presence of an equimolar amount of iodine.

Innovative solvent-free redox electrolytes for DSSCs were developed based on binary mixtures of ILs. The investigation of the physicochemical properties of the IL blends as a function of temperature has shown that they strongly depend on the type of the alkyl cation in the ILs' chemical structure. The Raman spectra of the corresponding electrolytes were dominated by the vibrational modes of the IL components in an additive way and confirmed the absence of any specific interaction, independent of the number of carbon atoms along the alkyl chain [38].

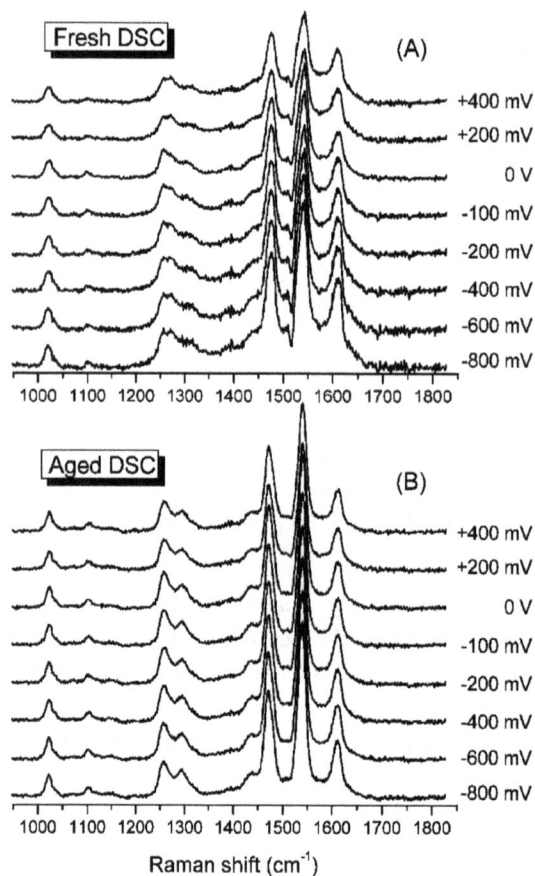

Figure 2.2.9: In situ micro-Raman spectra of the fresh (a) and aged (b) DSSCs in the dye vibration region, as a function of the polarization bias at 514.5 nm. Reprinted with permission from [36].

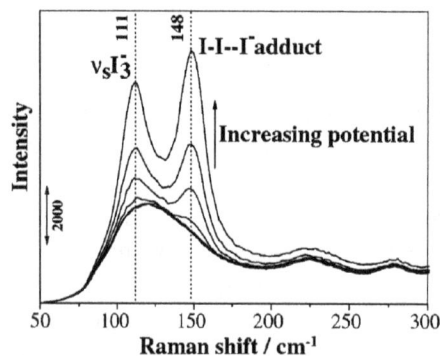

Figure 2.2.10: In situ Raman spectra of MPIm$^+$I$^-$ (x = 1) after applying anodic potentials from +0.2 V to 1.4 V in steps of 0.2 V. Reprinted with permission from [37].

Moreover, the behavior and properties of organic solvents typically used in redox electrolyte media for DSSCs have been systematically investigated by resonance Raman spectroscopy in combination with EIS, intensity-modulated photovoltage (IMVS), and photocurrent (IMPS) spectroscopies. Resonance Raman spectra reveal appreciable shifts in the vibrational frequency of the dye carboxyl anchoring groups as well as intensity variations of high- and low-frequency modes of dye/redox species by varying the electrolyte solvent and the polarization bias. These results are related to the variable surface coverage of the dye–TiO$_2$ photoelectrode by solvent molecules determined by their donor number, the concomitant change in the concentration of dye–redox couple intermediate adducts, and the dye–TiO$_2$ electronic coupling [39].

In order to improve the durability of DSSCs at high temperatures, ethyl isopropyl sulfone (EiPS) high-boiling point solvent was employed in the electrolyte. New species were formed by aging the electrolyte, which may modify the sensitized titania surface and reduce the photovoltage over the first stages of stressing at high temperatures. Related spectroscopic changes in the sensitized photoelectrode/electrolyte interface were detected via micro-Raman spectroscopy, the most prominent being the countervariation of the relative intensity ratio of the triiodide–dye coupling modes and the dye ones that depend quasi-linearly on the cell photocurrent in opposite directions [40].

Finally, it has also been shown that DSSCs using TG as a nonnitrile, high-boiling point, low-cost organic solvent for the iodide/triiodide redox shuttle can pass a harsh accelerated thermal aging test of 3000 h light soaking followed by an additional 2000 h of thermal aging at 85 °C. A combination of experimental techniques, namely linear sweep voltammetry, UV-Vis absorption, EIS, and micro-Raman spectroscopy, were applied to identify the underlying degradation mechanisms. This revealed a favorable conduction band edge shift for the TG-based cells effectively compensated for the diffusion limitations of the cell photocurrent induced by a moderate triiodide loss at high temperatures [41].

2.2.4.4 Quantum dot solar cells

This category of solar cells has emerged mainly from binary semiconducting compounds such as CdS, CdTe, CdSe, PbS, and PbSe [42, 43, 46, 47]. While using inorganic QDs PCEs of more than 15 % have been achieved [44], the current record value is 18.1 %, using perovskite QDs [45, 7]. CdSe nanocrystals are effective visual aids to demonstrate quantum mechanics. An interesting example of a nanospectroscopy study is the mapping of the performance of CdTe solar cells with nanoscale resolution [48]. Another study combines scanning electron microscopy, Raman spectroscopy, and UV-Vis spectroscopy for the ZnS/CdSe/CdS-ZnS/TiO$_2$ system (Fig. 2.2.11) [49].

In addition, the fast regeneration of colloidal CdSe QDs by polypyridyl ruthenium dyes employed as cosensitizers of mesoporous TiO$_2$ electrodes has been confirmed by

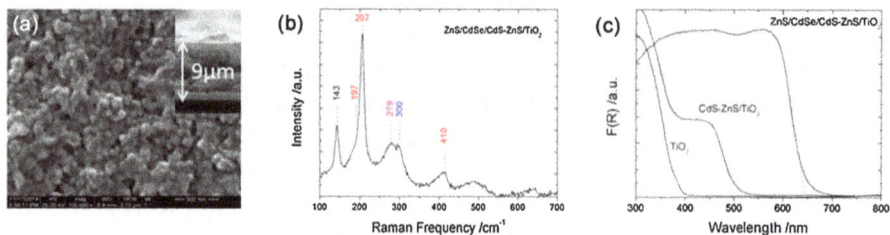

Figure 2.2.11: (a) Top view SEM image of the $ZnS/CdSe/CdS$-ZnS/TiO_2 photoanode surface and cross-section image as an inset. (b) Raman spectrum of $ZnS/CdSe/CdS$-ZnS/TiO_2. Bands are marked as follows: sensitized photoelectrode TiO_2 (in black), CdSe (in red), CdS (in blue), CdSSe (in magenta). (c) Absorption spectra (Kubelka–Munk units) of bare TiO_2 photoanode, TiO_2 film with CdS-ZnS layer, and TiO_2 film with $ZnS/CdSe/CdS$-ZnS layer (dashed curves are used to estimate the corresponding band-gaps). Reprinted with permission from [49].

Mora-Seró et al. [50]. Following a systematic investigation using PL, Raman, and transient absorption spectroscopies the authors observed that direct adsorption of QDs on TiO_2 leads to an efficient electron injection, whereas the use of colloidal ZnS-CdSe core-shell nanoparticles results in passivation of the QDs' surface and quantum confinement effects. Furthermore, cosensitization with a ruthenium molecular dye is accompanied by a strong photoinduced charge transfer between the photoexcited CdSe QDs and the dye molecules. Such a regenerating action of molecular dyes for QD sensitizers can have important implications in the development of efficient PV devices based on the synergistic action of dye–QD–TiO_2 heterostructures (Fig. 2.2.12). Thus, it has been confirmed that these nanospectroscopic techniques (PL, Raman, and transient absorption spectroscopy) qualify as sensitive probes of the individual cell components as well as of interfacial interactions in the TiO_2/QD/dye hybrid system.

The effect of light and thermal treatment (in air or in an inert N_2 atmosphere) on the characteristics of QDSCs and their stability was investigated by UV-Vis and Raman spectroscopies. Photoanodes bearing nanocrystalline titania and CdS, CdSe, and ZnS QDs present enhanced photon absorption in a large range of the visible spectrum. Studies using micro-Raman scattering as a function of light soaking confirmed that ZnS plays a combined passivation and stabilization role. Annealing of photoanodes resulted in increased QD size as verified by both UV-Vis diffuse reflectance and Raman spectroscopy. It was found that a composite CdS_xSe_{1-x} species was formed at the interface of the CdS and CdSe layers, which constitutes a stabilization factor and is further strengthened by annealing, which influences the electric characteristics of the sensitized photoanodes [51].

By combining micro-Raman, PL, and photocurrent measurements, Sfaelou et al. also confirmed that the most crucial component is the top ZnS layer, which passivates defects, prevents electron leakage, and results in photoanodes that resisted photodegradation for long illumination periods [52].

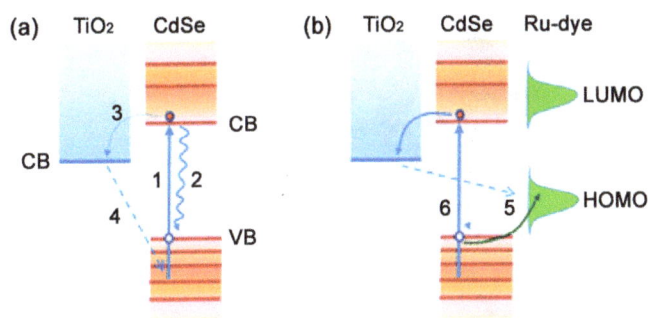

Figure 2.2.12: Schematic energy diagram of carrier photogeneration, transfer, and recombination in (a) QD/TiO$_2$ and (b) dye/QD/TiO$_2$. Arrows show the main processes occurring in sensitized electrodes after light illumination. The arrow thickness indicates the strength of the process. Process 1: After light illumination electron–hole pairs are photogenerated. Process 2: The pairs can recombine internally in the QD before any carrier is transferred outside the QD. This internal recombination can be nonradiative or radiative; radiative recombination results in the PL signal detected. Process 3: In addition, photogenerated electrons can be transferred to TiO$_2$, and they will recombine with the holes left in the QD (process 4). When the dye is attached to the QD the internal recombination (process 2) is strongly reduced (PL quenching) due to the fast hole injection from the QD into the dye (process 5). In this situation, the recombination of photoinjected electrons in the TiO$_2$ with photoinjected holes in the dye (process 6) is significantly slower than the recombination process QD/TiO$_2$, as has been characterized by transient absorption spectroscopy measurements. Reprinted with permission from [50].

2.2.4.5 Perovskite solar cells

Although perovskite materials have been extensively studied for over a century, their science is undergoing rapid expansion over the last few years due to their use in third-generation PV devices [53]. On the one hand, perovskite compounds of the general formulas ABX$_3$ and A$_2$BX$_6$ are incorporated as HTMs in DSSCs. Regarding their crystal structure, the cubic Cs$_2$SnX$_6$ (X = Cl, Br, I) perovskite structure is a defect variant of the archetype with the nominal composition of ABX$_3$ (e. g., SrTiO$_3$) [54], where half of the B sites are void (Fig. 2.2.13). The interstitial sites among the isolated [SnX$_6$] octahedra are filled with Cs atoms in a regular 12-fold coordination environment of X atoms. In contrast to CsSnX$_3$, Cs$_2$SnX$_6$ perovskites show an ionic character and may be considered as molecular salts. The use of CsSnI$_3$ as HTM has been investigated by M. Kanatzidis et al., reaching efficiencies of up to 10 % [55]. The same group also reported that in the case of Cs$_2$SnI$_6$ as HTM, a maximum PCE of 7.8 % has been reached [56]. Despite its lower efficiency, Cs$_2$SnI$_6$ is a much more promising material for solar cell applications thanks to the chemical stability of Sn^{4+} in ambient air [57].

On the other hand, hybrid organic–inorganic perovskites of the composition ABX$_3$ (A = CH$_3$NH$_3$, NH$_2$CHNH$_2$, B = Sn, Pb, and X = Cl, Br, I) can be used as photosensitizers in another category of solar cells, similar to DSSCs and QDSCs, known as PSCs. Depending on the presence of a porous semiconductor (usually TiO$_2$) as well as the

(a)

(b)

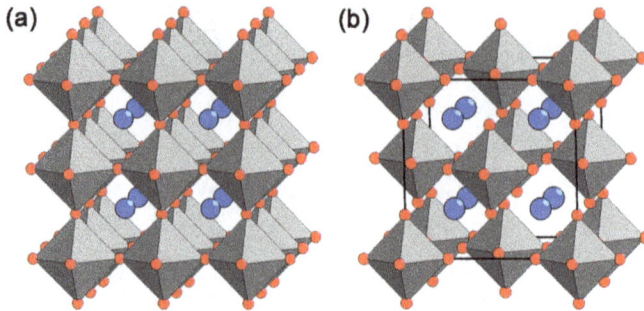

Figure 2.2.13: Crystal structure of the cubic perovskite CsSnX$_3$ (a) and the cubic defect perovskite Cs$_2$SnX$_6$ (b). Sn atoms reside in the octahedra formed by X atoms (red) and Cs atoms (blue) occupy the interstitial sites.

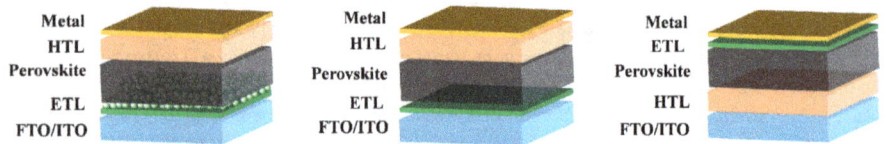

Metal	Metal	Metal
HTL	HTL	ETL
Perovskite	Perovskite	Perovskite
ETL	ETL	HTL
FTO/ITO	FTO/ITO	FTO/ITO

Figure 2.2.14: Regular PSC with mesoporous TiO$_2$ scaffold (left), planar (middle), and inverted (right) architecture in PSCs. Reprinted with permission from [62].

sequence of the electron-transporting layer (ETL)/perovskite layer/hole-transporting layer (HTL), three main architectures are studied (Fig. 2.2.14). Currently, one of the best-performing perovskite compounds is MAPbI$_3$ (MA = CH$_3$NH$_3$) and it yields efficiencies of over 20 % in solid-state mesoscopic PSCs [58, 59]. Noteworthily, Cs$_2$SnI$_6$ may also be used as absorber although the efficiency is as low as 1.5 % so far [60]. A record certified efficiency as high as 25.2 % has been achieved recently [61, 7].

The formation of ABX$_3$ solid solutions, namely mixing different atom types or molecular ions in each crystallographic site retaining the symmetry of the structure, is the main method of altering the optical and electronic properties in this class of materials. It is also well known that ABX$_3$ halide perovskites undergo various temperature-induced phase transitions that affect their physical properties [63]. In addition to their increased affinity and reactivity with humidity and oxygen, perovskite materials are also very sensitive to light-induced heating, which results in structural changes or even degradation. Raman spectra of pristine CH$_3$NH$_3$PbI$_3$ thin films were presented in [64, 65], where the structural changes under 514.5 nm laser illumination were thoroughly studied, as CH$_3$NH$_3$PbI$_3$ degraded into its precursors PbI$_2$ and CH$_3$NH$_3$I. In order to avoid this Raman laser-induced degradation, excitation above 800 nm is required [66]. As seen in Fig. 2.2.15, the band-gap of CH$_3$NH$_3$PbX$_3$ is finely tuned by altering and mixing the halogen composition, resulting in red shifts of both the low-frequency PbX and the MA cation internal vibrations towards heavy halide derivatives

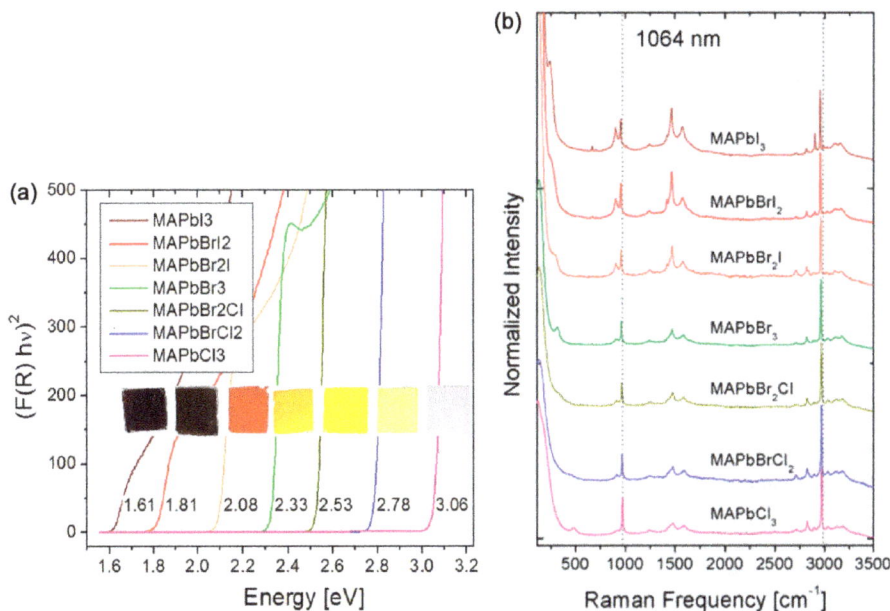

Figure 2.2.15: Spectral characteristics of MAPbX$_3$ (X = Cl, Br, I, and their mixtures). (a) UV-Vis spectra (where $F(R)$ is the Kubelka–Munk function) and (b) room temperature Raman spectra of mixed halide perovskite films. The films change color (from dark brown to white) and show a shift in band-gap energy (from 1.61 to 3.06 eV) as lighter halides are introduced into the structure. Reprinted with permission from [66].

indicative of strengthening the interaction between halides and the organic cation inside the inorganic cage. Amongst different MA vibrations, the torsion along the C-N axis is the most sensitive to the halide interaction with the NH$_3$ head of MA, spanning a broad frequency range from 249 to 488 cm^{-1}. Therefore, Raman shifts can be used in mixed halide perovskite layers to predict compositional ratios and study local inhomogeneities of the halide concentrations on a submicrometer scale [65, 63]. Furthermore, the high lateral resolution and sensitivity of Raman spectroscopy allows for the local study of the degradation product PbI$_2$.

A further study using multitemperature Raman and terahertz absorption spectra for MAPbX$_3$ (X = Cl, Br, I) has shown the key role of two types of disorder: (a) dynamic disorder caused by the unlocking of the rotation of the methylammonium ions in their cavities, resulting in homogeneous linewidth broadening, and (b) statistical disorder caused by the various possible cation orientations, which leads to inhomogeneous linewidth broadening. Theoretical analysis of these data supports domination of the optical phonon scattering for the low lattice thermal conductivity of these materials at room temperature [67].

Photoinduced phase segregation with iodide-rich and bromide-rich regions as well as nonhomogeneous distribution of methylammonium cations and iodide seg-

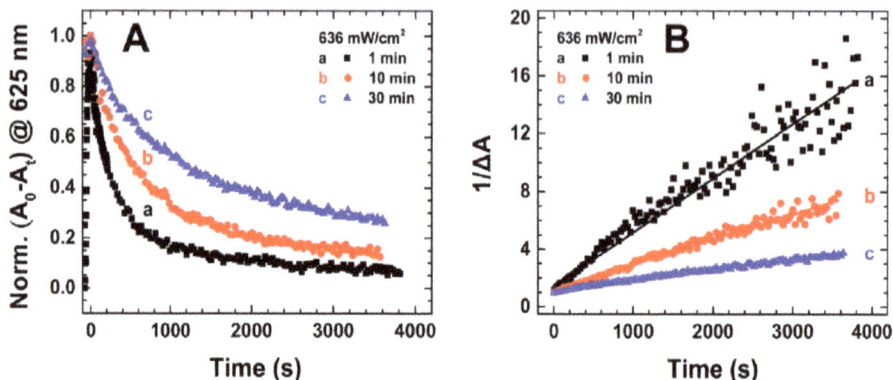

Figure 2.2.16: (a) Absorption changes representing recovery of a MAPbI$_{3-x}$Br$_x$ film following CW laser irradiation. (b) Second-order kinetic analysis of the recovery traces in (a). Reprinted with permission from [68].

regation stimulated by electron beams have been also observed in MAPbI$_{3-x}$Br$_x$ by PL [68] and CL [69–71]. Phase segregation in CH$_3$NH$_3$PbI$_{3-x}$Br$_x$ is shown in Figure 2.2.16 by the transient bleaching of the light absorption at 625 nm along with increased absorption in the low- and high-energy regions following a 387 nm laser pulse excitation. The photoinduced instability in the above structures has been attributed to spinodal decomposition of the mixed system with I- and Br-rich phases by studying the thermodynamics of the alloy formation with density functional theory calculations [72]. Furthermore, recent CL studies recorded ion segregation in MAPbI$_3$ and impacts of ion segregation on local optical properties in mixed halide perovskite films [73]. Local stoichiometric variations in iodide content may have a significant impact on carrier properties in CH$_3$NH$_3$PbI$_3$ films and mirror the effects of ion migration during PSC operation, which is considered responsible for the observed hysteresis in the J-V curves.

Similar to the DSSCs, there are long-term stability issues for PSCs which limit further exploitation in commercial devices [74]. Main stability issues for the lead perovskites are: (a) degradation due to humidity and (b) temperature-induced phase transformations. In order to elucidate fundamental decomposition pathways in organolead halide perovskites, in situ spectroscopic techniques are used. As an example we mention the work of Yang et al. [75] with absorption spectroscopy and in situ grazing incidence X-ray diffraction in a CH$_3$NH$_3$PbI$_3$ film exposed to flowing N$_2$ gas with controlled humidity conditions (Fig. 2.2.17). In situ vibrational and electronic nanospectroscopic characterization of PSCs under stressed conditions and DC or AC polarization is expected to become prominent in the future.

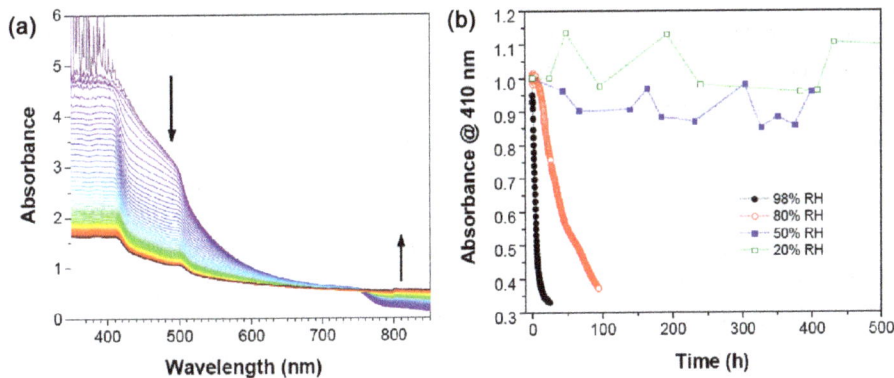

Figure 2.2.17: (a) UV-Vis spectra acquired at 15-min intervals of a $CH_3NH_3PbI_3$ film exposed to flowing N_2 gas with a relative humidity (RH) of $98 \pm 2\%$. (b) Normalized absorbance at 410 nm as a function of time for perovskite films exposed to various relative humidity levels. Data at RH values of 50 % and 20 % were acquired once per 24 h. The temperature was measured to be $22.9 \pm 0.5\,°C$ for all measurements. Reprinted with permission from [75].

2.2.5 Some challenges and solutions

In general, research in the field of third-generation PV devices aims at the following targets: (a) to go beyond the Shockley–Queisser limit and optimize the PV performance, (b) to use inexpensive and environmentally friendly materials, and (c) to improve efficiency and long-term stability as well as scale up production of the solar cells for commercial use.

Regarding the improvement of the solar cell efficiency, the main route is via tandem devices, which so far were mostly applicable on silicon tandem and 13–15 multijunction solar cells. The main problem though is the processing incompatibility such as junction deterioration due to high temperature or sputtering damage. As the efficiency of PSCs is expected to reach values above 25 % [76], and a value of 25.2 % has actually been achieved [61], it seems likely that a tandem PSC would surpass Si- or GaAs-based technology in the near future. In fact, a PCE of >30 % has been achieved with a perovskite/Si tandem cell [77, 7]. Nanophotonic concepts, involving nanostructures with typical length scales equal to or smaller than the wavelength of light that are incorporated in solar cell fabrication, can serve to reach these goals [5]. Moreover, one can exploit the large miscibility range in the perovskite structure on all three crystallographic positions, e. g., $FA_{1-x}Cs_xPb_{1-y}Sn_yI_{3-z}Br_z$, that leads to perovskite–perovskite tandem devices with PCE > 15 % and FF > 70 % and enhanced long-term stability [78]. Significant advances in the microstructure study are expected via the AFM-IR technique, which has already been applied to OSCs and recently to PSCs [70, 71]. The advantage is the direct correlation of the chemical composition in perovskite solid solutions to the frequency shifts of IR-active vibrational modes.

Among the alternative ways to improve overall efficiency, the hybridization of solar cell technology with artificial photosynthesis (water splitting) is particularly interesting and with huge environmental impact, if successful. According to current studies, a solar-to-hydrogen efficiency up to 15 % is possible by combining water photolysis with PSCs [79–82]. Secondly, the use of solar converters that incorporate fluorescent materials and QDs can improve the PCE of solar cells [83]. This approach also relies heavily on optical spectroscopy, as the fluorescent material absorbs UV photons and emits photons in the visible range. Lastly, the concept of nanophotonic solar thermophotovoltaic devices is investigated. Such systems could increase the PCE to higher values by exploiting simultaneously solar radiation and waste heat from the surface of the solar cell [84]. Obviously, this class of renewable energy sources is less intermittent (e. g., on cloudy days) as the temperature gradient of the device will be harvested by the thermoelectric unit beneath the PV unit of the hybrid device. In this respect, novel categories of environmentally friendly and abundant thermoelectric materials, such as tetrahedrites of the composition $M_{12}Sb_4S_{13}$ (M = Fe, Co, Ni, Cu, Zn) with high thermoelectric efficiency over the temperature range 25–100 °C [85, 86], have already reached commercialization for industrial-scale energy harvesting [87]. It has also been demonstrated that DSSCs can operate extremely efficiently under ambient lighting conditions [88]. In fact, PCEs as high as 34 % were obtained [89], opening great perspectives and high practical interest of the DSSC technology for indoor applications.

Regarding the physical and chemical properties of the PV materials, a significant leap forward towards this aim is the development of robust lead-free perovskites of reduced toxicity (e. g., based on tin) and the use of nonharmful solvents in the application of green, solution-based deposition procedures. Even more important issues are the scale-up of solar cell production and long-term maintenance of their efficiency. Improved stability of PSCs has been reported either by using less hygroscopic perovskites with mixed anions and cations [78] or by encapsulation of a $MAPbI_3/TiO_2$ heterojunction using a carbon counter electrode and polymethylsiloxane [90]. Nanospectroscopy techniques can strongly contribute to successfully address these challenges through materials characterization and investigation of operation/degradation mechanisms. Moreover, it has been demonstrated that expensive platinum counter electrodes can be successfully replaced by low-cost carbon nanomaterials. In fact, colloidal graphite and reduced graphene oxide thin-film electrodes, optimized following detailed characterization by a variety of experimental microscopic and spectroscopic techniques (including micro-Raman), show excellent electrocatalytic activity towards the redox couple regeneration [91, 92]. As a result, the corresponding DSSC devices surpassed in efficiency their analog using the very expensive platinum (Pt) counter electrode, thus considerably reducing the cell fabrication costs. These carbon nanomaterials can be also incorporated in PSCs, in order to improve charge transport and electron collection/hole extraction at the corresponding interfaces. Further developments can be expected by fine-tuning the semiconductor morphology via the introduction of 1D self-organized nanostructures, where the observed antenna Raman effect promotes vectorial electron transport along the titania nanotubes [93].

2.2.6 Summary and impact

Third-generation solar cells have the potential to become the dominant renewable energy source. The current chapter gives an overview and the main contributions in the development of solar cells with respect to nanospectroscopy. As most modern solar cells consist of multilayered structures, a larger number of characterization methods is needed. Examples of spectral analysis are given for all types of emerging solar cells. Optical nanospectroscopy offers a nondestructive analysis that is capable of reaching resolution in the nanometer range. By combining electron microscopy (SEM/EDX) and spectroscopic data, scientists working in the field are able to characterize in detail the cell components and fine-tune and optimize the corresponding interfaces at the nanoscale.

In the case of DSSCs significant information can be obtained for the semiconductor, the dye, the electrolyte (including the redox species and the corresponding solvents and additives), and the resulting interfaces. Additional evidence can be obtained on the sensitization mechanism (comprising the dye–redox couple interaction and the recombination processes) and the device behavior following prolonged aging. The existing know-how can be extended from the field of DSSCs to PSCs, where the adequate use and adaptation of the nanospectroscopy tools could provide the proper solutions to the investigation and optimization of the perovskite absorber phase transformations, the carrier transport in the mesoporous layer and the HTM, and the extraction of the photogenerated charges at the electrodes.

The use of these state-of-the-art nondestructive techniques enables the investigation of complex phenomena, elucidating the operation mechanisms and achieving significant scientific and technological developments. As a result, they may contribute to the enhancement of the cells' efficiency, lifetime, and stability and provide breakthroughs that will permit real advances in scale-up, industrialization, and commercialization of novel and highly performing, low-cost PV devices.

Bibliography

[1]   Hamann C, Hamnett A, Vielstisch W. Electrochemistry. 2nd ed. Wiley; 2007.
[2]   https://iea-pvps.org/wp-content/uploads/2022/04/IEA_PVPS_Snapshot_2022-vF.pdf. Accessed on September 19, 2022.
[3]   Renewables 2022 Global Status Report. http://www.ren21.net/status-of-renewables/global-status-report/. (https://www.solarpowereurope.org/press-releases/world-installs-a-record-168-gw-of-solar-power-in-2021-enters-solar-terawatt-age, as of 05/24/2022).

[4] http://www.sciencemag.org/news/2013/12/sciences-top-10-breakthroughs-2013. Accessed on September 19, 2022.

[5] Polman A, Knight M, Garnett EC, Ehrler B, Sinke WC. Photovoltaic materials: present efficiencies and future challenges. Science. 2016;352:4424.

[6] https://www.nrel.gov/news/program/2021/new-design-strategy-increases-organic-pv-efficiency.html. Accessed on September 19, 2022.

[7] National Renewable Energy Laboratory, NREL Efficiency Chart. https://www.nrel.gov/pv/cell-efficiency.html.

[8] Huth F, Govyadinov A, Amarie S, Nuansing W, Keilmann F, Hillenbrand R. Nano-FTIR absorption spectroscopy of molecular fingerprints at 20 nm spatial resolution. Nano Lett. 2012;12:3973–8.

[9] Zeng ZC, Huang SC, Wu DY, Meng LY, Li MH, Huang TX, Zhong JH, Wang X, Yang ZL, Ren B. Electrochemical tip-enhanced Raman spectroscopy. J Am Chem Soc. 2015;137:11928–31.

[10] Battaglia C, Cuevas A, De Wolf S. High-efficiency crystalline silicon solar cells: status and perspectives. Energy Environ Sci. 2016;9:1552–76.

[11] https://www.ise.fraunhofer.de/content/dam/ise/en/documents/press-releases/2022/1322_PR_ISE_World_Record_47,6Percent-SolarCell.pdf. Accessed on September 19, 2022.

[12] Muthmann S, Köhler F, Meier M, Hülsbeck M, Carius R, Gordijn A. In-situ Raman spectroscopy used to study and control the initial growth phase of microcrystalline absorber layers for thin-film silicon solar cells. J Non-Cryst Solids. 2012;358:1970–3.

[13] Alvarez-García J, Rudigier E, Rega N, Barcones B, Scheer R, Pérez-Rodríguez A, Romano-Rodríguez A, Morante JR. Growth process monitoring and crystalline quality assessment of CuInS(Se)$_2$ based solar cells by Raman spectroscopy. Thin Solid Films. 2003;431–432:122–5.

[14] Izquierdo-Roca V, Shavel A, Saucedo E, Jaime-Ferrer S, Álvarez-García J, Cabot A, Pérez-Rodríguez A, Bermudez V, Morante JR. Assessment of absorber composition and nanocrystalline phases in CuInS$_2$ based photovoltaic technologies by ex-situ/in-situ resonant Raman scattering measurements. Sol Energy Mater Sol Cells. 2011;95:S83–8.

[15] Oehlschläger F, Müller J, Künecke U, Hölzing A, Schurr R, Hock R, Wellmann P. Determination of material inhomogeneities in CuIn(Se,S)$_2$ solar cell materials by high resolution cathodoluminescence topography. Energy Proc. 2010;2:183–8.

[16] Chan EM, Mathies RA, Alivisatos AP. Size-controlled growth of CdSe nanocrystals in microfluidic reactors. Nano Lett. 2003;3:199–201.

[17] Wang X, Zhang D, Braun K, Egelhaaf HJ, Brabec CJ, Meixner AJ. High-resolution spectroscopic mapping of the chemical contrast from nanometer domains in P3HT:PCBM organic blend films for solar-cell applications. Adv Funct Mater. 2010;20:492–9.

[18] Dazzi A, Prater CB, Hu Q, Chase DB, Rabolt JF, Marcott C. AFM–IR: combining atomic force microscopy and infrared spectroscopy for nanoscale chemical characterization. Appl Spectrosc OA. 2012;66:1365–84.

[19] Marcott C, Awatani T, Ye J, Gerrard D, Lod M, Kjoller K. Review of nanoscale infrared spectroscopy applications to energy related materials. Spectrosc Eur. 2014;26:18–22.

[20] Veerender P, Saxena V, Chauhan AK, Koiry SP, Jha P, Gusain A, Choudhury S, Aswal DK, Gupta SK. Probing the annealing induced molecular ordering in bulk heterojunction polymer solar cells using in-situ Raman spectroscopy. Sol Energy Mater Sol Cells. 2014;120:526–35.

[21] Tsoi WC, Zhang W, Razzell Hollis J, Suh M, Heeney M, McCulloch I, Kim J-S. In-situ monitoring of molecular vibrations of two organic semiconductors in photovoltaic blends and their impact on thin film morphology. Appl Phys Lett. 2013;102:173302.

[22] Soultati A, Douvas AM, Georgiadou DG, Palilis LC, Bein B, Feckl JM, Gardelis S, Fakis M, Kennou S, Falaras P, Stergiopoulos T, Stathopoulos NA, Davazoglou D, Argitis P, Vasilopoulou M. Solution-processed hydrogen molybdenum bronzes as highly conductive anode interlayers in efficient organic photovoltaics. Adv Energy Mater. 2014;4:1300896.

[23] Singh R, Giussani E, Mroóz MM, Di Fonzo F, Fazzi D, Cabanillas-Gonzalez J, Oldridge L, Vaenas N, Kontos AG, Falaras P, Grimsdale AC, Jacob J, Mullen K, Keivanidis PE. On the role of aggregation effects in the performance of perylene-diimide based solar cells. Org Electron. 2014;15:1347–61.

[24] Stathopoulos NA, Palilis LC, Vasilopoulou M, Botsialas A, Falaras P, Argitis P. All-organic optocouplers based on polymer light-emitting diodes and photodetectors. Phys Status Solidi A. 2008;205:2522–5.

[25] Oregan B, Grätzel M. A low-cost, high-efficiency solar-cell based on dye-sensitized colloidal TiO_2 films. Nature. 1991;353:737–40.

[26] Wu JH, Lan Z, Lin JM, Huang ML, Huang YF, Fan LQ, Luo GG. Electrolytes in dye-sensitized solar cells. Chem Rev. 2015;115:2136–73.

[27] You P, Liu Z, Tai Q, Liu S, Yan F. Efficient semitransparent perovskite solar cells with graphene electrodes. Adv Mater. 2015;27:3632–8.

[28] Falaras P, Grätzel M, Hugot-Le Goff A, Nazeeruddin M, Vrachnou E. Dye sensitization of TiO_2 surfaces studied by Raman spectroscopy. J Electrochem Soc. 1993;140:L92–4.

[29] Hugot-Le Goff A, Falaras P. Origin of new bands in the Raman spectra of dye monolayers adsorbed on nanocrystalline TiO_2. J Electrochem Soc. 1995;142:L38–41.

[30] Falaras P, Hugot-Le Goff A, Bernard MC, Xagas A. Characterization by resonance Raman spectroscopy of sol-gel TiO_2 films sensitized by the $Ru(PPh_3)_2(dcbipy)Cl_2$ complex for solar cells application. Sol Energy Mater Sol Cells. 2000;64:167–84.

[31] Hugot-Le Goff A, Joiret S, Falaras P. Raman resonance effect in a monolayer of polypyridyl ruthenium(II) complex adsorbed on nanocrystalline TiO_2 via phosphonated terpyridyl ligands. J Phys Chem B. 1999;103:9569–75.

[32] Bernard MC, Cachet H, Falaras P, Hugot–Le Goff A, Kalbac M, Lukes I, Oanh NT, Stergiopoulos T, Arabatzis I. Sensitization of TiO_2 by polypyridine dyes. Role of the electron donor. J Electrochem Soc. 2003;150:E155–64.

[33] Stergiopoulos T, Bernard MC, Hugot-Le Goff A, Falaras P. Resonance micro-Raman spectrophotoelectrochemistry on nanocrystalline TiO_2 thin film electrodes sensitized by Ru (II) complexes. Coord Chem Rev. 2004;248:1407–20.

[34] Kontos AG, Stergiopoulos T, Tsiminis G, Raptis YS, Falaras P. In situ micro- and macro-Raman investigation of the redox couple behavior in DSSCS. Inorg Chim Acta. 2008;361:761–8.

[35] Kontos AG, Stergiopoulos T, Likodimos V, Milliken D, Desilvesto H, Tulloch G, Falaras P. Long-term thermal stability of liquid dye solar cells. J Phys Chem C. 2013;117:8636–46.

[36] Likodimos V, Stergiopoulos T, Falaras P, Harikisun R, Desilvestro J, Tulloch G. Prolonged light and thermal stress effects on industrial dye-sensitized solar cells: a micro-Raman investigation on the long-term stability of aged cells. J Phys Chem C. 2009;113:9412–22.

[37] Jovanovski V, Orel B, Jerman I, Hočevar SB, Ogorevc B. Electrochemical and in-situ Raman spectroelectrochemical study of 1-methyl-3-propylimidazolium iodide ionic liquid with added iodine. Electrochem Commun. 2007;9:2062–6.

[38] Bidikoudi M, Stergiopoulos T, Likodimos V, Romanos GE, Francisco M, Iliev B, Adamova G, Schubert TJS, Falaras P. Ionic liquid redox electrolytes based on binary mixtures of 1-alkyl-methylimidazolium tricyanomethanide with 1-methyl-3-propylimidazolium iodide and implication in dye-sensitized solar cells. J Mater Chem A. 2013;1:10474–86.

[39] Stergiopoulos T, Kontos A, Likodimos V, Perganti D, Falaras P. Solvent effects at the photoelectrode/electrolyte interface of a DSC: a combined spectroscopic and photoelectrochemical study. J Phys Chem C. 2011;115:10236–44.

[40] Perganti D, Kontos AG, Stergiopoulos T, Likodimos V, Farnell J, Milliken D, Desilvestro H, Falaras P. Thermal stressing of dye sensitized solar cells employing robust redox electrolytes. Electrochim Acta. 2015;179:241–9.

[41] Stergiopoulos T, Kontos AG, Jiang N, Milliken D, Desilvestro H, Likodimos V, Falaras P. High boiling point solvent-based dye solar cells pass a harsh thermal ageing test. Sol Energy Mater Sol Cells. 2016;144:457–66.

[42] Kamat PV. Quantum dot solar cells. Semiconductor nanocrystals as light harvesters. J Phys Chem C. 2008;112:18737–53.

[43] Kwizera P, Angela A, Wekesa M, Uddin J, Mobin Shaikh M. Synthesis and characterization of CdSe quantum dots by UV-vis spectroscopy. In: Uddin J, editor. Macro to Nano Spectroscopy. ISBN: 978-953-51-0664-7. InTech; 2012. Available from http://www.intechopen.com/books/macro-tonano-spectroscopy/synthesis-and-characterization-of-cdse-quantum-dots-by-uv-vis-spectroscopy.

[44] Song H, Lin Y, Zhang Z, Rao H, Wang W, Fang Y, Pan Z, Zhong X. Improving the Efficiency of Quantum Dot Sensitized Solar Cells beyond 15 % via Secondary Deposition. J Am Chem Soc. 2021;143:4790–800.

[45] Li J, Wang H, Chin XY, Dewi HA, Vergeer K, Goh TW, Lim JW, Lew JH, Loh KP, Soci C, Sum TC, Bolink HJ, Mathews N, Mhaisalkar S, Bruno A. Highly efficient thermally co-evaporated perovskite solar cells and mini-modules. Joule. 2020;4:1035–53.

[46] Baranov AV, Rakovich YP, Donegan JF, Perova TS, Moore RA, Talapin DV, Rogach AL, Masumoto Y, Nabiev I. Effect of ZnS shell thickness on the phonon spectra in CdSe quantum dots. Phys Rev B. 2003;68:165306.

[47] Rolo AG, Vasilevskiy MI. Raman spectroscopy of optical phonons confined in semiconductor quantum dots and nanocrystals. J Raman Spectrosc. 2007;38:618–33.

[48] Tennyson E, Leite M. Mapping the Performance of Solar Cells with Nanoscale Resolution. SPIE; 2015. http://leitelab.umd.edu/wp-content/uploads/2014/01/2015-Tennyson-SPIENewsroom.pdf.

[49] Givalou L, Antoniadou M, Perganti D, Giannouri M, Karagianni CS, Kontos AG, Falaras P. Electrodeposited cobalt-copper sulfide counter electrodes for highly efficient quantum dot sensitized solar cells. Electrochim Acta. 2016;210:630–8.

[50] Mora-Seró I, Likodimos V, Giménez S, Martínez-Ferrero E, Albero J, Palomares E, Kontos AG, Falaras P, Bisquert J. Fast regeneration of CdSe quantum dots by Ru dye in sensitized TiO$_2$ electrodes. J Phys Chem C. 2010;114:6755–61.

[51] Sfaelou S, Kontos AG, Givalou L, Falaras P, Lianos P. Study of the stability of quantum dot sensitized solar cells. Catal Today. 2014;230:221–6.

[52] Sfaelou S, Kontos AG, Falaras P, Lianos P. Micro-Raman, photoluminescence and photocurrent studies on the photostability of quantum dot sensitized photoanodes. J Photochem Photobiol A, Chem. 2014;275:127–33.

[53] Burschka J, Pellet N, Moon SJ, Humphry-Baker R, Gao P, Nazeeruddin MK, Grätzel M. Sequential deposition as a route to high-performance perovskite-sensitized solar cells. Nature. 2013;499:316–9.

[54] Glazer AM. Classification of tilted octahedra in perovskites. Acta Crystallogr, Sect B. 1972;28:3384–92.

[55] Chung I, Lee B, He J, Chang RP, Kanatzidis MG. All-solid-state dye-sensitized solar cells with high efficiency. Nature. 2012;485:486–9.

[56] Lee B, Stoumpos CC, Zhou NJ, Hao F, Malliakas C, Yeh CY, Marks TJ, Kanatzidis MG, Chang RPH. Air-stable molecular semiconducting iodosalts for solar cell applications: Cs$_2$SnI$_6$ as a hole conductor. J Am Chem Soc. 2014;136:15379–85.

[57] Kaltzoglou A, Perganti D, Antoniadou M, Kontos AG, Falaras P. Stress tests on dye-sensitized solar cells with the Cs$_2$SnI$_6$ defect perovskite as hole-transporting material. Energy Proc. 2016;102:49–55.

[58] Boix PP, Agarwala S, Koh TM, Mathews N, Mhaisalkar SG. Perovskite solar cells: beyond methylammonium lead iodide. J Phys Chem Lett. 2015;6:898–907.

[59] Jung HS, Park NG. Perovskite solar cells: from materials to devices. Small. 2015;11:10–25.
[60] Qiu X, Cao B, Yuan S, Chen X, Qiu Z, Jiang Y, Ye Q, Wang H, Zeng H, Liu J, Kanatzidis MG. From unstable $CsSnI_3$ to air-stable Cs_2SnI_6: a lead-free perovskite solar cell light absorber with band gap of 1.48 eV and high absorption coefficient. Sol Energy Mater Sol Cells. 2017;159:227–34.
[61] Jeong J, Kim M, Seo J, Lu H, Ahlawat P, Mishra A, Yang Y, Hope MA, Eickemeyer FT, Kim M, Yoon YJ, Choi IW, Darwich BP, Choi SJ, Jo Y, Lee JH, Walker B, Zakeeruddin SM, Emsley L, Rothlisberger U, Hagfeldt A, Kim DS, Grätzel M, Kim JY et al. Pseudo-halide anion engineering for α-$FAPbI_3$ perovskite solar cells. Nature. 2021;592:381–5.
[62] Marinova N, Valero S, Delgado JL. Organic and perovskite solar cells: working principles, materials and interfaces. J Colloid Interface Sci. 2017;488:373–89.
[63] Stoumpos CC, Malliakas CD, Kanatzidis MG. Semiconducting tin and lead iodide perovskites with organic cations: phase transitions, high mobilities, and near-infrared photoluminescent properties. Inorg Chem. 2013;52:9019–38.
[64] Antoniadou M, Siranidi E, Vaenas N, Kontos AG, Stathatos E, Falaras P. Photovoltaic performance and stability of $CH_3NH_3PbI_{3-x}Cl_x$ perovskites. J Surf Interfac Mater. 2014;2:323–7.
[65] Ledinsky M, Loper P, Niesen B, Holovsky J, Moon SJ, Yum JH, De Wolf S, Fejfar A, Ballif C. Raman spectroscopy of organic-inorganic halide perovskites. J Phys Chem Lett. 2015;6:401–6.
[66] Niemann RG, Kontos AG, Palles D, Kamitsos EI, Kaltzoglou A, Brivio F, Falaras P, Cameron PJ. Halogen effects on ordering and bonding of $CH_3NH_3^+$ in $CH_3NH_3PbX_3$ (X = Cl, Br, I) hybrid perovskites: a vibrational spectroscopic study. J Phys Chem C. 2016;120:2509–19.
[67] Leguy AMA, Goñi AR, Frost JM, Skelton J, Brivio F, Rodríguez-Martínez X, Weber OJ, Pallipurath A, Alonso MI, Campoy-Quiles M, Weller MT, Nelson J, Walsh A, Barnes PRF. Dynamic disorder, phonon lifetimes, and the assignment of modes to the vibrational spectra of methylammonium lead halide perovskites. Phys Chem Chem Phys. 2016;18:27051–66.
[68] Yoon SJ, Draguta S, Manser JS, Sharia O, Schneider WF, Kuno M, Kamat PV. Tracking iodide and bromide ion movement in mixed halide lead perovskites during photoirradiation. ACS Energy Lett. 2016;1:290–6.
[69] Dar MI, Jacopin G, Hezam M, Arora N, Zakeeruddin SM, Deveaud B, Nazeeruddin MK, Grätzel M. Asymmetric cathodoluminescence emission in $CH_3NH_3PbI_{3-x}Br_x$ perovskite single crystals. ACS Photonics. 2016;3:947–52.
[70] Zhao Y, Miao P, Elia J, Hu H, Wang X, Heumueller T, Hou Y, Matt GJ et al. Strain-activated light-induced halide segregation in mixed-halide perovskite solids. Nat Commun. 2020;11(1):1–9.
[71] Tang X, van den Berg M, Gu E, Horneber A, Matt GJ, Osvet A, Meixner AJ et al. Local observation of phase segregation in mixed-halide perovskite. Nano Lett. 2018;18(3):2172–8.
[72] Brivio F, Caetano C, Walsh A. Thermodynamic origin of photoinstability in the $CH_3NH_3Pb(I_{1-x}Br_x)_3$ hybrid halide perovskite alloy. J Phys Chem Lett. 2016;7:1083–7.
[73] Hentz O, Zhao Z, Gradečak S. Impacts of ion segregation on local optical properties in mixed halide perovskite films. Nano Lett. 2016;16:1485–90.
[74] Leijtens T, Eperon GE, Noel NK, Habisreutinger SN, Petrozza A, Snaith HJ. Stability of metal halide perovskite solar cells. Adv Energy Mater. 2015;5:1500963.
[75] Yang J, Siempelkamp BD, Liu D, Kelly TL. An investigation of $CH_3NH_3PbI_3$ degradation rates and mechanisms in controlled humidity environments using in situ techniques. ACS Nano. 2015;9:1955–63.
[76] Todorov T, Gunawan O, Guha S. A road towards 25 % efficiency and beyond: perovskite tandem solar cells. Mol Syst Des Eng. 2016;1:370–6.
[77] https://actu.epfl.ch/news/new-world-records-perovskite-on-silicon-tandem-sol/. Accessed on September 19, 2022.

[78] Eperon GE, Leijtens T, Bush KA, Prasanna R, Green T, Tse-Wei Wang J, McMeekin DP, Volonakis G, Milot RL, May R, Palmstrom A, Slotcavage DJ, Belisle RA, Patel JB, Parrott ES, Sutton RJ, Ma W, Moghadam F, Conings B, Babayigit A, Boyen H-G, Bent S, Giustino F, Herz LM, Johnston MB, McGehee MD, Snaith HJ. Perovskite-perovskite tandem photovoltaics with optimized bandgaps. Science. 2016;354:861–5.

[79] Luo JS, Im JH, Mayer MT, Schreier M, Nazeeruddin MK, Park NG, Tilley SD, Fan HJ, Grätzel M. Water photolysis at 12.3 % efficiency via perovskite photovoltaics and earth-abundant catalysts. Science. 2014;345:1593–6.

[80] Nordmann S, Berghoff B, Hessel A, Zielinsk B, John J, Starschich S, Knoch J. Record-high solar-to-hydrogen conversion efficiency based on a monolithic all-silicon triple-junction IBC solar cell. Sol Energy Mater Sol Cells. 2019;191:422–6.

[81] Kay A, Grätzel M. Artificial photosynthesis. 1. Photosensitization of TiO$_2$ solar-cells with chlorophyll derivatives and related natural porphyrins. J Phys Chem. 1993;97:6272–7.

[82] Grätzel M. Artificial photosynthesis – water cleavage into hydrogen and oxygen by visible-light. Acc Chem Res. 1981;14:376–84.

[83] Huang X, Han S, Huang W, Liu X. Enhancing solar cell efficiency: the search for luminescent materials as spectral converters. Chem Soc Rev. 2013;42:173–201.

[84] Lenert A, Bierman DM, Nam Y, Chan WR, Celanovic I, Soljacic M, Wang EN. A nanophotonic solar thermophotovoltaic device. Nat Nanotechnol. 2014;9:126–30.

[85] Lu L, Morelli DT, Xia Y, Zhou F, Ozolins V, Chi H, Zhou X, Uher C. High performance thermoelectricity in earth-abundant compounds based on natural mineral tetrahedrite. Adv Energy Mater. 2013;3:342–8.

[86] Vaqueiro P, Guelou G, Kaltzoglou A, Smith RI, Barbier T, Guilmeau E, Powell AV. The influence of mobile copper ions on the glass-like thermal conductivity of copper-rich tetrahedrites. Chem Mater. 2017;29:4080–90.

[87] http://www.freepatentsonline.com/y2017/0331023.html. Accessed on September 19, 2022.

[88] Freitag M, Teuscher J, Saygili Y, Zhang X, Giordano F, Liska P, Hua J, Zakeeruddin SM, Moser J-E, Grätzel M, Hagfeldt A. Dye-sensitized solar cells for efficient power generation under ambient lighting. Nat Photonics. 2017;11:372–8.

[89] Michaels H, Rinderle M, Freitag R, Benesperi I, Edvinsson T, Socher R, Gagliardi A, Freitag M. Dye-sensitized solar cells under ambient light powering machine learning: towards autonomous smart sensors for the internet of things. Chem Sci. 2020;11:2895–906.

[90] Liu Z, Sun B, Shi T, Tang Z, Liao G. Enhanced photovoltaic performance and stability of carbon counter electrode based perovskite solar cells encapsulated by PDMS. J Mater Chem A. 2016;4:10700–9.

[91] Giannouri M, Bidikoudi M, Pastrana-Martínez LM, Silva AMT, Falaras P. Reduced graphene oxide catalysts for efficient regeneration of cobalt-based redox electrolytes in dye-sensitized solar cells. Electrochim Acta. 2016;219:258–66.

[92] Perganti D, Giannouri M, Kontos AG, Falaras P. Cost-efficient platinum-free DSCs using colloidal graphite counter electrodes combined with D35 organic dye and cobalt (II/III) redox couple. Electrochim Acta. 2017;232:517–27.

[93] Likodimos V, Stergiopoulos T, Falaras P, Kunze J, Schmuki P. Phase composition, size, orientation and antenna effects of self-assembled anodized titania nanotube arrays: a polarized micro-Raman investigation. J Phys Chem C. 2008;112:12687–96.

Rosa Maria Montereali, Francesca Bonfigli, Enrico Nichelatti,
Massimo Piccinini, and Maria Aurora Vincenti

2.3 Luminescence of point defects in lithium fluoride thin layers for radiation imaging detectors at the nanoscale

2.3.1 Key messages

- Ionizing radiation generates point defects in the crystal lattice of insulating materials, among them color centers (CCs).
- CCs are localized over a few atomic sites and their luminescence properties allow applications to radiation detection and imaging.
- Broad-band, light-emitting, stable CCs in lithium fluoride are promising for radiation imaging at the nanoscale using optical nanoscopy as imaging tool.
- Lithium fluoride film-based radiation detector performance can be improved by selecting suitable substrates and 2D mapping of the absorbed radiation dose can be obtained at high spatial resolution.
- New methods exploiting CCs luminescence provide perspectives for applications and exciting possibilities for future work.

2.3.2 Pre-knowledge

CCs are point defects (imperfections in solids localized over a few atomic sites, among them vacancies and interstitials), whose electronic structure gives rise to the occurrence of energy levels in the forbidden gap between the conduction and valence bands. These levels drastically influence the electrical and optical properties of semiconductors and insulating materials. Charge capture at the lattice defects causes optical transitions which can absorb light in the previously transparent perfect crystal. CCs are named because their absorption bands are often located in the visible spectral range and give it a distinctive color.

Certain point defects emit light after the absorption of photons. The photons of the emitted light, known as photoluminescence (PL), are characterized by a lower energy – and, equivalently, a longer wavelength – than that of the absorbed photons. Such a difference is known as Stokes shift, named after the Irish physicist George G. Stokes. The Stokes shift is essentially due to energy transitions of the system (atoms, molecules, CCs, etc.) involved in the process: the absorbed photons cause these systems to enter an excited state; if part of the absorbed energy is lost in nonradiative ways (vibrations, dissipation, etc.), the remaining energy, released via photon emission, has a lower energy. The

Acknowledgement: The authors would like to dedicate this work to Prof. Anatoly Ya Faenov, a brilliant scientist and dear friend, who often visited our laboratories at ENEA C.R. Frascati for soft X-ray imaging experiments on lithium fluoride and joined stimulating discussions with positive optimism and great dynamism. The authors are indebted with many colleagues, researchers, technicians, students, and guests, whose contributions are cited in the reference list.

https://doi.org/10.1515/9783110442908-005

time needed to release the PL photon is known as decay time, because it corresponds to a transition (decay) of the excited system to a lower energy level. PL is a particular form of a class of phenomena known as luminescence (light emission following excitation due to the absorption of energy in different forms), in which excitation can be obtained also in other ways than excitation by photon absorption. Depending on the characteristic times of the phenomenon, PL can be classified as fluorescence or phosphorescence: if the release of the PL photon is immediate (decay times $\sim 10^{-7}$–10^{-9} s), one can speak of fluorescence; on the other hand, when the release of the PL photon is delayed, the phenomenon is known as phosphorescence.

A confocal laser scanning microscope (CLSM) is an optical microscope in which resolution and contrast are increased by placing a spatial pinhole at the confocal plane of the lens. In this way, out-of-focus light is stopped from reaching the detection system. For the theory of CLSM, see also Volume 1, Section 1 and Volume 2, Section 4. The fluorescent specimen is illuminated by light coming from a point source which is part of the microscope, and consequently all parts of the specimen become optically excited; however, the pinhole works as a selector of the fluorescence that is radiated only very close to the focal plane, because light coming from other planes is blocked. The resolution, especially the vertical one, is thereby increased beyond the limit obtainable in a conventional optical microscope. The price to pay is a lower signal intensity, so that often long exposures are needed. Because the specimen is illuminated only in one point, imaging of an area of it requires scanning over a pattern (parallel lines, a grid, etc.), an operation performed by servo-controlled mirrors in the instrument. The fluorescence coming from the specimen is collected by a detector and elaborated via software to reconstruct a 2D or 3D image – this latter is possible if several depth planes inside the specimen are analyzed. Usually, the detector is a photomultiplier tube (PMT), which is a vacuum tube very sensitive to light (ultraviolet, visible, and near-infrared) thanks to a process where the impinging photons strike a photocathode material, which ejects electrons due to the photoelectric effect. These electrons are then multiplied by multiple interactions with electrodes (dynodes) within the PMT by exploiting the phenomenon of secondary emission.

As far as other optical microscopy techniques are concerned, in particular scanning near-field optical microscopy (SNOM) and stimulated emission depletion (STED) microscopy, the reader should refer to Sections 3 and 4 in Volumes 1 and 2, respectively.

2.3.3 Introduction: luminescence and radiation detectors

Luminescence properties of point defects in pure and doped insulating materials – among which are oxides, fluorides, alkali halides, and glasses – are used for optically pumped lasers that operate in the visible and in the near-infrared, for solid-state light sources, as well as for radiation detectors, with a wide range of applications in lighting, imaging, and dosimetry [1, 2] for photonics, biomedical, and nuclear fields.

The most common radiation-sensitive materials that exhibit luminescence phenomena in the optical spectral range are doped aluminum oxide (Al_2O_3) [3], doped cesium halides, and pure and doped lithium fluoride (LiF) [4, 5], as well as Ag-activated phosphate glasses [6]; they recently attracted renewed scientific and industrial interest for radiation imaging and dosimetry based on luminescent point defects.

Thermoluminescence (TL, primarily in doped LiF) [7] and optically stimulated luminescence (OSL, mainly in doped Al_2O_3) [3] are quite complex light emission phenomena, which can be described in terms of the energy band model of electron–hole production in the high-energy gap between valence and conduction bands of an insulator. After the material is irradiated, light emission is obtained by heating and by light at effective wavelengths for TL and OSL, respectively. The TL and OSL signals are proportional to the absorbed radiation doses.

In optically transparent solids, also the PL properties of electronic point defects – such as impurities and intrinsic defects – are light emission phenomena stimulated by the absorption of photons at the proper wavelengths [8]. As a matter of fact, these localized defects generally possess specific spectral features, the optical absorption bands, which often give a distinctive color to the crystals. For this reason, they are also known as CCs [9]. For them, differently from OSL, the excited electron stays close to the defect that is created in the crystal lattice. The term radiophotoluminescence (RPL, mainly in doped glasses and LiF) [10] is often used when the PL phenomena involve radiation-induced defects, following the end of bombardment by ionizing radiation. Note that luminescence is also produced during irradiation (radioluminescence, RL) due to electron–hole pair recombination [11].

Even though scintillators and electronic radiation detectors allow for real-time dose measurement, the use of passive detectors relying on the luminescence of optically active electronic defects induced by irradiation is very common. In principle, such luminescent materials used for dosimetry are also good candidates to be employed as radiation imaging detectors; as a matter of fact, the atomic-scale size of such defects, which can be seen as minimum luminescent units, potentially provides an intrinsic high spatial resolution corresponding to the lattice spacing, that is, between about 0.5 and 2.0 nm for inorganic crystalline solids. The real spatial resolution is actually limited by the technique and the fluorescence microscope used to collect the luminescent latent image stored in the irradiated detector.

The diffusion of conventional and advanced optical microscopy instruments and methods, characterized by high versatility and novel spectroscopy approaches, enables high-spatial resolution fluorescence imaging combined with high sensitivity, especially in the visible spectral range. Obviously, the signal-to-noise (S/N) ratio hugely benefits from optical transparency of the hosting matrix at the luminescence wavelengths, thus enabling optical nanoscopy of the subtlest features. 3D volumetric imaging with subdiffraction-limited resolution can be also achieved in the presence of good transparency.

On the other hand, the areas of growth of thin films and synthesis of nanomaterials have seen a considerable development due to the need for a substantial reduction in the scale and costs of optical devices. The point defect dimensions and their active spectroscopic properties combined with the peculiar optical characteristics of the hosting matrix are promising for direct writing of nanostructures and for radiation imaging at the nanoscale.

In the next parts, the peculiarities and relevance of novel radiation detectors based on the PL properties of CCs in pure LiF crystals and thin films are presented and discussed together with some applications to X-ray imaging at the nanoscale and in proton beam advanced diagnostics in order to highlight the powerful opportunities that advanced fluorescence optical microscopy, spectroscopy, and nanoscopy offer in such investigations. The opportunity of PL enhancement in colored layered structures based on broad-band light-emitting radiation-induced point defects in optically transparent insulating LiF thin layers will be discussed. Finally, some conclusions and the future potential of nanoscopy of luminescent point defects in insulating materials in photonics and imaging are briefly addressed.

2.3.4 State-of-the-art

2.3.4.1 Radiation detectors based on photoluminescence of color centers in LiF

Laser-active radiation-induced CCs in LiF crystals [12] and thin films [13] are stable at room temperature (RT) and some of them efficiently emit broad-band PL in the visible spectral range under optical excitation [14]. They are investigated for their application in optically pumped solid-state tunable lasers [12], broad-band miniaturized light sources [4, 13], and radiation imaging detectors based on RPL [4, 15]. Optical spectroscopy and fluorescence microscopy [16], as well as SNOM [17] (see also Volume 1, Section 3) are very powerful tools for the study and development of these components.

LiF sensitivity to extreme ultraviolet (EUV) radiation [18], soft X-rays [15], and hard X-rays [19] enables advanced radiation imaging applications at the nanoscale, including microradiography of biological objects, even for in vivo samples [20].

X-ray imaging in the photon energy range from 20 eV to 30 keV is considered an important area in physics as well as in life sciences. Innovative LiF radiation imaging detectors, based on the peculiar broad-band spectral characteristics of CCs exploiting the optical properties and radiation sensitivity of the LiF material, have been proposed and tested and presently are under investigation for applications in X-ray microscopy of biological objects, even in vivo cells, in materials science and for the characterization of intense X-ray sources [15, 18]. The advantages of these components are their very high spatial resolution over a large field of view, wide dynamic range, versatility, and simplicity of use, combined with the high sensitivity of optical fluorescence microscopy reading techniques, and importantly giving a way to overcome some of the limits of standard X-ray imaging detectors [18].

Direct writing of micro- and nanostructures by focused low-energy electrons and photons [21, 22] and transfer of luminescent micropatterns by lithographic masks [18,

19] have been demonstrated for applications in the field of nanobiophotonics. Recently, LiF-based radiation detectors have been successfully utilized for advanced diagnostics and dosimetry of proton beams [23].

After exposure to radiation, the latent luminescent images stored in LiF can be read with conventional and advanced optical fluorescence microscopes. The atomic-scale defects that form these images are minimum luminescent units that ensure an intrinsic high spatial resolution. The technique and the microscope utilized to acquire the fluorescence images constitute the main bottleneck to the actually achievable resolution.

With the exception of exotic materials, such as rare gases in solid form, the bandgap of LiF is the largest one – greater than 14 eV – of any solid. The consequent high optical transparency allows investigation of features of the lattice defect by means of optical spectroscopy and advanced fluorescence microscopy techniques [24, 20], from which useful pieces of information regarding radiation effects on this material, either in bulk or thin-film form, can be also derived. As a matter of fact, optically transparent polycrystalline LiF thin films can be grown by physical vapor deposition methods on several amorphous and crystalline substrates [13]. Moreover, the absorption and emission spectral features of suitable radiation-induced CCs in this hosting material are located in the visible spectral range, where they can be interrogated with the standard instrumentation components for optical microscopy and nanospectroscopy, like high-performance laser and light sources, optical elements, and detectors.

2.3.4.2 Photoluminescence properties of radiation-induced color centers in LiF

In alkali halides, the primary defect is the F center, consisting of an anionic vacancy occupied by an electron. Its main absorption band in LiF is centered at about 248 nm [25, 26], but up to now its PL has not been detected unambiguously; a weak emission is theoretically expected at a wavelength of ~900 nm [27]. The F center plays a crucial role as primary defect because it can aggregate to form more complex defects. The laser-active F_2 and F_3^+ centers, consisting of two electrons bound to two and three adjacent anion vacancies, respectively, feature almost overlapping absorption bands at ~450 nm, which together form the so called M band [14]; under light excitation in this band, they show Stokes-shifted broad photoemission bands centered at 678 nm and 541 nm [14, 28], respectively. Because of their high emission efficiencies [12] and lasing characteristics [29], even at RT, many studies can be found in the literature regarding the optical properties of these two defects produced in LiF crystals and thin films by means of several irradiation sources.

Table 2.3.1 reports the peak energies (E_a, E_e), the full width at half maximum (FWHM$_a$, FWHM$_e$) of the absorption and emission bands of F, F_2, and F_3^+ CCs at RT and their luminescence decay time (τ) [30].

Table 2.3.1: Spectroscopic properties of F, F_2, and F_3^+ CCs at RT in LiF crystals.

Center	E_a (eV) λ_a (nm)	FWHM$_a$ (eV)	E_e (eV) λ_e (nm)	FWHM$_e$ (eV)	τ (ns)
F	5.00[a] 248	0.76[a]			
F_2	2.79[b] 444	0.16[b]	1.83[b] 678	0.36[b]	17[c]
F_3^+	2.77[b] 448	0.29[b]	2.29[b] 541	0.31[b]	11.5[c]

[a] [25, 26], [b] [28], [c] [30].

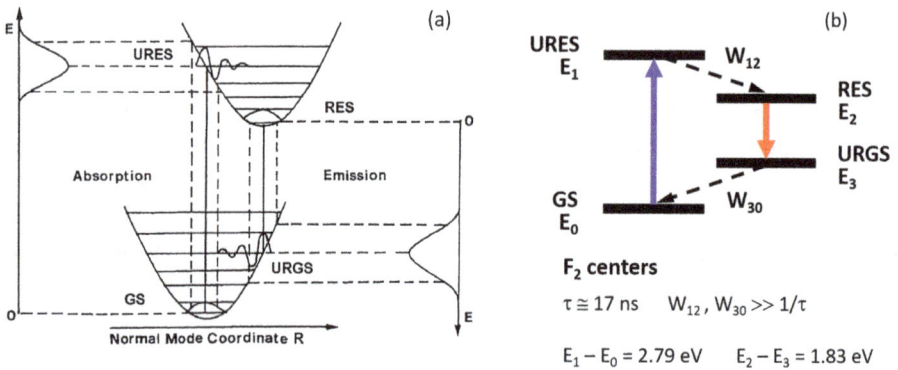

Figure 2.3.1: Configuration coordinate diagram of CCs (a) and energy level diagram of F_2 defects in LiF (b).

The theoretical treatment of the energetic levels of these CCs is based on the assumption that each defect could be considered as a molecular system interacting with the surrounding crystalline matrix. The detailed treatment of this interaction requires the use of a many-body Hamiltonian and a consequent elaborate theoretical approach. However, some relevant features of the interaction can be understood in terms of a simpler model, called the configuration coordinate model, lent from molecular physics, schematized in Figure 2.3.1a. This model can explain the Stokes shift (the fact that the absorption energy is higher than that of emission), the width of the absorption and emission bands, and their temperature dependence. In this model, derived from the Born–Oppenheimer approximation, the energies of the ground and excited states are functions of a normal coordinate R, representing the displacement of the nearby ions, which can be approximated by a parabola near the minimum (the equilibrium position of the ions). The energy levels for the vibrational states of the nuclei in both ground and excited electronic states are drawn as constant-energy lines. As shown in Figure 2.3.1a, the optical absorption, at fixed R according to the Franck–Condon principle, of a photon of appropriate energy causes the electron transition from a vibronic level of the electronic ground state (GS) to a vibronic level of the unrelaxed excited state (URES), from which the electron loses energy by phonon emission, reverting to the relaxed excited state (RES) in a characteristic time of 10^{-12} s. This transition is followed by a radiative transition, from the RES to the unrelaxed ground state (URGS), characterized by the luminescence decay time τ. A new fast nonradiative relaxation process, from the URGS to the GS, completes the optical cycle. The absorption and emission bands are the envelope of the various individual transitions between vibronic levels [31]. As an example, Figure 2.3.1b shows the characteristic four-level optical cycle of F_2 electronic defects in LiF.

In Figure 2.3.2a an RT absorption spectrum (left scale, optical density OD units) and a PL spectrum (right scale), obtained by exciting a LiF crystal (thickness = 2 mm), gamma-irradiated with an absorbed dose of 8.4×10^3 Gy, at a wavelength of 457.9 nm, are shown. In the absorption spectrum, one can identify two main features: the first one is the F band, centered at 248 nm, due to the primary F defects; the second one is the M band [14], located at ~450 nm, which consists of the overlap of the F_2 and F_3^+ absorption contributions. The PL spectrum is formed by the superposition of two broad emission bands arising from F_2 and F_3^+ CCs, which peak at 678 nm and 541 nm, respectively (see Table 2.3.1).

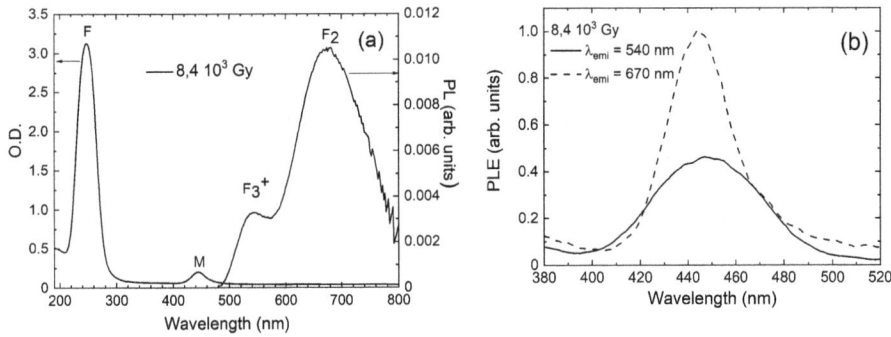

Figure 2.3.2: RT optical absorption and photoluminescence spectra, the latter one being obtained by exciting at the wavelength of 457.9 nm (a), and photoluminescence–excitation spectra, measured at emission wavelengths of 540 and 670 nm (F_3^+ and F_2 CCs, respectively) (b) of a gamma-colored LiF crystal (dose of 8.4×10^3 Gy). The crystal thickness is 2 mm.

Two RT PL–excitation (PLE) spectra of the same crystal, measured with a commercial spectrofluorometer with front-face detection geometry, are shown in Figure 2.3.2b. These two spectra represent the emission intensities at the wavelengths of 540 and 670 nm – i. e., the main emission contributions of F_3^+ and F_2 defects, respectively – as functions of excitation wavelengths spanning from 380 nm to 520 nm. The PLE spectra were suitably corrected for the effects of optical absorption inside the investigated sample [32]. According to the spectral features of the F_2 and F_3^+ absorption bands, the FWHM of the F_3^+ CCs is wider than that of the F_2 electronic defects, and the peak is located at a longer wavelength (see Table 2.3.1).

High-optical quality LiF crystals (pure or doped) and transparent polycrystalline LiF thin films can be grown by means of several methods. They play the role of host matrix for the broad-band visible-emitting F_2 and F_3^+ radiation-induced defects; these latter can be used to probe the host matrix characteristics [33]. LiF-based photonic microcomponents [34–36] and nanostructures [13, 20, 37] with particular optical PL properties can be produced by controlling the local formation, stabilization, and photothermal transformation of radiation-induced CCs. With respect to other alkali halides, radiation-induced point defects in LiF are stable even at RT, and high densities of F_3^+ centers are observed together with F and F_2 defects [38].

In irradiated LiF films the direct use of optical absorption spectra to identify the presence of electronic defects is often a hard task because of the presence of interference fringes due to the refractive index difference between film and substrate [39]. In this regard, PL measurements are more sensitive than the absorption ones in investigating the presence of visible-emitting electronic defects. Moreover, in colored LiF thin films, due to their limited thickness, it is possible to neglect the effects of pump absorption inside the investigated samples and to directly interpret PLE spectra as acquired, even at high CC volume concentrations.

Figure 2.3.3a shows the PL spectrum of a gamma-colored LiF film (~3 μm thick), irradiated with a dose of 10^6 Gy, deposited by thermal evaporation on a radiation-hard fused silica substrate, acquired at RT under laser excitation at 457.9 nm. The PL spectrum is formed by the superposition of two broad visible emission bands due to F_2 and F_3^+ CCs. By applying a two-Gaussian best-fit procedure, shown in Figure 2.3.3a, the peak positions – 678 and 541 nm, respectively – and the FWHMs of F_2 and F_3^+ CCs were obtained. These spectral features are very similar to those measured in bulk crys-

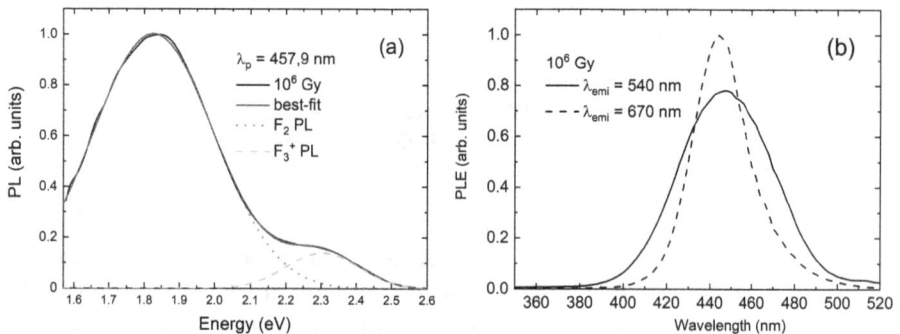

Figure 2.3.3: RT photoluminescence spectrum excited at 457.9 nm (a) and photoluminescence–excitation spectrum, measured at wavelengths of 540 and 670 nm (F_3^+ and F_2 CCs, respectively) (b) of a LiF thin film that was gamma-irradiated with a dose of 10^6 Gy. The film thickness is ~3 μm.

tals (see Table 2.3.1). In Figure 2.3.3b the PLE spectra of the same gamma-irradiated LiF film are shown; they were measured at RT at wavelengths of 540 nm (green emission, F_3^+) and 670 nm (red emission, F_2). Again, as expected, the PLE band of the F_3^+ defects is broader than the F_2 one, and the F_3^+ peak position is found at a longer wavelength.

The trend to miniaturization of photonic components has been increasingly stimulating the use of thin films as components of broad-band light-emitting devices. In this respect, LiF films are of great interest because they have the advantages of compatibility with amorphous and crystalline substrates, offering the possibility of growing multilayered structures with different materials [13] and functionalized emission properties, like in colored LiF film-based fully dielectric optical microcavities [40]; moreover, the control of dopant concentrations and interfaces by various versatile preparation techniques [41] can be achieved.

The modification of the LiF material local properties in a thin surface layer through the formation and stabilization of CCs depends on the nature and energy of the utilized radiation, on the irradiation parameters (dose, dose rate, etc.), and on the material growth conditions [33]. It has been observed that the formation and stabilization processes of CCs strongly depend on the characteristics of the LiF hosting material. From a practical point of view, one of the main aspects of the radiation–matter interactions that must be taken into account is the radiation penetration depth d in solids. For low-energy radiation, d can be comparable with, or smaller than, the typical thickness t of a film; in this case, the surface-colored layer can be hosted in a bulk single crystal matrix or in a polycrystalline LiF thin film, at the LiF–air interface (assuming it is the one exposed to radiation). In the case of energetic radiation, for which $d > t$, thin films become uniformly colored, thus providing 2D colored thin layers. This is the case, for instance, for the gamma-irradiated film of Figure 2.3.3.

2.3.4.3 Application of the photoluminescence properties of color centers in LiF to X-ray imaging

The principle of operation of a LiF film detector is schematized in Figure 2.3.4 for contact-mode operation [42]. Because no manipulation of X-rays is needed, contact mode is the simplest approach to X-ray microscopy. As shown in Figure 2.3.4a, the sample under investigation is placed in contact with the surface of the LiF detector surface in order to be exposed to X-rays; the irradiation causes the formation of a layer of stable CCs, whose local density is proportional to the dose absorbed in the film (assuming that fluence is low enough that saturation is not reached), which depends on the more or less transparent details of the sample. Indeed, the parts of the specimen that are more absorbing correspond to lower CC densities in the colored layer, because fewer photons reach it after passing through the specimen, while less absorbing parts induce higher CC densities.

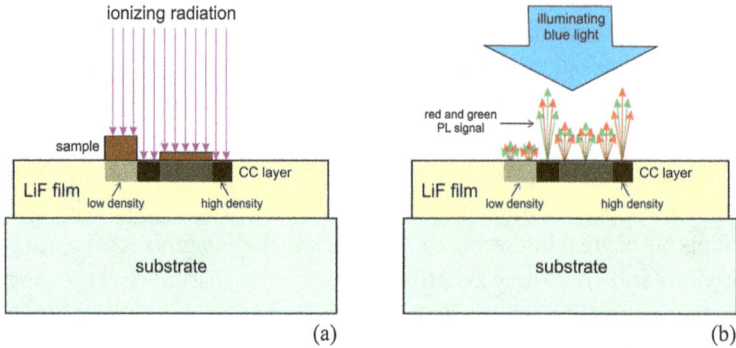

Figure 2.3.4: X-ray contact microscopy with a LiF film as radiation imaging detector. (a) Contact setup for X-ray imaging and formation of CCs in the LiF film during exposure. (b) Reading of the visible PL from the generated CCs under blue-light illumination. From Ref. [42] with kind permission of Società Italiana di Fisica.

The thickness of the colored layer depends only on the used X-ray energy. After exposure, the investigated sample is removed and the latent image, which has been generated and stored in the LiF film, can be read by illuminating it with blue light and detecting the red and green PL emitted by F_2 and F_3^+ CCs (see Fig. 2.3.4b). The optical readout process of LiF plates can be directly performed by means of a conventional optical fluorescence microscope, which has to be equipped with a blue-light pump spectrally centered around the peak wavelength of the M band to simultaneously excite the visible PL of F_2 and F_3^+ defects.

Since the size of a typical CC is less than 1 nm [43], the lattice constant of LiF being only 0.2013 nm, 2D images with high spatial resolution can be stored. High-quality, well-contrasted X-ray microradiographies of biological specimens and metallic meshes were achieved with spatial resolutions down to 250 nm across a wide field of view (>1 cm^2) by reading the irradiated detectors with a confocal laser fluorescence microscope [19]; moreover, resolutions down to 80 nm were demonstrated by means of SNOM [17, 44].

As an example, Figure 2.3.5a and b show contact X-ray microradiographies of two copper meshes with a typical wire dimension of 10 µm, characterized by 1000 and 2000 lines per inch (lpi), corresponding to periods of 25.4 and 12.7 µm, respectively [45]. They were placed in contact with the surface of LiF crystals in order to mask them from the incoming soft X-rays produced by an excimer-pumped laser plasma source with a copper target. The images were subsequently read by an optical microscope in fluorescence mode; the PL signal was detected under lamp illumination suitably filtered to excite in the M absorption band spectral region. The resulting pumping blue light was cut off from the detected image by another optical filter. The intensity of the emitted visible luminescence is proportional to the X-ray transmittance of the object. The uncolored black parts represent the dark shadow of the meshes, while the gray ones correspond to the luminescent colored areas.

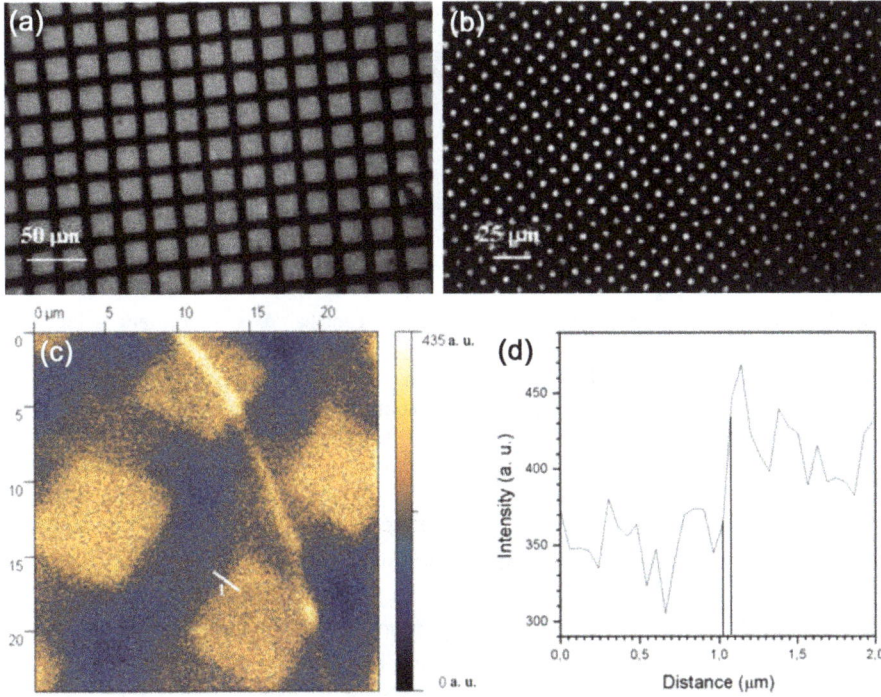

Figure 2.3.5: Fluorescence images of regular patterns of luminescent CCs in LiF crystals realized by masking LiF with copper meshes of different periods and irradiated by soft X-rays: (a) 1000 lpi and (b) 2000 lpi. The samples were observed with a fluorescence optical microscope at the same magnification. (c) 25 μm × 25 μm SNOM optical image of the masked LiF crystal in (b). (d) PL intensity profile, traced along the white line in (c). From [44, 45].

Thanks to the post-irradiation stability at RT of the generated F_2 and F_3^+ CCs, the X-ray microradiographies of the mask stored in LiF crystals and films could be read by an SNOM operating in collection mode: the light coming from an Ar laser, at a wavelength of $\lambda = 457.9$ nm, illuminated the whole area of the sample and the local fluorescence was revealed through a small aperture, consisting of the sharp tip at the end of an un-coated, tapered optical fiber, placed at a few nanometers from the illuminated surface. The near-field components were thus converted into far-field ones and propagated into the optical fiber. A high-pass filter, with a cut-off wavelength of 510 nm, was used to collect only the integrated PL signal coming from the active CCs in LiF by means of a PMT. Topographical (constant shear force) and optical (fluorescence) images were simultaneously acquired. Figure 2.3.5c shows the SNOM optical image of a portion of the irradiated LiF crystal. The PL intensity distribution arises from the locally patterned CCs produced during X-ray exposure. Figure 2.3.5d shows the intensity profile along the white line that crosses the edge of a square in Figure 2.3.5c. The profile shows an edge resolution of ~75 nm [44].

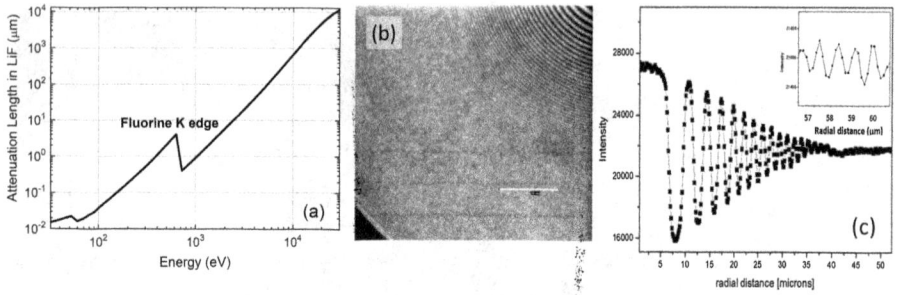

Figure 2.3.6: (a) Attenuation length of LiF in the EUV and soft and hard X-ray energy ranges. (b) Confocal fluorescence image of the PFZP X-ray microradiograph stored on a 1.4 μm thick LiF film grown on a glass substrate, under optical excitation at 458 nm. The brighter areas correspond to the zones most transparent to X-rays. Scale bar = 20 μm. (c) Spectrally integrated intensity as a function of the radial distance, magnified in the inset for a radius from 56 to 60 μm. From [19].

X-ray penetration in solids is strongly dependent on photon energy. Figure 2.3.6a shows the attenuation length of X-rays in LiF [46]: in the EUV (20–200 eV) it ranges from a few tens of nanometers to 200 nm, but it grows to several micrometers in the soft X-ray region (200 eV to 2 keV), depending on the energy, and goes up to 100 microns and more at higher energies. Due to their limited thickness, LiF film detectors allow to directly obtain thin colored layers also by using high-energy X-rays, because CC formation is limited in depth by the total film thickness.

As an example, a hard X-ray contact microradiography of a phase Fresnel zone plate (PFZP) – a diffractive lens for hard X-rays – on a 1.4 μm thick LiF film grown on glass substrate is shown in Figure 2.3.6b [19]. It was obtained by irradiation with a table-top X-ray microsource with Cu anode. During X-ray exposure, the PFZP was put in close contact with the LiF film surface. The PFZP radius was 120 μm and the width of its last zone was 203 nm. Fabrication of these opaque zones is achieved from electrodeposited gold, 0.9 μm thick, features that cause the absorption of 40 % of the X-ray Cu K_α radiation at 8 keV. The X-ray transparent zones consist of a 2 μm thick Si_3N_4 membrane. After X-ray exposure of the PFZP placed in contact with the LiF surface, the images stored in the LiF films were read by a confocal laser scanning microscope (CLSM) in fluorescence mode. The brighter areas correspond to the X-ray transparent zones of the PFZP. A modulation up to a radius of the order of about 90 μm, which corresponds to a zone width of about 270 nm, was observed [19]. Figure 2.3.6c shows the spectrally integrated intensity as a function of the radial distance for a radius from 56 to 60 μm. The FWHM of the single oscillation is of the order of 250 nm, and it is close to the limit of the confocal microscope in the used configuration. It can therefore be deduced that the final resolution is essentially dictated by the optical resolution of the reading instrument. The latent images generated in the LiF thin films by the local formation of active defects ensure an intrinsic high spatial resolution (<250 nm) also thanks to the limited thickness of the irradiated layer.

2.3.4.4 Confocal fluorescence microscopy: the ideal reading technique for 2D and 3D X-ray imaging luminescent detectors

The CLSM is an important evolution of the conventional wide-field fluorescence microscope. CLSM is a versatile and powerful optical instrument which is gaining interest in the scientific community for biological investigations and also for the characterization of materials, microstructures, and devices. A detailed description of the working principle of confocal microscopy is given in Volume 1, Chapter 1.5.

A CLSM has been used in fluorescence mode to read 2D X-ray microradiographies stored in irradiated LiF detectors and featuring high contrast and high spatial resolution. Regarding the latter, a value of the order of 250 nm, i. e., close to the limit of the confocal microscope, has been obtained for contact X-ray microradiography of a test pattern (see Fig. 2.3.6b and c).

One of the most critical factors to consider in wide-field fluorescence microscopy arises from the fact that regardless of the focal point, illumination from the objective produces fluorescence throughout the entire specimen volume. As a result, wide-field fluorescence images often contain a high level of background signal that obscures specimen details and dramatically reduces contrast. Historically, the only way to get a clear sharp image of a thicker sample was to physically cut it into thin sections, a technique which is time consuming, likely to introduce artifacts, and not practical for the imaging of live samples.

During the last few decades, a large number of techniques have been developed that allow optically removing the contributions from these out-of-focus regions, thus obtaining in the focal plane a clear image of the sample without having to physically harm it. This family of techniques is most commonly referred to as optical sectioning microscopy, including the confocal one. The basic key to the confocal approach is the use of spatial filtering techniques to eliminate out-of-focus light or glare in specimens whose thickness exceeds the immediate plane of focus. In addition to the possibility to observe in good contrast a single plane of a thick specimen, optical sectioning allows a great number of slices to be cut and recorded at different planes of the specimen, with the latter being moved along the optical axis by controlled increments. By computation, various aspects of the object can be generated from the 3D dataset (3D reconstruction, sections of any spatial orientation, etc.). The popularity of this technique has grown thanks to the 3D reconstruction of a great variety of biological and nonbiological samples with submicrometric resolution.

The confocal principle was first described by Minsky [47]; a simple explanation of it is found in the book by Lukosz [48]. It should be pointed out that the increase in lateral resolution with respect to a conventional microscope operating in the same working conditions can be quantified to be about 30 % [49]. The working scheme of a CLSM is illustrated in Figure 2.3.7a. A system of lenses and dichroic mirrors allows the illuminating laser (blue lines in the figure) to scan the region of interest of the

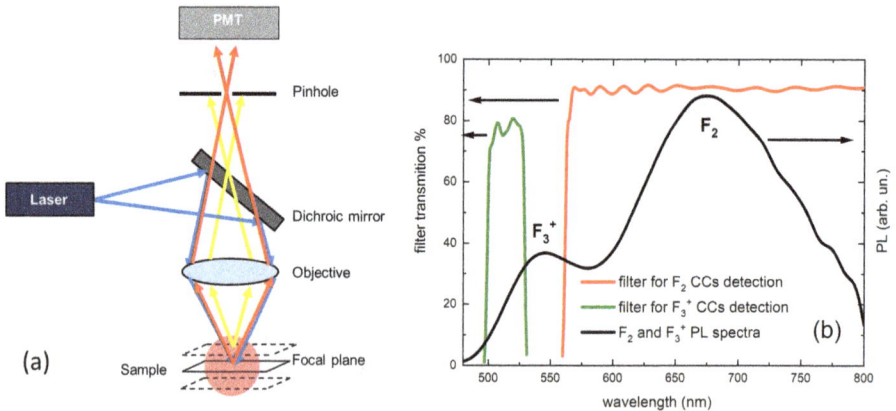

Figure 2.3.7: (a) Operating principle of a laser scanning confocal microscope. (b) RT PL spectra of a colored LiF crystal measured under laser excitation at the wavelength of 457.9 nm and transmission spectra of the optical filters mounted in the fluorescence collection system of the CLSM.

sample under observation. The light emitted from the sample (red and yellow lines) is collected by a system of PMTs. A small circular aperture (pinhole) is interposed on a plane conjugate to the focal plane of the objective lens along the optical path of this beam, whose function is to stop the light that does not come from the focal plane (yellow lines). The light from the focus (red line) passes through the pinhole and is collected by the PMT, delivering the photonic signals to form the final image. The result is that the final image is produced mainly by the light coming from a restricted vertical section of the sample, or from the planes close to the focus. By varying the focus plane, one can get images of further sections of the sample. Suitably assembling all the collected section images allows obtaining a full 3D reconstruction of the sample in a chosen vertical range of interest.

The CLSM used in our laboratory to read images stored in LiF-based detectors for X-ray imaging is equipped with an argon laser; the CCs excitation is performed using its 457.9 nm line. In fluorescence operating mode, the PL signal is detected by a system of two PMTs, which acquire separately and independently two different spectral ranges that are selected by an appropriate optical filtering system. The optics here is composed by a dichroic mirror that blocks the reflected excitation light and channels the emitted light in two optical paths directed to the filters of each detection stage. One filter is a low-pass filter and is reserved for the selection of the F_2 fluorescence broad band peaking at ~680 nm, while the second filter is a band-pass filter and it is dedicated to the selection of F_3^+ fluorescence broad band peaking at ~535 nm. Figure 2.3.7b reports typical F_2 and F_3^+ PL spectra of a colored LiF crystal measured under laser excitation at a wavelength of 457.9 nm together with the transmission spectra of the optical filters mounted in the fluorescence collection system of the CLSM.

As an example of CLSM applied to the 3D reading of a LiF detector, we report the detection of a quasi-parallel X-ray beam emitted by an X-ray tube (Cu anode) after emerging from a glass polycapillary semilens [50]. This beam uniformly irradiated the surface of a LiF crystal of dimensions $5 \times 5 \times 0.5 \, mm^3$. Figure 2.3.8a reports the 2D confocal images of an area of about $500 \times 500 \, \mu m^2$ of the colored LiF surface.

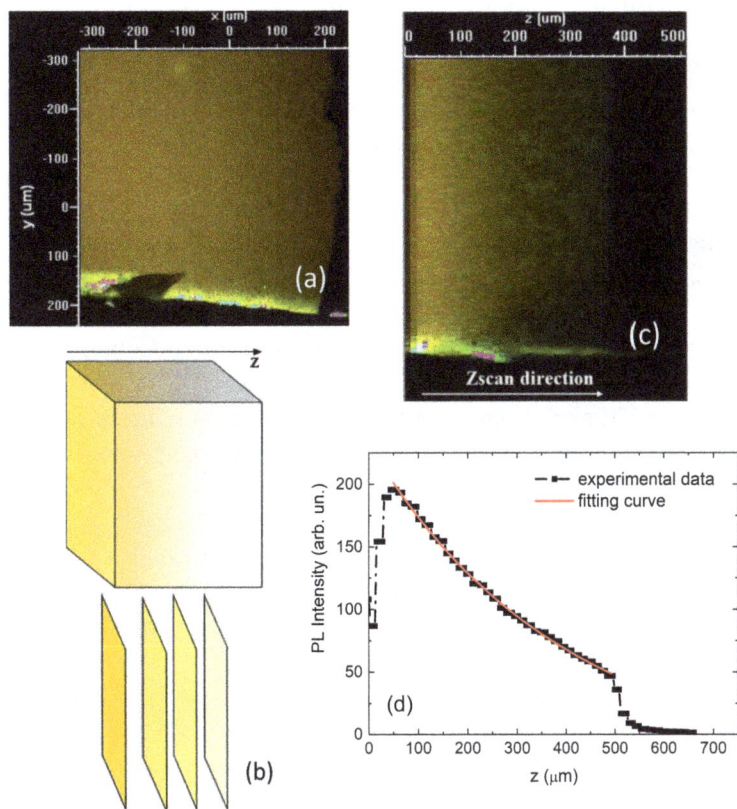

Figure 2.3.8: (a) CLSM 2D (X,Y) fluorescence images of the surface of a LiF crystal irradiated by a quasi-parallel X-ray beam emerging from a polycapillary semilens. (b) Scheme of how the CLSM optically sections the colored volume of the LiF crystal along the Z-axis. (c) Distribution along the Z-axis of the PL as obtained by the CLSM optical sectioning. (d) Experimental intensity profile along Z of the F_2 fluorescence (black points) as obtained from the distribution image in (c); the red solid curve is the best-fitting exponential decay curve. From [50].

According to the X-ray transmission properties in solids [46], the energy deposition of a monochromatic X-ray beam decreases in an exponential way along the penetration depth in LiF beyond a thin surface buildup layer. The emission radiation of a conventional tube mainly consists of an elastic component, Cu K_α, which is more intense than the "white" band emission, due to electron bremsstrahlung. If the concentration of CCs is assumed to be proportional to the absorbed energy, the CLSM proves to be a suitable technique for estimating its profile along the LiF crystal depth. Figure 2.3.8b shows the scheme of the optical sectioning that a CLSM can operate along the Z-axis in a crystal. By using this feature with controlled spatial increments, slices along the colored volume of the irradiated LiF crystal were obtained. Figure 2.3.8c shows the Z distribution of fluorescence intensity that was obtained with this approach. This characterization demonstrated that a single-exposure X-ray irradiation of LiF crystals combined with CLSM was enough to gather 3D information of the X-ray beam. Figure 2.3.8d reports the measured F_2 fluorescence intensity profile along the crystal thickness. These 3D imaging capabilities were recently extended to the visualization of a free electron laser (FEL) X-ray beam at 10.1 keV [51].

2.3.4.5 Peculiarities and enhancement of photoluminescence in thin irradiated LiF layers

The PL of CCs that are generated in a LiF film by exposure to ionizing radiation can be excited if a suitable optical pump is applied. In particular, F_2 and F_3^+ CCs luminesce in the red and green ranges of the spectrum, respectively, when they are optically excited in the so-called M absorption band, which is located in the blue and peaks at about 450 nm [9, 13, 14]. In this way, a PL pattern corresponding to the fluence cross-section of the previously applied ionizing radiation can be observed by means of an optical microscope, no development process being required [52–59]. Such a pattern replicates the radiation transmission properties of a sample, if this latter was placed on the LiF film during irradiation, so that either an absorption or phase contrast – depending on the irradiation setup – latent image is stored in the LiF film, ready to be "turned on" by means of blue illumination and detected. While the intrinsic spatial resolution of the stored image, thanks to the very low dimensions of the implied lattice defects, is below 250 nm [19], the actual spatial resolution of the detected image is bottlenecked by the optical characteristics of the instrument used for the PL observation.

The intensity of the detected PL image depends on the number of CCs that are stored, point by point, within the LiF film. If the film is irradiated with deeply penetrating ionizing radiation, it gets colored along its whole thickness. In such a case, the amount of emitted PL is larger for thicker films. On the other hand, if the ionizing radiation does not penetrate through all the film thickness, a surface-only layer of CCs is created at the irradiated face, and the film thickness becomes generally less important to the net amount of PL, unless significant optical interference phenomena take place.

It is possible, in either case, to amplify the emitted PL intensity by exploiting light confinement effects, in the presence of which the duration of the optical cycle of CCs gets shortened with a resulting increase of the photon emission rate [60–63]. Such light confinement effects were exploited, for instance, to study enhanced coherent anti-Stokes Raman scattering microscopy in a microcavity [64]. The desired improvement, in the presence of which a gain in efficiency of the device as luminescent radiation detector is achieved, can be designed by means of an analytical theoretical model [65]; it is practically attained, in its simplest configuration, by using an optically reflective material as substrate for the deposition of the LiF film, and also by choosing a film thickness that favors constructive interference effects of the optical pump and the PL at the same time – such a design reproduces that of a planar half-cavity, since the air–LiF interface is low reflecting in the visible range (reflectance of ~2.6 %) [4].

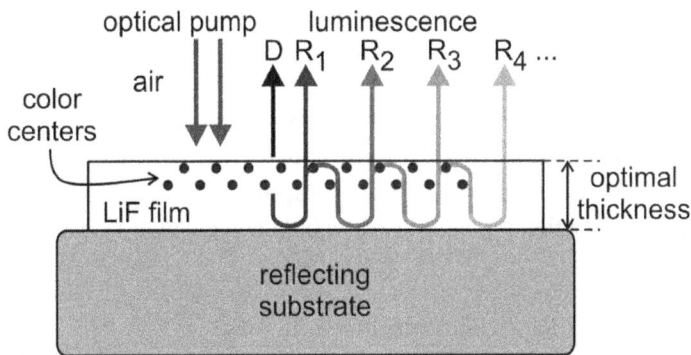

Figure 2.3.9: Scheme of how a half-cavity-enhanced LiF-film detector is read after exposition to ionizing radiation. The CCs created in the film get excited by the blue light of the optical pump. They emit into air a direct contribution to the PL signal (D in the figure) towards the observation device (e. g., a microscope objective). Thanks to the reflecting properties of the film–substrate interface, other multireflected contributions ($R_1, R_2, R_3, R_4, \ldots$), otherwise lost, can reach the observation device – however, because of the low LiF–air reflectance value (~2.6 %), multireflections beyond R_1 are generally negligible. If the thickness of the film is optimized, constructive interference contributes to provide an amplified PL signal. Additionally, constructive interference among the optical pump multireflections inside the film can increase the effective number of radiating CCs.

Figure 2.3.9 shows a scheme of the improved detector – this latter can provide, although containing fewer CCs, the same PL intensity as a nonimproved detector, such as a LiF film over a glass substrate, because the CCs in the improved detector emit photons at a higher rate. From a quantum theory viewpoint, the vacuum field modes – responsible for spontaneous emission from the CCs – are more intense in special locations (antinodes) within the structure. This peculiarity, once suitably exploited, allows recording in the LiF detector the latent image of the investigated samples by using a smaller amount of radiation, with all the beneficial effects that this fact implies, e. g., an exposure of the sample to a reduced radiation dose and/or shorter irradiation times, very useful for getting less blurred shots of in vivo biological samples.

Assuming that the irradiation process and the consequent PL detection take place from the air side, to obtain enhancement the substrate should reflect the PL radiated by the CCs towards air, because in this way it would be not lost through the substrate and would be redirected to the detection system; moreover, the reflection properties of the substrate should be ideally not affected by the irradiation with ionizing radiation if this latter is sufficiently penetrating to reach the substrate. In the design stage, one should also consider the phase of the LiF–substrate reflectance, interference between direct and substrate-reflected PL, the spectral distribution of PL, the wavelength, polarization state, angle of incidence of the optical pump, and the numerical aperture (NA) of the detection system. All these pieces of information should be taken into account to select a film thickness that maximizes both (i) the electric field intensity of the optical pump at positions where the CC distribution is denser and (ii) constructive interference among the direct and substrate-reflected PL components.

Next, the detection of the PL that is radiated from the stored latent image in the film detector is analyzed. If one considers a vertically impinging optical pump plane wave, of unitary in-air intensity and wavevector module k_P, onto a LiF film of refractive index n, it can be demonstrated that the intensity distribution of the pump inside the film is [65]

$$I_p(z) = T_{af} \frac{1 + R_{fs} + 2\sqrt{R_{fs}} \cos[k_P n(d + 2z) + \phi_{fs}]}{1 + R_{fa}R_{fs} - 2\sqrt{R_{fa}R_{fs}} \cos(2k_P nd + \phi_{fa} + \phi_{fs})}. \tag{2.3.1}$$

Here, the z-axis is perpendicular to the film interfaces, pointing towards the external air medium, its origin $z = 0$ being located at the film–substrate interface; as far as the used symbols are concerned, T_{af} is the air–film intensity transmittance, R_{fs} and R_{fa} are the film–substrate and film–air intensity reflectances, respectively, ϕ_{fs} and ϕ_{fa} are the film–substrate and film–air phase changes on reflection, respectively, and d is the film thickness.

If $N_{CC}(x, y, z)$ is the volume density of CCs that were generated inside the LiF film by the ionizing radiation, assuming the pump intensity to be everywhere in the film well below the saturation intensity of the active CCs, the density distribution of CCs that actually radiate PL photons under the action of the optical pump is a superposition of the two distributions [65],

$$N_{PL}(x, y, z) = qN_{CC}(x, y, z)I_P(z), \tag{2.3.2}$$

where q is a constant parameter. Such a quantity can be evaluated once $N_{CC}(x, y, z)$ is known and allows estimating the PL radiation pattern. If $w(x, y, z)$ is the power radiated by a single randomly oriented dipole in the stratified medium [4], if cooperative effects are excluded [66], the overall radiated power is simply [65]

$$W = \iiint N_{PL}(x, y, z)w(x, y, z)dxdydz. \tag{2.3.3}$$

In some cases, the above mathematical expression leads to a fully analytic expression of the radiated power [65].

In the estimation of the detected PL, other features of the detection setup have to be taken into account, among which important ones are the solid angle corresponding to the NA of the detection system (e. g., the objective of a microscope) and the spectral distribution of the PL. The actual detected signal is proportional to the integrals over those spanned angles and wavelengths. For our simulations, such a task is numerically accomplished by means of a MATLAB program that was specifically developed by us to calculate the PL radiated by an enhanced LiF film detector and detected by an optical system with certain characteristics.

In Figure 2.3.10, some example calculations are shown of the PL intensity, as detected by a NA = 0.13 microscope objective placed in the upper half-space (as in Fig. 2.3.9), versus the thickness of the LiF film for three distinct materials used as substrate: SiO_2, Si, and an ideal metal following Drude's law [67] with a plasma wavelength of 200 nm. Besides this Drude-like substrate, the dispersions of the optical constants of the other materials are those tabulated in Palik's handbook [68]. The LiF film is thought to be fully and homogeneously colored under the action of some kind of ionizing radiation. The optical pump, whose wavelength is set to 458 nm in the simulations, is assumed to be a vertically impinging plane wave from the air side. The spectral distributions of the PL radiated by F_2 and F_3^+ centers [9, 13, 14] have been properly taken into account. As far as the used PL intensity units are concerned, they are normalized not only to the LiF film thickness, but also in such a way that the PL values in Figure 2.3.10 would be exactly equal to 1 if the substrate were of the same material as

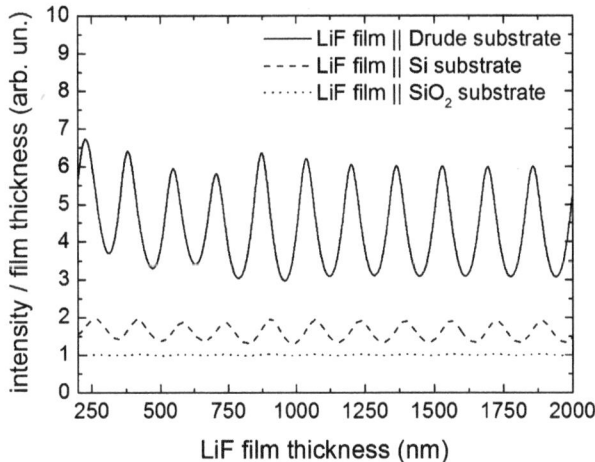

Figure 2.3.10: Theoretical PL, as a function of the film thickness, radiated by a homogeneous distribution of F_2 and F_3^+ color centers in the whole LiF film and detected by a microscope objective of NA = 0.13. Three distinct substrates are considered: SiO_2, Si, and an ideal metal following Drude's law with a plasma wavelength of 200 nm. See the text for other details of the simulations.

the LiF film. Improved amplification is predicted, at discrete thickness values, for the case of the ideal Drude metal, thanks to the constructive interference effects that act on both the optical pump and the PL multireflections, as previously explained. A similar, but less strong, beneficial effect is noticed for the Si substrate as well, while the SiO_2 substrate provides almost no amplification at all because of the small difference existing between its refractive index and that of LiF.

A similar situation as that described above is considered for the simulations shown in Figure 2.3.11. Here, however, the LiF film is considered to be homogeneously colored only in a thin surface layer of thickness 50 nm, as if poorly penetrating ionizing radiation was applied. The fact that in this second case the generated CCs are confined in a thinner layer allows for a better optimization of the device performance, because the nodes of the involved fields (optical pump and PL multireflections) can be better kept away from the active layer, thus minimizing effects that tend to lower the device performances due to destructive interference. Indeed, the plots of Figure 2.3.11 feature higher PL intensity peaks than in Figure 2.3.10. For the Drude metal case, amplifications as high as one order of magnitude can be expected at discrete thickness values of the LiF film. The Si provides amplification values that are close to 3, while in the SiO_2 case there is no appreciable amplification effect because of the poor reflectance of the LiF–SiO_2 interface. Note that the PL intensity units of Figure 2.3.11 are not normalized to the film thickness (as it happened in Figure 2.3.10), because the depth of the active layer is always set to 50 nm for any value of the film thickness.

By using LiF thin films thermally evaporated on glass and Si(100) substrates as radiation imaging detectors, soft X-ray contact microradiographies of very thin bio-

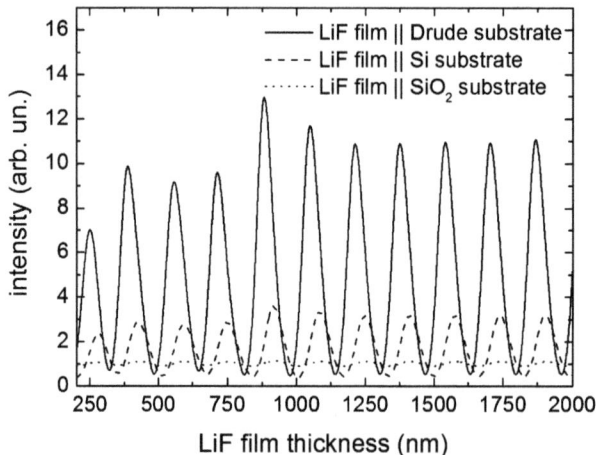

Figure 2.3.11: Theoretical PL, as a function of the film thickness, radiated by a 50 nm thin surface distribution of F_2 and F_3^+ color centers in the LiF film and detected by a microscope objective of NA = 0.13. Three distinct substrates are considered: SiO_2, Si, and an ideal metal following Drude's law with a plasma wavelength of 200 nm. See the text for other details of the simulations.

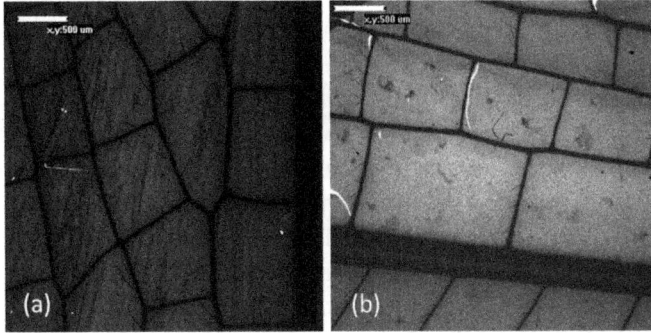

Figure 2.3.12: CLSM fluorescence images of soft X-ray contact microradiographies of a dried drag-onfly wing stored in 1 µm thick LiF films thermally evaporated on (a) glass and (b) Si(100) substrates obtained under the same irradiation conditions (single-shot, Cu target), scale: 500 µm. From [4].

logical objects (dried insect wings) were reported for the first time in [15]. Microra-diographies of a dragonfly wing on LiF thin films of equal thickness, but thermally evaporated on different substrates – (a) glass and (b) Si(100) – obtained by simul-taneously irradiating in vacuum with a single shot of a Nd-YAG laser plasma source with Cu target (X-ray fluence of ~350 µJ/cm^2/shot), are shown in Figure 2.3.12. The sig-nal intensities of these integrated fluorescence images are assumed to be proportional to the concentrations of active CCs, which depend on the transparency to X-rays of the exposed sample: the more absorbing parts of the biological object are darker in the flu-orescence image, while the less absorbing ones are lighter. Taking into account that the two pictures in Figure 2.3.12 share the same acquisition parameters, the higher re-sponse of the LiF film on Si substrate with respect to the LiF film on glass is clearly noticeable. If stronger signal intensities are involved, the higher S/N ratio allows for a well-contrasted observation of even subtler wing details, like ribs and hair. Finally, im-pressive results were also obtained for in vivo microscopy experiments [69, 56], which were mainly conducted in single-shot exposure to the soft X-rays (water window) emit-ted by a point-like Nd-YAG pumped laser plasma source [70].

2.3.4.6 Other applications of the photoluminescence of color centers in LiF for particle radiation detectors

TL dosimetry with doped LiF is currently used in clinical applications because of a LiF tissue equivalency (that is, absorbing and scattering properties that sufficiently match those of biological tissue) response. On this basis, an RPL method using LiF was de-veloped in [71] for gamma and X-ray dose measurements [72]. More recently, LiF film radiation imaging detectors were utilized for dosimetry [23] and advanced diagnostics of low-energy proton beams [73]. The formation by proton irradiation of aggregate F_3^+ and F_2 electronic defects in ~1 µm thick thin films of LiF grown on glass substrates was studied, at the nominal energies of 3 and 7 MeV and in the fluence range from 10^{11} to 10^{15} protons/cm^2, by means of PL spectroscopy for the beam of a linear accelerator at ENEA C.R. Frascati [23]. Proton energy loss continuously increases with depth, reaching its maximum value – known as the Bragg peak – at about the end of the implantation path. The depth of the Bragg peak is strongly dependent on the proton energy, and at the considered values of this latter it is deeper than the typical thickness of the grown LiF films. The PL signal was found to be proportional to the fluence, with linear response up to a value of ~8×10^{13} protons/cm^2, above which saturation takes place. Because of such a linearity, these radiation imaging detectors are also able to store a map of the transverse proton beam intensity; the map is obtained by simply exposing in air the surface of the LiF film perpendicularly to the proton beam propagation direction. Figure 2.3.13a shows the PL image of a 3-MeV proton beam stored in a LiF film, as read by a conventional fluorescence microscope equipped with a charge-coupled device (CCD) camera under blue-light excitation provided by a Hg lamp. The analysis of the acquired optical image allows one to record the transversal proton beam intensity profile with a high spatial resolution and shows even subtle intensity differences. The 8-bit camera used to acquire the image is likely to be adequate to show, with an optimal intensity scale accuracy, all the information stored in LiF, as reported in Figure 2.3.13b, where a 3D intensity representation is shown.

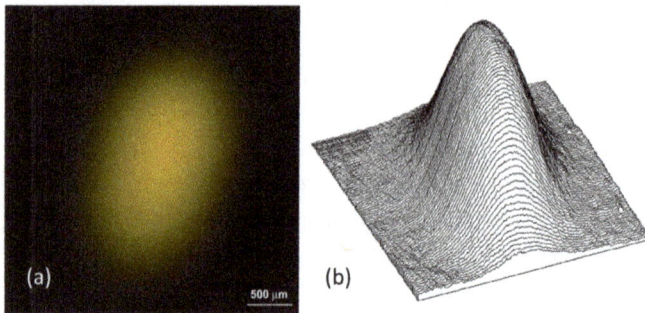

Figure 2.3.13: Photoluminescence image map (a) of the transversal intensity of a 3-MeV proton beam stored by color centers in a ~1 µm thick LiF film, thermally evaporated on a glass substrate, after irradiation at a fluence of 5.6×10^{13} protons/cm^2 and (b) its 3D intensity plot.

2.3.5 Conclusions and future perspectives

In conclusion, novel radiation imaging detectors that use the visible PL of radiation-induced laser-active CCs in LiF thin films have been developed. Thanks to the atomic-scale dimensions of the CCs, these detectors are sensitive to soft and hard X-rays with intrinsic submicrometric spatial resolution, which is limited only by the reading optical fluorescence microscope/technique utilized to acquire the latent images stored after exposure to ionizing radiation. Recently, large-area high-spatial resolution dose maps of low-energy proton beams covering at least three orders of magnitude of doses were demonstrated for the first time [74]. The sensitivity of the optical reading technique and the high emission efficiency of the F_2 and F_3^+ centers, combined with the high optical quality of the thermally evaporated films of this radiation-sensitive material, allows one to obtain accurate quantitative measurements of integrated visible PL intensity by using conventional and advanced fluorescence microscopy techniques. Spatial resolutions of 250 nm have been achieved using CLSM, and values down to 80 nm have been demonstrated by using SNOM for the detection of fluorescent CCs in LiF [17]. Even smaller resolution values, well below the diffraction limit and closer to the intrinsic size of these point defects, could be expected if other nonlinear optical microscopy techniques were employed. A method with potential for this task could be STED microscopy. This technique was successfully utilized to obtain the high-resolution imaging of individual CCs in diamond. The fluorescent nitrogen vacancy (NV) point defect, consisting in a substitutional nitrogen atom and a neighboring charged vacancy, emits with a significantly Stokes-shifted red light due to high phononic coupling with the lattice. These kinds of defects were mapped by STED microscopy with nanoscale (below 10 nm) resolution and subnanometer precision [75] by using focused laser light. The main reason for this success is that NV centers in diamond are not susceptible to photobleaching, providing a noninvasive quantitative characterization method for both the concentration and distribution of defects that is capable of working even a few microns beneath the sample surface, a region that would be unreachable by near-field optics-based nanoscopy technologies. In this case, the application of optical nanoscopy to transparent solids offers new insights into the physics underlying solid-state emitters. Near-atomic resolution of individual NV centers would open doors to new exciting approaches to their interactions with their nanoenvironments, indicating fluctuations in the local crystal fields in the case of diamond nanocrystals. However, reaching atomic resolution would mean also that the commonly used dipole approximation for treating the interaction between the excitation light and the emitter no longer holds, and there are other similar complications here to address. On the other hand, nanodiamonds are highly biocompatible and have been used as bioimaging agents; recently the STED imaging of green fluorescent nanodiamonds in a cell containing nitrogen vacancy nitrogen (NVN) defects was reported with resolution down to 70 nm with a commercial microscope [76], even if with low emission efficiency.

The potential is considerable to exploit the nanometric optical microscopy resolution and versatility in applications based on the luminescent properties of point defects in optically transparent insulating materials. Here, developments of solid-state emitters for photonics and fluorescent detectors for radiation imaging will surely bring advances in novel and simplified approaches to nanoscopy as well as lead to the development of faster and more powerful instruments. The full exploitation of nanotechnologies in material science to modify the host-matrix physical characteristics and to tailor the photophysical properties of intrinsic and extrinsic defects will play a fundamental role in the advancements in the growing field of luminescence of point defects in insulating thin layers; it presents an inspiring challenge for any young researcher or indeed even the experienced battle-hardened scientist.

Bibliography

[1]   Shionoya S, Yen WM. Phosphor Handbook. Boca Raton, FL, USA: CRC Press; 1998.
[2]   Itoh N, Stoneham M. Materials Modification by Electronic Excitation. Cambridge, UK: Cambridge University Press; 2001.
[3]   Yukihara EG, McKeever SWS. Optically stimulated luminescence (OSL) dosimetry in medicine. Phys Med Biol. 2008;53:R351–79.
[4]   Montereali RM, Bonfigli F, Piccinini M, Nichelatti E, Vincenti MA. Photoluminescence of colour centres in lithium fluoride thin films: from solid-state miniaturised light sources to novel radiation imaging detectors. J Lumin. 2016;170:761–9.
[5]   Bilski P. Lithium fluoride: from LiF:Mg,Ti to LiF:Mg,Cu,P. Radiat Prot Dosim. 2002;56:199–206.
[6]   Kurobori T, Yanagida Y, Chen YQ. A three-dimensional imaging detector based on nano-scale silver-related defects in X- and gamma-ray-irradiated glasses. Jpn J Appl Phys. 2016;55:02BC01-1-5.
[7]   McKeever SWS, Moscovitch M, Townsend PD. Thermoluminescence Dosimetry Materials: Properties and Uses. Ashford, UK: Nuclear Technology Publishing; 1995.
[8]   Schulman JH, Compton WD. Color Centers in Solids. Oxford, UK: Pergamon Press; 1963.
[9]   Fowler WB. Physics of Color Centers. New York, NY, USA: Academic Press; 1968.
[10] Kurobori T, Matoba A. Development of accurate two-dimensional dose-imaging detectors using atomic-scale color centers in Ag-activated phosphate glass and LiF thin films. Jpn J Appl Phys. 2014;53:02BD14.
[11] Quaranta A, Valotto G, Piccinini M, Montereali RM. Ion beam induced luminescence analysis of defect evolution in lithium fluoride under proton irradiation. Opt Mater. 2015;49:1–5.
[12] Basiev TT, Mirov SB, Osiko VV. Room-temperature color center laser. IEEE J Quantum Electron. 1988;24:1052–69.
[13] Montereali RM. Point defects in thin insulating films of lithium fluoride for optical microsystems. In: Nalwa HS, editor. Ferroelectric and Dielectric Thin Films, Handbook of Thin Film Materials, vol. 3. San Diego, CA, USA: Academic Press; 2002. p. 399–431.
[14] Nahum J, Wiegand DA. Optical properties of some F-aggregate centers in LiF. Phys Rev. 1967;154:817–30.
[15] Baldacchini G, Bonfigli F, Faenov A, Flora F, Montereali RM, Pace A, Pikuz T, Reale L. Lithium fluoride as a novel x-ray image detector for biological μ-world capture. J Nanosci Nanotechnol. 2003;3:483–6.
[16] Montereali RM, Bonfigli F, Menchini F, Vincenti MA. Optical spectroscopy and microscopy of radiation-induced light-emitting point defects in lithium fluoride crystals and films. Low Temp Phys. 2012;38:779–85.
[17] Ustione A, Cricenti A, Bonfigli F, Flora F, Lai A, Marolo T, Montereali RM, Baldacchini G, Faenov A, Pikuz T, Reale L. Scanning near-field optical microscopy images of microradiographs stored in lithium fluoride films with an optical resolution of λ/12. Appl Phys Lett. 2006;88:141107–9.
[18] Baldacchini G, Bollanti S, Bonfigli F, Flora F, Di Lazzaro P, Lai A, Marolo T, Montereali RM, Murra D, Faenov A, Pikuz T, Nichelatti E, Tomassetti G, Reale A, Reale L, Ritucci A, Limongi T, Palladino L, Francucci M, Martellucci S, Petrocelli G. Soft X-ray submicron imaging detector based on point defects in LiF. Rev Sci Instrum. 2005;76:113104-1-12 (also selected for the Virtual Journal of Biological Physics Research, December 1, 2005, issue).

[19] Almaviva S, Bonfigli F, Franzini I, Lai A, Montereali RM, Pelliccia D, Cedola A, Lagomarsino S. Hard X-ray contact microscopy with 250 nm spatial resolution using a LiF film detector and table-top microsource. Appl Phys Lett. 2006;89:054102.

[20] Bonfigli F, Faenov A, Flora F, Francucci M, Gaudio P, Lai A, Martellucci S, Montereali RM, Pikuz T, Reale L, Richetta M, Vincenti MA, Baldacchini G. High-resolution water window X-Ray imaging of in vivo cells and their products using LiF crystal detectors. Microsc Res Tech. 2008;71:35–41.

[21] Adam P, Benrezzak S, Bijeon JL, Royer P, Guy S, Jacquier B, Moretti P, Montereali RM, Piccinini M, Menchini F, Somma F, Seassal C, Rigneault H. Fluorescence imaging of submicrometric lattices of colour centres in LiF by an apertureless scanning near-field optical microscope. Opt Express. 2001;9:353–9.

[22] Larciprete R, Gregoratti L, Danailov M, Kiskinova M, Montereali RM, Bonfigli F. Direct writing of fluorescent patterns on LiF films by X-ray microprobe scanning. Appl Phys Lett. 2002;80:3862–4.

[23] Piccinini M, Ambrosini F, Ampollini A, Picardi L, Ronsivalle C, Bonfigli F, Libera S, Nichelatti E, Vincenti MA, Montereali RM. Photoluminescence of radiation-induced color centers in lithium fluoride thin films for advanced diagnostics of proton beams. Appl Phys Lett. 2015;106:261108.

[24] Montereali RM, Bigotta S, Piccinini M, Giammatteo M, Picozzi P, Santucci S. Broad-band active channels induced by electron beam lithography in LiF films for waveguiding devices. Nucl Instrum Methods Phys Res B. 2000;166–167:764–70.

[25] Hughes E, Pooley D, Rahman HU, Runciman WA. A survey of radiation damage, Harwell Atomic Research Establishment R-5604, UK, 1967.

[26] Fastampa R, Missori M, Braidotti MC, Conti C, Vincenti MA, Montereali RM. Temperature behaviour of optical absorption bands in colored LiF crystals. Results Phys. 2016;6:74–5.

[27] Baldacchini G, Cremona M, Grassano UM, Kalinov V, Montereali RM. Emission properties of gamma irradiated LiF crystals excited in the F absorption band by an excimer laser. In: Kanert O, Spaeth JM, editors. Defects in Insulating Materials. Singapore: World Scientific; 1993. p. 1103–6.

[28] Baldacchini G, De Nicola E, Montereali RM, Scacco A, Kalinov V. Optical bands of F_2 and F_3^+ centers in LiF. J Phys Chem Solids. 2000;61:21–6.

[29] Gellermann W. Color center lasers. J Phys Chem Solids. 1991;52:249–97.

[30] Baldacchini G, De Matteis F, Francini R, Grassano UM, Menchini F, Montereali RM. Emission decay times of F_3^+ and F_2 color centers in LiF Crystals. J Lumin. 2000;87–89:580–2.

[31] Agullo-Lopez F, Catlow CRA, Townsend PD. Point Defects in Materials. New York, NY, USA: Academic Press; 1988.

[32] Skuja L. In: Pacchioni G, Skuja L, Griscom DL, editors. Defects in SiO_2 and Related Dielectrics: Science and Technology. NATO Science Series, Series II, vol. 2. Dordrecht: Kluwer Academic Publisher; 2000. p. 73–116.

[33] Baldacchini G, Cremona M, d'Auria G, Martelli S, Montereali RM, Montecchi M, Burattini E, Grilli A, Raco A. Influence of LiF film growth conditions on electron induced color center formation. Nucl Instrum Methods Phys Res B. 1996;116:447–51.

[34] Ter-Mikirtychev VV, Tsuboi TT. Stable room-temperature tunable color center lasers and passive Q-switchers. Prog Quantum Electron. 1996;20:219–68.

[35] Chiamenti I, Bonfigli F, Gomes ASL, Michelotti F, Montereali RM, Kalinowski HJ. Optical characterization of femtosecond laser induced active channel waveguides in lithium fluoride crystals. J Appl Phys. 2014;115:023108-1-7.

[36] Vincenti MA, Almaviva S, Montereali RM, Kalinowski HJ, Nogueira RN. Permanent luminescent micro-patterns photoinduced by low power ultraviolet irradiation in lithium fluoride. Appl Phys Lett. 2006;89:241125.

[37] Mussi V, Granone F, Marolo T, Montereali RM, Boragno C, Buatier de Mongeot F, Valbusa U. Surface nanostructuring and optical activation of lithium fluoride crystals by ion beam irradiation. Appl Phys Lett. 2006;88:103116-8 (also selected for publication in the Virtual Journal of Nanotechnology, December 1, 2006 issue).

[38] Voitovich AP, Kalinov VS, Mikhnov SA, Ovsechuk SI. Investigation of spectral and energy characteristics of green radiation generated in lithium fluoride with radiation color centers. Sov J Quantum Electron. 1987;17:780.

[39] Nichelatti E, Montereali RM, Montecchi M, Marolo T. Optical properties of rough, inhomogeneous lithium fluoride films with colour centres. J Non-Cryst Solids. 2003;322:117–21.

[40] Bonfigli F, Jacquier B, Menchini F, Montereali RM, Moretti P, Nichelatti E, Piccinini M, Rigneault H, Somma F. In: Di Bartolo B, editor. Spectroscopy of Systems with Spatially Confined Structures. NATO Science Series. The Netherlands: Kluwer Academic Publisher; 2003. p. 697–703.

[41] Voitovich A, Goncharova O, Kalinov V, Stupak AP. Spectral-luminescent properties of gamma-irradiated fluoride crystals and film structures. J Appl Spectrosc. 2003;70:130–7.

[42] Montereali RM, Bonfigli F, Nichelatti E, Vincenti MA. Versatile lithium fluoride thin-film solid-state detectors for nano scale radiation imaging. Nuovo Cimento C. 2013;36:35–42.

[43] Sekatskii SK, Letokhov VS. Single fluorescence centers on the tips of crystal needles: first observation and prospects for application in scanning one-atom fluorescence microscopy. Appl Phys B. 1996;63:525–30.

[44] Oliva C, Ustione A, Almaviva S, Baldacchini G, Bonfigli F, Flora F, Lai A, Montereali RM, Faenov AY, Pikuz TA, Francucci M, Gaudio P, Martellucci S, Richetta M, Reale L, Cricenti A. SNOM images of X-rays radiographs at nano-scale stored in a thin layer of lithium fluoride. J Microsc. 2008;229:490–5.

[45] Bonfigli F, Faenov AY, Flora F, Marolo T, Montereali RM, Nichelatti E, Pikuz TA, Reale L, Baldacchini G. Point defects in lithium fluoride films for micro-radiography, x-ray microscopy and photonic applications. Phys Status Solidi A. 2005;202:250–5.

[46] X-ray interactions with matter. Accessed May 22, 2017, at http://henke.lbl.gov/optical_constants/.

[47] Minsky M. U.S. Patent No. 3,013,467 (19 December 1961).

[48] Lukosz W. Optical systems with resolving powers exceeding the classical limit. J Opt Soc Am. 1966;56:1463.

[49] Wilson T. Confocal microscopy: basic principles and architectures. In: Diaspro A, editor. Confocal and Two-Photon Microscopy. Foundations, Applications and Advances. New York, NY, USA: Wiley-Liss; 2002. p. 22.

[50] Bonfigli F, Hampai D, Dabagov SB, Montereali RM. Characterization of X-ray polycapillary optics by LiF crystal radiation detectors through confocal fluorescence microscopy. Opt Mater. 2016;58:398–405.

[51] Pikuz T, Faenov A, Matsuoka T, Matsuyama S, Yamauchi K, Ozaki N, Albertazzi B, Inubushi Y, Yabashi M, Tono K, Sato Y, Yumoto H, Ohashi H, Pikuz S, Grum-Grzhimailo AN, Nishikino M, Kawachi T, Ishikawa T, Kodama R. 3D visualization of XFEL beam focusing properties using LiF crystal X-ray detector. Sci Rep. 2015;5:17713.

[52] Baldacchini G, Bonfigli F, Faenov A, Flora F, Montereali RM, Murra D, Nichelatti E, Pikuz T. Luminescent patterns based on color centers generated in lithium fluoride by extreme ultraviolet radiation and soft X-rays. Radiat Eff Defects Solids. 2002;157:569–73.

[53] Gasilov SV, Faenov AY, Pikuz TA, Fukuda Y, Kando M, Kawachi T, Skobelev IY, Daido H, Kato Y, Bulanov SV. Wide-field-of-view phase-contrast imaging of nanostructures with a comparatively large polychromatic soft x-ray plasma source. Opt Lett. 2009;34:3268–70.

[54] Bonfigli F, Cecilia A, Heidari Bateni S, Nichelatti E, Pelliccia D, Somma F, Vagovic P, Vincenti MA, Baumbach T, Montereali RM. In-line X-ray lensless imaging with lithium fluoride film detectors. Radiat Meas. 2013;56:277–80.

[55] Heidari Bateni S, Bonfigli F, Cecilia A, Baumbach T, Pelliccia D, Somma F, Vincenti MA, Montereali RM. Optical characterization of lithium fluoride detectors for broadband X-ray imaging. Nucl Instrum Methods Phys Res, Sect A. 2013;720:109–12.

[56] Reale L, Bonfigli F, Lai A, Flora F, Poma A, Albertano P, Bellezza S, Montereali RM, Faenov A, Pikuz T, Almaviva S, Vincenti MA, Francucci M, Gaudio P, Martellucci S, Richetta M. X ray microscopy of plant cells by using LiF crystal as detector. Microsc Res Tech. 2008;71:839–48.

[57] Tomassetti G, Ritucci A, Reale A, Arizza L, Flora F, Montereali RM, Faenov A, Pikuz T. Two-beam interferometric encoding of photoluminescent gratings in LiF crystals by high-brightness table top soft x-ray laser. Appl Phys Lett. 2004;85:4163–5.

[58] Hampai D, Dabagov SB, Della Ventura G, Bellatreccia F, Magi M, Bonfigli F, Montereali RM. High resolution X-ray imaging by polycapillary optics and lithium fluoride detectors combination. Europhys Lett. 2011;96:60010.

[59] Nichelatti E, Bonfigli F, Vincenti MA, Cecilia A, Vagovič P, Baumbach T, Montereali RM. Broadband X-ray edge-enhancement imaging of a boron fibre on lithium fluoride thin film detector. Nucl Instrum Methods Phys Res, Sect A. 2016;833:68–76.

[60] Purcell EM. Spontaneous emission probabilities at radio frequencies. Phys Rev. 1946;69:681.

[61] Milonni PW, Knight PL. Spontaneous emission between mirrors. Opt Commun. 1973;9:119–22.

[62] De Martini F, Marrocco M, Mataloni P, Crescentini L, Loudon R. Spontaneous emission in the optical microscopic cavity. Phys Rev A. 1991;43:2480–97.

[63] Björk G, Machida S, Yamamoto Y, Igeta K. Modification of spontaneous emission rate in planar dielectric microcavity structures. Phys Rev A. 1991;44:669–81.

[64] Marrocco M, Nichelatti E. Coherent anti-Stokes Raman scattering microscopy within a microcavity with parallel mirrors. J Raman Spectrosc. 2009;40:732–40.

[65] Nichelatti E, Montereali RM. Photoluminescence from a homogeneous volume source within an optical multilayer: analytical formulas. J Opt Soc Am A. 2012;29:303–12.

[66] Nichelatti E, Marrocco M, Montereali RM. Cooperative optical effects in volumes embedded in layered media. J Raman Spectrosc. 2010;41:859–65.

[67] Ashcroft NW, Mermin ND. Solid State Physics. Philadelphia, PA, USA: Holt-Saunders International Editions; 1981.

[68] Palik ED. Handbook of Optical Constants of Solids. London, UK: Academic Press; 1985.

[69] Reale L, Bonfigli F, Lai A, Flora F, Albertano P, Di Giorgio ML, Mezi L, Montereali RM, Faenov A, Pikuz T, Almaviva S, Francucci M, Gaudio P, Martellucci S, Richetta M, Poma A. Contact X-ray microscopy of living cells by using LiF crystal as imaging detector. J Microsc. 2015;258:127–39.

[70] Martellucci S, Bellecci C, Francucci M, Gaudio P, Richetta M, Toscano D, Rydzy A, Gelfusa M, Ciuffa P. Soft x-ray generation by a tabletop Nd:YAG/glass laser system. J Phys Condens Matter. 2006;18:2039–44.

[71] Levita M, Schlesinger T. LiF dosimetry based on radiophotoluminescence (RPL). IEEE Trans Nucl Sci. 1976;23:667–74.

[72] Villarreal-Barajas JE, Piccinini M, Vincenti MA, Bonfigli F, Khan R, Montereali RM. Visible photoluminescence of colour centres in LiF crystals for absorbed dose evaluation in clinical dosimetry. IOP Conf Ser, Mater Sci Eng. 2015;80:12020-1–5.

[73] Piccinini M, Ambrosini F, Ampollini A, Carpanese M, Picardi L, Ronsivalle C, Bonfigli F, Libera S, Vincenti MA, Montereali RM. Solid state detectors based on point defects in lithium fluoride for advanced proton beam diagnostics. J Lumin. 2014;156:170–4.

[74] Piccinini M, Nichelatti E, Ampollini A, Picardi L, Ronsivalle C, Bonfigli F, Libera S, Vincenti MA, Montereali RM. Proton beam dose-mapping via color centers in LiF thin film detectors by fluorescence microscopy. Europhys Lett. 2017;117:37004-1–5.

[75] Rittweger E, Han KY, Irvine SE, Eggeling C, Hell SW. STED microscopy reveals crystal colour centres with nanometric resolution. Nat Photonics. 2009;3:144–7.

[76] Laporte G, Psaltis D. STED imaging of green fluorescent nanodiamonds containing nitrogen-vacancy-nitrogen centers. Biomed Opt Express. 2016;7(1):34–44.

Part 3: **Sensing**

Sensors have seen a steep, even exponential rise in their distribution over the last few decades [1]. Ambient monitoring of everything from water quality to smart homes and personal health parameters has experienced a stellar rise, further driven by the evolution of Industry 4.0, the Internet of Things, and the Internet of Everything [2]. Massive data collection of vital parameters regarding the global state of our environment, but also on a local or personal level, combined with machine learning-based data evaluation is envisaged to support intelligent decision making [3, 4]. This development is supported by the miniaturization of sensors for on-chip, point-of-care, or point-of-need application, combined with a simultaneous decrease in prices for individual sensing elements [5, 6]. Optical sensors have the potential of being noncontact and nondestructive, highly sensitive, little affected by electromagnetic interference, small, light-weight, and selective, and of providing high data rates [7]. A number of the sensing approaches are directly based on spectroscopic techniques. Many of these are currently further evolved to nanospectroscopic varieties for higher sensitivity in trace detection and higher spatial resolution, e. g., to make use of multiplexing. A prime example is surface-enhanced Raman spectroscopy (SERS), which plays a prominent role in the present applications volume [8, 9]. In SERS, vibrational information from molecules that are located in the nanovolume electric near-fields of nanoparticles serves to gain their unique chemical fingerprint. SERS has found entry into investigating such societally relevant issues as environmental pollution, traces of explosives, water contamination, or a variety of disease markers. Likewise, approaches such as localized surface plasmon resonance (LSPR) sensing or nonlinear sensing techniques are on the rise [10]. In the present collection, the emerging technique of LSPR shift sensing is illustrated with examples in gas sensing and medically relevant tasks such as glucose sensing. The principles and commercialization of current explosive detectors that exhibit chemical specificity instead of just detecting the shape or casing of explosive devices are discussed. Chapter 3.3 has a somewhat different character. It offers a perspective on differences between micro- and nanospectroscopy in view of relevant issues with respect to hardware and data classification and evaluation for a whole set of nanospectroscopic techniques and a wide field of applications, placing them into a particular historical and philosophical context.

Bibliography

[1]   Turner APF. Perspective – An age of sensors. ECS Sensors Plus. 2022;1(1):011601.
[2]   Yin MJ et al. Recent development of fiber-optic chemical sensors and biosensors: mechanisms, materials, micro/nano-fabrications and applications. Coord Chem Rev. 2018;376:348−92.
[3]   Zhu C, Huang J. Machine learning boosts performance of optical fiber sensors: a case study for vector bending sensing. Opt Express. 2022;30(14):24554−64.
[4]   Liang J et al. Applying machine learning with localized surface plasmon resonance sensors to detect SARS-CoV-2 particles. Biosensors. 2022;12(3):173.

https://doi.org/10.1515/9783110442908-006

[5]   Narayan RJ, editor. Medical Biosensors for Point of Care (POC) Applications. Elsevier Ltd.; 2017.

[6]   Roy S et al. Recent developments towards portable point-of-care diagnostic devices for pathogen detection. Sensors & Diagnostics. 2022;1:87–105.

[7]   Lobnik A et al. In: Wang W, editor. Advances in Chemical Sensors. IntechOpen. 2012. p. 3.

[8]   Perumal J et al. Towards a point-of-care SERS sensor for biomedical and agri-food analysis applications: a review of recent advancements. Nanoscale. 2021;13(2):553–80.

[9]   Xie LP et al. State of the art in flexible SERS sensors toward label-free and onsite detection: from design to applications. Nano Res. 2022;15(5):4374–94.

[10] Lopez GA et al. Recent advances in nanoplasmonic biosensors: applications and lab-on-a-chip integration. Nanophotonics. 2017;6(1):123–36.

Florian Laible, Anke Horneber, and Monika Fleischer

3.1 Localized surface plasmon resonance shift sensing

3.1.1 Key messages

- The dependence of the plasmon resonances of optical antennas on the refractive index of their surrounding medium can be harnessed for high-sensitivity sensing.
- The evolution, potential, and challenges of localized surface plasmon resonance (LSPR) shift sensing are considered on the background of other established optical sensing techniques.
- Nanofabrication plays an essential role for preparing suitable nanoantenna structures.
- Examples of state-of-the-art sensors that are made application-specific to sense particular analytes through surface functionalization are presented. Sensitivities down to single molecules have been demonstrated.
- Compared to the more established SPR sensing, LSPR shift sensing is still at the brink of commercialization.

3.1.2 Pre-knowledge

Plasmonic nanostructures with dimensions on the order of 100 nm can act as antennas for visible light, as discussed in Volume 1, Chapters 1.6 and 3.2. Under illumination with an electromagnetic wave, collective oscillations of the free electron density (i. e., localized surface plasmon polaritons) are induced, which exhibit geometry-dependent resonances. This LSPR depends on the polarizability α of the particle,

$$\alpha = 4\pi r^3 \frac{\varepsilon - \varepsilon_m}{\varepsilon + 2\varepsilon_m},$$

(3.1.1)

according to the quasi-static dipole approximation for small spherical particles, with the radius r, dielectric function $\varepsilon(\omega)$, and the dielectric constant of the surrounding medium ε_m. Resonances thus occur when the Fröhlich condition $\varepsilon(\lambda) = -2\varepsilon_m$ is fulfilled. At resonance, the particles exhibit a strong scattering and absorption cross-section, where the scattering in the limit of small nanospheres can be approximated as

$$C_{scat} = \frac{k^4}{6\pi}|\alpha|^2$$

(3.1.2)

and therefore also a high extinction cross-section $C_{ext} = C_{scat} + C_{abs}$ [1].[1] Strong local electrical evanescent near-fields are excited near the nanoantenna surface, which exponentially decay away from the surface over a few tens of nanometers, cf. Volume 1, Chapter 1.6.

---

[1] For a short discussion, see also equations (4.1.4) and (4.1.5) and the related footnote in Chapter 4.1.

https://doi.org/10.1515/9783110442908-007

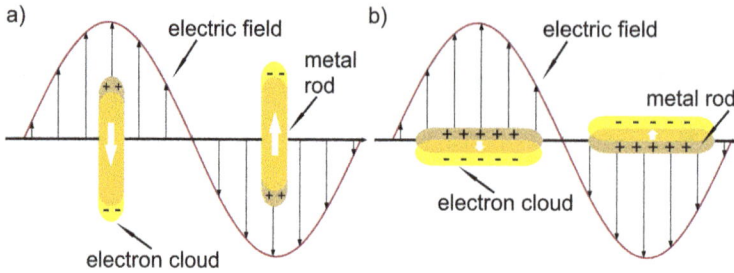

Figure 3.1.1: Excitation of (a) longitudinal and (b) transversal localized surface plasmons in nanorods (courtesy of P. Bieschke, after [2]).

For nanostructures with symmetry breaking, different resonances are observed depending on the polarization of the incident light relative to the antenna axes. This way, e. g., for nanorods a dichroism is observed, with a shorter-wavelength resonance for excitation of the short axis mode (Fig. 3.1.1b) and a longer-wavelength resonance for excitation along the long axis (cf. Fig. 3.1.1).

As can be deduced from the above condition, the resonance wavelengths are directly influenced by the dielectric constant, and thus the refractive index of the surrounding medium. They are defined by the respective wavelength at which the dielectric function of the respective metal matches the Fröhlich condition. This effect is exploited in LSPR shift sensors.

3.1.3 Importance of the application

In the broadest sense, a sensor is "a device that responds to a physical stimulus (such as heat, light, sound, pressure, magnetism, or a particular motion) and transmits a resulting impulse (as for measurement or operating a control)" [3]. In modern society, sensors are nearly ubiquitous, with a rising tendency. Sensing devices are employed in such varied application areas as medicine and patient monitoring, industry and manufacturing, metrology, transportation, food and agriculture, environmental monitoring, temperature control, health and safety, security, and the military. Most recently, interconnected sensor networks emerge for smart homes and cities, industry 4.0, connected health/wearable sensors, and the Internet of Things. In optical sensors, light is chosen as the physical stimulus. This can have several advantages. Optical sensors can be versatile, highly sensitive, noninvasive and nondestructive, immune to electromagnetic interference, small, light-weight, and selective, and provide high data rates. The working principle is typically based on electromagnetic radiation being detected by a receptor interacting with a transducer element, which transforms the optical signal into a measurable output signal [4]. Optical sensors account for a steadily growing multibillion US$ market, as "highly integrated, accurate and fast photonic sensors with multisensor data fusion are the sense organs of the digital society" [5]. The Web of Science in 2022 lists nearly 10^6 entries on the topic of sensors, out of which nearly 150,000 are categorized as optical and about 20,000 as plasmonic.

Plasmonic sensors form a subcategory of optical sensors. Here, surface plasmon polaritons are excited at a metal–dielectric interface by an incoming electromagnetic wave [6]. The plasmon resonance conditions are modified by changes in the permittivity of the metal or the environment. As a result, the SPR shift can be monitored as a sensor signal with ultrahigh sensitivity. SPR sensors based on metallic thin films have been commercially available for a few decades and are routinely used for chemical detection and molecular adsorption monitoring. By surface functionalization, they are rendered specific to the binding and recognition of particular analytes.

In a more recent development, the concept of SPR sensing has been extended to metallic nanoparticles. Here, plasmon propagation is restricted, and the plasmon oscillations are localized to the particle geometry, leading to the observation of LSPRs [7, 8]. Although not widely commercially available yet, LSPR shift sensors show a high potential in view of the ongoing demand for further miniaturization of optical biosensors. They operate on small volumes and enable easy read-out. The optical nature of the LSPR sensors allows for fast testing of a large number of samples, often on tabletop devices not requiring a full-scale lab. This could facilitate tests to be run routinely, enabling incidental and secondary findings of disease markers. Early disease diagnosis, e. g., for cancer or neurodegenerative pathologies, drastically improves the survival rate, thus strongly motivating the continuous development of better diagnostic techniques. LSPR biosensors could be used in handheld point-of-care (POC, typically used for tests on a patient in a medical context) and point-of-need (PON, more broadly used for tests run outside of a dedicated lab environment) applications as well as for high-end high-sensitivity measurements. For high-precision measurements, strongly advanced nanostructures and optical equipment are used. In contrast, for POC and PON solutions such as, e. g., smartphone attachments, usability and speed are essential. The optical part of the sensor is kept simple, and the nanostructure fabrication cost is factored in. Particularly in current times, the need to contain and control spreading diseases has become obvious. Fast and sensitive measurements are one of the keys to achieve this. The inner workings of the detector can be complex and sophisticated as long as the result is reliable, and the handling of the device is easy enough to mitigate user errors. While the development of the active area of the sensor, i. e., nanoantennas with functionalization, is not trivial, the read-out of the sensor is straightforward and can be achieved with limited equipment. The information on the presence of the analyte in the active area is nearly instant. Another time-saving aspect common to LSPR sensors is that the analytes are detected in a label-free manner, and they do not have to be pre-processed before the analysis. This brings the typical times for a full LSPR sensing cycle to the order of minutes. LSPR biosensors can be adapted in a timely manner to new analytes. For this purpose, "only" a suitable functionalization must be developed (which can be challenging enough), while leaving the rest unchanged. To further increase the usability of LSPR biosensors and bring down costs, efforts are undertaken to make them reusable.

In addition to accurate and relatively fast detection, LSPR sensors also allow for sensing with high spatial resolution. This offers the possibility of multiplexing, i. e., the simultaneous detection of more than one analyte. Parallel biosensing can be achieved, e. g., by a segmented functionalization of different areas of particles on the sensor with recognition structures for different analytes, since the detection sites are extremely small. Thus, multiple assays can be prepared in close proximity, e. g., on a single test strip. Multiplexing and statistical analyses are commonly used to improve the results of the sensors [9, 10].

Beyond medical applications, LSPR sensors may be used in a variety of other situations, such as detecting toxins in the environment or explosives in an airport setting. Another helpful potential application of LSPR sensors could be found in gas sensors. For the safety of residents of buildings heated with gas or workers working with flammable gases, the detection of leakages is of the highest importance.

The positive aspects of LSPR sensors are for now offset by their relatively high cost per test, making them niche products at best. The growing need for faster and user-friendly biosensors alongside the scientific effort of the LSPR community may however soon produce larger-scale applications for this promising technology.

3.1.4 State-of-the-art

3.1.4.1 History of plasmonic sensing

Recently, different sensor techniques have been developed that exploit plasmonic properties [11, 12]. Surface-enhanced Raman spectroscopy (SERS) and SPR sensing are widely known techniques, whereas LSPR sensing so far is less common. It is worth noting that there are also other nano-optical sensing approaches [13], like metal-enhanced fluorescence, where the electric fields of plasmonic particles are exploited for fluorescence enhancement, tip-enhanced Raman spectroscopy (TERS) with an illuminated metal probe, plasmonic Förster resonance sensors, where the distance-dependent energy transfer of two fluorophores is detected, or plasmonic ruler-based sensors, where the distance-dependent coupling of plasmonic modes is sensed; see also Volume 1, Sections 2 and 3. SERS and SPR sensing underwent quick development in the last century, whereas LSPR sensing methods mainly developed over the last decades. SERS is able to provide a chemical fingerprint of all Raman-active components present in a sample (together with added information, e. g., about their strain state, etc.) that can be decomposed using databases, principal component analysis, or further machine learning techniques. In contrast, SPR and LSPR sensing allow for the targeted screening for a specific analyte, or parallel sensing of different components in a multiplexed configuration. These techniques are briefly introduced below.

SERS sensing

Molecular bond vibrations can be detected by Raman spectroscopy, where light is inelastically scattered by molecules. Sharp lines in Raman spectra can be observed, which are characteristic of the vibrational modes of the molecule. They are often referred to as molecular "fingerprints" and allow for label-free qualitative analysis. For sensing applications, the limitation is the low signal intensity caused by extremely small scattering cross-sections, which can be overcome by enhanced Raman scattering techniques. Raman spectroscopy is described in more detail in Volume 1, Chapter 2.3. The group of Fleischmann in 1974 managed to observe the Raman signal of a pyridine monolayer [14]. Even though they recorded the first SERS signal, the description of SERS and the explanation of its origin were published in the following years by other groups [15]. The enhancement effect is nowadays attributed to an electromagnetic excitation and emission enhancement as well as a chemical effect via additional excitation paths. Usually, the electromagnetic effect is contributing to the total enhancement of up to 10^8 or more times [10], which is more significant than the chemical enhancement. More details on SERS can be found in Volume 1, Chapter 3.3.

SERS has a broad range of applications, e. g., in the detection of illegal drugs or toxic substances, chemical and biological analysis, medical diagnosis, and many more [10]. In this volume, Chapter 4.2, the application of SERS in cancer detection is presented. SERS is a well-established analytical technique. Raman microscopy is a standard method in commercial microscopes, and different kinds of SERS substrates can be purchased as well. Details on SERS substrates can be found in Volume 2, Chapter 6.2.3.

SERS offers great advantages like single molecular sensitivity, spectral fingerprints, narrow spectral peaks, multiplexing, and the simultaneous detection of different molecules. However, it also comes with several disadvantages. SERS spectroscopy is a rather slow method and the data analysis is complex. It also struggles with the low reproducibility of SERS measurements due to variations on the nanoscale of the plasmonic substrates.

SPR sensing

The first plasmonic sensing technique that was commercially available was SPR sensing. After the theoretical work on SPRs of Ritchie [16] in 1957 experiments followed quickly in 1968 by Kretschmann [17], Otto [18], and Raether [19], exciting a surface plasmon in a gold film by attenuated total reflection in a prism. The first publications of SPR sensors followed in the 1980s in publications of Gordon [20] (sensing in electrochemical reactions) and Nylander [21] (gas detection). Soon studies for the development of a commercial instrument started, and in 1990 the first SPR machine called

BIACore entered the market [22]. It allowed for real-time biospecific interaction sensing without the need for markers. Nowadays a broad range of instruments is available. SPR sensing can offer quantitative measurements and information about binding kinetics. Furthermore, differently functionalized biorecognition sensor chips offer sensitivity to many different analytes, like proteins, DNA, or even whole cells.

In SPR sensors a surface plasmon is excited in a thin metal layer. If the energy and momentum of the excitation photon (along the surface plane) and those of the surface plasmon match, the surface plasmon is resonantly excited. An evanescent wave is created at the metal–dielectric interface, which decays exponentially over the metal surface and interacts with the surrounding. The theoretical background of the surface plasmon can be read in Volume 1, Chapter 1.6. The SPR is sensitive to refractive index changes at the metal–dielectric interface. The sensors detect changes in either SPR angle, phase, or intensity at a fixed wavelength. The most common method is detecting the SPR angle dependence of the reflectance of monochromatic light with a prism. A dip occurs at the resonance angle due to the energy being transferred to the SPR, which shifts with a refractive index change [23].

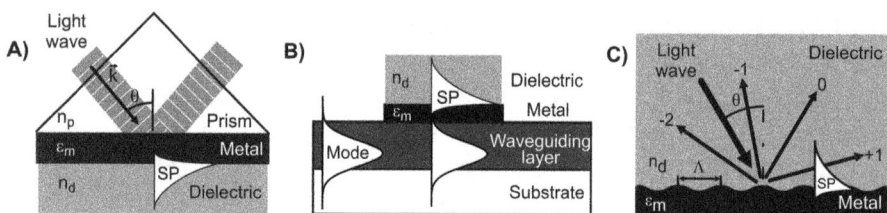

Figure 3.1.2: Illustration of different SPR sensing schemes. In (a) the SPR is excited by attenuated total reflection in a prism. (b) A setup using the metal–dielectric interface on top of a waveguide. (c) A sensor consisting of a metallic grating embedded in a dielectric medium. Adapted with permission from [23]. Copyright 2008 American Chemical Society.

SPRs in sensing are still mostly excited by a prism coupler. Nevertheless, two other methods are also well established: the excitation of the plasmon by a grating coupler and the excitation via a waveguide. A schematic illustration of these methods can be seen in Figure 3.1.2. Even though prism coupling is still dominating the field of SPR sensing, the other approaches have their advantages. Grating coupling devices can be easily mass produced, whereas waveguide coupling in optical fibers allows for small sizes. In optical fibers the total internal reflection of a fundamental mode is coupled to a metallic film on a polished face of the fiber core, creating a surface plasmon. Many different modifications of the fiber geometry are possible as well as different functionalization of the sensors [24]. A detailed description of the different surface coupling sensors with their respective measurement methods can be found elsewhere [23].

Up to now thousands of papers have been published on SPR sensing. An increasing interest in SPRs for biosensing applications can be seen. The application ranges from medical diagnostics and biological, pharmaceutical, and chemical analysis to environmental monitoring and food safety. Quantitative analysis can provide information on intermolecular interactions, structural changes in adsorbed molecules, and kinetics.

The development of SPR sensing led to many different instruments with their own specific applications. There are many approaches to enhance the sensors, like using metamaterials, 2D materials, multilayer structures, nanoholes, or nanoparticles [25–28]. The last-mentioned approaches show a direct transition to LSPR sensing.

LSPR sensing

The development of LSPR sensing took part parallel to that of SPR sensing [29, 30], but an increased interest appeared in recent years. Amongst others, the establishment of LSPR shift nanosensors relied on the availability of suitable nanofabrication techniques. In general, the surface plasmon has a larger decay length of the evanescent field (on the order of the wavelength) than the localized surface plasmon (a few tens of nanometers) and is therefore better suited for larger analyte structures, whereas smaller molecules can be better detected by LSPRs. LSPR sensing offers several advantages, such as easy colorimetric sensing [31, 32], which is a method where an in situ color change can be detected visually without any instruments. The change can be induced by refractive index changes, by aggregation, or less commonly, by a reaction-driven size change of particles [33]. Colorimetric sensing can be implemented at minimal cost in miniaturized devices and is easy, fast, and very sensitive. LSPR sensing in contrast to SPR sensing also allows for simple optical setups, multiplexing, and easy sensor adjustment via the particle or nanohole size and shape. With LSPR sensing, likewise single-molecule sensitivities can be reached [34–36]. The concept, realization, fabrication, properties, and applications of LSPR sensing will be discussed in the following.

3.1.4.2 Concept of LSPR shift sensors

In LSPR shift sensing, the spectrum of the optical antennas is recorded as the sensor signal, and the resonance wavelength is read out. Upon either a refractive index change of the medium within the range of the near-field or the binding of an analyte layer to the antenna surface, the resonance condition is modified, and the resonance wavelength is shifting to satisfy the new condition. For immersion in a medium the refractive index probed by the near-field is homogeneous. For adsorption of a thin layer, the indices of both the layer and the medium are present within the range of the near-field, which can be modeled by an effective refractive index with an intermediate

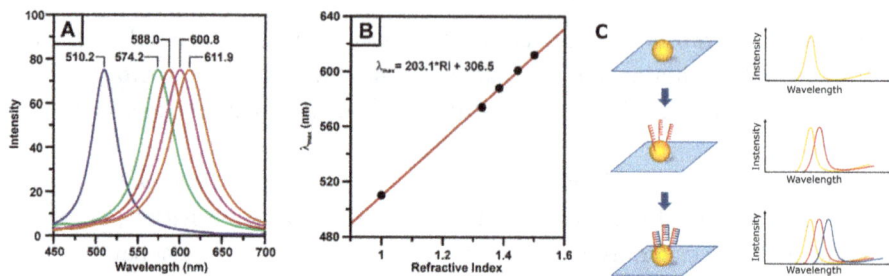

Figure 3.1.3: (a) Scattering spectrum from a single silver nanoparticle in various solvent environments (left to right): nitrogen gas, methanol, propan-1-ol, chloroform, and benzene. (b) Linear dependence of the LSPR wavelength on the refractive index. (c) Schematic of the biosensor principle (nanoantenna (yellow spectrum) with molecular recognition structure (red)) that indicates specific binding of an analyte (blue) through measuring the spectrum of the scattered signal. (a, b) Adapted with permission from [37], Copyright 2003 American Chemical Society; (c) adapted with permission from [38].

value. In the case of an increase of the (effective) refractive index, the resonance wavelength typically shifts to longer wavelengths, as schematically outlined in Figure 3.1.3a and c.

If the bulk index of the medium changes over a limited range, a roughly linear dependence of the shift $\Delta\lambda_{max}$ on Δn ensues (see Fig. 3.1.3b), from which the sensitivity of the sensor may be extracted. For larger index changes, correction terms need to be taken into account [39].

In analogy to SPR sensing, the spectral shift $\Delta\lambda_{max}$ through the adsorption of a thin layer can be approximated by

$$\Delta\lambda_{max} \approx m\Delta n\left(1 - \exp\left(\frac{-2d}{l_d}\right)\right), \tag{3.1.3}$$

with m being the nanostructure-dependent sensitivity factor in nm/refractive index unit (RIU), Δn the refractive index difference between that of the adsorbate or medium and the original medium, d the effective layer thickness of the adsorbate, and l_d the decay length of the electromagnetic field [39]. (For $d \to \infty$, the linear behavior $\Delta\lambda_{max} \approx m\,\Delta n$ is recovered.) From the observed shift, the main defining properties of the sensor can be determined:

- the sensitivity $S = \Delta\lambda_{max}/\text{RIU}$, and
- the figure of merit FOM = S/FWHM, with FWHM being the full width at half maximum of the resonance.

Thus, the further the wavelength is shifted per refractive index change and the sharper the resonance, the higher the quality and the lower the detection limit of the sensor. The challenge is therefore to design optical antennas that provide a strong inherent sensitivity factor m and a narrow linewidth.

LSPR shift sensors as such do not exhibit any specificity. The sensor signal is elicited solely by a change in the local permittivity within the near-field range of the optical antenna. In order to add chemical specificity to the detection, which is required for most practical applications (such as for detecting the presence of antibodies or toxins in a liquid or of explosives in the gas phase), the antenna surface can additionally be functionalized with a molecular recognition structure (cf. Fig. 3.1.3c). A typical example can be an antigen for binding of antibodies (cf. Fig. 3.1.4).

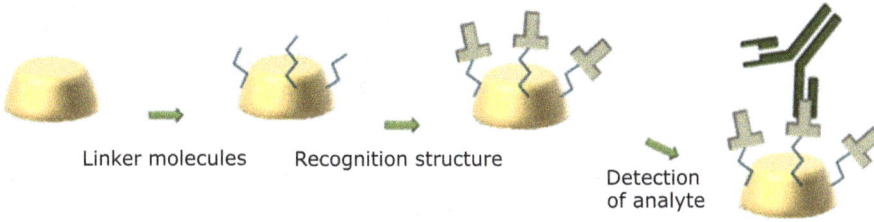

Linker molecules   Recognition structure

Detection
of analyte

Figure 3.1.4: Schematic of a LSPR sensor that is rendered specific by grafting a molecular recognition structure onto the surface. Adapted from [38].

In this case the time-dependent resonance shift is recorded, reflecting the kinetics of the binding process. Under specific boundary conditions, the change of the relative amount of bound analyte N with time t can be approximated by a simple kinetic equation as

$$\frac{dN}{dt} = k_a c(1 - N) - k_d N, \tag{3.1.4}$$

with c being the analyte concentration and k_a, k_d the kinetic rate constants for the association and dissociation [40]. For an initially pristine sensor, this results in the behavior

$$N(t) = \left[\frac{k_a c}{k_a c + k_d}\right](1 - e^{-(k_a c + k_d)t}). \tag{3.1.5}$$

After saturation of the binding sites with the analyte, the sensor can be regenerated for reuse. The characteristic ensuing curve shape is illustrated in Figure 3.1.5a.

After calibrating the LSPR shift for varying analyte concentrations, quantitative analyses are enabled. Assuming the resonance shift to be proportional to the amount of adsorbed analyte, the calibration curve can be reasonably well described by the Langmuir isotherm in line with equation (3.1.5) as $\frac{\Delta\lambda(c)}{\Delta\lambda_{max}} = \frac{k_a c}{k_a c + k_d} = \frac{K_a}{K_a + 1}$, with $K_a = k_a c / k_d$ (cf. the dotted line in Fig. 3.1.5b). Another characteristic of the sensor that can be deduced from the noise level in its nonlinear calibration curve (Fig. 3.1.5b) is the limit of detection (LOD), which indicates the minimum concentration of the analyte that can still be detected. Additionally, the dynamic range of the sensor (the range of

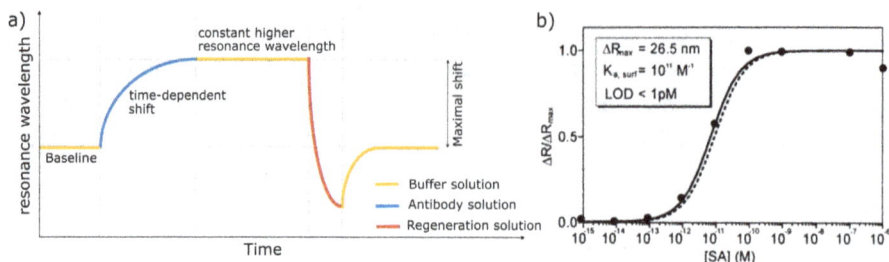

Figure 3.1.5: (a) LSPR shift over time displaying the binding kinetics and regeneration of an LSPR sensor. Adapted after [38]. (b) Normalized LSPR shift versus streptavidin concentration response curve for the specific binding to a biotinylated Ag nanobiosensor with fitting curves. Adapted with permission from [30]. Copyright 2002 American Chemical Society.

concentrations corresponding to a high slope in the calibration curve, where small changes in concentration lead to comparatively large wavelength shifts) is a key figure. In the case of LSPR biosensors the dynamic range indicates the range of analyte concentrations that can be distinguished. Other relevant properties are the cross-sensitivity to nonspecific binding of other analytes as well as features such as size, reusability, response time, or lifetime of the sensor [6].

In analogy to the above biosensing concepts, LSPR shift sensing can likewise be employed for sensing gases (H_2, CO, H_2S, NO_x, H_2O_2, Hg vapor, chloroform, etc.) [41]. Here, the adsorption of, incorporation of, or reaction with the gases can change the material properties of the host material. In the case of hydrogen sensing with palladium particles, e. g., upon saturating the surface with chemisorbed hydrogen atoms, the atoms diffuse into the nanoparticle host. Thus the material properties are changed by an expansion of the lattice and hydride formation. The modified permittivity can be detected either directly through a spectral shift of the LSPR of the nanoparticles themselves, or indirectly through that of an adjacent optical antenna [42].

3.1.4.3 Realization of LSPR shift sensors

Preparing ultrahigh-sensitivity, highly specific, artifact-free LSPR shift sensors is an art that offers many challenges. However, sensors with reasonable sensitivities to illustrate the basic sensor principle are rather straightforward to achieve. LSPR sensors can thus also be recommended, e. g., as demonstrators in class rooms or at open days. They likewise offer attractive training objects for short-term nanofabrication projects or internships in suitably equipped institutions.

Figure 3.1.6 shows the comparatively easy-to-prepare example of gold nanodisc antennas with different diameters fabricated by nanosphere lithography (NSL) (see next section). Polystyrene nanospheres are self-assembled in a hexagonal order on a thin gold film and used as an etch mask to create nanodiscs. The sensor is then char-

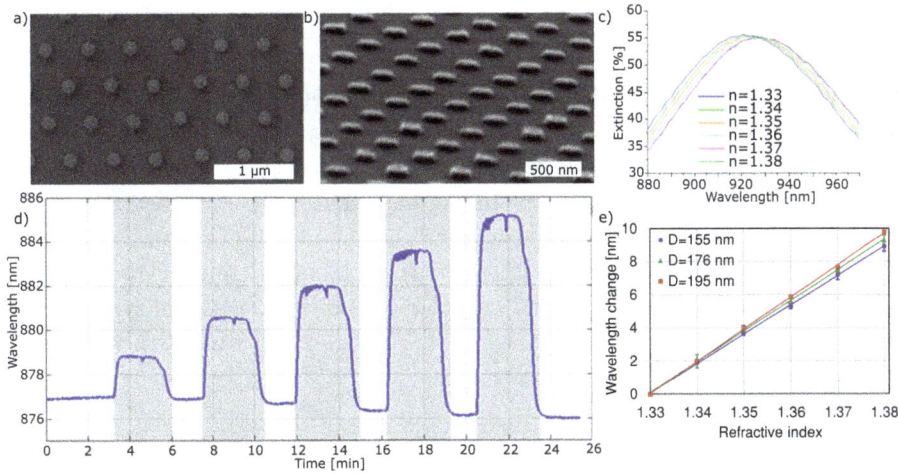

Figure 3.1.6: (a) Polystyrene spheres that self-assembled on a gold film in a hexagonal order, after reducing their diameter by reactive ion etching. (b) Hexagonally ordered gold nanodiscs after etching the film, using the nanospheres as an etch mask. (c) Extinction spectra showing the LSPR shift when the refractive index of the medium is increased in steps of 0.01. (d) Time series of alternating immersion in water and in water–glycerin mixtures with increasing refractive indices. (e) Plots of the LSPR shift over the refractive index with linear fits for different nanodisc samples to determine their sensitivity [43].

acterized by a typical index series using water–glycerin mixtures with the refractive index increasing in steps of $\Delta n = 0.01$. The corresponding resonance shifts are plotted over time, and the sensitivity is determined from a linear fit.

In the following, different techniques that have been employed for preparing advanced LSPR sensor platforms are detailed.

3.1.4.4 Fabrication of LSPR shift sensors

The plasmonic nanoantennas are the cornerstone of LSPR sensors [44]. For the fabrication of such antennas, different techniques are used depending on what requirements are placed on the structures or on the process itself. In general, the fabrication techniques are divided into top-down or bottom-up methods. Typical top-down approaches are based on lithography, defining the shape and location of the nanostructures. Bottom-up techniques rely on statistical processes such as self-assembly and wet-chemical synthesis to create the nanostructures on a surface. While the size and position of the nanoantennas are less well controlled in bottom-up processes, it is generally easier to generate a high number of them or to cover a large and irregular surface compared to top-down techniques. The following fabrication strategies are being used to create nanostructures for LSPR sensing. For further information on the individual fabrication techniques, the reader may also refer to Volume 2, Chapters 6.2 to 6.4.

Bottom-up

In view of LSPR sensing, the most common bottom-up process is to synthesize nanoparticles in solution and then disperse them on a substrate. The used chemicals dictate the material of the resulting nanoparticles. Common materials for synthesizing nanoparticles are silver, gold, platinum, palladium, aluminum, and copper. By systematically changing the parameters, the shape and size of the resulting particles can be controlled. The temperature and process time determine mainly the size of the grown particles [46]. Some of the more common geometries of synthetically grown nanostructures are shown in Figure 3.1.7. With the help of process engineering, specifically formed structures can be created.

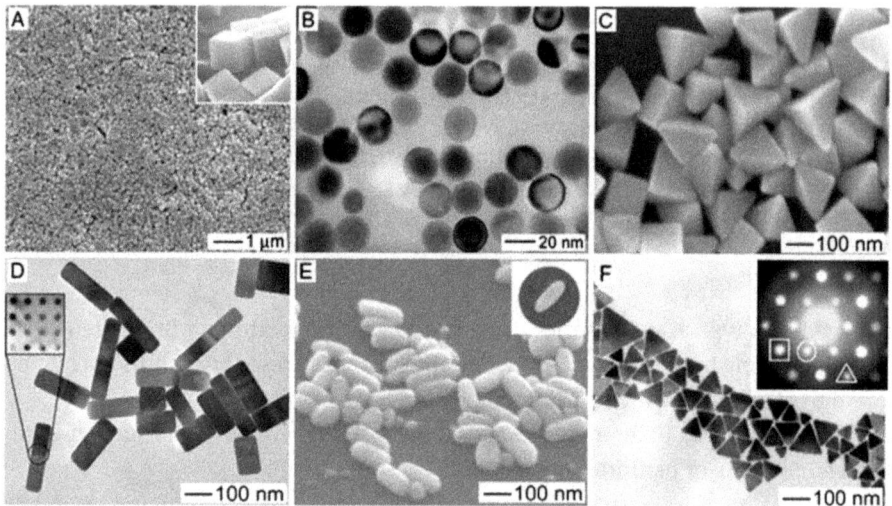

Figure 3.1.7: Ag nanostructures for sensing applications synthesized from solution. Reprinted with permission from Springer Plasmonics [45] © 2009.

Nanostructures that are synthesized in a solution can be suspended in a liquid. Alternatively, the nanostructures can be applied to a designated substrate by different coating techniques. This enables the coverage of a large area with nanoantennas of various shapes, sizes, and materials. For example, silver nanocarrots have been produced by Liang et al. and employed for refractive index sensing [47] (Fig. 3.1.8a). In Hedge et al., synthesized gold nanostars, nanorods, and nanocubes are used as LSPR sensors for glucose detection, either in solution or randomly immobilized on a glass substrate [48] (see Fig. 3.1.8b).

As an example for an LSPR assay, in the work of Endo et al. nanoparticles formed a self-assembled layer on top of a gold-coated glass substrate (see Fig. 3.1.9). This creates a large active area, enabling multiple tests on one substrate [49].

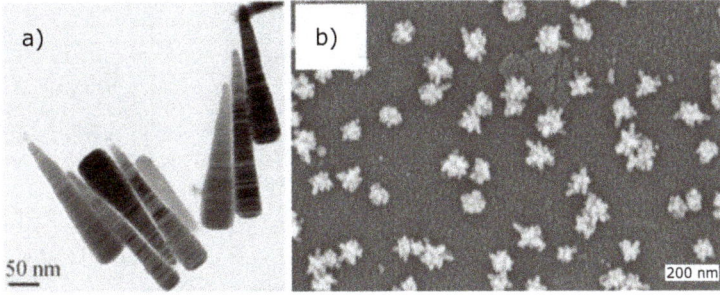

Figure 3.1.8: (a) Transmission electron microscopy image of silver nanocarrots grown in solution and used as refractive index sensors. Adapted with permission under CC BY 4.0 from [47]. (b) SEM image of gold nanostars used for LSPR detection of glucose. Adapted with permission under CC BY 4.0 from [48].

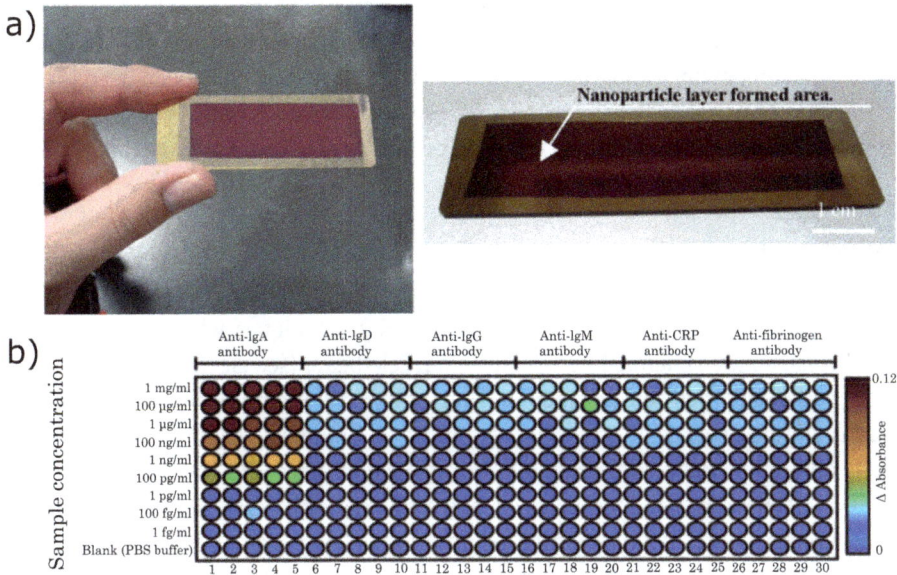

Figure 3.1.9: (a) Gold-coated glass substrate with a self-assembled monolayer of spherical gold nanoantennas forming (b) parallel LSPR assays for four immunoglobulins, C-reactive protein, and fibrinogen. Reprinted (adapted) with permission from [49]. Copyright 2006 American Chemical Society.

Lithography-based fabrication for LSPR sensing

As already mentioned, top-down techniques rely on lithography. In comparison to the previously discussed bottom-up techniques, lithography tends to be a relatively time- and labor-intensive process since multiple steps must be performed to realize the nanostructuring of a surface. Typically, there is a trade-off in lithography between

Table 3.1.1: Comparison of lithography techniques [50, 51].

Lithography technique	Minimal feature size	Throughput
Photolithography	~1 µm	Very high
Electron beam lithography	<5 nm	Very low
Nanoimprinting lithography	<10 nm	Medium to high
Laser interference lithography	~35 nm	High
Nanosphere lithography	<100 nm	Very high
Nanostencil lithography	~20 nm	Medium

the throughput and the minimal feature size. Pimpin et al. and Kassani et al. have compiled tables of common lithography techniques. Information from these sources has been combined and updated in Table 3.1.1 [50, 51].

In comparison to bottom-up techniques, top-down fabrication has higher requirements for the surface of the substrate in terms of flatness, and in some cases conductivity. In the following lithography techniques that have especially been used for the fabrication of nanostructures for LSPR biosensors are showcased. They are ordered from higher to lower throughput. It should be noted that this list is not exhaustive.

Nanosphere lithography

To achieve a middle ground between lithography techniques where each individual nanostructure is defined by a focused particle beam (see electron and ion beam lithography) and their self-assembly, a combination of bottom-up and top-down techniques can be used: NSL, which is introduced in Volume 2, Chapter 6.4. There are many different workflows to achieve a variety of geometries with NSL [52, 53].

One of the most common ways is shown by Chen et al. [54]. Negatively charged nanospheres are statistically distributed on a PMMA layer. A metal film is evaporated, and the spheres are removed by a strip-off process. With the film acting as a hole mask, the PMMA layer is etched all the way to the substrate, and a second metal layer is evaporated. The nanostructures remaining after a consecutive lift-off process have approximately the diameter and location of the nanospheres. With the NSL hole mask process, it is possible to structure large areas of flat surfaces with relatively homogeneous nanoparticles. The downside is that the shapes are limited.

Another common NSL process is using the nanospheres directly as an evaporation mask. This process is used for an LSPR sensor developed by Haes et al. [55]. In contrast to the previous case, the nanospheres are ordered in a closely packed monolayer. After the metal is evaporated on the substrate and the spheres are removed, the substrate is covered with hexagonally ordered triangles as shown in Figure 3.1.10. The triangular shape is beneficial to the performance of sensors, since corners and edges show

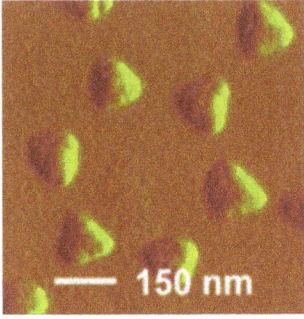

Figure 3.1.10: Ag triangles fabricated using NSL for the detection of Alzheimer disease markers. Adapted with permission from [55]. Copyright 2005 American Chemical Society.

enhanced near-fields. The greatest challenge of this technique is to achieve densely packed nanosphere monolayers over larger areas.

Laser interference lithography

Another way to achieve large-scale ordered nanostructures by a top-down process is using laser interference lithography [56]. Shen et al. used a laser interference pattern to expose a layer of photoresist on a substrate. The resist is exposed at the areas of constructive interference, which are in this case nearly rectangular. After the development resist pillars with rectangular footprints are left on the substrate. Now a gold layer is evaporated on top capping the pillars. The resulting disc-like structures on pillars over a metal-coated surface are shown in Figure 3.1.11. The structures resemble mushrooms and are ordered in a square grating, and are therefore called gold mushroom arrays. Here they are used for protein detection. Laser interference lithography can create a

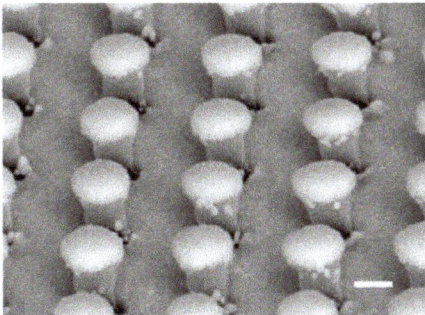

Figure 3.1.11: Tilted SEM image of a gold mushroom array fabricated via laser interference lithography for cytochrome *c* and alpha-fetoprotein detection. The scale bar represents 200 nm. Reprinted with permission from Nature, *Nature Communications* [56] © 2013.

limited number of nanoscale geometries in large-scale ordered arrays. For minimal structure size, UV lasers are used. Compared to NSL the technique is more demanding because of the optical interference setup.

Nanoimprint lithography

Nanoimprint lithography is a technique to structure larger areas with nanostructures. There are a lot of variations; the basic principle is to press a structured mold into a resist layer on a substrate to transfer the geometry of the stamp into the resist. After the imprinting, the structured layer can be used for nanofabrication by, e. g., metallization, dry etching, or lift-off processes. Jiang et al. present an imprinting process to create large numbers of nanostructures on a chip that is then used for LSPR biosensing [57]. Nanoimprint lithography is a highly scalable process. The molds are reusable and can even be designed for roll-to-roll imprinting, cf. Volume 2, Chapter 6.4.2.

Nanostencil lithography

Nanostencil lithography is another top-down approach for creating arrays of plasmonic nanostructures. Vazquez-Mena et al. used this type of lithography to create a nanostructure array for LSPR refractive index and biotin sensing (see Fig. 3.1.12) [58]. The stencils are made from silicon wafers onto which a silicon nitride layer is deposited. The nitride layer is then structured by electron beam lithography (EBL) and dry etched to create a hole pattern. After that, the backside of the wafer is structured via UV lithography. Afterward, the wafer is wet etched. This process aims to create windows of freestanding and patterned SiN films. The stencil is then fixed to the designated substrate, and a metal layer is evaporated. The stencil–substrate distance is crucial for this process. The smaller this distance, the more closely the nanostructures resemble the pre-defined pattern of the stencil. Also, the minimal distance achievable is limited by this factor. Since the stencil can be reused multiple times, this technique is more time-effective than direct EBL lift-off processes.

Electron beam lithography

One of the techniques with the highest control over the geometry and location of the fabricated nanostructures is EBL. In most cases, the electron beam is used to create a hole pattern in a resist. Afterwards, a material is evaporated on top, and a lift-off is performed. The resulting structure can either be the final nanoantenna or an etch mask, which is then transferred into a metal layer by dry etching. EBL is used to create either single very well-defined structures or complex geometries. In Volume 2, Chapter 6.4 more information on EBL can be found.

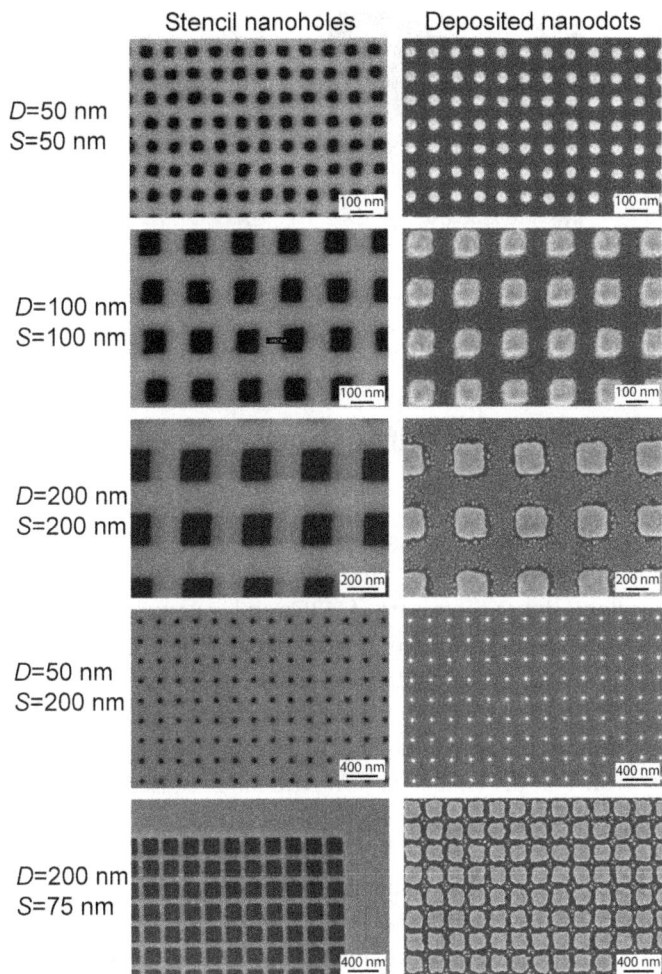

Figure 3.1.12: Nanostencils and the resulting metal nanoantennas. The nanostructures are used for bulk refractive index sensing and the detection of PLL-g-PEG-biotin. Reprinted with permission from [58]. Copyright 2010 American Chemical Society.

The main reason to use EBL for the fabrication of nanostructures is to generate geometries that are not achievable with other techniques. An example for this was brought forward by Lassiter et al. [59]. The structure used for LSPR sensing is a broad-band antenna shown in Figure 3.1.13. The structure is defined by EBL in a PMMA resist layer. After the development of the resist, a gold layer is evaporated, and a lift-off is performed. The outline of this antenna can only be achieved with devices that define shapes using focused beams of charged particles. The downside to EBL-based techniques is the time-consuming process and low throughput.

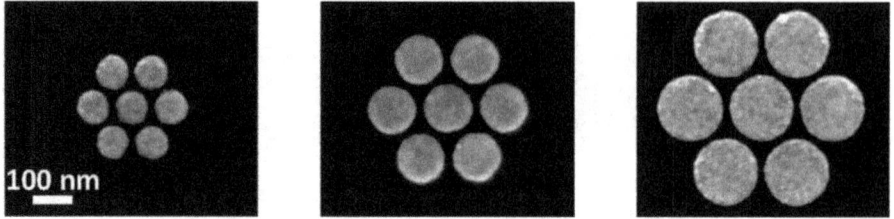

Figure 3.1.13: Heptamers of silver nanoparticles fabricated via an EBL lift-off process for Fano resonance-based refractive index sensing. The structures from the middle panel were tested for bulk refractive index changes and reached a FOM of 5.7. Adapted with permission from [59]. Copyright 2010 American Chemical Society.

3.1.4.5 Functionalization of LSPR shift sensors and choice of analytes

Independently of the fabrication technique and specific shape, plasmonic nanostructures can work as a sensor for refractive index changes of the medium that is surrounding them. This behavior is due to the nature of LSPRs and is described in equation (3.1.3) for the resonance frequency of plasmonic nanoantennas. For LSPR sensors the refractive index change is caused by the binding of analytes to the nanostructures. The refractive index value of DNA is ~1.58 in the visible range. A 60 % glucose solution in water has a refractive index of ~1.44. Compared to the refractive index of water ($n_{H_2O} = 1.33$) it becomes clear that the change in refractive index caused by adding organic molecules to a solution is not very big [60, 61], which poses a challenge for their detection.

Therefore, it is of advantage to introduce a functionalization layer to the nanoantennas of the LSPR biosensor. The functionalization can cause an accumulation of analyte molecules in the proximity of the nanoparticle, increasing the effective change in refractive index. It also introduces selectivity to the LSPR sensor if the functionalization primarily binds the analyte of interest. However, the analyte layer forming on the functionalization is thin, and therefore the rules of bulk refractive index sensing do not translate directly to this case. For a thin layer of analyte coating a spherical plasmonic nanoparticle, taking into account the fast-changing electric field across the layer thickness, an effective refractive index can be calculated as follows:

$$n_{eff} = n_b + (n_l - n_b) \cdot \left(1 - \left(1 + \frac{d}{r}\right)^{-(2p+1)}\right), \tag{3.1.6}$$

where n_l is the refractive index of the layer, n_b is the bulk refractive index, d is the thickness of the layer, r is the radius of the nanoparticle, and p indicates the order of the plasmon mode. In most LSPR sensors $p = 1$ since the dipolar mode is used. The near-field distribution is accounted for in this equation by the power law [62]. With the use of n_{eff} it is possible to calculate the wavelength shift through the equation

$\Delta\lambda_{max} \approx m\,\Delta n$, where $\Delta n = n_{eff} - n_b$. With this more rigorous approach, equation (3.1.3) would change to

$$\Delta\lambda_{max} \approx m(n_l - n_b) \cdot \left(1 - \left(1 + \frac{d}{r}\right)^{-(2p+1)}\right). \qquad (3.1.7)$$

A common functionalization approach is to use short molecules covering the nanoantenna, which selectively bind to the substance of interest. The length of the functionalization molecules is important, since the resonance frequency shift of the antenna is a near-field effect, meaning the shift is bigger if the refractive index change happens closer to the nanostructure. The functionalization effectively introduces a spacer between the analyte and the nanostructure, and thus a trilayer system [30, 39].

Other important aspects are the surface coverage of the functionalization and the reaction kinetics between the functionalization and analyte molecules. The main groups used for functionalization in the short-molecule approach are antibodies, aptamers (DNA and RNA parts), or other molecules that bind with specific substances, e. g., heavy metal ions. For a successful functionalization, the molecules must not only bind to the analyte, but also to the nanostructures. There they should form a monolayer. Therefore, they are outfitted with end groups that form strong bonds with the material of the nanoantennas. To functionalize gold nanostructures, thiol groups are added to the functionalization molecules. For a functional biosensor, also parasitic signals have to be minimized. Parasitic signals occur when other molecules from the examined solution nonspecifically bind to the functionalization or directly to the nanoantenna, or if the analyte itself binds to the surface of the antenna and not to the functionalization. A strategy to minimize the binding of any substance directly to the nanoantenna surfaces is to coat them with an additional inert layer. A commonly used material for this task is polyethylene glycol (PEG) [63].

The field of possible analytes and functionalization is vast. In the following the most common groups of analytes and functionalization will be discussed.

For environmental monitoring, LSPR sensors can be used, e. g., to detect toxins in water. LSPR sensors are useful to detect heavy metal ions. As functionalization for this task, molecules are used that form complexes with the metal ions. Especially for colorimetric sensing molecules are used that aggregate when the heavy metal ions are present. In solution, this leads to accumulations of nanoparticles [64, 65]. The sensors can also be used to detect other harmful substances in water such as urea or hormones, and vitamins [66]. The sensitivity can be further improved by combining the LSPR sensors with enzymatic signal amplification as in the established enzyme-linked immunosorbent assay (ELISA) technique [67].

LSPR sensors have been utilized to detect biomarkers that are linked to human health. These biomarkers can be cancer markers [68, 69], hormones, illness-specific enzymes, viruses, or bacteria. The functionalization for the detection of cancer markers consists in most cases of aptamers. For hormones, viruses, or bacteria, antibody functionalization is used first and foremost [70–73].

As another example for health-related LSPR sensors, Yang et al. have shown a system to detect glucose in human blood [74]. Enzymes are used as functionalization in this case. Since diabetes mellitus is a global health problem, sensitive and low-cost sensors are highly relevant. Another application of LSPR sensors is brought forward by Funari et al., who developed SARS-CoV-2 spike protein-specific sensors. An antibody functionalization is used, and the reported detection limit is 0.08 ng/mL, which is below the clinically relevant range. The test is with 30 min quite fast for its high precision [75]. The sensitivity of LSPR sensors will be discussed in the next section.

3.1.4.6 Sensitivity of LSPR shift sensors

Typically, LSPR sensors are aiming for high sensitivities and FOMs (cf. the section "Concept of LSPR shift sensors" above). The "nanobowl" structures reported by Jiang et al. for example exhibit sensitivities of 1600 nm/RIU [76]. This is one of the highest reported sensitivities in the context of LSPR biosensors. By, e. g., harnessing narrow Fano resonances in nanohole arrays, FOMs of over 100 have been demonstrated [28]. To indicate the LOD of selective biosensors, the minimal analyte weight per solution volume [g L^{-1}], the minimal concentration (e. g., mol/L [M]), or a weight-to-weight ratio (e. g., parts per billion [ppb]) can be given. A list of LOD values for different LSPR biosensors can be found in Guo et al. [77]. Typical LODs for LSPR sensors reported in the literature are in the order of pg mL^{-1} for the detection of cytokines (small proteins) and fM for cancer markers. The LOD of glucose using colorimetric approaches is in the micromolar to millimolar regime [78]. DNA can be detected at a concentration of 50 fM. LSPR sensors for detecting lead ions have shown LOD values of 1 to 10 ppb [79].

In comparison, SPR sensors, e. g., report single-digit nanomolar LOD values for respiratory viruses [80] and femtomolar LOD values for cancer markers [81]. SPR sensors for lead and mercury ions showed LOD values of 0.5 to 12 ppm [82].

In summary, it appears that the LSPR and SPR sensors do not significantly differ in sensitivity and LOD values for most similar tasks [83–85]. For the colorimetric detection of lead ions LSPR sensors have a significantly lower LOD than SPR ones. In many other cases the sensitivity of LSPR biosensors is in the same range as that for SPR sensing and SERS. The numbers vary for different applications, analytes, and methods. None of the techniques clearly outperforms the others comprehensively, and the advantages and drawbacks depend on the specific aims w. r. t. chemical fingerprinting, throughput, multiplexing, etc. SERS for instance gives an insight into the vibrational modes of the analytes and therefore their structure while detecting their presence. It thus provides chemical specificity without the need for additional recognition structures. Both LSPR and SERS sensors are well suited for multiplexing. Different sensing areas can be geometrically separated and can be read out individually, as in the example shown in Figure 3.1.9. The confinement of the metal to nanoparticles for LSPR sensing leads to a fast response time, as well as a less active sensing area that could be covered unspecifically, and there is no need for lasers, prims, or gratings to excite the plasmons. SPR sensors on the other hand often have a stronger response to bulk changes of the refractive index and tend to yield a better signal-to-noise ratio. An in-depth comparison of the techniques can be found, e. g., in the work of Yesudasu et al. [25].

A common goal of ultimate LSPR sensing is the detection of single analyte molecules in a solution [35]. This is achieved if the spectral shift of a plasmonic nanoantenna that is caused by the binding of single molecules can be detected. It can be done by monitoring the spectrum of individual functionalized plasmonic nanoantennas with high temporal and spectral resolution. Such a characteristic step-wise change in the measured signal can be seen in Figure 3.1.14b. In the work of Beuwer et al., the signal of many antennas is read out simultaneously to optimize the quantitative read-out and the dynamic range of the sensor. The time between binding events is the basis for the statistical analyses. The lower the concentration, the longer the time between binding events [86].

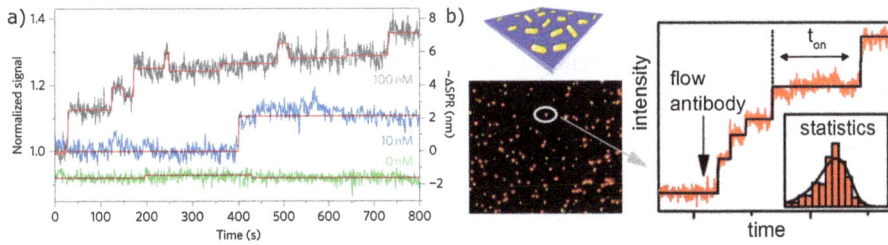

Figure 3.1.14: (a) Optical detection of single-molecule binding events of streptavidin–R-phycoerythrin conjugate to biotin-functionalized nanorods. Reprinted with permission from Nature, *Nature Nanotechnology* [36] © 2012. (b) Left: Schematic and field of view of the optical setup. Every dot represents a plasmonic nanoantenna functioning as an LSPR biosensor. Right: Time trace of the resonance position of a single nanoantenna. The time between steps in the signal is statistically analyzed. Adapted with permission from [86]. Copyright 2015 American Chemical Society.

Another experiment capable of live single-molecule detection was shown by Zijlstra et al. [36]. Exemplary measurement curves are shown in Figure 3.1.14a. The data show the spectral position of the plasmonic peak of a single gold nanorod antenna in real-time and in solutions of different analyte concentrations. The antenna is functionalized with biotin, which binds to the analyte, a streptavidin–R-phycoerythrin conjugate.

The optical setup monitoring the nanoantennas has to be sophisticated in order to detect the plasmonic shifts with high temporal resolution in a timely manner.

The feat to achieve single-molecule sensing does not only rely on a defined concentration limit. Another aspect that contributes to the sensitivity is the transportation of analyte molecules to the detection sites, i. e., the functionalized nanoantenna. One common way to deliver a solution containing an analyte to the nanoantennas is through microfluidic channels and flow cells. The functionalized nanoantennas are fabricated on a substrate that is then refitted with microfluidic channels or incorporated into a flow cell. In both cases, pumps are being used to drive solutions through and deliver the analyte to the nanoantennas. The binding of an analyte to the functionalized antennas is a statistical process [35]. To increase the chance of analyte binding, e. g., a strategy using dielectrophoresis to generate a flux of analyte particles to the nanostructures can be pursued; see also the next section and Figure 3.1.20 [87–89].

The sensitivity of the LSPR sensors increases with the volume and strength of the enhanced near-field of the nanoantennas. The near-field volume can be enlarged by using more nanoantennas. The near-field strength can be increased by using nanoparticles with sharp features, for example, the nanostars in Figure 3.1.8b. Another way to generate extra high near-fields is by using nanoantennas containing narrow gaps, as in Figure 3.1.7 or Figure 3.1.13. The high near-field areas of the antennas have to be accessible for the analyte to successfully bind it to the functionalization. For planar nanostructures like the ones typically fabricated by lithography, it is beneficial to slightly etch the substrate to open up the lower edge of the nanoantenna, which is typically the area with the highest near-field leading to an improved sensitivity (cf. Fig. 3.1.15) [90, 91].

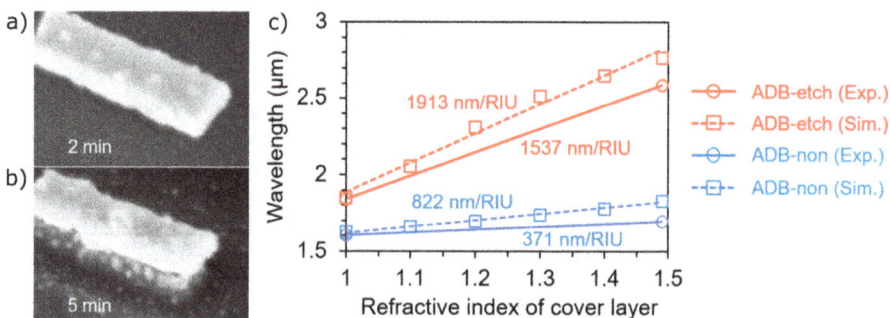

Figure 3.1.15: Left: EBL defined nanostructures with 2 and 5 min selective isotropic etching of the substrate. Right: Measurement of the resonance shift due to an increasing refractive index of the medium. The underetched particles show strongly increased sensitivities. Reprinted from [90].

Another related option to increase the sensitivity is called site-selective binding. With this strategy, only parts of the nanoantennas are functionalized that are located in high near-field areas, which leads to more efficient use of the analyte. This approach

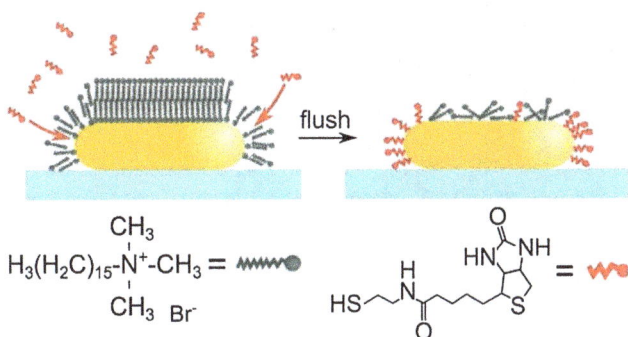

Figure 3.1.16: Site-specific functionalization of nanorods with a two-step functionalization process. Adapted with permission from the supporting information of [86]. Copyright 2015 American Chemical Society.

has been utilized in the work of Beuwer et al., who specifically functionalized the end-caps of rod-like nanostructures. The near-field at these locations is significantly higher than along the rest of the rods. A schematic of the site-specific functionalization is shown in Figure 3.1.16.

3.1.4.7 Some examples of own work

In recent work, we focused on two aspects of LSPR sensing: simultaneous sensing with the bonding and antibonding mode of a vertical dimer [38, 92] and miniaturizing an LSPR sensing setup by using a gradient index (GRIN) lens as the substrate [93]. Additional accumulation of the analyte can be achieved by making use of the dielectrophoretic forces created by the inhomogeneous electric field near the nanoantennas [87–89].

LSPR shift refractive index sensing and antibody detection by vertical dimers

In the first study [92], arrays of vertically stacked gold double-discs with a thin oxide spacer in between served as the detection elements (cf. also [94]). In this arrangement, three characteristic resonances can be observed. Due to the close proximity of the two nanodiscs, the electric near-fields of the discs couple across the spacer layer, and hybridization of the localized surface plasmon polaritons takes place [95] (see Volume 2, Chapter 6.2). If the dipoles that are excited in the two discs oscillate in phase (antibonding mode), the interaction energy is increased, and the corresponding resonance appears at shorter wavelengths than that of an individual disc. If the dipoles oscillate out-of-phase (bonding mode), the interaction energy it lowered due to attractive forces, and the resonance occurs at longer wavelengths. Due to the regular arrange-

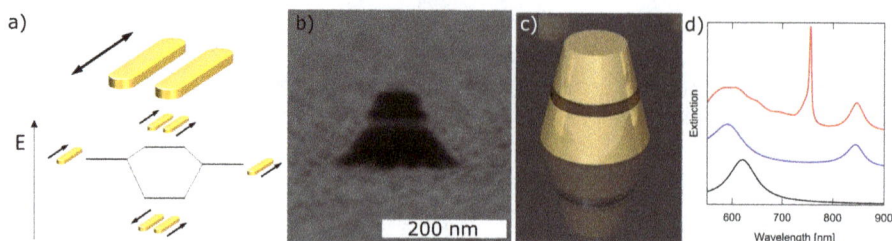

Figure 3.1.17: (a) Plasmon hybridization scheme of parallel (or vertically stacked) metal antennas. (b) SEM image and (c) schematic of vertical dimer. (d) Offset extinction spectra of a single nanodisc (black), a single vertical dimer (blue), and an array of vertical dimers (red). (a) and (b) Adapted from [38], (c) and (d) reprinted with permission from Springer, *Analytical and Bioanalytical Chemistry* [92] © 2015.

ment of the sensing elements in a grating structure, an additional narrow resonance appears, indicating the excitation of surface lattice resonances (SLRs) [96, 97] (see Fig. 3.1.17).

The SLR can be independently spectrally designed through the periodicity of the antenna structure, whereas the spectral positions of the bonding and antibonding mode depend on the geometry of the discs and the thickness of the spacer layer. The narrower the gap between the discs, the stronger the coupling, and the larger the energy splitting between the modes. As an additional degree of freedom, also the material of the nanodiscs can be tuned, fabricating either homodimers of the same, or heterodimers from different metals.

First, the sensitivity of the different modes is studied by numerical simulations [92], and by systematically varying the refractive index of the immersion liquid in a microfluidic channel (cf. Fig. 3.1.18a). The three resonances are clearly visible in the extinction spectrum in water (n = 1.33). Now the refractive index of the liquid was systematically varied in steps of Δn = 0.01 by adding different fractions of glycerin (n = 1.47). All resonances are shifted to longer wavelengths (by different degrees), as seen in Figure 3.1.18a. The shifts are quantified in the time series in Figure 3.1.18b, where the liquid is alternated between water and water–glycerin mixtures, and the corresponding shifts are monitored for all three modes. As a result, sensitivities of the modes of ~110 nm/RIU (antibonding), ~326 nm/RIU (SLR), and ~248 nm/RIU (bonding) and moderate FOMs of <1, ~9, and ~4, respectively, were obtained.

In a second step, the sensor is employed as an immunoassay for specific testosterone antibody sensing by grafting a molecular recognition structure (aminohexanethiol linker with testosterone oxime binding) on the surface that exposes the corresponding antigen. A binding and regeneration cycle is shown in Figure 3.1.19a. The resulting time traces for addition of different concentrations of testosterone antibody along with the resulting calibration curve are depicted in Figure 3.1.19b and c.

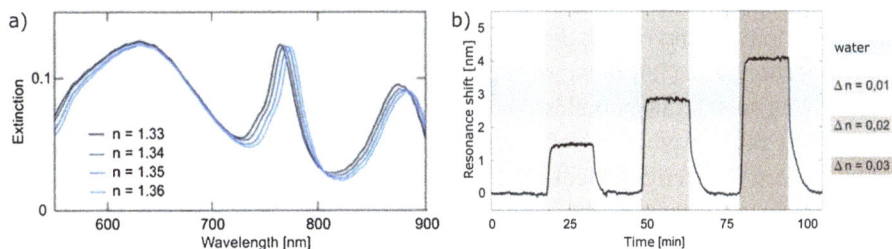

Figure 3.1.18: (a) Spectra of the vertical dimer LSPR sensor in water and in water–glycerin mixtures with increasing refractive index (n = 1.33 to 1.36). The shift of the (from left to right) antibonding (~630 nm), lattice (~770 nm), and bonding mode (~880 nm) can be seen. (b) Corresponding time-dependent refractive index sensing. The LSPR shift is recorded while water is alternated with the higher index mixtures in a microfluidic channel. Reprinted with permission from Springer, *Analytical and Bioanalytical Chemistry* [92] © 2015.

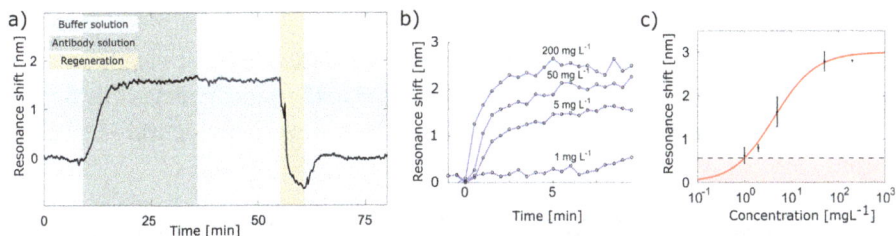

Figure 3.1.19: (a) Cycle of kinetic binding of testosterone antibodies to a testosterone oxime-functionalized vertical dimer sensor with regeneration. (b) LSPR shift for different concentrations. (c) Resulting calibration curve and LOD. Adapted from [38].

The dashed line in Figure 3.1.19c marks the LOD of this sensor configuration, which is estimated to ~6 nmol/L.

It is apparent that the SLR as an effect of far-field coupling between the dipoles of the dimers in the array reacts more sensitively to bulk refractive index changes. The LSPR modes on the other hand depend on interactions within the range of the near-field and are thus more sensitively influenced by the molecular binding. This way, multiscale sensing can be realized within one sensing structure.

Accumulating analyte near the nanoantennas by dielectrophoresis

A fundamental difficulty in LSPR shift sensing is presented by the fact that (i) only refractive index changes within the range of the decay length l_d of the electric near-field can be detected and (ii) l_d is extremely small, e. g., compared to the dimensions of a typical microfluidic channel. Therefore, most analyte molecules (with complex dielectric function ε_p^*) within the gas or liquid flow (with $\varepsilon_m^* = \varepsilon_m - i\sigma_m/\omega$, electric conductivity σ, and ω the frequency of the applied AC voltage) would go undetected

since they do not interact with the nanoantennas. This drawback can be partly reme-
died by accumulating any polarizable particles within the channel at the antenna locations. One such strategy builds upon dielectrophoretic forces that are exerted by
the strongly inhomogeneous electric fields near the nanoparticles due to their high
curvature [88, 98]. In the present example, gold nanocones are chosen as the optical
antennas, since particularly strong field gradients can be achieved near their tip apex.
The antennas are placed on a conductive indium tin oxide (ITO) layer as a bottom electrode, while a second transparent ITO layer is integrated in the microchannel lid acting
as the counter electrode. An AC bias is applied to the electrodes. Nano-objects within
the channel are then polarized by the electric field $\vec{E_0}$ and attracted to the antenna tips
following the force (calculated for spherical particles of radius a)

$$\vec{F_{dep}} = 2\pi\varepsilon_m \, \text{Re}\left(\frac{\varepsilon_p^* - \varepsilon_m^*}{\varepsilon_p^* + 2\varepsilon_m^*}\right) a^3 \vec{\nabla} |\vec{E_0}|^2. \tag{3.1.8}$$

The principle is schematically outlined in Figure 3.1.20a. In [87] it was shown that
fluorescent dye molecules could indeed be attracted to the nanoantennas this way as
shown in Figure 3.1.20b, where they would then be accessible for optical LSPR shift
sensing.

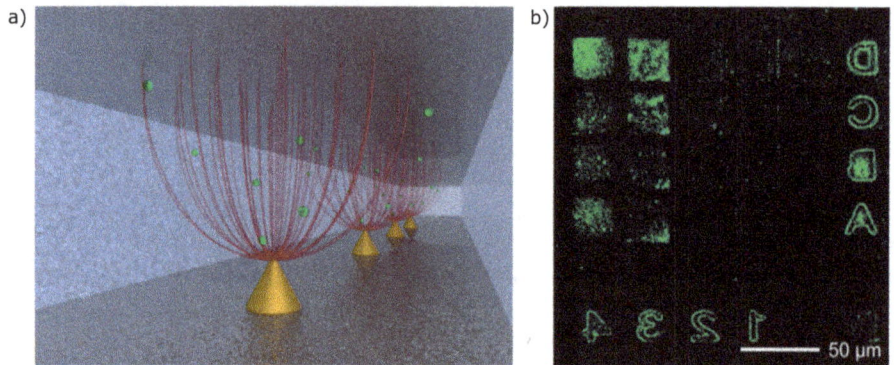

Figure 3.1.20: (a) Schematic of gold nanocone antennas in a microfluidic channel creating a gradient of the electric field between the cone tips and a top electrode (red lines) to attract polarizable
molecules (green). (b) Resulting image of fluorescently labeled bovine serum albumin proteins that
are selectively attached to square arrays of gold nanocone antennas with their pitch decreasing from
left to right. Used with permission of the Royal Society of Chemistry, from [87]; permission conveyed
through Copyright Clearance Center, Inc.

Miniaturized sensors by integration on a GRIN lens

In an attempt to miniaturize the detection unit and move away from spacious and
expensive optical setups, plasmonic nanoantennas were integrated on the surface of

a GRIN lens with a numerical aperture of NA = 0.5 [93]. GRIN lenses are glass cylinders that are doped with metal ions in such a way that the refractive index increases parabolically towards the center (see Fig. 3.1.21b). As a result, light follows a sinusoidal path within the lens and is periodically focused and collimated. If a lens with a pitch length of 0.25 is chosen, parallel light incident on one surface is focused on the opposite surface, or point sources on one surface are turned into a collimated beam on the other side. Arrays of metal nanodiscs as the LSPR sensors were nanofabricated on one face using a transfer process (Fig. 3.1.21a). The light scattered by the point-like nanoantennas leaves the lens as a collimated beam and can be focused onto a spectrometer for detection. The nanostructured lens is then directly integrated in a microfluidic channel that is formed in a polymeric disc and sealed with a coverslip. The full sensing device has a diameter of only about 2 cm (see Fig. 3.1.21d and e). The overall optical configuration including read-out is sketched in Figure 3.1.21c.

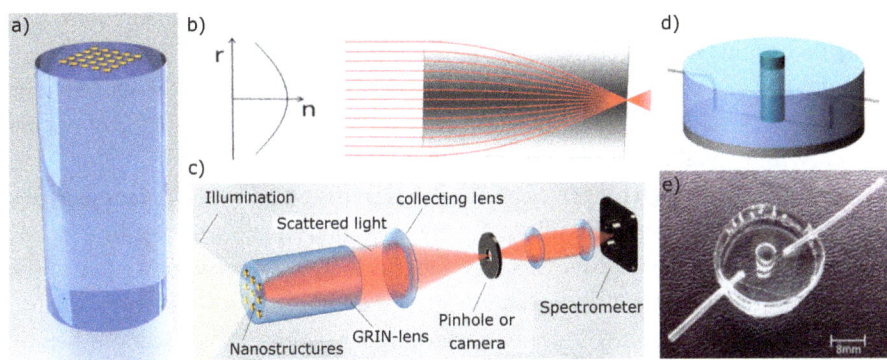

Figure 3.1.21: (a) GRIN lens with nanoantennas structured on one face. (b) Principle of the GRIN lens; for a 1/4 pitch lens a collimated beam is focused on the opposite face and vice versa. (c) Schematic of the miniaturized read-out configuration; for measurements in liquid, the GRIN lens can be integrated in a microfluidic channel. (d) Schematic of a nanostructured GRIN lens embedded in a polymer disc with an integrated microfluidic channel at the bottom. (e) Final device. (a)–(c) Adapted from [38]; (d) and (e) adapted from [99].

The final fluidic LSPR sensor cell was again characterized as a refractive index sensor in a dark-field microscope. Refractive index series with water–glycerin mixtures were evaluated on sensors with two different disc sizes, and sensitivities of 139 nm/RIU and 372 nm/RIU were extracted for 80 nm and 150 nm discs, respectively. The sensitivity scales with the resonance wavelength [100]. A proof-of-principle for testosterone antibody sensing was demonstrated. Combined with a simple grating spectrometer, the configuration could offer a step towards the development of portable low-cost LSPR sensing devices.

3.1.4.8 Applications of LSPR shift sensors

As discussed above, SPR sensing systems in which binding processes on gold films are monitored have been well established for some time now, with numerous suppliers providing the required commercial systems. In contrast, only a few companies that are working on LSPR-based sensing platforms have been founded, and even fewer instruments are commercially available yet (e. g. Nicoya [101], Insplorion [102], Lambdagen [103], or LSPR AG [104]).

Nevertheless, applications of LSPR in many important areas have been indicated for example in medical diagnosis (e. g., diabetes diagnosis by glucose sensing [105]), environmental sensing [65], and gas detection [41, 42], just to name a few. In the area of gas detection one of the few commercially realized devices available is manufactured by Insplorion. Furthermore, many interesting applications and their instrumentations have already been demonstrated as proofs-of-principle in the literature, waiting for entry into the market. Selected examples of such applications in different areas will be highlighted in this section. Many of these are exploiting the following specific advantages: low colloidal particle production costs, easy and inexpensive instruments, feasibility of miniaturized versions as no complicated optics with movable parts are necessary, and therefore portability.

The easiest approach for sensor applications is the use of colloidal nanoparticles monitoring a color change. This allows for the direct detection of analytes by a visible colorimetric change in solution. This wavelength change can occur either due to the local change in the refractive index or due to the coupling of multiple nanoparticles. Some examples for colorimetric systems were already outlined in the sections before. Here three examples will be highlighted: the use of functionalized Au particles for Hg^{2+} ion detection and two colorimetric lateral flow assays [65, 67, 106].

The detection of environmental pollution by heavy metal ions can be carried out under laboratory conditions with DNA-functionalized gold nanoparticles. These nanoparticles carry two DNA strands that form aggregates up to a transition temperature T_M. By coordination with the Hg^{2+} ion the transition temperature rises, which can be monitored for quantitative analysis. The change from blue to red is even visible to the bare eye [65] (cf. Fig. 3.1.22). This method has also been extended to other ions and to room temperature [107].

The colorimetric analysis under laboratory conditions is time consuming and costly and requires specialized staff. These problems were circumvented by developing lateral flow cells with the metal nanoparticles. These fast-measurement LSPR sensors are very well suited for POC sensors, which allow for medical sensing on-site close to the patients. They usually work by interaction of the analyte and a recognition system using the gold particles as fluorescent labels to form a visible stripe in the detection window (see Fig. 3.1.23). Lateral flow assays with gold nanoparticles are successfully used to identify a vast variety of analytes such as toxins, drugs, cortisol,

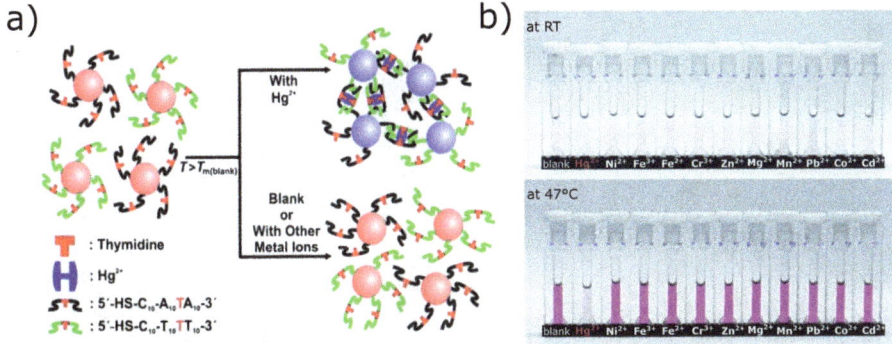

Figure 3.1.22: Colorimetric detection of Hg^{2+} ions with colloidal nanoparticles. Adapted from [65].

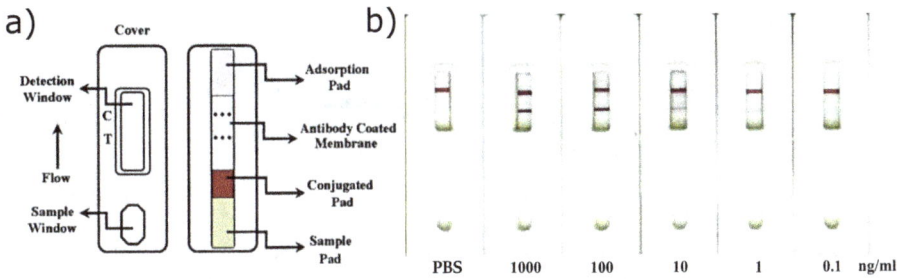

Figure 3.1.23: Lateral flow assay of staphylococcal enterotoxin B. (a) Schematic illustration of a lateral flow cell. (b) Measurements with the buffer solution (left) and different staphylococcal enterotoxin B concentrations (right). Adapted from [106].

or pregnancy hormones as well as diseases ranging from bacterial and viral infections to cardiovascular diseases, just to name a few [108]. Some devices even allow for semiquantitative analysis. An example of a lateral flow cell with staphylococcal enterotoxin B is shown in Figure 3.1.23b [106].

Nevertheless, the easy and fast sensor concepts with colloidal nanoparticles in solution have several disadvantages. Especially sensors read out by eye show less sensitivity than more complex approaches. Additionally in solution the particles may aggregate even without analyte and show interference with other molecules.

As an example for a more complex flow cell with immobilized nanoparticles an immunosorbent assay system for horseradish peroxidase enzyme detection by enzymatic signal amplification is shown. It is based on refractive index change sensing, and Chen et al. even showed single-molecule sensitivity [67]. The system consists of functionalized gold particles fabricated by hole mask colloidal lithography and a spectral readout (Fig. 3.1.24).

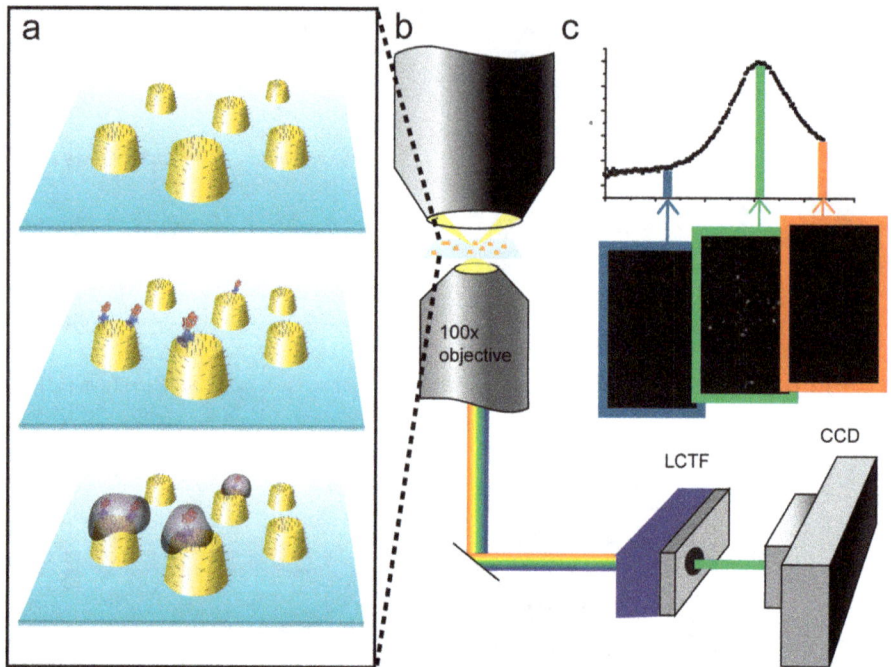

Figure 3.1.24: Immunosorbent assay based on refractive sensing. (a) Sensing principle. (b) Setup, LCTF: liquid crystal tunable filter. (c) Example spectrum. Adapted with permission from [67]; Copyright 2011 American Chemical Society.

Another group of devices are chip-based sensors, quite often coupled with microfluidics. There are two main approaches. On the one hand some systems add nanoparticles or nanostructures on chips and use SPR-like systems with wavelength, interferometric, or intensity readout, for example for increasing the SPR sensing sensitivity in cancer detection [109]. Some approaches are hybrid systems exploiting SPR and LSPR, others are relying solely on the LSPR effect. LSPR-based systems have the advantage that a stable signal is achievable without temperature control [13]. Chip-based flexible biosensing platforms based on LSPR are also commercially available, e. g., in the openSPR setup developed by Nicoya Lifescience, which is commercially available since 2018 as a benchtop system.

On the other hand, there are devices taking advantage of the possibility to build compact systems and POC devices. Those devices are based around cost-effective, simple, and compact optics, for example only a LED as the light source and a CCD chip as the detector, but without any lenses [110], or even with a mobile phone camera as the detector [111]. Two compact solutions for portable/handheld instruments are shown in Figure 3.1.25. In Figure 3.1.25a a device for food control by analysis of casein in milk is shown [112]. It consists of a compact light source and a compact spectrometer, both connected to an optical probe, which is situated on top of the biosensor. The sensor it-

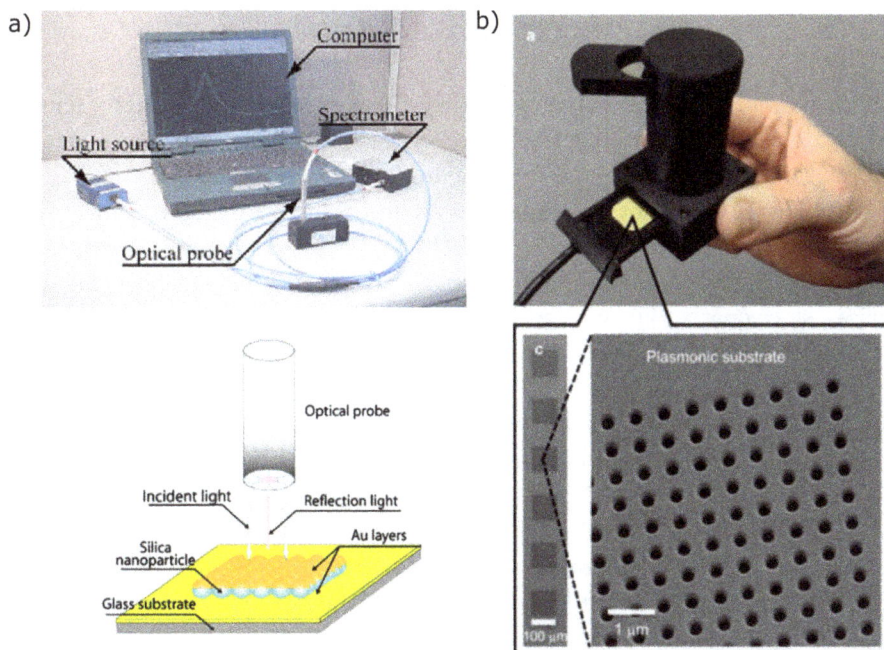

Figure 3.1.25: Examples of compact plasmonic sensing devices. (a) Device for analysis of casein in milk; adapted after [112]. (b) Handheld sensor based on nanoholes; reprinted with permission from Nature, *Light: Science & Applications* [110] © 2014.

self is a monolayer of gold-coated silica nanoparticles functionalized with anti-casein antibodies on a gold layer. This device showed results with high accuracy comparable with standard SPR or ELISA techniques but with lower costs, and is reported to be fast and easy to use [112].

A second example of a handheld device by the Altug group is depicted in Figure 3.1.25b [110]. The setup consists of a battery, a LED, a nanohole sensor chip, and a CMOS detector for readout. The whole device is 7.5 cm tall and weighs 60 g. The diffraction pattern of the nanohole array is imaged, and intensity variations in the transmission are recorded. The system was demonstrated on mono- and bilayers of proteins and protein quantification. With the small dimensions and the low-cost optical setup, this device serves as an excellent example for a POC device.

The examples in this chapter show that LSPR sensing is not only applicable to a vast area of applications, but may also be realized in numerous ways for instrumentation. Some devices are already on the market, and many ideas may follow over the coming years.

3.1.5 Some challenges and solutions

The still very limited number of commercial LSPR sensors on the market at the time of the release of this book indicates that there are challenges remaining that need to be overcome to generate viable products from this technology.

Ongoing challenges for the sensing process are the increase of the selectivity, sensitivity, and dynamic range.

The selectivity must be maximized to create a sensor that is suitable for end-user devices, since only sensors with good selectivity can give meaningful results. To increase selectivity suitable recognition elements need to be developed, and the functionalization must be optimized to reduce cross-sensitivity to other molecules. Parasitic signals from any molecules binding directly to the antennas or substrates bypassing the functionalization must be minimized as well. This can be done by an additional inert coating applied to the antennas, substrates, and channels.

Another challenge for LSPR sensors is to develop their selectivity to avoid pretreatment for composite solutions. For emulsions, like blood, this is more complicated since the solid particles in the liquid can interrupt the sensing process by cloaking the microfluidic channels or by covering the nanoparticles. For now, these emulsions must be at least diluted or even separated before the test can be performed. There are however interesting developments of tabletop optical sensors with incorporated centrifugal microfluidic platforms called lab-on-a-disc [113].

The specific binding of the analyte to the molecular recognition structure opens the door for reusable biosensors. The core idea is to regenerate the sensor after measurement to use it again. Regeneration solutions are chosen for offering a high solubility of the analyte while not reacting with the rest of the sensor. To successfully clear the sensor from the analyte it is highly beneficial if the analytes are only bound to the functionalization, so they are accessible for the purging solution, and the bond strength is known and roughly constant.

The pursuit of high sensitivity is universal for all types of sensors. In view of LSPR biosensors, it is a more pressing issue since they compete with other optical biosensors like SPR and SERS approaches, which show comparable sensitivity [84, 85]. In order to decrease the LOD, ideally down to single-molecule sensitivity, the signal-to-noise ratio has to be improved. This can be achieved by using better optical equipment to reduce the noise. To increase the signal, nanostructures with stronger and more accessible near-fields can be used. With these measures or by combining them the LOD can be significantly improved, but the complexity of the system increases correspondingly.

Another approach to decrease the LOD is to use statistics to extract more information from a given signal. More nanoantennas or longer measurement cycles are beneficial for this strategy, since more data lead to higher statistical accuracy.

This approach also helps with the dynamic range of the sensor. For a high dynamic range, the number of potential binding spots must be high, so the sensor does not get saturated too fast. But the sensitivity must be high as well to confidently measure small analyte concentrations. Many individual nanostructures are monitored in parallel, and all signals are included in statistical analysis.

The successful and efficient transport of the analyte to the nanoantennas is likewise a challenge concerning LSPR biosensors. Since the active area of the sensor is the near-field volume of nanoantennas it is relatively small, and the probability of analyte reaching this area has to be increased to aid the effectivity of the sensor. In many cases, microfluidic channels are used to transport a solution that potentially includes the analyte to the active area of the sensors, i. e., the nanoantennas. To increase the probability of an analyte encountering the antennas the channel volume around the active sensor area should be as small as possible. A decreasing channel cross-section leads to higher pressure needed to drive the fluid, which has to be taken into account when designing the flow cell. The probability of contact of the analytes with the nanoantennas can be further increased by using measures like, e. g., dielectrophoresis, as shown in Figure 3.1.20.

LSPR biosensors perform poorly in the detection of small molecules forming thinner adsorption layers (cf. equations (3.1.3) and (3.1.7)), even though LSPR is still better suited than SPR. The change of the effective refractive index caused by these molecules is small due to their size. This leads directly to a lower sensitivity of the sensors. Since this is an inherent problem of the core mechanics of LSPR sensors, the sensor has to be improved in every aspect to enhance the performance for small molecules. Alternatively, another sensing mechanisms like SERS can be chosen for this kind of analytes.

LSPR biosensors can set themselves apart from other biosensing approaches by the distinguishing feature that the sensor can be built with a small footprint to the point where the whole device including any optical components can be made handheld. Due to the working mechanisms of the sensor, it is relatively fast, and the binding of the analyte can be detected nearly instantaneously. This renders LSPR biosensors potentially powerful candidates for PON and POC use.

3.1.6 Summary and impact

Optical sensing technologies have undergone tremendous development over the last decades, and research on LSPR shift sensors has seen a rapid increase since the start of the twenty-first century. The development of nanoantenna-based detection techniques occurred hand-in-hand with the progress of nanofabrication techniques to prepare suitable transducer elements and the advancement of suitable molecular recognition elements. LSPR shift sensors represent a direct translation of the concept of optical nanospectroscopy into the field of (bio)sensing. They offer label-free, versatile, highly sensitive sensing devices, which are extremely well suited for miniaturization and multiplexing. Signal read-out is obtained in a straightforward manner by following the spectral shifts of nanoantenna resonances. The suitability of LSPR shift sensors for the detection of relevant markers has been proven in scientific research and lab-based demonstrations. The most important application fields include healthcare, POC diagnostics, and environmental monitoring, as well as refractive index and gas sensing.

The demand for sensors is growing and projected to increase even further in importance in the context of the Internet of Things. Instrumentation harnessing surface plasmons for sensing purposes in the shape of interface-based SPR sensors has reached the marketplace in the 1990s. The development of nanostructure-based LSPR shift sensors started with a temporal delay, and they have not yet reached the same maturity. Such sensors are currently at the brink to commercial realization, with the first specimens becoming available. The near future will show whether their advantages prevail and remaining challenges can be sufficiently overcome to make them viable competitors for existing technologies.

Bibliography

[1]   Maier SA. Plasmonics: Fundamentals and Applications. Springer; 2007.

[2]   Xavier J, Vincent S, Meder F, Vollmer F. Advances in optoplasmonic sensors – combining optical nano/microcavities and photonic crystals with plasmonic nanostructures and nanoparticles. Nanophotonics. 2018;7:1.

[3]   Merriam-Webster dictionary site on sensors, Springfield, MA, USA: Merriam-Webster Inc., (20.05.2022 at https://www.merriam-webster.com/dictionary/sensor).

[4]   Lobnik A, Turel M, Korent Urek S. Advances in chemical sensors. In: Wang W, editor. IntechOpen. 2012. p. 3.

[5]   Europe's age of light! How photonics will power growth and innovation, Düsseldorf, Germany: Photonics21 Executive Board, 2019. (23.05.2022 at https://www.photonics21.org/download/ppp-services/photonics-downloads/Europes-age-of-light-Photonics-Roadmap-C1.pdf).

[6]   Dahlin AB, Jonsson MP. Performance of nanoplasmonic biosensors. In: Dmitriev A, editor. Nanoplasmonic Sensors. New York, NY: Springer New York; 2012. p. 231.

[7]   Estevez M-C, Otte MA, Sepulveda B, Lechuga LM. Trends and challenges of refractometric nanoplasmonic biosensors: a review. Anal Chim Acta. 2014;806:55–73.

[8]   Lopez GA, Estevez M-C, Solera M, Lechuga ML. Recent advances in nanoplasmonic biosensors: applications and lab-on-a-chip integration. Nanophotonics. 2017;6(1):123–36.

[9]   Ruemmele JA, Hall WP, Ruvuna LK, Van Duyne RP. A localized surface plasmon resonance imaging instrument for multiplexed biosensing. Anal Chem. 2013;85:4560.

[10]   Denizli A, editor. Plasmonic Sensors and Their Applications. Wiley; 2021.

[11]   Sannomiya T, Voros J. Single plasmonic nanoparticles for biosensing. Trends Biotechnol. 2011;29:343.

[12]   Byun KM. Development of nanostructured plasmonic substrates for enhanced optical biosensing. J Opt Soc Korea. 2010;14:65.

[13]   Hill RT. Plasmonic biosensors. Wiley Interdiscip Rev Nanomed Nanobiotechnol. 2015;7:152.

[14]   Fleischmann M, Hendra PJ, Mcquillan AJ. Raman-spectra of pyridine adsorbed at a silver electrode. Chem Phys Lett. 1974;26:163.

[15] Moskovits M, Piorek BD. A brief history of surface-enhanced Raman spectroscopy and the localized surface plasmon Dedicated to the memory of Richard Van Duyne (1945–2019). J Raman Spectrosc. 2021;52:279.

[16] Ritchie RH. Plasma losses by fast electrons in thin films. Phys Rev. 1957;106:874.

[17] Kretschmann E, Raether H. Notizen: radiative decay of non radiative surface plasmons excited by light. Z Naturforsch A. 1968;23:2135.

[18] Otto A. Excitation of nonradiative surface plasma waves in silver by the method of frustrated total reflection. Z Phys A Hadrons Nucl. 1968;216, 398.

[19] Raether H. Surface-plasmons on rough surfaces. Nuovo Cimento B. 1977;39:817.

[20] Gordon JG, Ernst S. Surface-plasmons as a probe of the electrochemical interface. Surf Sci. 1980;101:499.

[21] Nylander C, Liedberg B, Lind T. Gas-detection by means of surface-plasmon resonance. Sensor Actuator. 1982;3:79.

[22] Liedberg B, Nylander C, Lundstrom I. Biosensing with surface-plasmon resonance – how it all started. Biosens Bioelectron. 1995;10:R1.

[23] Homola J. Surface plasmon resonance sensors for detection of chemical and biological species. Chem Rev. 2008;108:462.

[24] Duan Q, Liu Y, Chang S, Chen H, Chen JH. Surface plasmonic sensors: sensing mechanism and recent applications. Sensors. 2021;21:5262.

[25] Yesudasu V, Pradhan HS, Pandya RJ. Recent progress in surface plasmon resonance based sensors: a comprehensive review. Heliyon. 2021;7:e06321.

[26] Eftekhari F, Escobedo C, Ferreira J, Duan X, Girotto EM, Brolo AG, Gordon R, Sinton D. Nanoholes as nanochannels: flow-through plasmonic sensing. Anal Chem. 2009;81:4308.

[27] Brolo AG, Gordon R, Leathem B, Kavanagh KL. Surface plasmon sensor based on the enhanced light transmission through arrays of nanoholes in gold films. Langmuir. 2004;20:4813.

[28] Yanik AA, Cetin AE, Huang M, Artar A, Mousavi SH, Khanikaev A, Connor JH, Shvets G, Altug H. Seeing protein monolayers with naked eye through plasmonic Fano resonances. Proc Natl Acad Sci USA. 2011;108:11784.

[29] Raschke G, Kowarik S, Franzl T, Sonnichsen C, Klar TA, Feldmann J, Nichtl A, Kurzinger K. Biomolecular recognition based on single gold nanoparticle light scattering. Nano Lett. 2003;3:935.

[30] Haes AJ, Van Duyne RP. A nanoscale optical biosensor: sensitivity and selectivity of an approach based on the localized surface plasmon resonance spectroscopy of triangular silver nanoparticles. J Am Chem Soc. 2002;124:10596.

[31] Piriya VSA, Joseph P, Daniel SCGK, Lakshmanan S, Kinoshita T, Muthusamy S. Colorimetric sensors for rapid detection of various analytes. Mater Sci Eng C Mater Biol Appl. 2017;78:1231.

[32] de la Rica R, Stevens MM. Plasmonic ELISA for the ultrasensitive detection of disease biomarkers with the naked eye. Nat Nanotechnol. 2012;7:821.

[33] Zhang Z, Wang H, Chen Z, Wang X, Choo J, Chen L. Plasmonic colorimetric sensors based on etching and growth of noble metal nanoparticles: Strategies and applications. Biosens Bioelectron. 2018;114:52.

[34] Sannomiya T, Hafner C, Voros J. In situ sensing of single binding events by localized surface plasmon resonance. Nano Lett. 2008;8:3450.

[35] Taylor AB, Zijlstra P. Single-molecule plasmon sensing: current status and future prospects. ACS Sens. 2017;2:1103.

[36] Zijlstra P, Paulo PM, Orrit M. Optical detection of single non-absorbing molecules using the surface plasmon resonance of a gold nanorod. Nat Nanotechnol. 2012;7:379.

[37]   McFarland AD, Van Duyne RP. Single silver nanoparticles as real-time optical sensors with zeptomole sensitivity. Nano Lett. 2003;3:1057.

[38]   Horrer A. Brechungsindexsensoren basierend auf metallischen Nanostrukturen – Untersuchung der Sensitivität, Anwendung als Biosensor und Integration in einen kompakten Aufbau. PhD thesis. Germany: University of Tübingen; 2018.

[39]   Jung LS, Campbell CT, Chinowsky TM, Mar MN, Yee SS. Quantitative interpretation of the response of surface plasmon resonance sensors to adsorbed films. Langmuir. 1998;14:5636.

[40]   Homola J. Present and future of surface plasmon resonance biosensors. Anal Bioanal Chem. 2003;377:528.

[41]   Tittl A, Giessen H, Liu N. Plasmonic gas and chemical sensing. Nanophotonics. 2014;3:157.

[42]   Wadell C, Syrenova S, Langhammer C. Plasmonic hydrogen sensing with nanostructured metal hydrides. ACS Nano. 2014;8:11925.

[43]   El-Asfar S. Brechungsindex–Sensitivität selbstangeordneter plasmonischer Golddisk–Nanostrukturen hergestellt mit Nanosphere–Lithographie. Internship report. Germany: University of Tübingen; 2016.

[44]   Gordon R. Nanostructured metals for light-based technologies. Nanotechnology. 2019;30:212001.

[45]   Cobley CM, Skrabalak SE, Campbell DJ, Xia YN. Shape-controlled synthesis of silver nanoparticles for plasmonic and sensing applications. Plasmonics. 2009;4:171.

[46]   Thanh NT, Maclean N, Mahiddine S. Mechanisms of nucleation and growth of nanoparticles in solution. Chem Rev. 2014;114:7610.

[47]   Liang H, Rossouw D, Zhao H, Cushing SK, Shi H, Korinek A, Xu H, Rosei F, Wang W, Wu N, Botton GA, Ma D. Asymmetric silver "nanocarrot" structures: solution synthesis and their asymmetric plasmonic resonances. J Am Chem Soc. 2013;135:9616.

[48]   Hegde HR, Chidangil S, Sinha RK. Refractive index sensitivity of Au nanostructures in solution and on the substrate. J Mater Sci, Mater Electron. 2022;33:4011.

[49]   Endo T, Kerman K, Nagatani N, Hiepa HM, Kim DK, Yonezawa Y, Nakano K, Tamiya E. Multiple label-free detection of antigen-antibody reaction using localized surface plasmon resonance-based core-shell structured nanoparticle layer nanochip. Anal Chem. 2006;78:6465.

[50]   Pimpin A, Srituravanich W. Review on micro- and nanolithography techniques and their applications. Eng J. 2012;16:37.

[51]   Kasani S, Curtin K, Wu NQ. A review of 2D and 3D plasmonic nanostructure array patterns: fabrication, light management and sensing applications. Nanophotonics. 2019;8:2065.

[52]   Haynes CL, Van Duyne RP. Nanosphere lithography: a versatile nanofabrication tool for studies of size-dependent nanoparticle optics. J Phys Chem B. 2001;105:5599.

[53]   Ai B, Yu Y, Mohwald H, Zhang G, Yang B. Plasmonic films based on colloidal lithography. Adv Colloid Interface Sci. 2014;206:5.

[54]   Chen S, Svedendahl M, Kall M, Gunnarsson L, Dmitriev A. Ultrahigh sensitivity made simple: nanoplasmonic label-free biosensing with an extremely low limit-of-detection for bacterial and cancer diagnostics. Nanotechnology. 2009;20:434015.

[55]   Haes AJ, Chang L, Klein WL, Van Duyne RP. Detection of a biomarker for Alzheimer's disease from synthetic and clinical samples using a nanoscale optical biosensor. J Am Chem Soc. 2005;127:2264.

[56]   Shen Y, Zhou J, Liu T, Tao Y, Jiang R, Liu M, Xiao G, Zhu J, Zhou ZK, Wang X, Jin C, Wang J. Plasmonic gold mushroom arrays with refractive index sensing figures of merit approaching the theoretical limit. Nat Commun. 2013;4:2381.

[57]   Jiang S, Saito M, Murahashi M, Tamiya E. Pressure free nanoimprinting lithography using ladder-type HSQ material for LSPR biosensor chip. Sens Actuators B, Chem. 2017;242:47.

[58]   Vazquez-Mena O, Sannomiya T, Villanueva LG, Voros J, Brugger J. Metallic nanodot arrays by stencil lithography for plasmonic biosensing applications. ACS Nano. 2011;5:844.

[59]   Lassiter JB, Sobhani H, Fan JA, Kundu J, Capasso F, Nordlander P, Halas NJ. Fano resonances in plasmonic nanoclusters: geometrical and chemical tunability. Nano Lett. 2010;10:3184.

[60]   Lide DR. CRC Handbook of Chemistry and Physics. 85th ed. Taylor & Francis; 2004.

[61]   Refractive index database, Polyanskiy MN. 2008–2022. (13.04.2022 at https:// refractiveindex.info).

[62]   Jatschka J, Dathe A, Csáki A, Fritzsche W, Stranik O. Propagating and localized surface plasmon resonance sensing — A critical comparison based on measurements and theory. Sens Biosensing Res. 2016;7:62.

[63]   Chandradoss SD, Haagsma AC, Lee YK, Hwang JH, Nam JM, Joo C. Surface passivation for single-molecule protein studies. J Vis Exp. 2014;86:e50549.

[64]   Ratnarathorn N, Chailapakul O, Dungchai W. Highly sensitive colorimetric detection of lead using maleic acid functionalized gold nanoparticles. Talanta. 2015;132:613.

[65]   Lee JS, Han MS, Mirkin CA. Colorimetric detection of mercuric ion (Hg^{2+}) in aqueous media using DNA-functionalized gold nanoparticles. Angew Chem, Int Ed Engl. 2007;46:4093.

[66]   Gupta BD, Pathak A, Shrivastav AM. Optical biomedical diagnostics using lab-on-fiber technology: a review. Photonics. 2022;9:86.

[67]   Chen S, Svedendahl M, Duyne RP, Kall M. Plasmon-enhanced colorimetric ELISA with single molecule sensitivity. Nano Lett. 2011;11:1826.

[68]   Acimovic SS, Ortega MA, Sanz V, Berthelot J, Garcia-Cordero JL, Renger J, Maerkl SJ, Kreuzer MP, Quidant R. LSPR chip for parallel, rapid, and sensitive detection of cancer markers in serum. Nano Lett. 2014;14:2636.

[69]   Yavas O, Acimovic SS, Garcia-Guirado J, Berthelot J, Dobosz P, Sanz V, Quidant R. Self-calibrating on-chip localized surface plasmon resonance sensing for quantitative and multiplexed detection of cancer markers in human serum. ACS Sens. 2018;3:1376.

[70]   Liu J, Jalali M, Mahshid S, Wachsmann-Hogiu S. Are plasmonic optical biosensors ready for use in point-of-need applications? Analyst. 2020;145:364.

[71]   Caucheteur C, Guo T, Albert J. Review of plasmonic fiber optic biochemical sensors: improving the limit of detection. Anal Bioanal Chem. 2015;407:3883.

[72]   Miranda B, Rea I, Dardano P, De Stefano L, Forestiere C. Recent advances in the fabrication and functionalization of flexible optical biosensors: toward smart life-sciences applications. Biosensors. 2021;11:107.

[73]   Peltomaa R, Glahn-Martinez B, Benito-Pena E, Moreno-Bondi MC. Optical biosensors for label-free detection of small molecules. Sensors (Basel). 2018;18:4126.

[74]   Yang QS, Zhang X, Kumar S, Singh R, Zhang BY, Bal CL, Pu XP. Development of glucose sensor using gold nanoparticles and glucose-oxidase functionalized tapered fiber structure. Plasmonics. 2020;15:841.

[75]   Funari R, Chu KY, Shen AQ. Detection of antibodies against SARS-CoV-2 spike protein hy gold nanospikes in an opto-microfluidic chip. Biosens Bloelectron. 2020;169:112578.

[76]   Jiang J, Wang XH, Li S, Ding F, Li NT, Meng SY, Li RF, Qi J, Liu QJ, Liu GL. Plasmonic nano-arrays for ultrasensitive bio-sensing. Nanophotonics. 2018;7:1517.

[77]   Guo LH, Jackman JA, Yang HH, Chen P, Cho NJ, Kim DH. Strategies for enhancing the sensitivity of plasmonic nanosensors. Nano Today. 2015;10:213.

[78]   Gao Y, Wu Y, Di J. Colorimetric detection of glucose based on gold nanoparticles coupled with silver nanoparticles. Spectrochim Acta, Part A, Mol Biomol Spectrosc. 2017;173:207.

[79]   Qiu GY, Law AHL, Ng SP, Wu CML. Label-free detection of lead(II) ion using differential phase modulated localized surface plasmon resonance sensors. Proc Eng. 2016;168:533.

[80]   Shi L, Sun Q, He J, Xu H, Liu C, Zhao C, Xu Y, Wu C, Xiang J, Gu D, Long J, Lan H. Development of SPR biosensor for simultaneous detection of multiplex respiratory viruses. Biomed Mater Eng. 2015;26(Suppl 1):S2207.

[81] Liu R, Wang Q, Li Q, Yang X, Wang K, Nie W. Surface plasmon resonance biosensor for sensitive detection of microRNA and cancer cell using multiple signal amplification strategy. Biosens Bioelectron. 2017;87:433.

[82] Abdi MM, Abdullah LC, Sadrolhosseini AR, Mat Yunus WM, Moksin MM, Tahir PM. Surface plasmon resonance sensing detection of mercury and lead ions based on conducting polymer composite. PLoS ONE. 2011;6:e24578.

[83] Yonzon CR, Jeoung E, Zou S, Schatz GC, Mrksich M, Van Duyne RP. A comparative analysis of localized and propagating surface plasmon resonance sensors: the binding of concanavalin A to a monosaccharide functionalized self-assembled monolayer. J Am Chem Soc. 2004;126:12669.

[84] Svedendahl M, Chen S, Dmitriev A, Kall M. Refractometric sensing using propagating versus localized surface plasmons: a direct comparison. Nano Lett. 2009;9:4428.

[85] Haes AJ, Van Duyne RP. A unified view of propagating and localized surface plasmon resonance biosensors. Anal Bioanal Chem. 2004;379:920.

[86] Beuwer MA, Prins MW, Zijlstra P. Stochastic protein interactions monitored by hundreds of single-molecule plasmonic biosensors. Nano Lett. 2015;15:3507.

[87] Schäfer C, Kern DP, Fleischer M. Capturing molecules with plasmonic nanotips in microfluidic channels by dielectrophoresis. Lab Chip. 2015;15:1066.

[88] Barik A, Cherukulappurath S, Wittenberg NJ, Johnson TW, Oh SH. Dielectrophoresis-assisted Raman spectroscopy of intravesicular analytes on metallic pyramids. Anal Chem. 2016;88:1704.

[89] Barik A, Otto LM, Yoo D, Jose J, Johnson TW, Oh SH. Dielectrophoresis-enhanced plasmonic sensing with gold nanohole arrays. Nano Lett. 2014;14:2006.

[90] Moritake Y, Tanaka T. Impact of substrate etching on plasmonic elements and metamaterials: preventing red shift and improving refractive index sensitivity. Opt Express. 2018;26:3674.

[91] Acimovic SS, Sipova H, Emilsson G, Dahlin AB, Antosiewicz TJ, Kall M. Superior LSPR substrates based on electromagnetic decoupling for on-a-chip high-throughput label-free biosensing. Light Sci Appl. 2017;6:e17042.

[92] Horrer A, Krieg K, Freudenberger K, Rau S, Leidner L, Gauglitz G, Kern DP, Fleischer M. Plasmonic vertical dimer arrays as elements for biosensing. Anal Bioanal Chem. 2015;407:8225.

[93] Horrer A, Haas J, Freudenberger K, Gauglitz G, Kern DP, Fleischer M. Compact plasmonic optical biosensors based on nanostructured gradient index lenses integrated into microfluidic cells. Nanoscale. 2017;9:17378.

[94] Dmitriev A, Pakizeh T, Kall M, Sutherland DS. Gold-silica-gold nanosandwiches: tunable bimodal plasmonic resonators. Small. 2007;3:294.

[95] Nordlander P, Oubre C, Prodan E, Li K, Stockman MI. Plasmon hybridization in nanoparticle dimers. Nano Lett. 2004;4:899.

[96] Auguie B, Barnes WL. Collective resonances in gold nanoparticle arrays. Phys Rev Lett. 2008;101:143902.

[97] Humphrey AD, Barnes WL. Plasmonic surface lattice resonances on arrays of different lattice symmetry. Phys Rev B. 2014;90:075404.

[98] Jose J, Kress S, Barik A, Otto LM, Shaver J, Johnson TW, Lapin ZJ, Bharadwaj P, Novotny L, Oh SH. Individual template-stripped conductive gold pyramids for tip-enhanced dielectrophoresis. ACS Photonics. 2014;1:464.

[99] Haas J. Transfer von Goldnanostrukturen auf eine GRIN-Linse und Integration in eine Flusszelle mit Mikrofluidikkanal zur Verwendung als LSPR-Sensor. Bachelor Thesis. Germany: University of Tübingen; 2015.

[100] Miller MM, Lazarides AA. Sensitivity of metal nanoparticle surface plasmon resonance to the dielectric environment. J Phys Chem B. 2005;109:21556.

[101] Homepage of Nicoya, Kitchener, ON, Canada: Nicoya, (13.04.2022 at https://nicoyalife.com).

[102] Homepage of Insplorion, Göteborg, Sweden: Insplorion, (13.04.2022 at https://www.insplorion.com/).

[103] Homepage of Lambdagen, Menlo Park, USA: LambdaGen Cooperation, 2021-2022. (13.04.2022 at https://lamdagen.com/).

[104] Homepage of LSPR AG, Wallisellen, Switzerland: LSPR AG, (13.04.2022 at https://lspr.swiss/).

[105] Serra A, Filippo E, Re M, Palmisano M, Vittori-Antisari M, Buccolieri A, Manno D. Non-functionalized silver nanoparticles for a localized surface plasmon resonance-based glucose sensor. Nanotechnology. 2009;20:165501.

[106] Rong-Hwa S, Shiao-Shek T, Der-Jiang C, Yao-Wen H. Gold nanoparticle-based lateral flow assay for detection of staphylococcal enterotoxin B. Food Chem. 2010;118:462.

[107] Li M, Cushing SK, Wu N. Plasmon-enhanced optical sensors: a review. Analyst. 2015;140:386.

[108] Posthuma-Trumpie GA, Korf J, van Amerongen A. Lateral flow (immuno)assay: its strengths, weaknesses, opportunities and threats. A literature survey. Anal Bioanal Chem. 2009;393:569.

[109] Law WC, Yong KT, Baev A, Prasad PN. Sensitivity improved surface plasmon resonance biosensor for cancer biomarker detection based on plasmonic enhancement. ACS Nano. 2011;5:4858.

[110] Cetin AE, Coskun AF, Galarreta BC, Huang M, Herman D, Ozcan A, Altug H. Handheld high-throughput plasmonic biosensor using computational on-chip imaging. Light Sci Appl. 2014;3:e122.

[111] Dutta S, Saikia K, Nath P. Smartphone based LSPR sensing platform for bio-conjugation detection and quantification. RSC Adv. 2016;6:21871.

[112] Hiep HM, Endo T, Kerman K, Chikae M, Kim DK, Yamamura S, Takamura Y, Tamiya E. A localized surface plasmon resonance based immunosensor for the detection of casein in milk. Sci Technol Adv Mater. 2007;8:331.

[113] Strohmeier O, Keller M, Schwemmer F, Zehnle S, Mark D, von Stetten F, Zengerle R, Paust N. Centrifugal microfluidic platforms: advanced unit operations and applications. Chem Soc Rev. 2015;44:6187.

Roy Aad, Suzanna Akil, and Safi Jradi

3.2 Chemical nanosensors for the detection of explosives

Explosives detection has become a major societal issue over the past few years. Accordingly, the fabrication of selective ultrasensitive nanosensors able to detect small traces of explosives molecules is a great challenge for the scientific community. This chapter provides an overview of the progress made in the development of chemical nanosensors and their use in explosives detection. The chapter mainly focuses on the advances in nanospectroscopic methods applied for the detection of 2,4,6-trinitrotoluene (TNT) and 2,4-dinitrotoluene (DNT) vapor traces. More particularly, the use of fluorescent polymers and surface-enhanced Raman spectroscopy (SERS) in this field is developed. Examples of explosives detector devices which have been spurred by nanospectroscopic techniques are given.

3.2.1 Key messages

- Conventional techniques that are used for explosives detection, such as X-ray screening and landmine detectors, do not detect explosive traces but rather rely on shape, object density, and the use of metallic casings to identify possible explosive materials. With recent advances in explosives, these techniques have been rendered obsolete.
- Explosive trace detection techniques are being researched and developed. Currently, ion mass spectrometry has gained wide acceptance as an explosive trace detector and is already in use in major US and European airports. Ion mass spectrometry however is not suitable for on-site explosive trace detection. In that regard, chemosensors, such as SERS substrates and fluorescent sensing polymers (FSPs), have shown to be promising candidates for explosives trace detection.
- SERS substrates can detect explosive traces down to a concentration of few attomoles; however, due to lack of standardization, SERS substrates are still not widely accepted as a technique for explosives detection, and no commercialized explosive trace detectors which use this technique exist at this time.
- FSPs can detect explosive traces down to a concentration of few femtomoles. SpectraFluidics is currently commercializing explosive trace detectors which use this technique under the brand name Fido X.

Acknowledgement: Financial support of the "Fonds Européen de Développement Régional (FEDER)" for the project "Nanoassemblage 3D," NanoMat (www.nanomat.eu) by the "Ministère de l'enseignement supérieur et de la recherche", and the COST action MP1302 Nanospectroscopy is acknowledged.

https://doi.org/10.1515/9783110442908-008

3.2.2 Pre-knowledge

The current chapter focuses on the application of optical nanospectroscopic techniques for the detection of explosives. It is therefore vital for the reader to be familiar with the basics of fluorescence and Raman spectroscopy (Volume 1, Section 2). A background is required in the fabrication and characterization of plasmonic nanostructures (Volume 2, Section 6) and their use in surface-enhancement Raman spectroscopy (Volume 1, Section 3).

3.2.3 Introduction

According to a study by the UK-NGO Action on Armed Violence [1], in the decade from 2011 to 2020 more than 12,000 improvised explosive device (IED) events occurred in at least 100 countries, causing more than 135,000 civilian casualties. Numbers initially increased drastically, reaching more than 22,000 casualties in 2013, and then decreased gradually, levelling off in 2020/21 at just under 5,000. IEDs account for 52 % of civilian injuries and deaths by explosives overall. Of all incidents with IEDs, 60 % occurred in populated areas, and 90 % of the casualties in these areas were civilian. This global and drastic development in the use of IEDs comes as no surprise as the know-how, instruments, and techniques to synthesize certain explosives, such as the most commonly used ammonium nitrate and fuel oil (ANFO) explosive mixture (a. k. a. fertilizer bomb), are rather basic compared to current technological standards, and the materials used to synthesize them are easily accessible and purchasable [2]. Thus, it is clear why explosives detection has become a major societal issue over the past few years [3].

Importance of nanospectroscopy methods for the detection of explosives

The most common techniques in use nowadays for the detection of explosives are X-ray security screening systems, which are widely used in airports and attraction centers, and landmine detectors (i. e., metal detectors), which are mainly used by armies to detect buried landmines. These techniques, however, present a large number of false positives as they lack the required selectivity and accuracy for effective explosives detection. X-ray screening systems, for example, rely on shape and object density to identify explosives hidden in bags and suitcases. Landmine detectors, on the other hand, rely on the metallic nature of the landmine casing to detect buried mines. None of these features however, whether it is the shape and object density or the metallic casing, can be considered as a distinct feature, or in other words a fingerprint, of explosive compounds. The development of plastic chemistry and the success and dominance of plastics in the early twentieth century eventually led to the development of plastic explosives. This development made the detection of explosives using X-ray screening and conventional landmine detectors a difficult, if not impossible, task. Plastic explosives, which can be cleverly disguised as they can be molded into almost any shape and possess similar density as other plastic components, are hard to identify with X-ray screening systems. Nonmetallic mines, on the other hand, which contain no metal whatsoever, cannot even be found using conventional landmine (i. e., metal) detectors.

The failure of conventional techniques to comply with recent advances in explosives technology has sparked a great interest to research and develop more accurate explosives and complementary detection techniques to counter the rising illicit use of IEDs. To achieve higher accuracy, a great number of these techniques, whether stand-off or contact techniques, use vibrational or mass spectroscopy to detect explosives, as they offer a full insight into the chemical composition of the probed chemical compounds, allowing for an accurate detection of any explosive trace. In addition, with recent development in chemosensors, fluorescence/luminescence spectroscopy has shown to be an equally interesting and promising technique for explosives trace detection.

Explosives detection techniques are usually classified into two categories, non-contact stand-off techniques and near-field/contact techniques, according to the detection distance. Stand-off detection techniques are, in principle, techniques that are capable of detecting and identifying threats at a long-range distance, often between 5 and 25 m, which is, of course, an important issue when it comes to explosives. These techniques mainly consist of sending electromagnetic waves to a suspected area and afterwards recovering and analyzing the scattered signal in order to determine the presence/absence of explosives. Such stand-off techniques include Raman light detection and ranging (Raman-LIDAR), laser-induced breakdown spectroscopy (LIBS), coherent anti-Stokes Raman spectroscopy (CARS), and terahertz (THz) spectroscopy. However, the detection of explosives at a distance is a complicated issue. Stand-off techniques often require the use of high-intensity lasers and a laborious analysis of the detected signal, in order to overcome the atmospheric attenuation and noise and achieve the required sensitivity. As a consequence, stand-off explosives detection devices are extremely bulky, which requires them to be mounted on on-road vehicles for on-site explosives detection. Moreover, while some studies have reported on environmental pollution monitoring using vehicle-mounted LIDAR systems [4], stand-off systems, in general, fail to be accepted as effective techniques for explosives detection.

On the other hand, near-field/contact detection techniques, as their name suggests, are techniques that detect explosive traces by contacting, or more practically by being in the vicinity, of the suspected sample or area. Compared to stand-off techniques, contact techniques are smaller, cheaper, simpler, and more promising, with some techniques already in use in airports and by armies around the world. Near-field/contact techniques include mass spectroscopy, ion mass spectroscopy (IMS), fluorescence/luminescence spectroscopy, Fourier transform infrared spectroscopy, and SERS. With contact detection techniques proving to be more useful, analytical contact detection techniques, especially IMS, have gained wide acceptance as explosive trace detectors, with security companies, such as Thermo Scientific, Smiths Group plc, and Safran S. A. already supplying major US and European airports with more and more IMS screening systems over the past few years [5]. While capable of detecting explosives down to the picogram in a few seconds, IMS systems, nevertheless, do not offer a solution for on-site explosives detection, such as landmine detection. Popular IMS systems, such as IONSCAN 500DT, detect explosives traces by wiping a swab sampler over the items and then inserting the swab sampler into a slot where the sampler will be heated in order to release and, afterwards, analyze vapor traces that were caught by the sampler. The use of a swab sampler makes these technologies not suitable for many explosives detection scenarios, since direct vapor trace detection is required; therefore, their use remains limited to cargo and passenger screening in airports. Moreover, the IONSCAN 500DT, as an example, has dimensions of $40 \times 31 \times 40\,cm^3$, a weight of 19 kg, and a price in the ten-thousands USD. IMS systems that detect direct explosive vapor traces are usually gigantic walk-through portals, such as the EntryScan 4 [6], equipped with a venting system or a self-contained air collection system in order to collect the ambient air surrounding a passenger or a package for explosives trace detection. The EntryScan systems, as an example, have dimensions of $235 \times 138 \times 102\,cm^3$, a weight of 341 kg, and a price of over 100,000 USD.

Recent research focused on optical chemosensing techniques in an effort to find an answer to the ever-growing need for amenable and affordable explosives detection techniques. Optical chemosensing devices are mainly composed of a laser, a transducer (i. e., chemosensor), and a spectrometer. With recent advances in diode lasers and mini-spectrometers, optical spectroscopy setups can be drastically reduced in size and cost. The chemosensor, however, is the most important part of the device. The chemosensor is a material that reacts to the presence of a chemical compound by changing its optical (e. g., fluorescence, reflectance, etc.) or electrical (i. e., conductance properties, impedance, etc.) properties.

Studies on explosives chemosensing are usually realized in solution state, by dissolving explosives in a solvent, or in gas state, and detecting the generated vapor traces. Vapor trace detection is, of course, favored for more practical uses. The detection of explosives vapor traces, however, is not an easy task. Explosives are majorly solid-state compounds and, therefore, exhibit very low vapor pressures (Fig. 3.2.1).

TNT, for instance, exhibits a very low equilibrium vapor pressure of 8.02×10^{-6} mmHg (i. e., 5–10 ppb) at 25 °C [8]. In the head space over a landmine, the TNT concentration is of course an order of magnitude lower than the saturated vapor pressure. In other words, for a real-time explosives vapor trace detection, the chemosensor needs to be sensitive to vapor concentrations 10 times smaller than the saturated

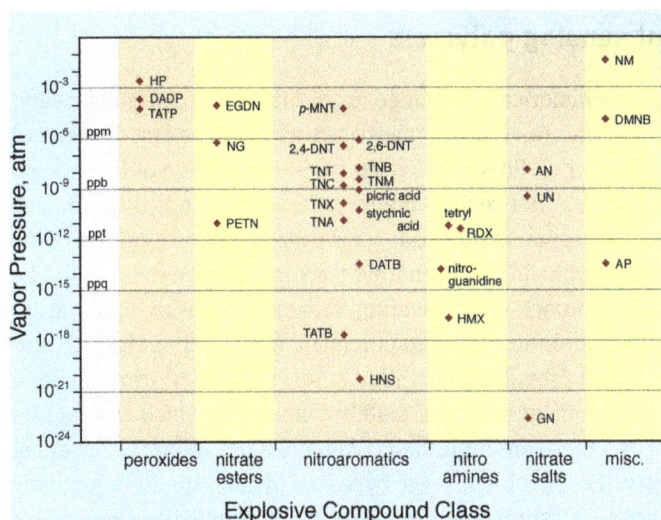

Figure 3.2.1: Vapor pressure of different explosive compound classes. Reprinted from [7]. Copyright 2013, with permission from Elsevier.

vapor concentration of the targeted explosive. In this regard, nanomaterials have received major interest due to their high surface-to-volume ratio, as well as their high surface energy. These characteristics enable their surface functionalization with biological/chemical molecules, creating recognition events upon interaction with target molecules or compounds that cause measurable physical changes. Chemosensors can thus be nanoengineered to achieve the desired sensitivity and chemically engineered to achieve the desired selectivity. Optical chemosensing, thus, paves the way for an explosives trace detection system that can be functional, fast, and accurate while simultaneously being portable and affordable. In the following sections, we focus on two optical chemosensing techniques: FSPs and SERS.

3.2.4 State-of-the-art of the nanospectroscopy methods used for explosives detection

We focus on FSP and SERS as they have proven to be very promising with prototypes already being tested and products already are being sold. For the sake of simplicity and clarity, the upcoming sections will solely focus on the detection of TNT and DNT, as they compose the majority of the studies. DNT is a natural breakdown product of TNT. Due to its high vapor pressure (25 times higher than that of TNT), DNT vapor will dominate the headspace over stored and left-over TNT. DNT is thus a predominant and important marker of TNT.

3.2.4.1 Fluorescent sensing polymers

Fluorescent indicators can be historically regarded as the first fluorescent sensors and the stepping stone for today's fluorescent polymer-based sensors. The first fluorescent indicator dates back to 1876, when Friedrich Krüger proposed the use of fluorescein for the detection of free acid [9]. In a general manner, fluorescent indicators consist of two chemical moieties: "a receptor responsible for the molecular recognition of the analyte and a fluorophore responsible for signaling the recognition event" [10]. The easiest and most common approach for fabricating fluorescent sensing material involves the casting of fluorescent indicators into polymeric matrices in order to cause their physical entrapment and provide them with a solid support [11]. Nonetheless, the approach suffers from inhomogeneity and stability problems, which reduce lifetime and reproducibility and therefore limit their commercial use. With the advancement of polymer chemistry, these problems were bypassed by the exploitation of new polymerization techniques (e. g., click polymerization [12]) that allowed the synthesis of functionalized fluorescent polymers, which can be easily covalently attached into polymeric matrices, leading to a stable fluorescent sensing material. Fluorescent polymers thus emerged as a new trend of chemical sensors which have been receiving lots of interest due to their many advantageous features and numerous applications. One of the most successful examples of such fluorescent polymeric sensors is probably that of the fluoroionophore currently commercialized in the Roche OPTICCA portable blood optical analyzer [13].

Fluorescent polymer solutions and thin films have been recently applied to detect explosives. The detection method mainly consists on monitoring quenching (i. e., intensity decrease) of the polymer fluorescence in the presence of nitroaromatic compounds, such as TNT [14]. The quenching mechanism is attributed to an electron transfer (Fig. 3.2.2) between the electron-rich polymer and electron-deficient nitroaromatic compounds that effectively deactivates the polymer excitation. Upon optical excitation, an electron from a ground state of the polymer (e. g., valence band) is promoted to a higher-energy excited state (e. g., conduction band). If the fluorescent polymer is free of analytes, the excited electron will recombine to its initial ground state by releasing a photon. However, if an electron-deficient analyte is present in the vicinity of the fluorescent polymer, it acts as an electron trap which can spontaneously capture the electron from the excited state of the polymer [14]. Thus, an electron transfer occurs from the excited state of the polymer towards the lowest unoccupied molecular orbital (LUMO) of the analyte, followed by a nonradiative reverse electron transfer from the quencher LUMO state to the polymer ground state [14]. As a consequence, the analyte presence enhances the rate of nonradiative recombination, leading to quenching of the polymer fluorescence.

Fluorescence quenching may arise from either a static or a dynamic process. The static process is characterized by a strong polymer–analyte bonding, which is ensured by strong electrostatic interactions and the formation of a polymer–analyte bound

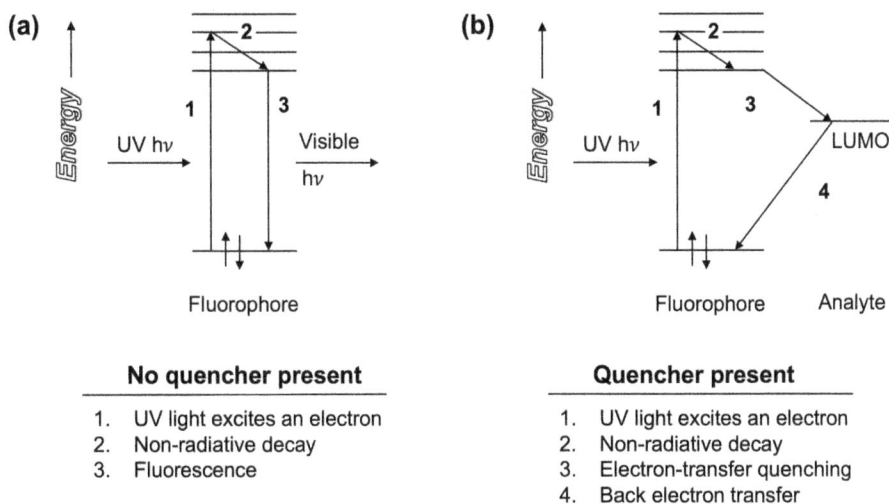

(a) No quencher present
1. UV light excites an electron
2. Non-radiative decay
3. Fluorescence

(b) Quencher present
1. UV light excites an electron
2. Non-radiative decay
3. Electron-transfer quenching
4. Back electron transfer

Figure 3.2.2: (a) Illustration of the fluorescence process of a polymer. (b) Illustration of the fluorescence quenching of a polymer in the presence of an analyte. Adapted from [14] with permission from The Royal Society of Chemistry.

complex. On the other hand, the dynamic process is characterized by a weak bonding, where the polymer–analyte interaction is purely collisional. The kinetics of the static and dynamic quenching process can be described by the Stern–Volmer relationship. In a Stern–Volmer plot (i. e., ratio of emission intensities I_0/I vs. quencher concentration), the Stern–Volmer relationship is the equation of a straight line with a slope that is usually denoted as K_{SV} (M^{-1}) known as the Stern–Volmer constant. The Stern–Volmer constant is commonly used for fluorescent polymers in solution to indicate the strength of interaction, i. e., the sensitivity to quenchers. High Stern–Volmer constants indicate a high sensitivity to quenchers, while low Stern–Volmer constants indicate a low sensitivity to quenchers.

Among existing fluorescent polymers, conjugated polymers are the most promising for sensing applications, due to their semiconductor properties. The semiconductive nature of conjugated systems results from the alternating single and double bonds (Fig. 3.2.3), which creates a delocalized electronic structure (i. e., energy bands) through a continuous overlapping of π orbitals along the polymer backbone (i. e., a semiconductive "molecular wire") [15]. A classical chemosensor approach (Fig. 3.2.4a) usually consists of individual receptor-fluorophore molecules (i. e., nonconjugated recognition sites), where a single analyte molecule can only quench one fluorophore. However, in the case of conjugated polymers, the receptor sites are "molecularly wired in series" (Fig. 3.2.4b), through the polymer backbone. This "wiring" results in a "photoinduced electron transfer quenching" mechanism (Fig. 3.2.4c), which leads to a self-amplified ultrahigh sensitivity. In other words, the quenching of a fluorophore in a conjugated polymer does not necessarily involve the direct binding of analyte

Resonance   **Electron delocalization**

Aromatric ring

Conjugated double bonds

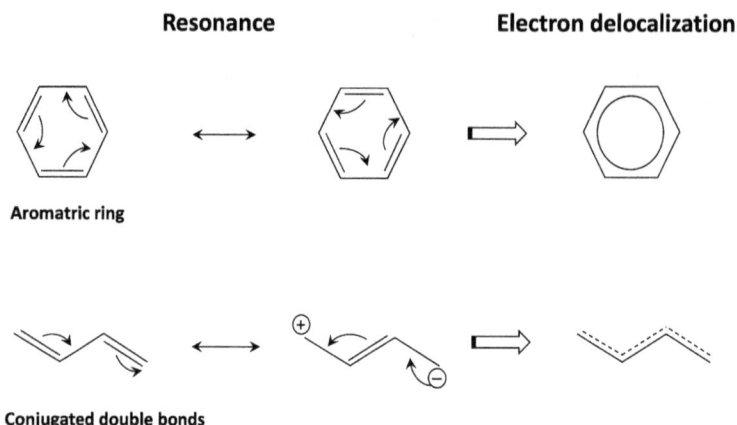

Figure 3.2.3: Comparative schematics showing the analogy between delocalized electrons in aromatic rings and conjugated bonds.

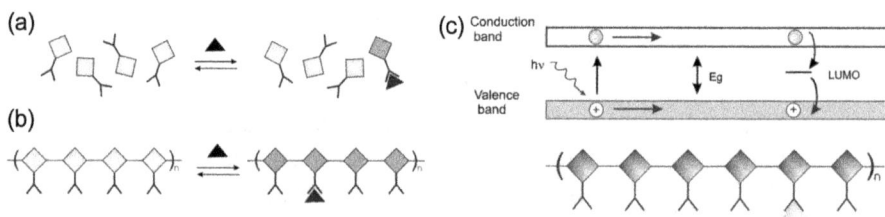

(a)

(c)

(b)

Figure 3.2.4: (a) Schematic illustration of a situation of partial quenching where individual receptor molecules interact with a small quantity of analyte acting as a quencher. (b) Schematic illustration of a conjugated polymer interacting with a small quantity of analyte resulting in a complete quenching of the polymer chain. (c) Band diagram illustrating the "photoinduced electron transfer quenching mechanism" through energy migration in a conjugated polymer exposed to a single analyte. Adapted from [10] with permission from The Royal Society of Chemistry.

to its related receptor site. In delocalized electronic structures, the fluorescence of a fluorophore can be quenched by analytes bound to other, even distant, receptor sites on the polymer chain. This collective behavior is emphasized by the easy energy migration (i. e., exciton transport) along the polymer backbone over large distances (Fig. 3.2.4c). Thus, in conjugated systems, a single analyte molecule can quench a large number of fluorophores, leading to an amplified quenching signal.

Conjugated polymers for TNT sensing applications have been studied and developed by the groups of Swager and Trogler. In 2003, the Trogler group, of the University of California at San Diego, reported on photoinduced amplified quenching properties of poly(tetraphenyl-1,1-silole) (Fig. 3.2.5) in solution in the presence of TNT [16]. Poly(tetraphenyl-1,1-silole) is a polysilole copolymer that has a Si-Si backbone which induces the delocalization of σ electrons along the main chain. To study the quenching

Green light out

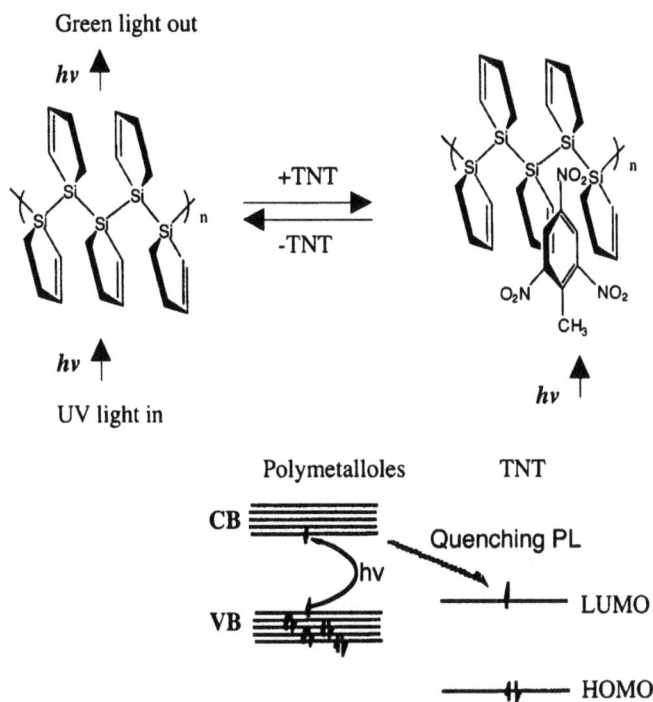

Figure 3.2.5: Diagram of an electron transfer process leading to photoluminescence quenching from polymetallole in the presence of TNT under UV light. Reprinted with permission from [16]. Copyright © 2003 American Chemical Society.

response, TNT was successively added to a tetrahydrofuran (THF) solution containing a concentration of the copolymer of 2×10^{-6} M. The THF did not interfere with the copolymer fluorescence, which presented a high K_{SV} of 4340 M^{-1} for TNT, the highest K_{SV} reported for a fluorescence-quenching polymer in solution. However, the copolymer did not perform as well in the solid state for TNT vapor detection, due to aggregation and self-quenching issues.

On the other hand, Swager's group, of the Massachusetts Institute of Technology, focused on the development of poly(phenyleneethynylene) (PPE)-based polymer thin films for the detection of TNT vapor. In solution state (i. e., in organic solvent), the polymers investigated by Swager's group presented much lower K_{SV} values to the presence of explosives, compared to the polysilole studied by the Trogler group. However, the chemical structure design of the polymers studied by Swager's group prevented their aggregation and self-quenching in the solid state (i. e., thin films), which allowed for superior vapor sensing capabilities.

In 1998, Yang and Swager studied the quenching response to various nitroaromatic analytes of a 25-Å pentiptycene-derived polymer thin film spin-cast onto a glass substrate [17]. The incorporation of the rigid pentiptycene moiety into the polymer

(a)

Cavities

Pentiptycene Groups

Polymer Backbone

TNT

(1)

(b)

Fluorescence Intensity

Quenching (%)

Time (s)

Wavelength (nm)

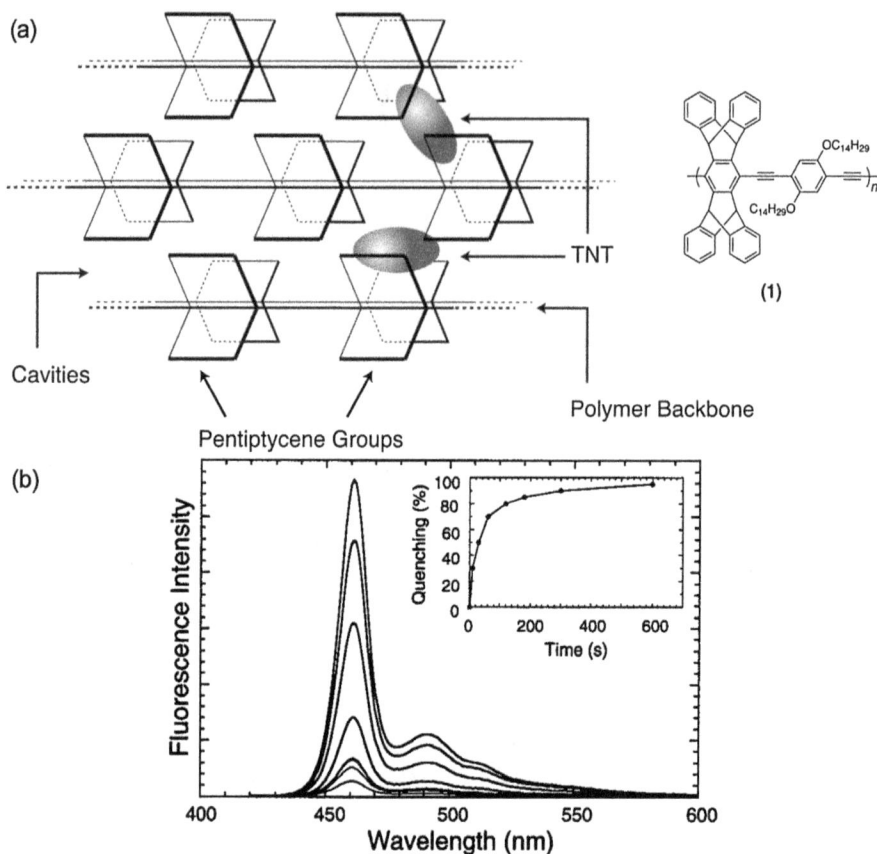

Figure 3.2.6: (a) Pentiptycene groups preventing the π-stacking of the polymer backbone. The polymer chemical structure is shown on the right [18]. (b) The fluorescence intensity measured at room temperature of 25-Å PPE film at different times of exposure to TNT vapor. The exposure time varies from 0 s (top) to 600 s (bottom). The inset shows the percentage of fluorescence quenching as a function of time. Reprinted with permission from [17]. Copyright © 1998 American Chemical Society.

backbone prevented π-stacking or excimer formation (Fig. 3.2.6a), which could lead to self-quenching of the polymer chains in solid state. To ascertain the quenching response, the PPE thin film was inserted into sealed vials at room temperature that contained solid analyte (e. g., DNT, TNT) crystals. Upon exposure to saturated TNT vapor (10 ppb at 25 °C), the PPE thin film fluorescence reached 75 % quenching after 60 seconds (Fig. 3.2.6b).

The film thickness of 25 Å in Yang and Swager's study is of extreme importance to the high quenching efficiency they obtained. The dense nature of polymer matrices can prevent the rapid diffusion of analytes into the polymeric matrix, leading to low quenching efficiencies (Fig. 3.2.7). In fact, the best reported quenching efficiencies were obtained with porous nanolayers that are 25 Å thin [19, 20].

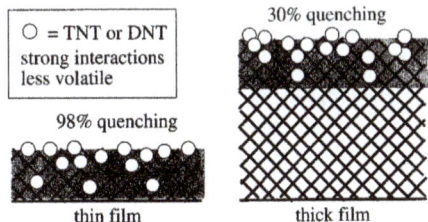

Figure 3.2.7: A schematic illustration of the film thickness effect. Reprinted with permission from [17]. Copyright © 1998 American Chemical Society.

Nano- and microporous polymeric matrices can further enhance analyte diffusion leading to an enhanced quenching process. In 2002, Zahn and Swager further enhanced the fluorescence quenching using a triphenylene-containing PPE [20]. The helical grid structure of triphenylene-based PPE in its aggregated (solid) state (Fig. 3.2.8) increased its excited-state lifetime and quantum yield. Under similar experimental conditions as [17], the triphenylene-based 25-Å PPE film showed 75 % fluorescence quenching after 10 s of exposure to TNT vapor, which corresponds to a 4-fold increase in sensitivity [20].

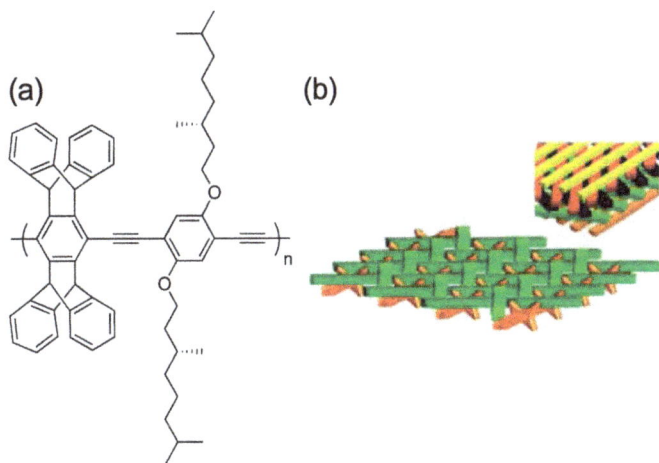

Figure 3.2.8: (a) Chemical structure of the triphenylene-based PPE. Reprinted with permission from [21]. Copyright © 2007, American Chemical Society. (b) Helical grid structure of the PPE thin film. Copyright 2002 Wiley. Used with permission from [20].

One of the problems of fluorescent polymer-based chemosensors is their limited selectivity. Effectively, fluorescent polymers can quench a wide variety of explosive and nonexplosive electron-withdrawing compounds (e. g., TNT, DNT, nitrobenzene, picric acid, NO$_2$, etc.). In [16], Trogler noted that "because the reduction potential of TNT

(−0.7 V vs. NHE (normal hydrogen electrode)) is less negative than that of either DNT (−0.9 V vs. NHE) or nitrobenzene (−1.15 V vs. NHE), it is detected with highest sensitivity." Thus, the polymer sensitivity (i. e., electron transfer) is strictly linked to the energy gap separating the polymer excited state and the analyte LUMO state. The electron transfer preferably occurs from the excited polymer state towards the highest analyte LUMO state. Che et al. increased the selectivity and sensitivity of TNT detection using a carbazole-based tetracycle (Fig. 3.2.9a) [19]. In a reverse strategy, Che introduced a carboxyl group into the side chain linker in order to lower the energy of the highest occupied molecular orbital (HOMO) state (i. e., excited state) of the studied polymer. The lowering of the HOMO-state energy improved the polymer stability towards oxygen, which enhanced the polymer selectivity and sensitivity to TNT (Fig. 3.2.9b). The polymer exhibited 75 % of fluorescence quenching after 10 s of exposure to TNT saturated vapor. However, no fluorescence quenching was observed in the presence of other oxidizing reagent vapors. The polymer fluorescence quenching was significant even at ppt concentrations of TNT (Fig. 3.2.9c).

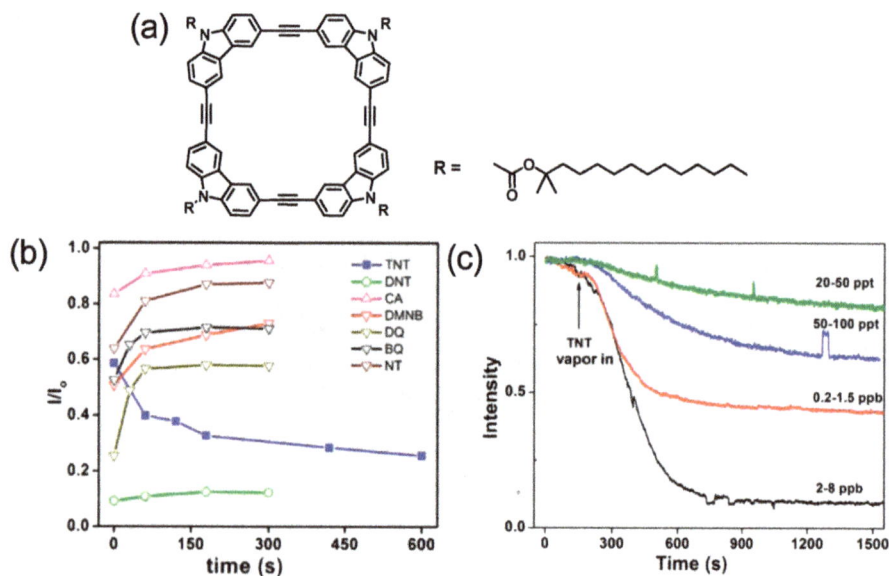

Figure 3.2.9: (a) Chemical structure of the carbazole-based tetracycle. (b) Fluorescence intensity (I_0/I) of the tetracycle showing a selective quenching to TNT. (c) Fluorescence quenching of the tetracycle when exposed to various TNT concentrations. Adapted with permission from [19]. Copyright © 2012 American Chemical Society.

Eventually, Nomadics, Inc. developed and tested a portable landmine detector (Fig. 3.2.10) based on the PPE thin films studied by Swager [22, 23]. The developed detector was reported to detect femtogram quantities of TNT (56 fg/mL, 6 ppt) in the

Figure 3.2.10: Photo of a soldier field-testing a portable landmine detector based on self-amplifying polymers. Reprinted with permission from Springer Nature: Springer MRS BULLETIN [18] Self-Amplifying Semiconducting Polymers for Chemical Sensors, Timothy M. Swager and Jordan H. Wosnick, Copyright © 2011, The Materials Research Society.

air in near real-time (1 s exposure). Field studies showed that this detector is more likely to detect a landmine and less likely to give a false detection than a TNT-sniffing dog team (currently the most used method for explosives detection in the United States). In addition, Teledyne FLIR (formerly FLIR Systems) is advertising explosives detection devices based on the amplified fluorescent polymers developed in Swager's group at MIT. According to the company website, "The FLIR Fido® X products [24] are the only commercially available handheld explosives trace detectors (ETD) to use TrueTrace technology" and "is the lightest and most sensitive handheld explosives trace detector in its class." The Fido X3, for example, has dimensions of $37 \times 12 \times 7$ cm^3, a weight of 1.36 kg, and a detection time of 10 s (limits of detection and price are not available). Moreover, Fido X3 has two detection modes: trace mode, where detection is performed via a sampler, and vapor mode, where detection is performed via direct vapor emission.

3.2.4.2 Surface-enhanced Raman spectroscopy

Most explosive compounds exhibit extremely low Raman scattering cross-sections. TNT, for example, is Raman-inactive and shows no spectral features in conventional Raman measurements. Explosive compounds are therefore impossible to detect using conventional Raman spectroscopy. SERS offers an enhanced light–matter interaction, which overcomes the low scattering cross-sections and allows the detection of explosives down to practical detection concentrations.

The first use of SERS for explosives trace detection dates back to 1995, when Kneipp et al. detected TNT at the $100 \, \mathrm{nmol.L^{-1}}$ (1 pg) level using gold and silver colloidal suspensions [25]. Shortly thereafter, Spencer and colleagues demonstrated the detection of DNT vapor traces down to the 5 ppb level using a flow-through probe head (Fig. 3.2.11) design and electrochemically etched gold and a principal component analysis (PCA) algorithm to improve the limits of detection [26, 27].

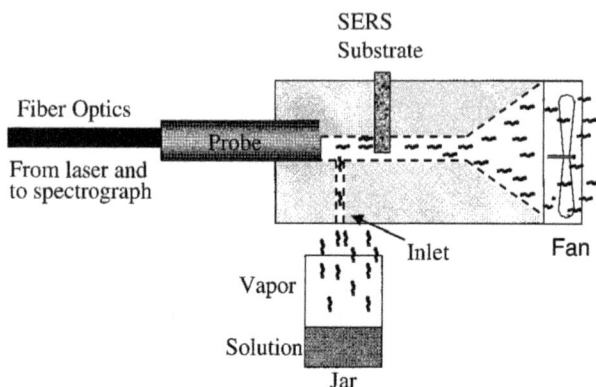

Figure 3.2.11: Illustration of the SERS adapter sleeve designed for detection of vapor signals. The stainless steel sleeve slides over the fiber-optic Raman probe. A fan draws the vapor through the sleeve at a rate of 1.2 L/s. The gold SERS substrate is positioned at the focal point of the Raman probe and partially blocks the air flow, creating turbulent flow at the substrate face. Figure and figure caption reprinted with permission from [27]. Copyright © 2000 American Chemical Society.

Within a year, a field-portable device based on the design proposed by Spencer and colleagues was subjected to field tests and proved successfully to have the ability to locate buried landmines [28].

Despite these encouraging results, SERS use for explosives detection witnessed a slow growth period, where for about 5 years (2004 to 2009) only single-digit numbers of studies on explosives detection using SERS were reported. However, SERS use for explosives detection saw renewed interest, with more than 200 studies published since 2010. The SERS studies investigated the detection of various explosive substances using a number of methodologies and substrates. Hakonen et al. [29] published a comprehensive review in 2015 focusing on SERS techniques and substrates for explosive and chemical threat detection. In his review, Hakonen compiled the numerous studies into a table that gives a full insight into the various SERS substrates that were used and the sensitivity toward TNT and other explosives that were attained. Table 3.2.1 is largely inspired by Table 4 presented in [29]. For simplicity however, Table 3.2.1 only considers publications on TNT and DNT.

Table 3.2.1: SERS substrates for TNT and DNT demonstrated to date.

SERS substrate	Sensitivity TNT (M)	Sensitivity DNT (M)	Reference
Klarite™	20 pg		[30]
Klarite™	20 pg		[31]
Klarite™ MIPS	3×10^{-6}		[32]
Ag NPs		3×10^{-7}	[33]
Ag NPs	10^{-10}		[34]
Ag NPs	4×10^{-10}		[35]
Ag NPs	10^{-12}	10^{-8}	[36]
Ag NPs in agarose	10^{-8}		[37]
Ag NPs on carbon nanotube grid	10^{-3}		[38]
Ag NPs on Q-tip		5 ng	[39]
PABT complex Ag NPs	1.5×10^{-17}		[40]
μ-LC and Ag Qdots		3×10^{-9}	[41]
Colloidal Ag NPs on paper	94 pg	7.8 pg	[42]
Ag-Fe$_3$O$_4$ nanohybrid	10^{-8}		[43]
M-comp 4-ATP-AgNP	5×10^{-9}	10^{-7}	[44]
TNT-SO$_3$-CP Ag NP	5×10^{-11}		[45]
Au nanostars on Ag film		10^{-17}	[46]
Au NPs		10 nmol	[48]
Au NPs	10^{-6}		[49]
Au NP polymers		0.4 ag (vapor)	[50]
Au NP array	10^{-9}		[51]
Au NPs in porous alumina	3×10^{-16}	5×10^{-16}	[52]
Fe$_3$O$_4$/AuNPs	10^{-12}		[53]
Peptides-Au nanorods	10^{-10}		[54]
Au nano bowtie	7×10^{-6}		[55]
Au nano sapphire		1.1×10^{-7} (vapor)	[56]
Au nanopopcorn	10^{-13}		[57]
Au nanoprism β-CD		10^{-11}	[58]
Ni/Au nanocarpet	10^{-7}	10^{-7}	[59]
Roughened Au Foil		5 ppb (vapor)	[27]
M-comp L-cys-AuNP	10^{-12}		[60]
M-comp L-cys-AuNP	10^{-9}		[61]
M-comp L-cys-AuNP	2×10^{-12}		[62]
M-comp cysteamine-AuNP	10^{-10}		[63]
Graphene oxide-Au nanocage	10^{-14}		[64]
GNP graphene M-comp	10^{-11}		[65]
Microfluidic NPs		5.5×10^{-12} (vapor)	[66]
Nanoclusters		10 ppt (vapor)	[67]
Cu laser ablation nanocavities	10^{-4}		[68]

From Table 3.2.1, we can conclude that there is not a clear idea on what type of SERS substrate is best for explosives detection. Research on SERS for explosives detection is rather governed by trial and error than by a well-set research path, mainly due to the lack of knowledge on the various mechanisms behind the SERS effect and the weight of their contributions to the final surface-enhanced Raman signal. Numerous materials (Au, Ag, Cu, graphene, etc.) and geometries (nanoparticles (NPs), nanostars, nanorods, bowtie, etc.) were thus investigated. Nevertheless, we can realize that, amongst the investigated SERS substrates, Au and AgNPs are most popular.

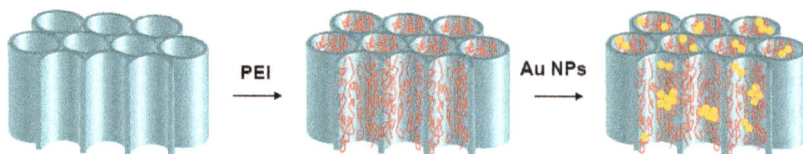

PEI   Au NPs

Figure 3.2.12: Fabrication procedures for 3D nanoporous membranes decorated with aggregated gold nanoparticle clusters. The surfaces of porous alumina membranes are functionalized with positively charged amine groups by modification with PEI, and CTAB-capped gold colloids are then passed through the PEI-modified porous alumina membranes resulting in the formation of aggregated gold nanoparticle clusters inside the pores. Figure and figure caption reprinted with permission from [52]. Copyright © 2009, American Chemical Society.

In 2009, Ko et al. [52] reported on label-free molecular level detection of TNT and DNT. The SERS substrate investigated in [52] consisted on a 3D nanoporous alumina membrane decorated with aggregated gold NP clusters (Fig. 3.2.12). The surface of the nanoporous membrane was modified with positively charged polyethylenimine in order to prevent the attachment of gold NPs on the outer surface, resulting in the formation of aggregated gold clusters on the inner walls of the membrane by partial replacement with the cetyltrimethylammonium bromide ligands on the gold NP surface. Ko et al. drop-evaporated TNT and DNT solutions in ethanol and, using the NO_2 stretching mode at $1342\,cm^{-1}$, repeatedly detected TNT and DNT levels down to 5–10 zeptograms (3×10^{-16} to 5×10^{-16} M), or about 15–30 molecules,[1] respectively. Ko et al. attributed this exceptional Raman enhancement to (1) "hotspots" in the vicinity of gold NP clusters, (2) the larger specific surface area of the porous alumina membrane, and (3) an optical waveguide effect caused by multiple total internal reflections at the pore air–alumina interface [52].

In 2011, Zhou et al. [40] reported on the possibly most sensitive SERS substrate to date with a TNT limit of detection of 1.5×10^{-17} M.[2] However, in order to achieve

1 The Raman spectra were recorded with laser excitation at a wavelength of 784 nm, a power of 20 mW, and an integration time of 20 s.

2 The maximum Raman intensity was recorded at ~100 s after the addition of the liquid droplet using a laser power of 1 mW. The integration time was not mentioned.

this high sensitivity, Zhou et al. detected the presence of TNT molecules via *p*-amino-benzenethiol (PABT), which can form a charge transfer complex with TNT. Thus, instead of using the Raman signal of the NO$_2$ stretching mode at 1342 cm^{-1}, Zhou et al. used the Raman signal of PABT at 1434 cm^{-1} to detect TNT. The SERS substrate consisted of free-standing top-closed flexible silver nanotubes, whose surface was modified with PABT. When the TNT/ethanol solution was drop-evaporated onto the SERS substrate, the PABT-TNT complexes formed PABT-TNT-PABT molecular bridges leading to the self-approach (i. e., bending) of the flexible silver nanotubes (Fig. 3.2.13). The self-approached silver nanotubes present Raman "hotspots" at the intimate interspace between the nanotubes. In addition, while TNT and PABT molecules exhibited no absorption at 532 nm, the extended conjugation system of PABT-TNT-PABT absorbed light at 532 nm, making PABT resonant with both the laser line and the surface plasmon resonance of the silver nanotubes. The modified PABT absorption, shifting from an off-resonant to an on-resonant state, coupled with the "hotspots" caused by self-approached silver nanotubes, strongly enhanced the Raman signal of PABT, leading to a very high enhancement factor of ~1 × 10^8, very close to that of resonant dyes.

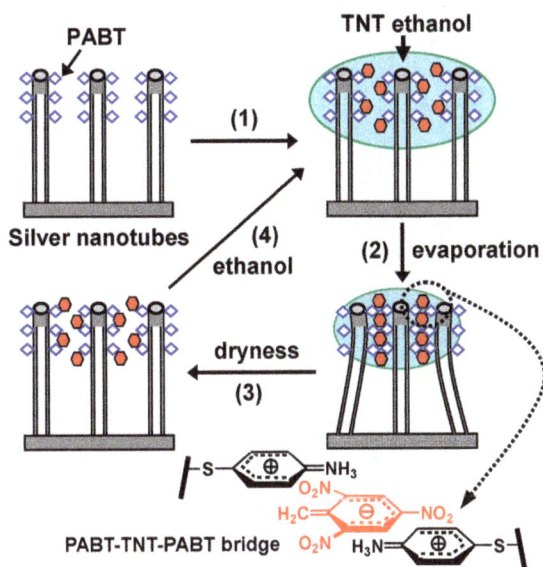

Figure 3.2.13: Mechanism for TNT-induced resonance Raman enhancement of PABT on the top-closed silver nanotube array. Figure and figure caption reprinted with permission from [40]. Copyright © 2011, American Chemical Society.

While the sensitivity reported by Zhou et al. [40] is 20 times higher than that reported by Ko et al. [52], the approach suffers from a drawback. One of the main strong points of Raman spectroscopy is that it is a label-free technique. Due to the strong PABT selectivity to TNT, the SERS substrate studied by Zhou et al. [40] exhibited no Raman enhancement when tested with structurally similar molecules. In other words, the SERS substrate in [40] only functioned with TNT. On the other hand, the SERS substrate tested by Ko et al. was label-free and could be applied to detect TNT, DNT, and other explosive molecules.

In 2014, Lee et al. [46] reported the detection of attomole levels of DNT. Lee et al. investigated the sensitivity of a high-density gold nanostar (GNS) assembly (Fig. 3.2.14). The GNSs were electrostatically assembled on top of a 200-nm silver film, separated by a 1-nm layer of poly(diallyl dimethylammonium). The authors used the GNS assembly surface density to tailor the interplay between interparticle (GNS–GNS) and particle–film (GNS–Ag film) plasmon coupling to provide maximum SERS effects. With a GNS surface density of ~43 GNSs per μm^2, the authors achieved a SERS enhancement factor of 4.4×10^7 and were able to observe the 834 cm^{-1} (NO_2 out-of-plane bending) and 1355 cm^{-1} (NO_2 symmetric stretching) Raman peaks of DNT after immersion in a DNT–methanol solution with a concentration of 10 aM.[3]

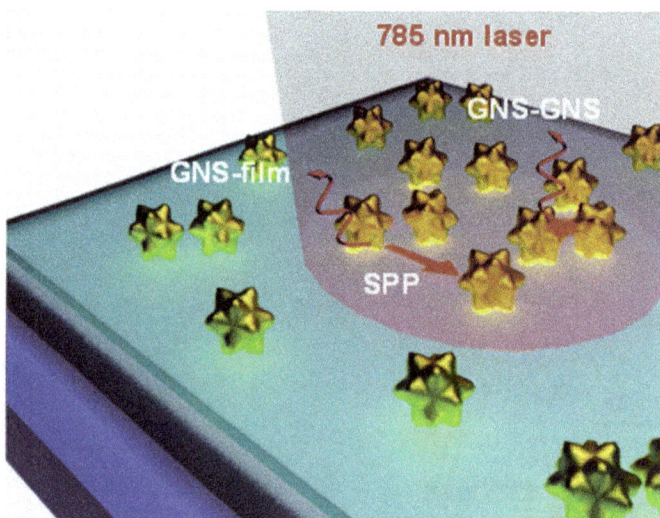

Figure 3.2.14: Schematic illustration of electromagnetic field enhancement in a gold nanostar (GNS) layer assembled on a continuous silver film (Ag) used as SERS substrate for the detection of DNT. The enhancement results from coupling between GNS–Ag and GNS–GNS, in combination with the propagating surface plasmon polariton modes on the silver film. Reprinted from [46] with permission from The Royal Society of Chemistry.

---

3 The Raman spectra were recorded with laser excitation at 785 nm, a power of 15 mW, and an integration time of 0.5 s. The SERS substrates were immersed in DNT solution in methanol for 2 h, rinsed in methanol, and blown dry for the SERS measurement of DNT.

It is important to mention that most studies on SERS substrates for TNT and DNT detection were performed in liquid phase. The extension of SERS measurements to microfluidic systems thus becomes a natural direction since they offer some advantages that are necessary for real-life in situ applications, such as convenient sample delivery and separation/pre-concentration of complex sample matrices [29].

In 2002, Keir et al. demonstrated the detection of surface-enhanced resonance Raman scattering (SERRS) of a TNT derivative [69] using a microfluidic cell (Fig. 3.2.15). Using the chip-based SERRS system, Keir et al. were able to achieve a limit of detection of 10 fmol. However, the study was conducted using an azo dye of TNT rather than TNT.

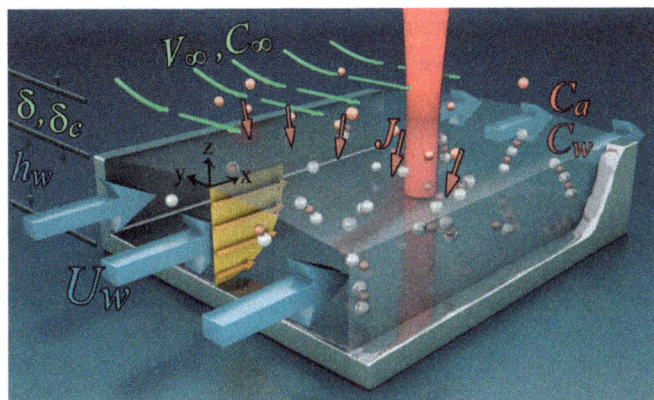

Figure 3.2.15: Cutaway illustration of material flows in the free-surface microfluidic channel. The aqueous microfluidic phase flows from left to right (blue arrows). The gas phase flows from back to front (green arrows). Analyte molecules (red spheres) diffuse from the gas phase into the liquid phase (red arrows). Nanoparticles (white spheres) suspended in the aqueous phase adsorb to suspended analyte molecules before interrogation by 658 nm laser light (red vertical beam) for detection by SERS. Figure and figure caption reprinted with permission from [66]. Copyright © 2012, American Chemical Society.

In 2012, Piorek et al. [66] demonstrated real-time vapor trace detection of DNT using a microfluid/surface-enhanced Raman spectroscopy system. The microfluidic system consisted of a 4–40 μm deep free-surface microchannel which mimics the canine olfactory system. The structure allowed for efficient transfer of DNT molecules from the air into the liquid region, which confined and concentrated the airborne DNT molecules and thereby enhanced their subsequent detection [66]. Thus, a vapor concentration of 1 ppb (less than 1 % of the saturated DNT vapor concentration) of DNT emanating from a solid-phase DNT crystal resulted in an aqueous concentration of DNT of 11.7 μM. Silver NPs (35 nm diameter) continuously flowed in the microchannel, which allowed for continuous, real-time, long-term gaseous-headspace chemical detection monitoring due to aggregation control and continuous SERS substrate refreshing by the flow [66]. Piorek et al. were thus able to repeatedly detect intense SERS signals[4] (1350 cm^{-1} NO$_2$ symmetric stretch and 1600 cm^{-1} NO$_2$ asymmetric stretch) due to dimerized silver NP "hotspots" and the increased aqueous analyte concentration due to the microfluidic channel design. The real-time vapor trace detection of DNT at 1% of the saturated DNT vapor concentration is one of the most interesting results to date. It is important to stress that this technology was then commercialized by Spectra Fluidics.

3.2.5 Summary and impact

Research in the area of design and fabrication of sensors and chemosensors for explosives detection is of great interest to a variety of scientific communities ranging from optical and chemical sciences to engineering. Several tools and techniques for explosives detection based on the latest trends in nanomaterials and micro- and nanotechnologies are being developed [70]. Chemical nanosensors based on optical spectroscopy are one of the most promising sensors for small-trace explosives detection as they allow selective and rapid detection of low concentrations, while simultaneously the device could be portable and affordable.

Among the different nanospectroscopic methods, fluorescent polymers and SERS have made big advances over the past two decades. Limits of detection of explosives using SERS have evolved from 100 nM in 1995 [25] to attomolar levels [40, 46, 47]. The integration of microfluidic systems, which allows for a more efficient delivery, separation, and pre-concentration of explosive traces, has made SERS a more reliable and a more adapted technique for real-life applications, which eventually led to a commercialized product by SpectraFluidics Inc. However, SERS has not yet succeeded in being accepted as a technique fit for explosives detection, and SpectraFluidics, Inc. has ceased to exist. This could be mainly due to the fact that the control and reproducibility of SERS on a commercial and research basis is still challenging/difficult. Nevertheless, significant improvements in performance and standardization can lead to a better acceptance of SERS as a technique for explosives detection. Fluorescent polymers, on the other hand, have a completely different success story. Although the lowest reported limit of detection for fluorescent polymers (56 fg/mL, 246.6 fM) is 24,660 times higher than the lowest reported limit of detection for SERS (10 aM), Teledyne FLIR seems to be having significant success in commercializing and developing various explosives detection devices, such as the Fido X2, Fido X3, Fido X4, and Fido X80.

As described on their website, Fido X "utilizes amplifying fluorescent polymer technology licensed from the Massachusetts Institute of Technology (MIT) as part of its

---

4 The maximum Raman intensity was recorded at ~10 min after exposure to DNT vapor using a laser power of 35 mW. The intergration time was not mentioned, estimated to be 4 s.

ultrasensitive suite of detection materials" for screening personal belongings, parcels, and cargo. Fido X seems to be a fierce competitor to more established and widespread techniques such as IMS, as it is gaining ground and being more and more used in airports and seaports.

Bibliography

[1]   Dathan J. A decade of explosive violence harm, action on armed violence. 2021. https://aoav.org.uk/wp-content/uploads/2021/05/A-Decade-of-Explosive-Violence-Harm.pdf. Griffith E. Explosive violence monitor 2021, action on armed violence. 2022. https://aoav.org.uk/wp-content/uploads/2022/04/Explosive-Violence-Monitor-2021_v5.pdf.
[2]   IED attack fact sheet: improvised explosive devices. U.S. Department of Homeland Security.
[3]   Initiative for explosives detection, Pacific Northwest National Laboratory, http://www.pnl.gov/nationalsecurity/explosives.detection.
[4]   Sachse G, Steele T, Rana M. Application of a new gas correlation sensor to remote vehicular exhaust measurements. In: Proceedings of the 8th CRC On-Road Vehicle Emissions Workshop. 1998.
[5]   Zolotov YA. Ion mobility spectrometry. J Anal Chem. 2006;61(6):519.
[6]   https://www.morpho.com/sites/morpho/files/entryscan-4-en-072016.pdf. Accessed March, 2017.
[7]   Bielecki Z et al. Sensors and systems for the detection of explosive devices – an overview. Metrol Meas Syst. 2012;19(1).
[8]   Howard PH, Meylan WM, editors. Handbook of Physical Properties of Organic Chemicals. Boca Raton, Fla: Lewis Publishers; 1997.
[9]   Krüger F. Fluoresceïn als Indicator beim Titriren. Ber Dtsch Chem Ges. 1876;9(2):1572.
[10] Basabe-Desmonts L, Reinhoudt DN, Crego-Calama M. Design of fluorescent materials for chemical sensing. Chem Soc Rev. 2007;36(6):993.
[11] Yoshimura I, Miyahara Y, Kasagi N, Yamane H, Ojida A, Hamachi I. Molecular recognition in a supramolecular hydrogel to afford a semi-wet sensor chip. J Am Chem Soc. 2004;126(39):12204–5.
[12] Li H et al. Facile synthesis of poly(aroxycarbonyltriazole)s with aggregation-induced emission characteristics by metal-free click polymerization. Sci China Chem. 2011;54(4):611–6.
[13] He H, Mortellaro MA, Leiner MJP, Fraatz RJ, Tusa JK. A fluorescent sensor with high selectivity and sensitivity for potassium in water. J Am Chem Soc. 2003;125(6):1468–9.
[14] Toal SJ, Trogler WC. Polymer sensors for nitroaromatic explosives detection. J Mater Chem. 2006;16:2871–83. 2006.
[15] Dai L, Soundarrajan P, Kim T. Sensors and sensor arrays based on conjugated polymers and carbon nanotubes. Pure Appl Chem. 2002;74(9).
[16] Sohn H, Sailor MJ, Magde D, Trogler WC. Detection of nitroaromatic explosives based on photoluminescent polymers containing metalloles. J Am Chem Soc. 2003;125(13):3821–30.
[17] Yang J-S, Swager TM. Fluorescent porous polymer films as TNT chemosensors: electronic and structural effects. J Am Chem Soc. 1998;120(46):11864–73.
[18] Swager TM, Wosnick JH. Self-amplifying semiconducting polymers for chemical sensors. MRS Bull. 2002;27(6):446–50.
[19] Che Y et al. Diffusion-controlled detection of trinitrotoluene: interior nanoporous structure and low highest occupied molecular orbital level of building blocks enhance selectivity and sensitivity. J Am Chem Soc. 2012;134(10):4978–82.

[20] Zahn S, Swager TM. Three-dimensional electronic delocalization in chiral conjugated polymers. Angew Chem Int Ed. 2002;41(22):4225–30.

[21] Thomas SW, Joly GD, Swager TM. Chemical sensors based on amplifying fluorescent conjugated polymers. Chem Rev. 2007;107(4):1339–86.

[22] Cumming CJ et al. Using novel fluorescent polymers as sensory materials for above-ground sensing of chemical signature compounds emanating from buried landmines. IEEE Trans Geosci Remote Sens. 2001;39(6):1119–28.

[23] la Grone MJ, et al. Detection of land mines by amplified fluorescence quenching of polymer films: a man-portable chemical sniffer for detection of ultratrace concentrations of explosives emanating from land mines. 2000. p. 553.

[24] https://www.flir.com/products/fido-x4/. Accessed: September, 2022.

[25] Kneipp K et al. Near-infrared surface-enhanced Raman scattering of trinitrotoluene on colloidal gold and silver. Spectrochim Acta, Part A, Mol Biomol Spectrosc. 1995;51(12):2171–5.

[26] Haas JW III, Sylvia JM, Spencer KM, Johnston TM, Clauson SL. Surface-enhanced Raman sensor for nitroexplosive vapors. 1998. p. 469.

[27] Sylvia JM, Janni JA, Klein JD, Spencer KM. Surface-enhanced Raman detection of 2,4-dinitrotoluene impurity vapor as a marker to locate landmines. Anal Chem. 2000;72(23):5834–40.

[28] Spencer KM, Sylvia JM, Marren PJ, Bertone JF, Christesen SD. Surface-enhanced Raman spectroscopy for homeland defense. 2004. p. 1.

[29] Hakonen A, Andersson PO, Stenbæk Schmidt M, Rindzevicius T, Käll M. Explosive and chemical threat detection by surface-enhanced Raman scattering: a review. Anal Chim Acta. 2015;893:1–13.

[30] Botti S, Cantarini L, Almaviva S, Puiu A, Rufoloni A. Assessment of SERS activity and enhancement factors for highly sensitive gold coated substrates probed with explosive molecules. Chem Phys Lett. 2014;592:277–81.

[31] Botti S, Almaviva S, Cantarini L, Palucci A, Puiu A, Rufoloni A. Trace level detection and identification of nitro-based explosives by surface-enhanced Raman spectroscopy: trace level detection of explosives by SERS. J Raman Spectrosc. 2013;44(3):463–8.

[32] Holthoff EL, Stratis-Cullum DN, Hankus ME. A nanosensor for TNT detection based on molecularly imprinted polymers and surface enhanced Raman scattering. Sensors. 2011;11(12):2700–14.

[33] Mbah J, Moorer K, Pacheco-Londoño L, Hernandez-Rivera S, Cruz G. A rapid technique for synthesis of metallic nanoparticles for surface enhanced Raman spectroscopy: rapid technique for synthesis of metallic nanoparticles. J Raman Spectrosc. 2013;44(5):723–6.

[34] Zhang C, Wang K, Han D, Pang Q. Surface enhanced Raman scattering (SERS) spectra of trinitrotoluene in silver colloids prepared by microwave heating method. Spectrochim Acta, Part A, Mol Biomol Spectrosc. 2014;122:387–91.

[35] Sil S, Chaturvedi D, Krishnappa KB, Kumar S, Asthana SN, Umapathy S. Density functional theoretical modeling, electrostatic surface potential and surface enhanced Raman spectroscopic studies on biosynthesized silver nanoparticles: observation of 400 pM sensitivity to explosives. J Phys Chem A. 2014;118(16):2904–14.

[36] Primera-Pedrozo OM, Jerez-Rozo JI, De La Cruz-Montoya E, Luna-Pineda T, Pacheco-Londono LC, Hernandez-Rivera SP. Nanotechnology-based detection of explosives and biological agents simulants. IEEE Sens J. 2008;8(6):963–73.

[37] Raza A, Saha B. In situ silver nanoparticles synthesis in agarose film supported on filter paper and its application as highly efficient SERS test stripes. Forensic Sci Int. 2014;237:e42–6.

[38] Sun Y et al. Highly sensitive surface-enhanced Raman scattering substrate made from superaligned carbon nanotubes. Nano Lett. 2010;10(5):1747–53.

[39] Gong Z, Du H, Cheng F, Wang C, Wang C, Fan M. Fabrication of SERS swab for direct detection of trace explosives in fingerprints. ACS Appl Mater Interfaces. 2014;6(24):21931–7.

[40] Zhou H et al. Trinitrotoluene explosive lights up ultrahigh Raman scattering of nonresonant molecule on a top-closed silver nanotube array. Anal Chem. 2011;83(18):6913–7.

[41] Zachhuber B, Carrillo-Carrión C, Simonet Suau BM, Lendl B. Quantification of DNT isomers by capillary liquid chromatography using at-line SERS detection or multivariate analysis of SERS spectra of DNT isomer mixtures: quantification of DNT isomers by capillary liquid chromatography using at-line SERS detection. J Raman Spectrosc. 2012;43(8):998–1002.

[42] Fierro-Mercado PM, Hernández-Rivera SP. Highly sensitive filter paper substrate for SERS trace explosives detection. Int J Spectrosc. 2012;2012:1–7.

[43] Bao ZY et al. Quantitative SERS detection of low-concentration aromatic polychlorinated biphenyl-77 and 2,4,6-trinitrotoluene. J Hazard Mater. 2014;280:706–12.

[44] He X, Wang H, Li Z, Chen D, Zhang Q. ZnO–Ag hybrids for ultrasensitive detection of trinitrotoluene by surface-enhanced Raman spectroscopy. Phys Chem Chem Phys. 2014;16(28):14706.

[45] Liu H, Lin D, Sun Y, Yang L, Liu J. Cetylpyridinium chloride activated trinitrotoluene explosive lights up robust and ultrahigh surface-enhanced resonance Raman scattering in a silver sol. Chem - Eur J. 2013;19(27):8789–96.

[46] Lee J et al. Tailoring surface plasmons of high-density gold nanostar assemblies on metal films for surface-enhanced Raman spectroscopy. Nanoscale. 2014;6(1):616–23.

[47] Verma AK, Soni RK. Ultrasensitive surface-enhanced Raman spectroscopy detection of explosive molecules with multibranched silver nanostructures. J Raman Spectrosc. 2022;53(4):694–708.

[48] Cecchini MP, Turek VA, Paget J, Kornyshev AA, Edel JB. Self-assembled nanoparticle arrays for multiphase trace analyte detection. Nat Mater. 2012;12(2):165–71.

[49] Zhang C, Li Z, Wu Z, Han D. Study of surface enhanced Raman scattering of trace trinitrotoluene based on silver colloid nanoparticles. Spectrosc Spectr Anal. 2012;32(3):686–90.

[50] Khaing Oo MK, Chang C-F, Sun Y, Fan X. Rapid, sensitive DNT vapor detection with UV-assisted photo-chemically synthesized gold nanoparticle SERS substrates. Analyst. 2011;136(13):2811.

[51] Liu X, Zhao L, Shen H, Xu H, Lu L. Ordered gold nanoparticle arrays as surface-enhanced Raman spectroscopy substrates for label-free detection of nitroexplosives. Talanta. 2011;83(3):1023–9.

[52] Ko H, Chang S, Tsukruk VV. Porous substrates for label-free molecular level detection of nonresonant organic molecules. ACS Nano. 2009;3(1):181–8.

[53] Mahmoud KA, Zourob M. Fe$_3$O$_4$/Au nanoparticles/lignin modified microspheres as effectual surface enhanced Raman scattering (SERS) substrates for highly selective and sensitive detection of 2,4,6-trinitrotoluene (TNT). Analyst. 2013;138(9):2712.

[54] Nergiz SZ, Gandra N, Farrell ME, Tian L, Pellegrino PM, Singamaneni S. Biomimetic SERS substrate: peptide recognition elements for highly selective chemical detection in chemically complex media. J Mater Chem A. 2013;1(22):6543.

[55] Hatab NA, Rouleau CM, Retterer ST, Eres G, Hatzinger PB, Gu B. An integrated portable Raman sensor with nanofabricated gold bowtie array substrates for energetics detection. Analyst. 2011;136(8):1697.

[56] Chou A et al. SERS substrate for detection of explosives. Nanoscale. 2012;4(23):7419.

[57] Demeritte T et al. Highly efficient SERS substrate for direct detection of explosive TNT using popcorn-shaped gold nanoparticle-functionalized SWCNT hybrid. Analyst. 2012;137(21):5041.

[58] Xu JY, Wang J, Kong LT, Zheng GC, Guo Z, Liu JH. SERS detection of explosive agent by macrocyclic compound functionalized triangular gold nanoprisms. J Raman Spectrosc. 2011;42(9):1728–35.

[59] Sajanlal PR, Pradeep T. Functional hybrid nickel nanostructures as recyclable SERS substrates: detection of explosives and biowarfare agents. Nanoscale. 2012;4(11):3427.

[60] Guo Z et al. Ultrasensitive trace analysis for 2,4,6-trinitrotoluene using nano-dumbbell surface-enhanced Raman scattering hot spots. Analyst. 2014;139(4):807–12.

[61] Zhou X, Liu H, Yang L, Liu J. SERS and OWGS detection of dynamic trapping molecular TNT based on a functional self-assembly Au monolayer film. Analyst. 2013;138(6):1858.

[62] Dasary SSR, Singh AK, Senapati D, Yu H, Ray PC. Gold nanoparticle based label-free SERS probe for ultrasensitive and selective detection of trinitrotoluene. J Am Chem Soc. 2009;131(38):13806–12.

[63] Jamil AKM, Izake EL, Sivanesan A, Fredericks PM. Rapid detection of TNT in aqueous media by selective label free surface enhanced Raman spectroscopy. Talanta. 2015;134:732–8.

[64] Kanchanapally R, Sinha SS, Fan Z, Dubey M, Zakar E, Ray PC. Graphene oxide–gold nanocage hybrid platform for trace level identification of nitro explosives using a Raman fingerprint. J Phys Chem C. 2014;118(13):7070–5.

[65] Liu M, Chen W. Graphene nanosheets-supported Ag nanoparticles for ultrasensitive detection of TNT by surface-enhanced Raman spectroscopy. Biosens Bioelectron. 2013;46:68–73.

[66] Piorek BD, Lee SJ, Moskovits M, Meinhart CD. Free-surface microfluidics/surface-enhanced Raman spectroscopy for real-time trace vapor detection of explosives. Anal Chem. 2012;84(22):9700–5.

[67] Wang J, Yang L, Boriskina S, Yan B, Reinhard BM. Spectroscopic ultra-trace detection of nitroaromatic gas vapor on rationally designed two-dimensional nanoparticle cluster arrays. Anal Chem. 2011;83(6):2243–9.

[68] Hamad S, Podagatlapalli GK, Mohiddon MA, Soma VR. Cost effective nanostructured copper substrates prepared with ultrafast laser pulses for explosives detection using surface enhanced Raman scattering. Appl Phys Lett. 2014;104(26):263104.

[69] Keir R et al. SERRS. In situ substrate formation and improved detection using microfluidics. Anal Chem. 2002;74(7):1503–8.

[70] Aragay G et al. Recent trends in macro-, micro-, and nanomaterial-based tools and strategies for heavy-metal detection. Chem Rev. 2011;111:3433–58.

Norman McMillan, Raul D. Rodriguez, Douglas McMillan,
Mark Heaton, Victor Hrymak, Simon Perry,
Enisa Omanović-Mikličanin, and Stavros Pissadakis

3.3 Converging optical micro- and nanospectroscopies via the development of surface science with forensic Shannon's measure of information applied to environmental applications

3.3.1 Key messages

- A review of the state-of-the-art is given, providing a perspective in the various fields of optical nanospectroscopy to reprise, freshen, and contextualize issues dealt with in Volumes 1 and 2 of "Optical Nanospectroscopy" as they relate to the key micro- and nanohardware requirements for a nanospectroscopic surface science.
- A literature review of the applications of micro- and nanovolume spectroscopy modalities is given, highlighting issues of nanoforensic analysis where maximizing Shannon's measure of information (SMI) has special and obvious practical importance to the fusion of micro- and nanotensiospectroscopy.
- A short primer (prelude) on each type of optical nanospectroscopy (modality) is presented in this chapter to help make the discussion of theoretical and hardware fusion understandable for tensiospectroscopy. The results of a comprehensive literature review of the existing state-of-the-art for micro- and nanospectroscopy are given in two tables and essential issues and advances are identified for the fusion of optical micro- and nanospectroscopy.

Acknowledgement: The chapter was supported by the COST Action MP1302 and specifically an outcome of the work of the Textbook subcommittee and the plan for Volumes 1 and 2 of the "Optical Nanospectroscopy" textbook from the lead authors who developed the AMRA method and developed the proposal for the content which was taken forward by Professors Meixner and Fleischer. The work of ANT Ltd. has been driven here by the lead author, but the work of Dr. Sven Riedel, Dr. Kavanagh, and Dr. John Nolan should be acknowledged here as coworkers in developing tensiospectroscopy and data mining/statistical work. Mr. Nathan Hulme, Director, Starna Scientific and ANT Group provided an input on cuvette nanodroplets accessories and Dr. Liam McDonnell provided some structural input to the chapter. The work of Professor Tatiana Perova has been important in this recent development and the ongoing work on SERS in liquids will hopefully mean that this work linking with Professor Iouri Gounko and Dr. Sarah McCarthy will prove of importance. The H2020 ACTPHAST Consortium is acknowledged as it forms the basis of the design work discussed here, in particular the final report coauthored by Dr. G. Violakis. The contribution by Dr. Teresa Madeira at the AMRA initial stage of the chapter should be acknowledged.

https://doi.org/10.1515/9783110442908-009

- A new metareview method used in the literature review has been focused specifically on tensiospectroscopy to objectively review the limitations of microtensiospectroscopy, to extend methods, hardware, and analysis by identifying issues and constraints necessary to bring this forward to the nanodomain.
- Some critical related challenges of optical nanospectroscopy are discussed: (i) the design optimization of the integrated nanoinstrumentation, (ii) the combination of the data from different optical nanospectroscopic modalities, (iii) the issues of connecting nanoinstrumental development to nanoproducts, and (iv) finally, methods exploring how to extend the analytical nanodeliverables of optical tensiospectroscopic products.
- Some concluding consideration on the importance of the role of Shannon information theory methods in optical nanospectroscopic product engineering tasks are described as these theoretical proposals are vital for calibration and the delivery of traceability for nanotensiospectroscopy. The use of data scatter theory is discussed and we connect "ray" to "wave" physics problems on either side of the diffraction limit.
- Drawing on the authors' experience, the integration of various micro- and nanospectroscopies is discussed, as this fusion process opens up new horizons of tensiospectroscopic applications for agricultural monitoring and farm payments, as well as addressing vital ecological challenges with mankind's future threatened by climate change.
- In the concluding discussion, a new synthesis designed in large part by the new metareview method is presented from concrete developments of hardware inventions, new nanotensiospectroscopic methods, nanoanalysis using data scatter theory, the STRAPS App, and the Shannon Depository Archive. These methods in combination facilitate the analysis, fusion, standardization/calibration, and, importantly, the correlation of data/information in the most universal way using SMI representation. The use of this approach to metrological information theory advance is discussed for delivering environmental health indices.

3.3.2 Pre-knowledge

The following content is a discussion on the developing integration of nanospectroscopies and their growth from microspectroscopy. The useful platform of knowledge from Volume 1 includes the following: Section 1: basics of light–matter interaction: optical microscopy and nanofocusing; Section 2: optical spectroscopies; Section 3: nanoscopy, super-resolution microscopies, near-field-enhanced spectroscopies and microscopies, near-field-enhanced microscopies, tip-enhanced spectroscopies, and nano-IR spectroscopy. Useful knowledge from Volume 2 includes the following: Section 4: instrumentation for optical microscopy and nanospectroscopy, light sources, optical detectors, spectrometers and optical filters, microscopy, and scanning probe microscopy; Section 6: nanomaterials for nanospectroscopy, nanostructures for enhanced microscopies, and methods for nanofabrication. In the present volume, all chapters dealing with surface-enhanced Raman spectroscopy (SERS) and sensing are relevant to this chapter.

Since the chapter makes extensive use of the SMI, which does not appear anywhere else in this book, it might be useful to at least provide the relevant definitions here. Let x_i, $i = 1, 2, \ldots, n$, be the discrete values of a random variable X, with associated probabilities $p(x_i)$. Then $H(X) = -\sum_{i=1}^{n} p(x_i) \cdot \log_b(p(x_i))$ is the SMI introduced originally for telecommunication signals. The logbase b is typically 2 (Shannon's preference), e, or 10; they are equivalent and the results can be mutually converted easily. As an example, in an ecological context, H has been used as a diversity index, with x_i being the different species of a population and $p(x_i)$ their relative abundance [1]. As such, H as SMI can serve as a measure to judge correspondingly digitized spectra. On a similar note, the data scatter plot, where, e. g., one variable is plotted against another (but not restricted to two variables), is a tool used to visualize and characterize data [2]. It is particularly useful for recognizing correlations, e. g., if the datapoints for different events are lying on a straight line, a linear correlation becomes obvious.

3.3.3 Importance of the application

(i) Miniaturization of spectroscopy focusing on nano

Despite his wonderful contributions to science, Archimedes was most proud of his development of burning beams technology based on enormous mirrors; this is relevant to mention regarding our discussion, as it produced incredibly intense light beams unrivaled until modern times with the advent of laser technologies. We also know that he was proud of being able to write down a number that was greater than the number of grains of sand ($\sim 10^{23}$) on a beach at Syracuse. The number of photons involved in delivering information to our eyes in optical nanospectroscopy experiments is even more staggering. A signal of 1 nW of green light delivers approximately $3 \cdot 10^9$ photons a second. Planck's constant is $6.62607004 \times 10^{-34}$ Js and the earliest time the known universe can be measured at is 5.4×10^{-44} s. The density of the universe as a whole is 10^{-27} kg/m^3. We can put these figures against a measure of DNA absorbance, which can typically be 50 absorbance (A) units, but we must remember that this is a logarithmic measure. Indeed, perhaps a Guinness Book of Records optical measurement of 203 ± 1 A-units for Parasol sunscreen [3], weakening down to 185 in 350 min of aging, was made in 2005 by McMillan et al., done then with some novel technology, although such spectroscopic accessories are today routine [4]. The content of this chapter also extends directly from the unique collaboration on an Advanced Meta-Review Analysis (AMRA) to address the training needs of early-stage researchers (ESRs) in optical nanospectroscopy fundamentals that resulted in this three-volume textbook on optical nanospectroscopy, created in the context of the EU-funded COST Action MP1302. A "coherent" content for this textbook project was proposed based on the AMRA, and this chapter itself can be considered a study on the utility of AMRA [5] in view of this textbook project.

For several decades, three trends have dominated the development of optical spectroscopy: miniaturization of the spectroscopy instrumentation, miniaturization of the analyzed sample, and, more recently, miniaturization of the probe. Below some innovations in nanovolume-sample handling for tensiospectroscopy are briefly discussed for probe, green, and space technologies [6]. The initial impetus to reduce the spectroscopy instruments' physical size or footprint was space saving within laboratories and clean rooms, and later, to perform spectroscopy at application sites. This trend was facilitated by adopting developments in microfabrication, photonics, and information technology. The key to current portability has been the commercial development of modular spectrometers from companies such as Hamamatsu, Ocean Optics, and Wasatch. Each of these products integrates key mechanical, optical, and electrical components within a small package that forms the core of a spectroscopy instrument. Importantly, modular spectroscopy instruments now have performances comparable to laboratory instruments. Tematys [7] estimated that 44,000 modular spectroscopy products were sold worldwide in 2016 with sales expected to quadruple by 2021. Three examples of the application of portable spectroscopy instruments – often called compact or handheld – illustrate their advantages: bringing diagnosis closer to the patient, roadside testing of drugs by law enforcement personnel, and identifying pollutants on-site during brown-field site remediation. One consequence of the migration of spectroscopy from the laboratory is the need to build analytical intelligence into the instrument, using on-board or cloud-based libraries to interpret spectra automatically. Today, networked, disposable smart sensors (the Internet of Things) are being deployed in field applications such as agribusiness and environmental monitoring. Further miniaturization in instrument size can be anticipated, for example, with integrated photonic circuit technology transferred from telecommunications applications into spectroscopy. There is no doubt that disposable spectrometers will join the Internet of Things in what we term analytical "globalization."

Reductions in sample size are demanded by applications where samples are either a priori small (such as forensics and neonatal care), costly (pharmaceutical drug development), or rare (art fraud). Of course, there are limitations to sample size reduction that arise from the practicalities of sample handling and presentation. There is also the fundamental barrier to sample size reduction that arises from analyte concentration. The latter is defined volumetrically, for example, as molarity, and clearly there is a minimum theoretical sample volume beyond which no detection is possible for a given analyte concentration due to sensitivity limits, noise, and other operational issues. However, there are sample handling and presentation methods that, to some extent, circumvent such limitations. One method uses solvent evaporation to increase analyte concentration. Another "captures" the analyte onto a surface or interface, converting a volumetric analysis to a real one. Of course, the latter method is only fully successful when the spectrometer's field of view encompasses all the captured analytes. When describing spectroscopic modalities, it is essential to distinguish between those operating on liquid samples, where the analyte is distributed throughout the volume probed, and situations where the analyte is solid, adsorbed at a solid interface or in the liquid/solid interface.

Optical probe dimensions are diffraction-limited when operated in far-field configurations. The development of near-field geometries overcomes this limit and has rev-
olutionized imaging and spectroscopy at the nanoscale. In this chapter, we include
within nanospectroscopy those methods and techniques that perform spectroscopy
with a nanometric spatial resolution in at least one dimension. The latter caveat is
important as it recognizes that analytical spectroscopies do not have to achieve 3D
resolution to be nanosensitive. Classic surface chemical analysis techniques such as
Auger electron spectroscopy and X-Ray photoelectron spectroscopy are recognized as
nanoscale in sample depth but with probing areas in the millimeter or micrometer
range.

The need for the development of micro- and nanometrology arises as elaborated
by Hansen et al. [8]. However, matters are complicated enormously below the diffraction limit as critical dimensions are scaled down and geometrical complexity of objects is increased, and the available technologies appear insufficient. Major research
and development efforts must be undertaken to answer these challenges that involve
the spectroscopist. Pathlength issues are a starting point in spectroscopic metrology
but these have not become critical, since sample volumes for UV-Vis, near-infrared
(NIR), and mid-infrared (MIR) quantitative spectroscopy are still in the microvolume
range above the diffraction limit. New measuring principles and instrumentation, tolerancing rules, and procedures need consideration. The issues of nanotraceability and
calibration for nanometrology await urgent development. Spectroscopic traceability
and calibration requirements will be identified in our discussion below in the relevant places. The development of laser sources, interferometry, and fabrication technologies transformed the potential for measurement technology, while advances in
SEM and nanomicroscopy are also closely related to spectroscopy. However, there still
remains a technological and fundamental gap. There appears to be a failure to sufficiently connect up the microscopic/mechanical advances and the needs in qualitative
nanospectroscopic standards, thus delineating a research opportunity for ESRs.

(ii) Moving from microvolume to nanovolume spectroscopy, keeping the UV-Vis yardstick in mind

As a reference point for this chapter, examples of commercial microvolume spectroscopy systems are reviewed for the most important molecular modalities. The
history of spectroscopy goes back to the nineteenth-century work of Kirchhoff and
Bunsen, but the departure point for this chapter is the twentieth-century development of commercial UV-Vis spectroscopy used by laboratories all around the world.
The "golden standard" of spectroscopy was largely quantitative, as this addressed the
large and growing number of industrial applications for quality assurance, process
control, and other commercial uses. The technological advance of these spectrom-

eters was developed in Europe and the USA most urgently during World War II and subsequently thereafter by market competition.

Let us begin the discussion of the state-of-play of today's spectroscopic laboratory capabilities with the technological review in Table 3.3.1. This table gives the various performance specifications of a range of commercial microspectrometers from Ocean Optics taken as an example. This important economic and volumetric downsizing of spectroscopy connecting with fiber geometries was developed to exploit the latest spectroscopic components. Below, a similar table provides comprehensive information on hardware and other issues for nanospectroscopy. The architecture of a typical system for microvolume spectroscopy is almost universal now and is based around the modular spectrometer type including a modular design of light sources, fiber spectrometers, and sample handling accessories, like e. g. the DMV Bio-cell from Starna Scientific to measure ultra-absorbing liquids such as DNA. The key to the success of the approach in the 1980s by McMillan with the first tensiomicrovolume spectroscopy system commercialized as the "Fiber Drop Analyser" was fiber connectivity and engineering based on Eon O'Mongain's University College Dublin multi-fiber fiber spectrometer.

There is now really a developing situation where spectroscopic standards, traceability, interlaboratory alignment, regulatory procedures, and standardization of data processing/analysis will be necessary in every professional laboratory. Starna Scientific's John Hammond contributed to "Standard Best Practice in Absorption Spectrometry" [9], which was very much the millennium spectroscopic bible for those in the British Isles and Ireland, but indeed beyond most English-speaking countries. The work of Hammond with K. Hulme and N. Hulme made subsequently important contributions to commercialize and indeed professionalize spectroscopy by developing UV-Vis standards. From there, other standards in molecular spectroscopic modalities appeared, such as in the National Institute of Standards and Technology (NIST). UV-Vis is the yardstick for microvolume spectroscopy from which we might say "a proper and formal development" of other spectroscopic methods has evolved. An elephant in this particular room has come from the explosion of nanospectroscopy that has not allowed for the niceties of a regulated development to occur. UV-Vis is the traditional spectroscopy against which all else is measured but is not an essential foundation stone of optical nanospectroscopy as it is in the microscale. While transmission UV-Vis techniques have relevance to SERS/tip-enhanced Raman spectroscopy (TERS), for confocal microscopy and super-resolution techniques, it is not the yardstick for the new field of optical nanospectroscopy. In this regard, Smith, McMillan et al. [10] developed a general quantitative method of "noncollimated light spectroscopy" (NCLS) that works above and below the diffraction limit. This transmission method for nanodrops and other surface science nanostructures may have established a theoretical advance that will allow nano-UV-Vis/NIR/MIR to become a standard for nanovolume methods.

Table 3.3.1: Examples of instrumental specifications of microvolume spectrometers/nephelometers that are useful in comparison of microvolume and nanovolume/nanospectroscopic capabilities.

	UV-Vis	Raman	NIR	Fluorescence	Nephelometers	Refractometers
Instrument	ANT UV-Vis	QE-PRO Raman[a]	NIRQuest512[b]	USB4000-FL[c]	Aurora 1000[d]	Abbe (DR-A1-Plus)[e]
Description	Quick, portable, NIST traceable, custom software	Portable; sensitive; powerful; stable	Fast; robust; modular; rugged	Modular; fast; portable; compatible	Automatic calibration; easy maintenance	High precision, bench-top
Resolution (nm)	1.5 nm FWHM	$7–11\ cm^{-1}$ over the range	~3.1 nm FWHM w/ 25-µm slit	~10.0 nm FWHM	$<0.3\ Mm^{-1}$ (scattering coefficient)	Refractive index (nD) 0.0001, Brix 0.1 %
Range	200–850 nm	185–1100 nm	900–2500 nm	200–1100 nm	0 to $20{,}000\ Mm^{-1}$ (scattering coefficient)	Refractive index (nD) 1.3000 to 1.7100; Brix 0.0 to 100.0 %
Comments	Operational principle: noncollimated light spectroscopy and real nanodrop capability 100 nL	Traditionally a low-sensitivity technique and principal utility for qualitative research	Beer–Lambert law quantitation with qualitative spectral library	Principally used as a quantitative technique, with phase and other signal modulation capable of nanodetection	Traditionally used for quantifying turbidity but with PM capable of nanodetection	Traditionally a technique partitioned from spectroscopy despite theoretically being an integral component

[a] https://www.oceaninsight.com/products/spectrometers/raman/qepro-raman-series/
[b] https://www.oceaninsight.com/products/spectrometers/near-infrared/nirquest2.5/
[c] https://www.oceaninsight.com/blog/alternatives-to-usb-spectrometers/
[d] http://ecotech-research.com/aurora-1000
[e] https://www.atago.net/USA/images/catalog/dr-a1-plus_us.pdf

3.3.4 State-of-the-art

3.3.4.1 Quantitating perspectives on forensic optical nanospectroscopy applications

This section looks at the state-of-the-art of various nanospectroscopic modalities as a prelude to the later discussions on an integrated vision of this field. It highlights some specific and crucial developments that distinguish nanoscience from the micro level. We begin by confronting the conundrum of UV-Vis spectroscopy being the golden standard of macro- and microspectroscopy, which is paradoxically a backmarker in the development of nanospectroscopy. The bar chart shown in the introduction (Part 1) reveals that the pathfinding nanospectroscopies are especially found in IR and Raman techniques. What has been revealed is a veritable explosion of research publications, commercial articles, and books for these pathfinding areas, where micro- and nanospectroscopy have connected profoundly and moved spectroscopic analysis fully down below the diffraction limit.

Forensic analysis is a field of subspecialties with different techniques to acquire criminal or further legitimate evidence. Nanospectroscopy has a growing influence in providing direction to forensic investigations. But nanoscience still has some distance to go before it can confidently be used in a court of law. This is especially true for nanospectroscopy, which is still in its infancy. Nanoforensics will become a major part of forensic analysis as its emergence is driven by developments in nanospectroscopy. This is accompanied by the emergence of nanosensor methods for real-time investigation of crime scenes and terrorist activity, explosive compounds detection, gases, biological mediators and filtrates, contributions to the analysis of toxic materials, and the old forensic staples, providing evidence about tissue, materials, and soil. Nanospectroscopy has low sensing capability and can be tuned to target specific molecules. Unsurprisingly there are growing applications for nanodetection at the crime scene and at airports for detecting counterfeit food, beverages, and fuel. Nanospectroscopy also has a growing role to play in the detection of art fraud. It complements many other forensic analytical techniques and can provide nanomaterial recognition in assembling evidence, which would not have been possible using traditional methods. Numerous novel approaches are emerging to ease the way for forensic scientists, for example, DNA extraction from palm prints, gun residues, fingerprints, explosives, and heavy metals, providing conclusive evidence. Here the fusion of optical nanospectroscopies and physical measurements discussed by the authors is hopefully a harbinger of profound future developments in forensic science.

The AMRA study used to define and develop the content of this textbook on optical nanospectroscopy was developed to plan this chapter. This analysis was delivered using the EBSCO Discovery Service (EDS), a research discovery tool that allows searching across a vast range of databases, e-journals, e-books, reports, multimedia items, and more. Figure 3.3.1, for instance, shows the published result of López and Guarcia-Ruiz [11] and provides some quantitative measures to deepen the understanding of the schematic classification of explosives, concentrating here on nitrogen detection examined for airborne or touch contact. Taking just one point to highlight the value of this study, the AMRA analysis provides a quantitative perspective to explain the rather small commercial drive for forensics with respect to the big beasts in this IR and Raman sector, namely industry, military, and environment, which are an order of magnitude greater than forensics. Table 3.3.2 presents the results of a library search in the forensic science fields, which, while being only a basic study, nevertheless constitutes a very

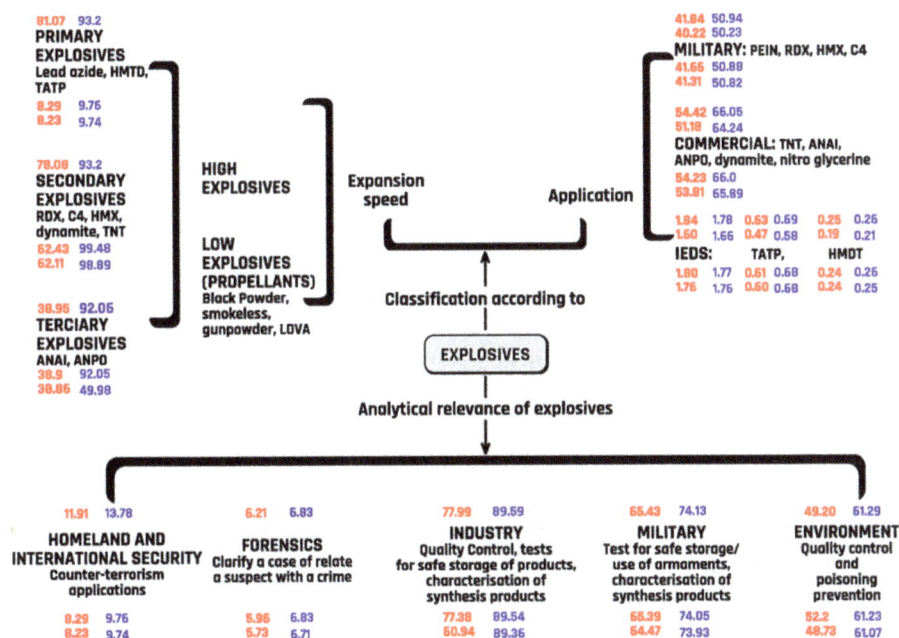

81.07 93.2
PRIMARY EXPLOSIVES
Lead azide, HMTD, TATP
8.29 9.76
8.23 9.74

78.08 93.2
SECONDARY EXPLOSIVES
RDX, C4, HMX, dynamite, TNT
62.43 99.48
62.11 98.89

36.95 92.05
TERCIARY EXPLOSIVES
ANAI, ANPO
38.9 92.05
38.86 49.98

HIGH EXPLOSIVES

LOW EXPLOSIVES (PROPELLANTS)
Black Powder, smokeless, gunpowder, LOVA

Expansion speed

Application

41.84 50.94
40.22 50.23
MILITARY: PEIN, RDX, HMX, C4
41.55 50.88
41.31 50.82

54.42 66.05
51.18 64.24
COMMERCIAL: TNT, ANAI, ANPO, dynamite, nitro glycerine
54.23 66.0
53.81 65.89

1.84 1.78 0.53 0.59 0.25 0.26
1.50 1.66 0.47 0.58 0.19 0.21
IEDS: TATP,   HMDT
1.90 1.77 0.51 0.58 0.24 0.25
1.76 1.75 0.50 0.58 0.24 0.25

Classification according to

EXPLOSIVES

Analytical relevance of explosives

11.91 13.78
HOMELAND AND INTERNATIONAL SECURITY
Counter-terrorism applications
8.29 9.76
8.23 9.74

6.21 6.83
FORENSICS
Clarify a case of relate a suspect with a crime
5.96 6.83
5.73 6.71

77.99 89.59
INDUSTRY
Quality Control, tests for safe storage of products, characterisation of synthesis products
77.38 89.54
60.94 89.36

65.43 74.13
MILITARY
Test for safe storage/ use of armaments, characterisation of synthesis products
65.39 74.05
64.47 73.93

49.20 61.29
ENVIRONMENT
Quality control and poisoning prevention
52.2 61.23
48.73 61.07

Figure 3.3.1: EDS Discovery Service Classification and numerical evaluation of relative importance of each entry in the diagram of explosives and subfields. The figures in red print are *percentages* for Raman and those in blue are *percentages* for SERS. The reference numbers of EDS Discovery Service hits (100 %) for explosives are respectively 95,734 and 65,151. The three sets of figures are for respectively the *percentages* for the EDS Discovery Service hits for the category (i) on their own above and (ii) below for High Explosives and (iii) below for Low Explosives.[1] We illustrate here for the "Forensics" entry (second entry from left on bottom line) the *percentages* for Explosives AND Forensics (6.21, 6.83) with High Explosives AND Forensics (5.96, 6.83) and Low Explosives AND Forensics (5.73, 6.71). Note, there are thus no entries for High Explosives and Low Explosives in the diagram as their percentages are distributed to appear among all other entries in the figure. Based on [11].

---

1 High explosives have a fuel and oxygen built into the same molecule, meaning they deflagrate, while low explosives are usually a mixture of fuels and oxidizers, such that they detonate.

Table 3.3.2: AMRA results (number of hits) for forensic applications in niche subfields for each nanospectroscopic modality with EDS "hits." NIR = near-infrared; MIR = mid-infrared; SERS = surface-enhanced Raman spectroscopy; SERRS = surface-enhanced resonant Raman spectroscopy; TERS = tip-enhanced Raman spectroscopy; Fluor. = fluorescence nanospectroscopy; DLS = dynamic light scattering; MALS = multiangle light scattering.

A = Nanomodality D = Applications	NIR	MIR	SERS	SERRS	TERS	Fluor.	DLS	MALS
Toxicology	6631	10,413	4977	429	11,208	13,218	3857	9196
Chemistry	12,319	19,223	9385	707	21,516	14,802	6337	14,321
Biology	12,400	19,330	9436	710	21,550	24,842	6352	14,361
Entomology	5033	8412	3609	192	8903	8616	3110	7248
Serology	5911	9316	4216	295	9704	11,069	3397	8137
Pharmacology	9829	15,269	7194	576	19,192	18,724	5418	12,292
Explosives	10,002	15,859	7518	555	17,450	18,256	5291	11,817
Clandestine drugs	1913	3309	1602	94	3318	2907	1180	2712
Paint	9155	14,657	6957	506	16,572	16,611	5076	10,952
Glass	11,884	18,593	8788	678	20,930	23,534	6176	13,812
Plastics	11,832	18,920	8684	614	20,877	22,857	6145	14,052

revealing dataset. This search has been done with the keyword "*nanospectroscopy*"; for example, the MIR "hits" are for publications linked to nanospectroscopic techniques. The emergence and importance of SERS here in the field of forensics is significant despite its recent emergence as a spectroscopic technique.

The main reference to cite is the overview of Aad, Akil, and Jradi in this volume's Chapter 3.2, "Chemical nanosensors for the detection of explosives". It is well known that there are serious issues in the acceptance of SERS, fluorescence, and other techniques discussed by these authors for such forensic monitoring tasks. This study highlighted the use of SERS in the liquid phase for explosives trinitrotoluene, or more specifically, 2,4,6-trinitrotoluene (TNT) and 2,4-dinitrotoluene (DNT). Drop evaporation techniques are applied, using a silver nanorod surface modified with *p*-aminobenzenethiol (PABT) that are forming a PABT-TNT-PABT complex by pulling together (self-approach), thus creating hotspots with enormous enhancement factors of up to eight orders of magnitude (10^8). The importance of microfluidic systems with delivery and separation/pre-concentration of complex sample matrices using microchannels is also discussed.

Chapter 5.1 by Omanović-Mikličanin on optical nanospectroscopic applications in food science deals with diagnostic issues relating to beverages and food which have enormous importance to customs control operations in Europe. Chapter 5.2 on SERS applications in cultural heritage by Shabunya-Klyachkovskaya is of vital importance for criminal and other legal prosecutions in the art world. Culha et al. in Chapter 4.2 on "SERS in label-free detection of cancer" address a topic of profound importance to this chapter because the method can be utilized as an alternative approach to existing label-based assays. Furthermore, this method reduces analysis time while increasing diagnostic accuracy. It is a method with perhaps enormous importance for forensic analysis in the future.

One of the latest developments is to deliver SERS simply by putting a metallized atomic force microscopy (AFM) or scanning tunneling microscopy (STM) tip working as a nanoantenna near the target molecule [12]. The presence of the nanoantenna enables the conversion of evanescent waves at the surface over a small area defined by the tip radius of curvature into propagating waves detectable in the far-field, typically reducing the enhancement volume to nanometers. This brings forward the vexed methodological issues still to be addressed before TERS can be widely used for forensic purposes. There is no established calibration sample that could provide the nanoscale structure for routine imaging, since only subdiffraction imaging can ensure that TERS has been successfully achieved. The TERS spectra also could differ from the microvolume Raman spectra due to statistical variations at the single-molecule level, molecule–probe interactions affecting the spectral signature, heating, and photocatalytic reactions on the plasmonic nanoantenna. Another challenge of TERS is the routine demonstration of nanoscale imaging on samples deposited on nonmetallic substrates due to lower field amplification obtained in that configuration.

Despite the promise of TERS, it remains confined to the laboratory but presents a considerable potential in forensics for the trace-level detection of illegal chemical substances, DNA sequencing, etc. TERS, as well as the very extraordinary commercial instrument nanoIR [13] developed by Anasys Instruments with resonance-enhanced AFM and tapping AFM-IR, a portable device enabling high-spatial resolution chemical imaging, has been developed working with lab-on-a-tip technology. Such developments are of utmost importance to forensic science, where the Fourier transform IR (FTIR) (cf. Volume 1, Chapter 2.3 and Volume 2, Chapter 4.6) is a workhorse technology with rich, interpretable spectra that directly correlates to nanoIR and is true nanoscale FTIR. There is a major point here to make with regard to tip technologies; this technique does not require high-vacuum technology for their operation, thus making portability a realizable objective. The power of IR has been beautifully presented by McDowell, Patterson, and Harter [14] from Los Alamos Science. This would be hard to outclass for the introduction of quantum applications.

Quantum dots (QDs) have importance in established fields of forensics for fingerprints and explosives. Perhaps a contributory reason is the necessity of a "problem solving" approach in forensics, where the required physical evidence or "questioned sample" is then to be compared with a known material in seeking to identify the origin of this sample. A current massive research effort exists for the imaginative use of QDs that has spurred something akin to a Gold Rush spilling over into forensics. This wild activity is not helpful in forensic research, given the need for properly established and traceable procedures and the strongest statistical proof. This undeveloped state of affairs for QD studies perhaps partly explains why Cantu's book [15] is a rather lonely volume. Forensics is certainly a field that highlights the cardinal importance of connecting spectroscopy and imaging, which is a primary advantage of analytical nanospectroscopy over microspectroscopy. The ability of QDs to fluoresce (with a high extinction coefficient that is combined with a comparable quantum yield to that of fluorescent dyes) gave nanomaterials entrée to a classical method of identification in improving the visualization of latent fingermarks, or poorly developed fingerprints [16]. Furthermore, the potential applications of QDs in detecting nitroaromatic explosives, such as TNT, based on directive fluorescence quenching of QDs, electron transfer quenching processes, or fluorescence resonance energy transfer have been paid attention to. DNA analysis is associated tightly with forensic applications in molecular diagnostics.

Raman spectroscopy is an excellent method with high potential for applications in food forensics. The main advantages of Raman spectroscopy over other analytical approaches applied in food analysis are the following: it can be used both in solids and liquids; the sample does not require preparation; the water content does not interfere like in FTIR; it is a nondestructive method; it is highly specific to the analyte's chemical fingerprint; the spectra are quickly acquired within seconds; and the samples can be analyzed through a glass container or a polymer packaging. Raman spectroscopy has been used to test honey adulterated with fructose, glucose, inverted sugar, hydrolyzed inulin syrup, and malt must [17]. Adulterant agent concentrations have been determined using partial least squares regression (PLSR) and principal component regression (PCR) methods. The major problem in this type of analysis is the sample's viscosity which significantly influences the peak distribution. This contributes to the error occurrence and makes chances for quantitative determination very low. On another side, Raman spectroscopy has a great potential for determining methanol levels in alcoholic drinks [18]. Methanol could cause poisoning when it is present in high concentrations. Unfortunately, this is a common way of fraud with strong health implications. Raman peaks obtained from those samples are sharp and clear, allowing precise and accurate qualitative and quantitative analysis.

3.3.4.2 Overcoming the diffraction limit

The optical systems in Table 3.3.1 are far-field and thus have a spatial resolution limit (the diffraction limit of light) set by the operating wavelength and numerical aperture of the optical system. Moving beyond the diffraction limit for a given spectroscopic modality is not straightforward and requires innovative approaches. Practical nanoscale imaging (nanoscopy), and thus nanospectroscopy, conceptually began in 1928 with the work of Synge [19, 20] in trying to overcome the diffraction barrier in optical microscopy. Synge conceived and described an instrument that would image in what today we term the near-field. His original idea was based upon using an intense light source behind a thin, opaque metal film with a small orifice of about 100 nm. The orifice was to remain within 100 nm of the surface, and information was to be collected by point-by-point scanning. He even foresaw the illumination and the detector movement as the greatest technical difficulties at the time. Indeed, the success of this approach required the development of modern computer control and feedback systems with piezoelectric scanners available in the 1980s. His work inspired many modern studies as reviewed by Kim and Song [21]. Whilst this chapter will not review nanoscopy per se, nanoscale imaging based on spectroscopic data such as fluorescence is included.

The present study reveals some major holes in nanospectroscopic advances, and here some useful insights are drawn from the authors' own historical research. The work aims at uniting elements from both the micro- and nanovolume multispectroscopic systems, the long-term development being the delivery of an omnispectroscopy that allows the fusion of the hardware and analytical analysis of any number of molecular spectroscopies. This architectural connectivity nightmare is realized with an elegant rotating load (TWIST) droplet system moving from measurement position to measurement position. The third and largely underdeveloped facet is the ability of nanospectroscopies to generate images such as offered by the coherent Raman (CoR) microscopy techniques that notably attempts at the optical equivalent of magnetic resonance imaging.

Today the massive explosion in optical nanospectroscopy means that there is a pressing need for a paradigm shift, the full shaping of which requires considerable research effort. Some progress is reported here on new micro- and nanospectroscopic metrology standards that establish a solid starting point for this work. Recognizing these metrological issues implies an immediate and direct challenge for regulatory and standard bodies such as the American Standards for Testing and Materials (ASTM) International. Some examples are flagged to raise the importance of what is indeed a pressing issue of developing relevant nanostandards. It is possible to anticipate many difficulties and probably considerable inertia from the body of established science that is not yet thinking openly about nanospectroscopy. It is true that many commercial spectroscopic products marketed with nanospectroscopic names alarmingly ignore in their marketing that the technologies face a fundamental physical barrier. This chapter openly reviews how these issues of micro- and nanospectroscopy connect and transform, and examples show how "diffraction limit" issues have been successfully addressed for various modalities of optical nanospectroscopy. One approach to UV nanospectroscopy has been the outcome of the authors' research based on surface science that directly addresses these fundamental concerns. One of the most ingenious ways to overcome the diffraction limit is with a plasmonic needle acting as a nanoscopic light source that localizes and enhances the incident light. This is behind the tip-enhanced optical spectroscopy methods that constitute an emerging field in nanospectroscopy. The obvious step from this point is to consider how far things have advanced towards a universal vision that we could describe as a new spectroscopic paradigm. Even if there is no consistent nanospectroscopic answer yet, solid advances have been identified and discussed.

3.3.4.3 Historical review

The diffraction limit applies to light which has traveled a distance significantly greater than its wavelength, as in far-field illumination. This topic is carefully covered in Volume 1, Section 3. Here we briefly touch on the remarkable development of Hutchinson and Synge. The near-field scanning optical microscopes (NSOMs/SNOMs) place the source or detection probe into the near-field by using a subwavelength-sized aperture engineered for the light hitting the surface. Figure 3.3.2 illustrates the three approaches (called aperture-NSOM in Fig. 3.3.2(a) and (b) and apertureless-NSOM in Fig. 3.3.2(c)) that are the modern realizations of Synge's radical concept. In the illumination mode of Figure 3.3.2(a), light is directed through a local probe placed near the sample to generate an evanescent field at the probe tip. Light is scattered from the probe and the sample system when the tip approaches and the near-field reaches the probe. An NSOM image is obtained by the detector by collecting the scattered light. This approach has practical difficulties leading to intensive research in probe development. In the collection mode illustrated in Figure 3.3.2(b) the sample is irradiated in the far-field regime. The dashed line indicates a critical angle, and illumination beyond that angle induces total internal reflection (TIR) below the sample and the excitation of an evanescent field. A bare dielectric probe tip located above the surface acts as a near-field-to-far-field converter, and the converted near-field in the proximity of the tip apex propagates to a photodetector. Figure 3.3.2(c) shows the apertureless probe geometry in which light is irradiated onto the sample and the evanescent field is generated by TIR illumination. By locating the probe tip near the sample to induce scattering of the evanescent field, this wave is converted into a propagating wave, carrying the optical signal with nanoscale resolution.

The working principle of plasmonic surface charge waves is shown schematically in the bottom half of Figure 3.3.2. These waves are generated when an electromagnetic field is confined and propagates at the interface between a noble metal thin layer and a dielectric. Surface plasmon resonance (SPR) can only be generated in transverse magnetic (TM) modes where the polarization of the magnetic field vector is in the plane of the metal–dielectric interface since transverse electric modes cannot excite SPR. A transversal field distribution with maximum field intensity at the metal–dielectric interface has associated evanescent waves that penetrate both adjacent media but decay exponentially in the near-field. Plasmonic excitation can be generated using grating couplers, by a dielectric waveguide, or as the most common way, by using a coupling prism, as shown in Figure 3.3.2A. Such an experimental arrangement limits the multiplexing possibilities. This configuration is not suitable for use in plasmonic sensors because of miniaturization requirements. Instead, miniaturization is better addressed by using metallic nanostructures in the subwavelength size range to generate localized SPR (LSPR). In this case, particular electronic modes can be excited when light strikes the metallic nanostructures so that free electrons oscillate collectively. LSPR has gained much interest as an alternative technique to the standard SPR, as

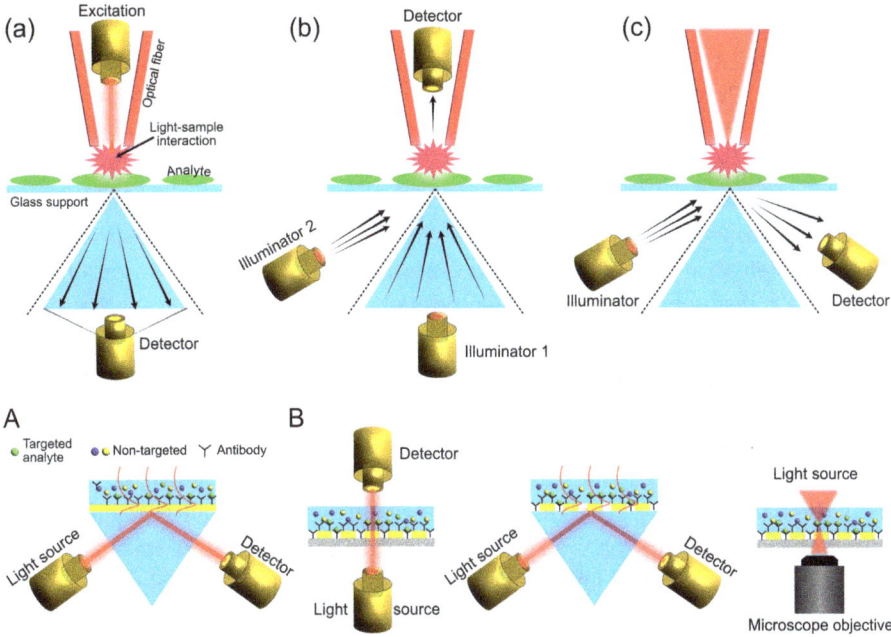

Figure 3.3.2: Top: Schematic of illumination mode (a), collection mode (b), and apertureless NSOM mode (c). Bottom: Schemes representing the most common configurations for (A) SPR and (B) LSPR sensors. Redrawn from [21] and [22].

metal nanoparticles (NPs) offer unprecedented opportunities for multiplexing since the sensing area is limited by the size of the nanostructure, which virtually expands the throughput to the nanostructure level. Moreover, since no bulky coupling methods are required, simple transmission or reflection configurations are sufficient to generate the LSPR effect shown schematically in Figure 3.3.2B.

The claimed enhanced sensitivity that nanoplasmonic sensor devices can reach has not yet been fully achieved, and there is some controversy regarding the real improvement that LSPR can provide over conventional SPR sensors (see also Chapter 3.1). The latter have proven their effectiveness in the monitoring and characterization of biomolecular interactions with a sensitivity (i. e., a shift of the SPR wavelength per refractive index unit (RIU)) that usually ranges between 10^5 and 10^7 nm/RIU. In the case of LSPR sensors, when using conventional nanostructures such as nanospheres, nanorods, or nanodisks, the sensitivity is in the same range or even lower (usually between 10^4 and 10^6 nm/RIU). Whereas in terms of bulk sensitivity, SPR clearly outperforms LSPR, a significantly better surface sensitivity can theoretically be obtained in LSPR. In general terms, although some publications have dealt with this controversy, there is still a lack of convincing studies that confirm whether nanoplasmonics is competitive enough with SPR in terms of surface sensing performance. However, a few recent works demonstrate that sensitivity levels are in the same order of magnitude, although the sensitivity seems to improve and be higher at low analyte concentration in the case of LSPR, both in a competitive assay and in direct approaches. This could be partially due to the strong LSPR field confinement of the nanostructures compared to SPR, which becomes more evident at low target concentrations, especially in direct assays.

Here some issues are presented briefly to provide an overview of hardware challenges of nanospectroscopy and related applications. This section deals with the concept of pathlength at the heart of quantitative issues and suggests the concept of NCLS might be a *transformative element* when posited as a theoretical concept that disregards the "diffraction limit," and applies equally well at either side of this barrier. The problem of taking the NCLS below this physical limit is, as will be seen, a practical one only. NCLS theory does not require rays but applies equally to wave solutions. Pathlength is a concept that does not apply to NCLS.

The analytical development of NPs and QDs was touched on above, but here the use of UV-Vis spectroscopy to characterize NPs should be mentioned. QDs and indeed other nanoentities were comprehensively detailed by Lin, Lin, and Sridrhar [23] in what is a wide-ranging and impressive overview of the physiochemical characterization of such nanomaterials. Spectroscopy, as we discover, is just a small part of an array of technologies for nanocharacterization, but specifically in this context, it is used in determining the physiochemical measure of size, concentration, and aggregation state. Also, the measurement/study of bioconjugation from UV-Vis used in conjunction with fluorescence and circular dichroism (CD) is important. However, UV-Vis spectroscopy here cannot be considered to be a nanospectroscopic method. Indeed, this optical characterization method is more than 150 years old and was authored in Tyndall's scattering studies [24], as was IR spectroscopy itself. They are both derivatives from his research on the quantification of the absorption and emission of constituents in the atmosphere. Molecular interpretation [25] of his spectroscopic results was demonstrated in Tyndall's differentiation of ozone from molecular oxygen. The development of particulate matter characterization is today very wide, but these are expansive applications [26]. One most important review article on dynamic light scattering (DLS) and the zeta potential is by Bhattacharjee, whose work was undertaken at University College Dublin [27]. This development comes from the older nephelometric (science of light scattering) method of multiangle light scattering (MALS) [28]. Wyatt's Ultimate Guide is an excellent start here for ESRs to get a basic introduction [29] to modern developments of nephelometry, introducing history, measurement principles of MALS and DLS, detector issues, number of angles for MALS, and type of laser and an overview of the full range of nephelometric instrument types including an on-line viscometer. A wide range of systems and techniques is presented, such as instruments for hollow fiber field flow, gradient systems, on-line configurations, plate readers, temperature systems, electrophoretic mobility measurements, protein instruments, and nanofilter kits for small-volume measurements. The tensiograph approach here offers perhaps now further enhanced measurements as nephelometric measurements are linked to absorption measurements, where it seems this may be a new measurement option for liquids, although such multiangle absorption photometers (MAAPs) are exploited in airborne monitoring [30]. The crucial point is that in liquid measurements a direct connection to UV-Vis spectroscopy via fundamental surface science is achieved. Optical tensiometry, as will be explained below, allows UV spectra to be modulated by nanoprocesses (e. g., local chemical reactions), while being recorded optically with an NCLS UV-Vis drop spectrometer. This suggestion is supported by connecting this development to well-established thermodynamic molecular methods in surface science.

3.3.4.4 UV-Vis seeking to go below the diffraction limit

An important initial role for UV-Vis spectroscopy is in characterizing NPs and QDs and other such nanomaterials. An excellent overview of its many facets can be found in Lin, Lin, Wang, and Sridhar [23]. The role of UV-Vis in this field is truly historic, but the use of the generalized Rayleigh light scattering equation is very much an art. The AZoNano publication gives a beautifully crafted explanation of UV-Vis spectroscopy in that context [31]. With so many diverse nanomaterials, characterization has become a complex issue, especially given the concerns for human health that are raised by the use of these new materials. QDs – semiconductor nanocrystals that exhibit intrinsic optical and electrical properties that are size-dependent due to the quantum confinement effect – have an important role in UV-Vis nanospectroscopy. The most comprehensive overview of their use in bioanalysis is perhaps that by Petryayeva, Algar, and Medintz [32] when judged on several fronts, not the least being that beautifully presented in Figure 3.3.3. A very good visual presentation of the physical and optical properties of luminescent NPs and organic fluorophores, including various types of QDs, is presented there. Consideration of the role of NPs and organic fluorophores reveals these are a "transformative element" in the full sense of what has been outlined in the previous section; the possibilities opened up by these entities transform for example tensiography from micro- to nanotechnology. Moving away from the home base, the initial field of medical diagnostics and therapeutics, again by way of perspective, many cutting-edge applications for QDs have emerged in the fields of information technology, electronics, biotechnology, energy, medicine, cellular imaging, and diagnostic biosensing, discussed by Hauppauge [33].

QDs can be tuned to give a desired fluorescence emission wavelength. Due to their unique properties, QDs have attracted considerable attention in different scientific areas. Their quantum thermodynamic properties are different from those of the bulk material from which they are manufactured. Most importantly, luminescence has an increasing wavelength moving towards that of the bulk by increasing the QD size. As the confinement energy depends on the QD size, both absorption onset and fluorescence emission can be tuned by changing the QD size during synthesis. In addition,

Quantum Dots — Semiconductor — 2-10 nm — 10 - 90 % — 25-35 nm — > 10 ns

Organic Fluorophores — Dye molecules, proteins — <1 nm, ~4 nm — N/A — Variable — Broad, red-tailed — < 10 ns

Material / Size / Functionalization / Quantum Yield / FWHM / Photostability / Lifetime / Multiplexing / 2PE

UCNPs — Lanthanide-doped matrix — 20-50 nm — Variable — < 15 nm — > 100 ms

Dye-Doped NPs — Silica — 20-200 nm — Variable — < 10 ns — > 60 nm — <100%

Metal Nanoclusters — Gold, silver — < 2 mm — < 20% — < 100 ns — < 5 ns — > 60 nm — < 25%

Nanodiamonds — Carbon — 5-20 nm — 10 - 20 ns — > 80 nm — <10%

Carbon Nanotubes — Carbon — Variable — < 10 ns — > 60 nm — 5 - 60%

Graphene Oxide — Carbon — Thickness 0.6 nm

C-dots — Carbon — <10 nm

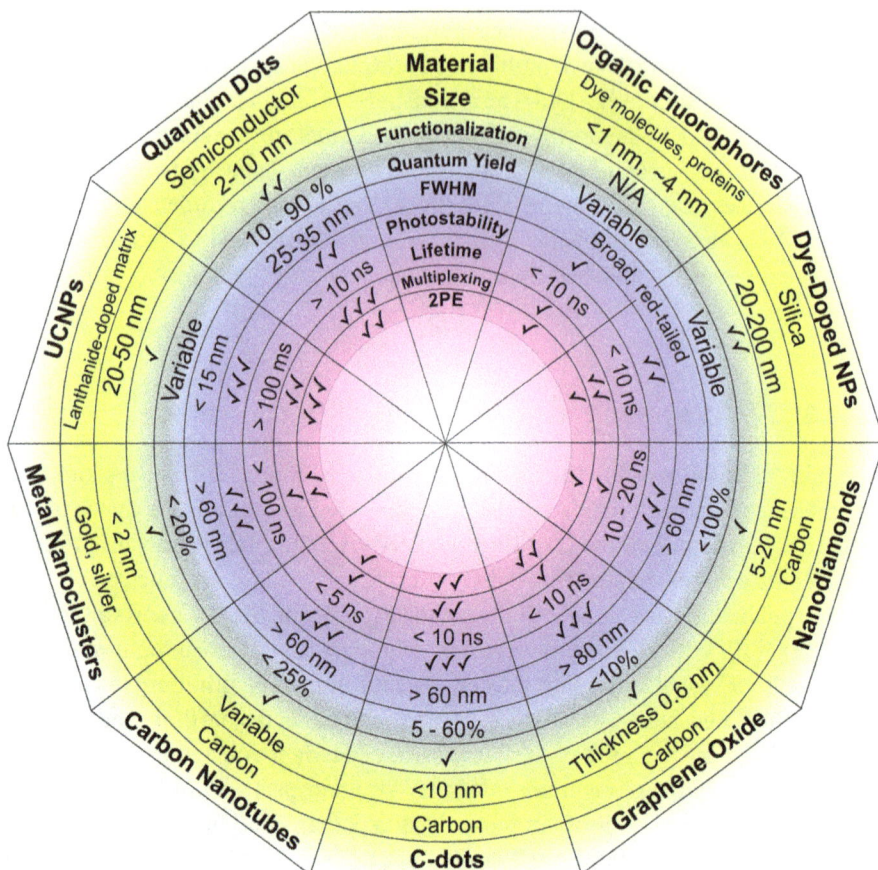

Figure 3.3.3: Comparison of physical and optical properties of luminescent NPs and organic fluorophores. The check marks (√) indicate the relative degree of favorability. Redrawn after [32].

the lifetime of fluorescence is also size-dependent. The engineering possibilities are not exhausted, e. g., to improve fluorescence quantum yield, QDs can be made with "shells" of a larger band-gap semiconductor material around them. The improvement is suggested to be due to the reduced electron and hole access to nonradiative surface recombination pathways in some cases, but also due to reduced Auger recombination in others.

As mentioned above, one very important spectroscopic application of QDs derives from the tiny and limited amounts of DNA at a crime scene. Human DNA has to be quantified accurately after the process of DNA extraction. Accordingly, highly sensitive detection of human genomic DNA is an essential issue for forensic study. QDs also have a variety of uses as emission probes in forensic DNA quantification. Some new developments suggest that medical breakthroughs could lead to forensic applications.

QDs can act as delivery vehicles for small interfering RNAs, which are powerful tools for silencing gene expression. There is now a new class of easily adjusted multifunctional NPs emerging from the conjugation of QDs with targeting agents and photosensitizers. The extraordinary photostability of QDs allows them to be used in real-time tracking of molecules and cells over extended periods, and antibodies, streptavidin, peptides, DNA, nucleic acid aptamers, or small-molecule ligands can be used to target QDs to specific proteins on cells. These materials containing NPs often appear as evidence and thus are subject to forensic analyses. These usually are spectroscopic identification or simple pass/fail tests for spectroscopic analysis, which extends to using luminescent QDs, up-converting nanoscale phosphors, and nonluminescent NPs as security tags to label products or to add security to documents, and as anticounterfeiting measures. These assist in determining if an item was fraudulently made or altered. There are forensic reagents that exploit gold NPs used in what is essentially a photographic process and are amplified by a silver developer. There are some good examples of straightforward applications of quenched oil-soluble CdSe QDs for the detection of TNT at concentrations of 1.0×10^{-7} to 5.0×10^{-5} mol/L by Ki [34]. Still, the difficulty of applying QDs in forensics is obvious, for example in proposals for detecting hemoglobin where specially manufactured QDs are then modified into a Nafion (fluorinated polymers functionalized with sulfonic acid moieties) membrane to be investigated with enterochromaffin-like (ECL) cells (ECL cells are a distinctive type of neuroendocrine cell in the gastric mucosa underlying the epithelium). This technique is used for high-sensitivity detection of hemoglobin from a characteristic quenching effect [35], with a sobering message that this requires high skill.

It would be remiss not to mention that there are many other nanocomponents apart from QDs that play a transformative role in moving microspectroscopy to the nanodomain [36], and here we simply refer to those categories of particles indicated in Figure 3.3.3. The reader is referred to Volume 2, Section 6 for more details.

3.3.4.5 Important nanospectroscopic UV-Vis development

There is little evidence in the literature of any substantive UV-Vis nanospectroscopic liquid research, although there is ongoing work on UV-Vis SERS substrates [37]. As noted above, UV-Vis is also used as a characterization method for nanomaterials [38] but this would be considered a micro-UV-Vis technique used in support of a nanoscience study. Some nanospectroscopic studies are reported on QDs in the visible spectrum and on TERS in the UV spectrum, and a pioneering study of UV-Vis spectroscopy of CO_2 gas absorption in a microporous polymer was reported [39]. Chapter 5.3 in this volume highlights J- and H-aggregation on the surface of noble metal NPs. This review's importance goes beyond UV-Vis nanospectroscopy as this development is important also for fluorescence spectroscopy, Raman spectroscopy, DLS, and the use of birefringence in polarization micrography. To avoid duplication, only reference to the important use of cyanine dyes in labeling proteins, antibodies, peptides, nucleic acids, and indeed any kind of biomolecules shall be given. The application of dye-NP assemblies for medical diagnostics, chemical sensing, and catalysis, together with the capping of these NPs or similar NP innovations, will be vital tools in the armory of nanospectroscopists developing UV-Vis research.

Tensiospectroscopy or NCLS [40] has been developed by the authors for both sessile and pendant drops underpinned by an absorption and turbidity coefficient [41], with the patent covering a new general quantitative method and a major advance since the Beer–Lambert law. As displayed in Figure 3.3.4, in tensiography, a liquid is introduced via a capillary tube running down the center of a drophead. A source fiber couples the light source with the drop, whilst a collector fiber delivers the output signal to a detector. The signal, plotted in a tensiotrace, i. e., detector signal versus time, arises from modulated rainbow peak couplings between the source and collector fibers as the drop grows under gravity. Molecular activity on the drop surface is encoded in the signal, but the drophead's surface significantly affects the signal. The tensiograph presents the opportunity to conduct several spectroscopic measurements simultaneously, and differential spectroscopy with one fiber pair measuring the reaction on the drop surface and a second pair measuring the spectroscopy of the bulk feeding the reactions on the surface. The first peak in the tensiotrace gives an extremely sensitive measurement of refractive index, and the trace length gives an improved measurement of surface tension using the standard drop period measure. The time to the last peak in the tensiotrace (called tensiopeak) gives an optical measure of surface tension. The trace obtained for each wavelength measures turbidity and then, using capacitive methods, drop volume and other measures are obtained. The scattering methods developed on the substrate of bovine serum albumin were used to deliver unique

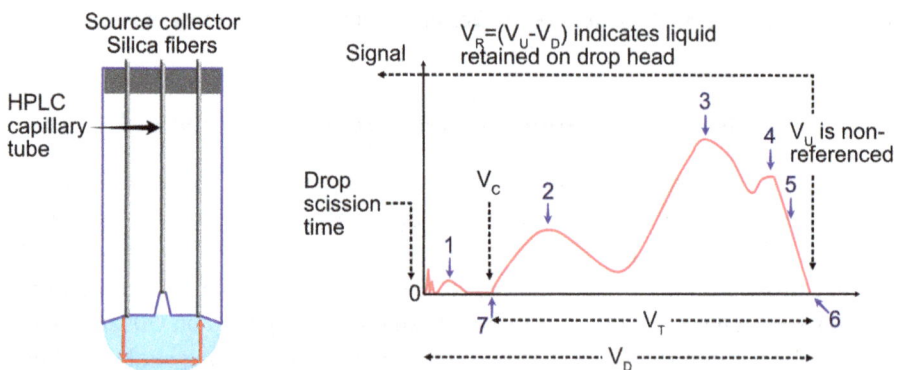

Figure 3.3.4: Example of a drophead configuration and a tensiotrace showing the important trace features and the times associated with drop mechanics. 0 = separation vibration; 1 = first-order peak (protopeak); 2 = rainbow (second-order deuteropeak); 3 = tensiopeak (third-order tritopeak); 4 = shoulder peak (fourth-order tetratopeak); 5 = separation peak (fifth-order pemptopeak); 6 = drop period; 7 = rainbow peak commencement.

nanospectroscopic monitoring and formation of monolayers and bilayers, but significantly reporting for the first time that of triple layers.

The tensiograph's commercial embodiment, the STASIS, is the first ever multimeasurand (*omni*) instrument designed for optical monitoring of complex liquids that has outperformed high-performance liquid chromatography (HPLC) detection levels in terms of limit of detection (LOD) and limit of quantification (LOQ) values in the tracking of complex polycyclic aromatic hydrocarbons (PAHs) and priority pollutant molecules in trial studies conducted by large European water companies in real waters with complex changing backgrounds. The multiplex of signals from the tensiospectrograph furthermore monitors surface properties, optical dispersion, and turbidity. The authors consider the STASIS, originally developed as a microvolume UV spectrophotometer technology, as perhaps the best prospect for general-purpose UV-Vis nanotensiospectroscopy [42].

Tensiography has been quantitatively modeled by Pringuet, Smith, and McMillan [43], which transforms the instrument into a bona fide UV-Vis spectrophotometer demonstrated with sub-ppb detection levels [44]. Using the nomenclature of Figure 3.3.4 there is a universal relationship between the drop period v_D (in s) and the tensiopeak period v_T: $v_T = k \cdot v_D$. The constant $k < 1.0$ connects the drop period with the tensiograph peak (an optically determined quantity) used to generate a UV-Vis spectrum from the tensiopeak heights [45]. The fundamental universal relationship transforms the mechanical v_T measurement into an optical peak height measurement of the tensiopeak and thus transforms mechanical tensiometry into an optical response. The measurement of v_T can therefore be used as a basic method of studying diverse biomolecular and chemical processes directly with UV-Vis spectroscopy. Proteins and other molecules adsorb to the drophead substrate, thus changing this period to a purely optically determined quantity, which this UV-Vis instrument can monitor. For example, a comparative contamination study has demonstrated real-time monitoring of protein contamination of polymeric surfaces [46–48], demonstrating that the instrument not only records accurate UV-Vis spectra for the drop under test, but also responds to nanoprocesses that modulate the UV-Vis spectra. The device is indeed an optical tensiometer, and here is the secret of its nanocapabilities. The dynamic tension is measured from the tensiopeak period, but this surface measurement can of course be made repeatedly by stepper-pump control with drop volumes cyclically scanning the moving peak maximum.

There are many possibilities of extending these tensiospectroscopic capabilities by exploiting hollow core fibers and microstructured optical fibers (MOFs) to controllably deliver micro- and nanovolume drops, where hybrid designs overlaid with ZnO nanocoatings [49] inside the optical fiber capillaries can be cleaned using all optical methods [50] with total removal of the fluidic sample. Advanced biocell surfaces overlaid with ZnO layers can provide in a single sample combined optofluidic, disinfecting [51], and sensing [52, 53] capabilities, boosting the functionalities, robustness, sensitivity, and potential selectivity of the developed tensiographic devices in the generic levels of systems or subsystems. The use of MOFs and hollow fibers for controlling the size, shape, and volume of liquid drops for tensiographic spectroscopy can lead to the emergence of a new type of miniaturized and robust tensiometers either hosted on the end face of standard [54] and MOFs [55] or floating on liquid interfaces [56] while exploiting whispering gallery mode (WGM) resonation.

Fluorescence at the nanoscale

It is perhaps true that fluorescence spectroscopy was the basic launchpad for nanospectroscopy, being the unchallenged nanobiosensing spectroscopy until the rise of SERS/TERS. Fluorescence nanospectroscopy has the sensitivity to deliver single-molecule detection, even at a single-cell level, but has evolved as a complex multifaceted method. Different fluorescence biosensing types will be considered here due to their importance in forensic science. The direct method exploits powerful fluorescent tags such as the green fluorescent protein (GFP), and Förster resonance energy transfer (FRET) offers the ability to study protein interactions with intermolecular separation.

Fluorescence has a sensitivity of typically three orders, but as much as six orders greater than UV-Vis, with lower detection limits for less sample volume. Fluorescence is generally more accurate and precise, although it has the obvious downside of being applicable to the limited number of molecules that fluoresce, which is a positive in terms of instrumental specificity. In nanovolume measurements, control of the source's intensity is a major issue due to the saturation of fluorophore absorption, which has led to the use of short pulses to combat photobleaching. For application technologies, having the options in source types for fluorescence is attractive. LEDs, lasers, and other microsources possibly connected to graded index fiber (GRIN) lenses to produce a collimated beam for both micro- and nanosystems are available. The detectors used in fluorescence spectrometer systems are varied; the Irish-made solid-state SensL avalanche photodiode (APD) provides an excellent option for micro- and nanoinstrument applications. CMOS and CCD detection are also useful options, but there are many others.

Discussions of some of the relevant technical issues in stepping from micro to nano are nicely illustrated in an impressive overview of fluorescence correlation spectroscopy (FCS) [57], which is now one of the major biophysical techniques for unraveling molecular interactions in vitro and in vivo. FCS, however, is just one example in optical nanospectroscopy of these new departures, which is based on powerful data analysis of ultrafast optical nanosensitive fluorescence data streams for the key innovation of the auto-correlation analysis. FCS not only monitors molecular processes, but beyond that, it also delivers information on a whole range of nanomolecular processes.

FRET, resonance energy transfer (RET), or electronic energy transfer (EET) describe energy transfer between two light-sensitive molecules known as chromophores. A donor chromophore, initially in its electronic excited state, may transfer energy to an acceptor chromophore through nonradiative dipole–dipole coupling [58]. This is an example of a photonic discovery leading to a new and distinct spectroscopic method. The schematic diagram of the FRET process in Figure 3.3.5 is instructive in helping this vital point to be succinctly explained, since the radius of interaction is much smaller than the wavelength of the light emitted, i. e., in the near-field. The excited chromophore emits a virtual photon that is instantly absorbed by a receiving chromophore. These virtual photons are undetectable, as their existence violates the conservation of energy and momentum; hence FRET is known as a *radiationless* mechanism. The efficiency of this energy transfer is inversely proportional to the sixth power of the distance between donor and acceptor, making FRET extremely sensitive to small changes in distance. There are various ways this process occurs, but the one between two fluorescence molecules is the most studied, being identified from the shortening of the fluorophore donor lifetime. FRET can determine if two fluorophores are within a certain distance of each other, for insights into molecular interactions and structural information. Such measurements are increasingly exploited in biology and chemistry. Here, an elegant optical approach involves key technologies from ingenious optical designs taking spectroscopies below the diffraction limit. The prism in TIR delivers an evanescent field that decays exponentially in the near-field to illuminate fluorophores in TIR fluorescence (TIRF), which restricts excitation to a thin region 100–200 nm be-

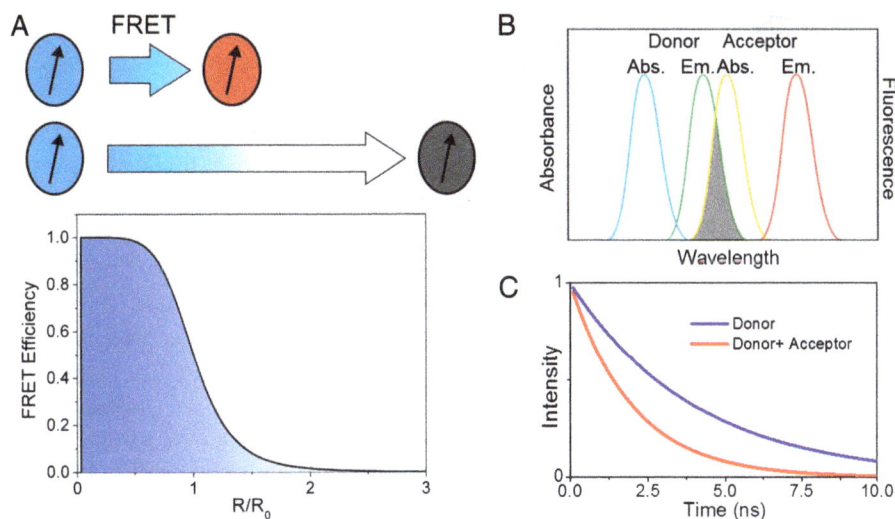

Figure 3.3.5: FRET concept. (a) Distance dependence of FRET efficiency. (b) Overlap of donor emission and acceptor absorption. (c) Shortening of donor fluorophore lifetime in the presence of a FRET acceptor. Reprinted with permission from [58].

hind the cover glass. This results in an essential gain in the signal-to-noise ratio because of the removal of the out-of-focus blur from far-field.

Some critical work on fluorescence nanospectroscopy is useful for the presentation of transformative deliberations starting with that of Cricenti et al., whose landmark review of the development of the near-field nanospectroscopy methods dates back to 2004 [59]. SNOM has augmented the usefulness of optical spectroscopy at a microscopic level in the region between 300 nm and 10 nm. Schermelleh et al. [60] produced a guide to super-resolution fluorescence microscopy.

Raman at the nanoscale: SERS, TERS, and beyond

A most comprehensive overview of SERS and the techniques beyond can be found in this volume's chapter on advanced nanospectroscopy in bioapplications by Kneipp et al., Chapter 4.3, which although addressing biomedical applications is accessible for ESRs dealing with important issues: labels, tags, and probes (basic and multimethod); spatial resolution and imaging issues; SERS substrate engineering; and multiphoton methods. Culha et al. in Chapter 4.2 on SERS in label-free detection of cancer from proteins, cells, and tissues thoroughly address a topic of profound importance in forensics, and hence to this chapter, because the method can be utilized as an alternative approach to existing label-based assays, reducing analysis time while increasing diagnostic accuracy. The chapter by Omanović-Mikličanin on optical nanospectroscopic applications to food science (5.1) deals with diagnostic issues relating to beverages and foods that have enormous importance to customs control operations in Europe. The chapter on SERS applications in cultural heritage (5.2) by Shabunya-Klyachkovskaya is of vital importance for criminal and other legal prosecutions in the art world.

A gentle introduction to Raman and advanced nano-Raman techniques is that by Dad and Agrawal [61] and a well-presented unified view is given by Lombardi and Birk [62], including the issue of SERS selection rules. Itoh et al. [63] have quantitatively evaluated the processes underlying SERS, providing also an informed theoretical overview pinpointing the advantage of SERS over fluorescence spectroscopy as "an intrinsically potential method for label-free detection of biomolecules." A chronological history of SERS substrate development is given by Yao and coauthors [64]. In general, SERS substrates can be divided into three main groups: colloidal, solid, and flexible substrates. A wide range of different solid SERS substrates is suggested at the present time [65, 66]. Deposition of metal films over 2D arrays of nanosphere-based structures [67] has been investigated, as well as so-called "printed SERS substrates" via inkjet and screen printing of different types of metal NPs on silicon, glass, or polymer substrates. Indeed, nanopillar substrates paved the way for the development of powerful new SERS technologies [68]. Confinement of light through plasmonic structures is utilized not only in biosensing applications as explained above, but also in optical trapping. Nanoplasmonic structures have been used for ultrasensitive biosensors with high multiplexing characteristics, spectroscopy systems with high enhancement factors, and nanometer-precision optical tweezers. There are gross variations on the SERS-active surface, showing the importance of the local environment, and here, e. g., even in the simplest and most widely used approach of metallic NPs or of magnetic NPs (MNPs) in colloidal suspension, aggregation makes reproducibility challenging.

The issue of combining SERS with surface science has been addressed in the devices shown in Figure 3.3.6a and b, with two instrumental variations of the Raman Flipper system developed by the author's company Advanced Nano Technologies Ltd. (ANT). The nanodroplet surface physics defining the sample are now able to deliver SERS measurements from what is called the "rear-read" excitation through the transparent base of the drop directly to concentrated droplets. This patented instrument [69] has many possible options. The first offers a novel approach for micro- and nanovolume samples (3 µL to 100 nL) that can be quickly pipetted onto the top of the instrument, but it can also take standard vials, cuvettes, and SERS sample loads for small commercial SERS substrates onto which droplets with the analyte are pipetted and allowed to dry. A modular instrument design for this ANT Raman Flipper has been developed with magnetic locking accessories for a complete range of samples, including packets of drugs that can be put into an accessory called the Large Sample Handler. The basic manual operation sample handling system is designed for SERS commercial substrates loaded with pipetted samples and left to dry. This manual loaded stub with the now activated sample is stuck to the plinth on the disposable sample stub. These plinths can hold a pipetted sample droplet deposited onto a steel plinth, which is in position on the stub holder. The ANT-patented accessory for replicating readings done on droplets is called the Nanodrop & SERS Accessory (N&SA), which involves no human handling and can be automated with robotic loading. The strict rotational droplet placement allows the laser excitation to be closed down to the smallest point and could deliver the most remarkable SERS enhancement of up to eight orders.

Figure 3.3.6b is a photograph of the Raman probe version, which uses the standard Raman probe that fits simply into the housing in a couple of minutes. The probe can be used as a general laboratory or field instrument in the normal way. In both modalities of the instruments, the droplet sample is loaded and then flipped over for an exact and reproducible measurement in an enclosed environment. The usual situation is that the Laplace pressure on the droplet produces a force that pins the sample onto a stainless stub and with a sample-holding plinth similar in design to that shown in Figure 3.3.6c usually 2 mm in diameter but without the cylindrical hole in the stub base. Such a design gives excellent Raman spectra and is one of the smallest possible volume sample systems, offering rapid replicate readings. Additionally, it is a green technology reducing sample waste to an absolute minimum. Replicate readings are essential for tensiospectroscopy. The disposable steel stubs can be wiped with negligible carry-over working with manual and automatic N&SA facilitating 3D raster scanning of droplets from the droplet substrate to the vapor corona.

(a)

(b)

(c)

Individual

Load

Read
Laser

MagSERS

Raman

Magnetic
NPs

SERS

Concentrated
MagNPs

Solenoid
OFF

Solenoid
ON

(d)

Simple Aliquant Plot

SERS signal [a.u.]

Pipette step

(e)

Evaporation and Concentration
Aliquant Plot

SERS signal [a.u.]

Time

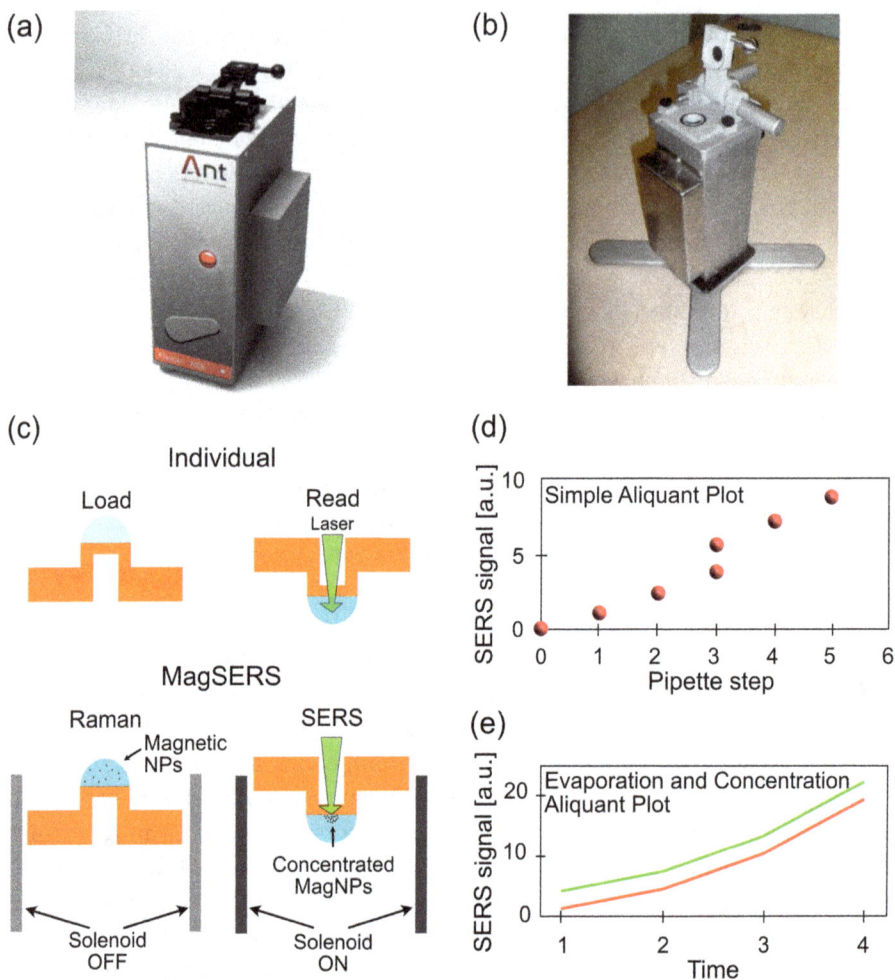

Figure 3.3.6: Raman Flipper Mark 1. (a) The spectrometers in their commercial form are the Dias Raman Flipper and the Auxiliary Raman Flipper for free-space laser systems or Raman probe operation. (b) There are additional formats for this product but two formats are presented to underline the versatility of this product concept. (c) Diagram showing Raman load and read and SERS read, magnetic concentration, and back-read for both Raman and SERS. The droplet does not need to be flipped of course for the unflipped read. (d) Simple "Aliquant" MagSERS Plot using traceable standard additions. The x-axis represents the pipetted steps and the y-axis the SERS signal. (e) Use of evaporation of two MNP solutions. The x-axis is time. The y-axis is the SERS signal. Signals from two droplets of the same solution but different analyte concentrations.

At this point, a development of conceivably major importance will be briefly explained. Figure 3.3.6c shows a design to deliver rapid replicate SERS measurements, which has been fully engineered. There is a problem with nanospectroscopy in droplets because of attenuation of the laser beam by NPs/MNPs, macromolecules, colloidal and other aggregational molecular structures, and many other nanostructured entities. This challenge can be addressed in two well-established ways. The first is pipetting droplets onto a substrate and allowing them to dry before using the laser read to record the nanospectra. The second is using the "flipping" innovation, which allows the pipetted sample to be held pinned to a transparent plinth when the magnet field is switched on to magnetically aggregate the MNPs to the base of the plinth to be read without attenuation as the plasmonic wave is at the base of the transparent plinth. The sample can be read as a Raman and then SERS signal simply by switching the magnetic field on and off. This method is illustrated here and patented in a droplet system patent by McMillan et al. [70]. This method is called "Flipping MagSERS" with 1- to 3-μL droplets on a 2 mm diameter plinth, but it is possible to access droplets as small as 300 nL for a top read with the laser on an unflipped 1 mm diameter plinth. These methods importantly are immediate and remove the need for drying. The cavity under the plinth makes the laser "back-read" for the laser excitation through the base of the plinth as short as practically possible. MagSERS, at this point, requires solenoid operation. The solenoid field is used to produce a concentration of the MNPs on the base of the plinth, but using a neodymium rare earth magnet mechanically pushed up into the plinth base with a stepper is an alternative method to aggregate the MNPs, but blocks the laser if the rear-read operation is required. This patent proposes also using a SlipSERS substrate on the plinth as an alternative method for achieving a similar flipping SERS method. The "Slips" stands for slippery liquid-infused porous surface(s), patented substrates described by Yang et al. [71], although this SlipSERS proposal does not allow instant back-read as it requires droplet drying and surface preparation of the plinth and thus is not recommended. This option is mentioned here, as there are a number of possible new SERS variations on the MagSERS theme approaches using flipping of droplets and back-read laser excitation.

The MagPLINTH innovations importantly also offer a way of removing the nano-spectroscopic baseline variabilities that are the other problem that bedevils SERS measurements. This innovatory approach based on traceable methods puts calibration at the heart of every measurement. The costs of such methods are made possible by the minute volume of 1 microliter/measurement despite the cost of standard solutions. The nanovolume sample economies arise from a routine calibration economic enough to make traceability a standard part of all measurement procedures. Furthermore, flipping these plinths makes it possible to produce replicate readings rapidly. For water droplets, merely touching a water-based sample surface with a pin dipped in oil produces an oil cap and allows kinematic measurements, e. g., of enzymatic reactions to be monitored, using SERS and Raman toggling between both measurements with the laser excitation in the same position.

The McMillan patent [72] proposes using a new method called "Aliquant," a variation on the method of standard additions. Figure 3.3.6(c) shows schematically pipette loading of transparent "back-drop-read" plinths for both standard aliquot-adapted replicate Raman measurements, but also for the magnetic concentration of MNPs when switching on the field for the SERS measurement, with the laser read through the cylindrical cavity at the back of the plinth. Let us look at just a couple of options of the method noting there are other possible variations on the Aliquant theme. Approach 1: Using a 1.00-µL droplet with a calibrated pipette: (i) make a standard with the analyte in solvents that evaporate; (ii) pipette a single drop of MNP solution and add one droplet of analyte; (iii) rear-read measure the signal with switching magnetic field on; (iv) add successively two pipetted drops of analyte and rear-read measure; (v) add sample droplet and rear-read; (vi) add perhaps three standard analyte droplets to complete the trend line shown in Figure 3.3.6d. The lift or fall in the trend line allows a quantitative sample measurement using standard curve fitting. Such quantification in SERS is difficult, so this reliable method has many advantages, although this approach depends on the evaporation of droplets to make the process work, so it is not instant. The problem of SERS measurement is associated with removing the scattered background and other issues, but the measurements follow a sequential and reproducible stepped method. The error estimates can be made by obtaining replicates by repeating several measurement series a number of times. The axes are SERS intensity versus steps.

Figure 3.3.6e shows the time increase in the intensity of the SERS signal with two different analyte concentrations but in a sample containing MNPs and using rear-read. The plinth engineering allows for disposable sample holders and excitation from below via flipping or a Raman probe. Variations in the design of the plinth, droplet sample, MNPs and sometimes volume additions, magnetic and environmental control of the droplet sample, and indeed other experimental control issues are the fundamental building block-steps for these new assays. The Aliquant is a variation in the Aliquot theme, but rather than volume additions with additions of MNPs the "quants" are additions of known increments of sample MNP numbers.

It should be noted that the additional droplet method with evaporation is not the preferred option as it takes time, and replicates are not easily obtained. If a rear-read is done with droplets mixed in a small vial, quite simply droplets of standard analyte solution and the NP solution are mixed and then measured on a plinth with rear-read. There are several options here, but suppose the following mixtures of standard known concentration and volume of analyte in a droplet (A), a standard NP droplet (N), and the standard volume droplet of sample (S) are made: A+N, 2A+N, 3A+N+S, 4A+S+N, 5A+S+N, and 6A+N+S. Then a plot of concentration against step will show a decreasing concentration curve, which will have a discontinuity of a lift or fall in the trend line from which the concentration can be obtained. These calculations are simple. The method allows immediate reading of pipetted droplets via the rear-read and thus replicates of the reading can be taken.

We summarize the advantages that have not been touched on in the above discussion of Aliquant. The system can be improved in volume measurements using a camera rather than the poor accuracy of microvolume deposition using the pipette volume settings, which is a limitation here. Advantages of the Aliquant method are as follows:

1. This analytical approach can deliver Raman and SERS measurement from a laser excitation that is not moved.
2. This approach of concentration of magnetic plasmonic NPs should increase the SERS sensitivity to below existing LODs. The current invention solves problems associated with maintaining sample integrity and the value of preventing contamination from carry-over, and ensures no other extraneous material enters the sample aliquot.
3. A succession of drops can be added, and importantly the magnetically concentrated signal does not depend on sample volume but on the volume of sample addition.
4. Vitally, the SERS sample does not need to dry on substrates with Flipping-MagSERS plinths with rear excitation/signal read Raman/fluorescence/other probe/etc. to obviate the need for waiting for a droplet to evaporate.

This approach can be used for nanospectroscopic measurements on SERS forensic plinths for total sample recovery. The instrument has the advantage of precise control of the scattering volume of a liquid drop as well as minimal losses of light during Raman measurements. The system offers an easy quantitative Raman data collection method from the large set of samples with variable concentration (or polymorphism, functionalization, etc.) of the compound under investigation, as well as easy quantitative comparison with the reference samples.

A current goal in producing a more "all-encompassing theory" is to quantify the SERS effect, which is a very difficult challenge, but one that is presently concentrating many minds. The confined nature of the hotspot enabled in tip-enhanced Raman scattering [73] allows one to obtain resolution below 10–50 nm without losing any spectral information (cf. Fig. 3.3.7). The enhanced area expands 10–30 nm from the tip apex, making the technique surface-sensitive. Regarding TERS instrumentation, it mandates precision control of the tip position determined by the AFM or STM feedback loops. The optical readout is obtained by focusing the excitation laser on the tip landed on the sample and detecting the scattered light by the same optics. Imaging requires the stability of the optical alignment in such a system. TERS has been demonstrated in various applications, from examining the purity and defects in carbon nanotubes to visualizing individual carbon-based nano-objects [74] and the compositional analysis of viruses and protein structures. In the case of our special focus on forensics, TERS has been demonstrated for determining the bases [75] and sequence [76] in single DNA strands and for identification of inks used in hand-written documents [77].

Besides system stability, image quality is largely determined by the contrast between the far-field signal (the signal obtained from all the volume illuminated by the

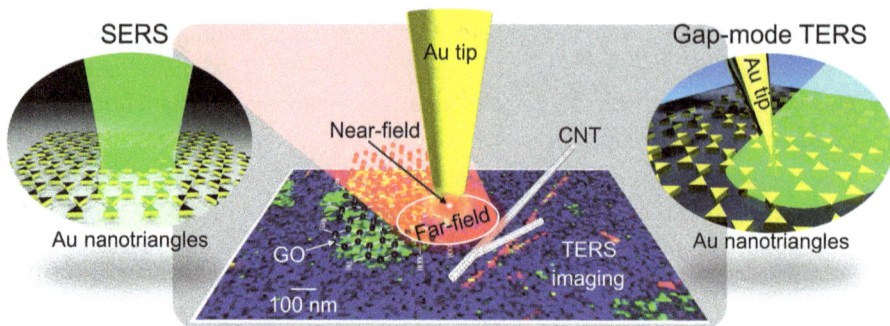

Figure 3.3.7: In SERS, plasmonic NPs offer the ultrasensitive detection of minute amounts of chemicals and the identification for forensic purposes, for example. TERS is the ultimate case of SERS, where a single hotspot is generated by the tip, allowing to obtain chemical images with nanoscale resolution. Gap-mode TERS is one of the most successful implementations of TERS that arises when a metallic substrate or plasmonic nanostructure is used, and the plasmon hybridization between the tip and metal substrate provides a further boost to the signal enhancement.

laser) and the near-field signal (the signal coming from the enhanced field region underneath the tip). The far-field can be reduced by selecting a thin sample and high-numerical aperture optics. The near-field intensity is determined by the enhancement of the tip and the enhanced volume. While the former intensity is well controlled by modern technology, the reproducibility of tip enhancement is low. The largest enhancement and confinement are achieved with very sharp tips, but the sharper the tip, the harder it is to attain reproducibility, which hampers their mass production. Many recipes in the literature describe the manufacture of good STM probes for TERS [78]. Still, many are fragile metal-coated AFM tips that delaminate due to the difference in thermal properties of the silicon tip and the metal coating. These temperature differences arise in TERS experiments due to photothermal heating upon laser illumination. Efforts are focused on extending current reproducible AFM tip batch manufacturing processes to TERS, the most common being the use of metallic substrates. Their optical properties support a strong image dipole in the substrate that provides higher enhancement and confinement, even for the low enhancement tips. That means that the easiest samples for TERS are nanometer thin samples on gold or silver substrates, which is widely reflected in the literature. Recently, full metal probes have been developed for AFM-TERS that, similarly to STM, are entirely made of silver or gold, with no metal coating whatsoever [79]. Even though probe development is on its way to be solved, there are methodological issues to be addressed before the technique can be widely used for forensic purposes, such as the lack of a calibration standard, and also the tendency of the TERS spectra to differ from the microvolume Raman spectra. Another challenge is the routine demonstration of nanoscale imaging on samples deposited on nonmetallic substrates. Thus, although the promise of TERS stays in the labs for the moment, there is considerable potential in forensics for the trace level detection of illegal chemical substances, DNA sequencing, etc.

Exploiting refractospectroscopy at the nanoscale – contributions to tensionanospectroscopy and nanodroplet sample preparation and handling

The chapter by Laible, Horneber, and Fleischer entitled "Localized surface plasmon resonance shift sensing" in this volume (Chapter 3.1) details the refractospectroscopy at the nanoscale. The size of the probed volume by laser is restricted by the diffraction limit as we have seen with TIR. The diffraction limit can be overcome using TIR and special near-field probes, which take advantage of evanescent fields. An important effect of what might be called molecular nanorefractometry is achieved from the refractive index sensitivity of metallic NPs to molecular binding, leading to a reversible shift in the localized surface plasmonic resonance. Horrer et al. [80], e. g., modeled this effect using the narrow antisymmetric resonance in coupled plasmonic vertical dimers. The resonance wavelength depends, among other parameters, on the refractive index of the particles' surroundings [81].

This work is part of a current research effort extending from the authors' pioneering work on tensiography and tensiospectroscopy in the late 1980s. It initially integrated imaging with fiber light path analysis in large pendant droplets of 80 to 120 µL, culminating in the invention of optical tensiography in 2006 [82]. That work later evolved to tensiospectroscopy on sessile droplets and the invention of NCLS to produce an accurate measurement system [83] to replace the Beer–Lambert law below the diffraction limit. Now a multimeasure and multimeasurand microcamera system, the Multimeasure Tensionanospectrometer, has been devised to monitor the refractive index using supernumerary interferometry in a sessile droplet working below the diffraction limit and so as a nanospectroscopic technique. The patented "Multimeasure Technology" not only measures the refractive index with great precision on pipetted droplets of typically 1 µL sitting on quartz plinths, but also measures surface tension, conducts dynamic measurements of volume to determine the complex evaporation rate, and calculates the UV-Vis absorption coefficient (α_f). It has been extended to measure the rheometric properties of a vibrating droplet. Other physical quantities can be obtained with this Multimeasure system, namely the rheometric, absorption, and turbidity coefficients. In Figure 3.3.8a, the refraction of rays coming through a droplet are shown in color to aid the eye track the ray paths. A stepper drive intercepting the rays systematically allows the engineering of a very sophisticated refractometer, but the simplest measurement possible with a camera system is for the minimum deviation ray. This measurement can also be determined from interference and diffraction patterns, and the latter is a true nanomeasurement being described by wave physics. Figure 3.3.8b shows a refraction measurement from the stepper motor-driven Multimeasure. This Multimeasure instrument is a partner in crime

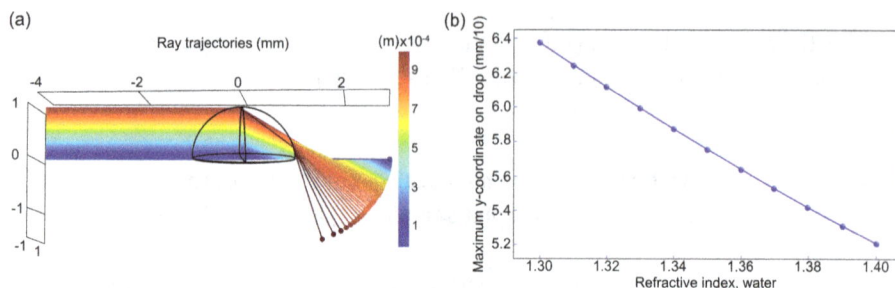

(a) Ray trajectories (mm)   (m)x10⁻⁴

(b) Maximum y-coordinate on drop (mm/10) vs Refractive index, water

Figure 3.3.8: (a) Modeled ray paths through droplet showing supernumerary interfering paths color-coded and not representing wavelengths. (b) Refraction measurement from the Multimeasure Tensionanospectrometer.

with all the ANT forensic spectroscopic innovations such as the Dual-spectroscopic NanoWand [84], a portable system developed for monitoring bogs, or the patented so-called HHOUDINIS [85]. This invention for drop sample processing uses partial hydrophobic coatings to ensure that the droplet can be removed simply by gravitational tipping, blowing/sucking, charging, or other ways. The reason for highlighting this invention is that tensionanospectroscopy is a field of surface science, and here the removal of all contamination from surfaces is not just "nice to achieve," but essential to the instrumental platform delivering the science.

3.3.5 Challenges and solutions

(i) Instrument challenges

The comparison of performance specifications for nanospectroscopy to those shown in Table 3.3.1, for microspectroscopy, is not too revealing of the differentiation of these two spectroscopy domains. There are some essential differences between micro- and nanospectrometer architectures, for example, nanofluorescence is made possible by advances in laser technologies that are not generally exploited in microspectroscopy. These differences, although vital, can be thought of more as variations on the established theme of microspectroscopy. Volume 2, Section 4 has highlighted some of the nanocomponents in the spectroscopic measurement position which is often in an image plane.

The tangible differentiation of spectroscopies above and below the diffraction limit relates to what will be called "nanospectroscopic issues," and below simply "issues" in Table 3.3.3, which have been compiled through a collective effort of the authors. It has been condensed to those issues which we believe are most important for ESRs, particularly the key research advantage. Table 3.3.3 must be considered in conjunction with Table 3.3.1 to inform and support conclusions. Table 3.3.3 is further

Table 3.3.3: Review of issues, performance factors, and major applications for various nanospectroscopic approaches. The details here are illustrated by just one transformative element (there are others and these are just examples of some main elements.)

ACROSS: Nanomodality DOWN: "Issue"	Nano UV-Vis	Nano UV-Vis/fluorescence/NIR	Dynamic light scattering/multi-angle light scattering (MALS/DLS-zeta potential)	SERS-IR	TERS-IR	Nano IR (NIR/MIR)	(i) Nanofluorescence (ii) Autocorrelation fluorescence (iii) FRET	Nanorefractometry
Essential or improved hardware component(s) for nanospectroscopy	NPs (with cyanine dyes) QDs Quartz tensiographic drophead designed for nano-UV-Vis	Molecular excitation from nano source or QD	New specific laser sources and detectors	NP (various) and plasmonic nanostructures, functionalization for specificity and sensitivity of target molecules	Reproducible plasmonic tip; objective lens; stable STM/optics coupling	Reproducible nanoaperture, tapered waveguide or nanoantenna, for radiation sources	Prism for near-field coupling; intense pulsed/laser sources, detector architectures, labels, NPs, etc.	Rayleigh scattering-based Aptasensors are a class of refractometric sensors/dimer arrays (here with sandwich of truncated cone with Au-SiO$_2$-Au), etc.

Table 3.3.3 (continued)

ACROSS: Nanomodality DOWN: "Issue"	Nano UV-Vis	Nano UV-Vis/ fluorescence/ NIR	Dynamic light scattering/multi-angle light scattering (MALS/DLS-zeta potential)	SERS-IR	TERS-IR	Nano IR (NIR/MIR)	(i) Nanofluorescence (ii) Autocorrelation fluorescence (iii) FRET	Nanorefractometry
Nanocontrol factor(s)	Vibration and damping and ultraleveling, environment chamber for sample	Tuned and engineered QDs (size)	Temperature typically −15 °C to 150 °C but ultra option to 210 °C. Laser mode-hopping hardware to improve their stability for both 658 and 785 nm.	pH; polarization; use of raster scan to locate "hotspots"	Feedback controlling tip sample separation, vibrational damping, alignment on tip, specialized scanning modes, especially in AFM	Aperture/surface feedback, vibrational damping, engineering towards portability and chip-based consolidation	Various but for one example SM fluorescence through control of sample volume using confocal system. There are many options.	Engineering of LPR and other factors such as those concerning excitation of this dimer structure with white light from top through PDMS
Key nano software algorithms for nanocontrol requirement(s)	Big Data capability; qualitative fingerprinting nanospectroscopic effect from multiplexed encoded complex signals	Laser control technologies for example in analytic applications but extending to commercial, e.g., display/TV/solar panels	Dedicated packages making possible multiple measurements and applications in biotherapeutics, protein, biopolymers, synthetic polymers, NPs, etc.	PCA and some standard data mining methods, Big Data capabilities with access to nano reference libraries	Synchronized STM/optical scanning, automated identification of localized "hotspots" on the tip, spectral database	Spectral database providing fingerprint capability with Big Data capabilities	Environmental (various) and surface passification protocols, etc.	Analysis of results requires integration of overall near-field and normalized electric field but depends on use of reference refractive index (RI) and absorption coefficient data

Table 3.3.3 (continued)

ACROSS: Nanomodality DOWN: "Issue"	Nano UV-Vis	Nano UV-Vis/fluorescence/NIR	Dynamic light scattering/multi-angle light scattering (MALS/DLS-zeta potential)	SERS-IR	TERS-IR	Nano IR (NIR/MIR)	(i) Nanofluorescence (ii) Autocorrelation fluorescence (iii) FRET	Nanorefractometry
Key theoretical/scientific tool	Beer–Lambert ditto Universal optical ten-siometer law connecting molecular thermodynamics using dynamic surface tension (DST) and UV-Vis from one optical measurand	Theoretical treatment of QDs and other nanosensors directed to opening up analytical applications	(i) Rayleigh equation has been developed (MALS) by recent researches and DLS theory (ii) Stokes–Einstein relationship (DLS)	Identification of plasmonic and chemical enhancement factors for local environment	Treatment of beam/probe/sample interaction as a point dipole, either oscillating or finite	EM theory to connect far-field to the near-field	Taking one example FRET nonradiative dipole–dipole coupling underpinned by theoretical description of interaction	Plasmonic theoretical substrate design and developments linked to near-field spectroscopy.

Table 3.3.3 (continued)

ACROSS: Nanomodality DOWN: "Issue"	Nano UV-Vis	Nano UV-Vis/fluorescence/NIR	Dynamic light scattering/multi-angle light scattering (MALS/DLS-zeta potential)	SERS-IR	TERS-IR	Nano IR (NIR/MIR)	(i) Nanofluorescence (ii) Autocorrelation fluorescence (iii) FRET	Nanorefractometry
Key research advantage	Richness and flexibility from multimeasurand, and, exploitation of new data mining capabilities.	A palette of opportunities with limits seemingly only bounded by the researcher's imagination given the proliferation of quantum particle types.	PEGylated protein drug products (where PEG is polyethylene glycol), biologically derived substances such as low-molecular-weight heparin (LMWH), and inherently complex pharmaceutical substances, etc.	An enormous range of possible SERS methods and assays are opening up one of the largest research fields in optical nanoscience	Single-molecule detection is possible but submolecular resolution is typical for both imaging and spectroscopy	SNOM can be applied in all environments, from ambient conditions to cryogenic temperatures, and for liquid samples.	Microscopic systems offer a research platform in a rapidly evolving field for which component and substrate engineering flexibility is key	A foundational study has been highlighted in this chapter as an example of several recent studies but which describes important novel measurement approaches using RI as a calibration tool and thus opening up new research opportunities.

Table 3.3.3 (continued)

ACROSS: Nanomodality DOWN "Issue"	Nano UV-Vis	Nano UV-Vis/ fluorescence/ NIR	Dynamic light scattering/multi-angle light scattering (MALS/DLS-zeta potential)	SERS-IR	TERS-IR	Nano IR (NIR/MIR)	(i) Nanofluorescence (ii) Autocorrelation fluorescence (iii) FRET	Nanorefractometry
Measurement time	Fast Fourier transform analysis. Minutes for full analysis result but real-time monitoring is done	From short times excited by laser pulses to very extended period of minutes	s	ms	Tens of minutes to hours	Tens of minutes to hours	ns up to hours	ms
Analysis issues	Drift a small problem but handled by new software of measurements	Need for chemical labeling of targeted medical diagnostics and with product engineering advancing to provide automation.	Absolute measurements of molecular weight and absolute dn/dc measurements are important issues to be considered.	Fluorescence is a major issue for weak Raman/SERS signals in quantitative measurements	Background signal, changes in the vibrational spectra as compared to classical methods, spectral changes with time	Interference from the background signal is a problem, with drift issues	Interferences are not significant	The modeling is not yet fully connected to the experimental results and thus cannot be used to optimize fully the measurement system

Table 3.3.4: Summon Discovery Service (library search algorithm) search and analysis of the applications for various nanospectroscopic modalities for publications from 2000 to 2018. Figures indicate the number of publications for this search to identify the number of nanospectroscopic studies in the 12 wide-ranging disciplines, while the five sets of methods across the top differentiate the number of hits using search information of the major technique used in this nanospectroscopic work. The five sets of methods are those supportive of the research in the discipline and contain grouped or related topics. The final column indicates the relative importance of the various nanospectroscopic works in the discipline but contains studies on consumer products which were 2.7 % of the total number of hits.

A = Nano-modality D = Applications	NIR/MIR	SERS/SERRS/TERS	Fluorescence/ Correlation Fluorescence	DLS/MALS	Refrac- tome- try	100 %
Medical	6525/2010	8321/306/2846	26,528/1035	8228/2132	6844	6.29 %
Veterinary	466/468	836/32/437	2262/1248	575/365	10	0.84 %
Bioscience	19,727/4164	26,469/818/8156	78,676/28,433	23,639/5288	2422	18.05 %
Chemical	25,933/5148	32,439/855/10,982	88,542/30,919	25,270/6827	31,377	24.18 %
Pharmaceutical	3894/1419	5236/175/1968	18,924/8380	7783/1727	4734	5.07 %
Environmental	8759/2141	11,786/343/3567	23,232/13,459	8416/2366	9500	8.60 %
Food	4523/1653	7274/252/2138	21,275/9248	7243/1736	6045	5.84 %
Agriculture	1014/545	1550/55/671	4088/2041	1051/461	1283	1.35 %
Material	25,490/5038	31,077/784/10,665	84,784/29,287	24,537/6681	31,833	23.46 %
Geology	649/379	1158/33/600	2961/1649	435/333	748	0.91 %
Built environment	2564/894	3485/113/1410	8131/4110	1469/880	3375	2.65 %
Consumer	2467/959	3622/57/1428	8890/4110	2702/1014	3081	2.77 %

contextualized by the "Summon" (a search algorithm used in this work) analysis of the number of publications in the various application fields of nanospectroscopy provided in Table 3.3.4. The apparent overarching conclusion from the analysis is that critical nanospectroscopy advances relate to the theoretical/experimental breakthroughs and hardware and software invented to address the new demands of the specific nanotechnique, operating below the diffraction limit. These innovations are varied but depend on the genius of researchers to break through the diffraction limit barrier. It also indicates that there is no major difference in prominence among all modalities of the various nanospectroscopies when MIR and NIR are taken together, as they should be, since many studies mentioning NIR also mention MIR, and vice versa. The readers are invited to consider these issues for themselves critically; there is not enough space here to further explain the details of the table, but we believe this table is a rich mine of valuable knowledge where the micro to nano issues arise.

The starting point for forensics today is to understand that IR absorption frequencies are the equivalent of chemical species fingerprints and provide identification of the unknown by comparison with IR spectral databases. It so happens that FTIR is also the reference method for carbon measurements and soil characterization. Thus, the "environmental" application study extending from the forensic backbone of this study (see below) lines up beautifully. The macro libraries are far and away more extensive and analytically powerful for microspectroscopy than for nanospectroscopy. Fourier transform IR (FTIR) spectroscopy is a powerful technology for obtaining spectral information from a small sample area. Focal plane array (FPA) detectors are an enabling technology here for chemical imaging, allowing an array of perhaps 65,000 location-specific IR spectra to be recorded in almost real-time and for most biological processes. Far-field IR microscope systems with every pixel enabling a high signal throughput to receive information from relatively large areas are limited in lateral resolution, so they are of limited utility and unable to deliver at the nanoscale.

Chemical and bioimages of samples are computer-generated from measurements of peak heights, areas, peak ratios, correlations, principal component analysis (PCA), and other spectral measures. These images are usually useful for monitoring chemical or biological changes. Images can be recorded by collecting point measures in mapping, scanning a surface, or parallel multichannel detection. Techniques such as PCA can be used to process these data. Microscopy is far-field, being diffraction-limited, and nanoscopy is near-field data collection with the spatial resolution being usually defined by the aperture size. IR nanoscopy overcomes the diffraction limit with nanometer resolution, being a nondestructive method, and does not require vacuum conditions. This technique requires sophisticated control and electronic feedback systems, but offers several methods of data collection. It places demands on data collection sensitivity as signal intensity is low.

Further nanoscale advantages can be delivered by connecting complementary nanoscale techniques. Centrone for example [86], based on the old workhorse AFM, has developed photothermal induced resonance (PTIR), resonance-enhanced AFM-IR (RE-AFMIR), and s-SNOM to provide overlapping and spatially correlated information. Lahiri, Holland, and Centrone showed a concrete example of the advantages of nanoscopy over microscopy [87].

(ii) Software challenges – visualizing nanospectroscopic signals

There are many possible software options to assist nanospectroscopies in both imaging and spectroscopic applications; for example, the data scatter method pioneered by ANT with some notable successes, for example, (i) in solving a century-old deconvolution problem of quantitating food color admixtures [88], (ii) in Raman silicon wafer measurements [89] matching the low-cost instrument's performance with that of the state-of-the-art Raman systems used in Intel Ireland, and (iii) in drop shape imaging analysis [90].

Data mining methods are essential tools in nephelometric spectroscopies (measuring the intensity of scattered light), but today are almost turnkey in their operation. These provide well-trodden paths and ones that do not need to be discussed here. The considerations here relate rather to philosophical issues of nanospectroscopy, addressed by new approaches designed specifically for these unique problems of analytically complex and interconnected issues. The software enables the removal of the complex and variable background signals, treatment of read laser beam attenuations and scattering, the investigation of plasmonic signal complexities, and the probing into the compromised statistical distributions of measurements from foregoing issues. The basic questions of chemical and biological quantitation from nanospectroscopy are at the heart of these discussions. The shape analysis of the spectral lines is connected fundamentally to the statistical analysis of these spectra. While the tensiographic shape analysis is accepted as the key to understanding surface kinetics in chemical and biological processes from measurements of drop shape variations that are defined by these processes, in nanospectroscopy, shape analysis of spectral lines is in itself an innovative departure whose potential contribution is largely ignored. The importance of thinking about these issues is however underlined by the well-known fact that turbid solutions from complex biological liquids cause variabilities in spectral heights. What we are addressing in tensionanospectroscopy is line shape variabilities from surface science droplet shape variations in addition to other nanospectroscopic variations. The way to connect these issues is using statistics traced back to fundamental information theory from statistics of energy levels.

This method was developed initially for drop shape analysis, which is a very challenging analysis problem starting with the fact that Fourier components are limited in value. Data scatter has also been used in optical telecommunications system diagnostics again with monitoring very subtle shape variations from small optical dispersions. The data scatter methods described here are a tool able to view the nanoscales of optics with reciprocal scatter measures of these signals which reciprocate, making the small large. Here, the discussion will be restricted to the direct data scatter approach which is itself almost a virgin field.

This approach is unique in being able to identify causation from the data scatter patterns. Figure 3.3.9c shows such a pattern of a digitized signal with a distinctive X and a distinct vertical line at 10° to the vertical. These scatter patterns are similar to X-ray patterns being arcs/loops, lines, and rows. There are pattern measurands that have Shannon measures for size, orientation, and indeed other measurands. The quantitative measures sensitively tune to give quantitative data scatter measures of many variabilities of great utility. Recently, this data mining tool has been fundamentally developed delivering the simple-to-use Cloud App STRAPS (**S**hannon **T**raceability **R**eference **A**nd **P**attern-Analytic for **S**pectra). The ultrafine scatter details as can be

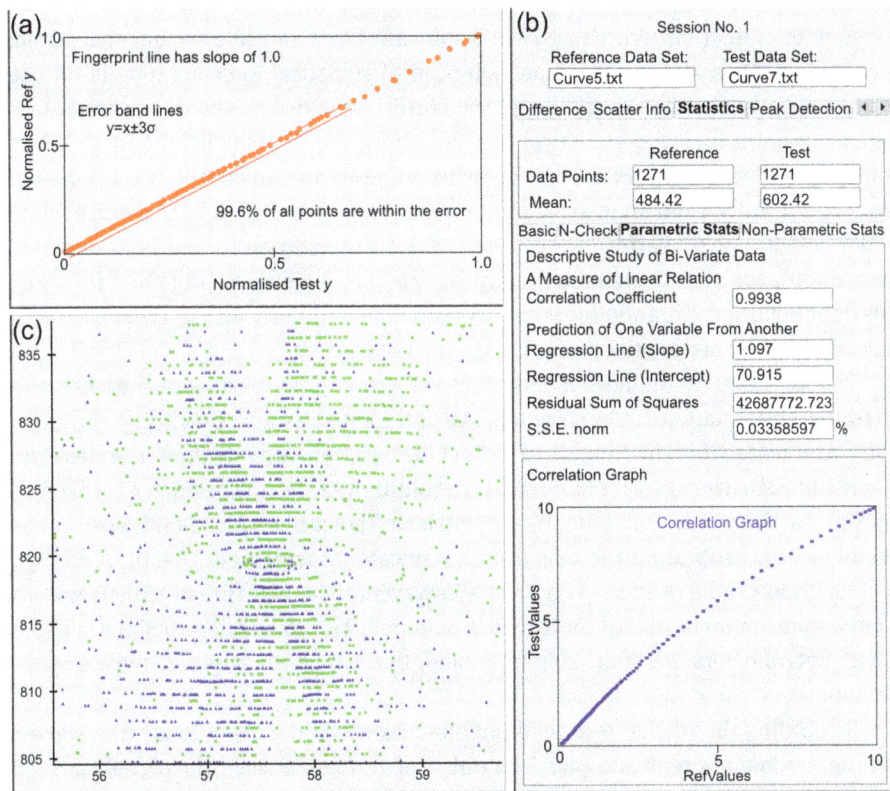

Figure 3.3.9: (a) Double-click (DC) 6-sigma noise reference measurement illustrated. (b) "Measurement signal" of STRAPS of a correlated deviation of signals at higher values in a statistical window. (c) Example of data scatter for digital signal showing considerable amount of "qualitative detail."

seen from this one scatter diagram are for optical distance measures that are nanodimensional. This is a data mining nanotechnique that is in its infancy.

The double-click 6-sigma noise reference measurement is illustrated in Figure 3.3.9a for two Raman spectra recorded by a rapid "double-click" acquisition. For perhaps most instruments, they have reproducible signals for two captures taken in quick successions, such as with Raman spectra. These two snapshots are effectively replicate signals. This plot displays such a signal pair of the one signal against its matching twin. It is a straight line of slope 1.0, and the difference between the two is the signals' noise. A bit of simple mathematics confirms the statistical fact that 99.6% of all plotted points sit inside the "$\pm 3\sigma$ error band." From this plot, it is simple to deconvolve the two orthogonal noise components for each signal and plot the noise distribution. This is a "fingerprint line plot" noise check, which is called the "signal fingerprint check" (SFC) plot. Given that the scatter signals usually have thousands of digitized points, the plots are a very good visual check on whether there is a normal or parametric distribution. This STRAPS software check (a useful visual check from the Toolbox) then allows the proper selection of statistical tools acceptable for use in the analysis. Having knowledge of the correct statistics tool is the all-important departure point for valid statistical analysis. Now, the double-click method allows "on-the-spot" visual inspection of the distribution to help make this vital decision. More importantly, from an instrumentation metrology point of view, this noise figure is the best possible "performance measure" for the instrument and here obtained immediately for this spectrometer analysis. The "noise measure" obtained from just this double-click calibration measurement provides a number for the LOD ($\pm 3\sigma$) and, indeed, the limit of quantitation and analytical sensitivity.

This preliminary analysis is important as $\pm 3\sigma$ can be taken as a measure of a Shannon "information bit" exploiting SMI interpreted as an uncertainty measure. The y-axis scale, from the "double-click" calibration, can be made dimensionless and labeled in SMI. The spectral bandwidth (resolution) of the spectrometer is taken as an SMI for the y-axis. For Raman, an "intrinsic" silicon standard ($520.5\,\mathrm{cm}^{-1}$) has been used for calibrating the axis from a replicate measurement. Again, a double-click method can be used to generate a $\pm 3\sigma$ wavenumber measure of SMI. From this a noise measurement plot of the x-axis is obtained. The conversion to SMI scales for other spectrometers is a short step from here, but will not be explained given space limitations.

Reminding the reader, the rapid double-click method is used to acquire signals that approximate a replicate pair. The only differences between the signals are the noise components that are different on each of these two matched digital signals. Figure 3.3.9c shows the scatter pattern from the signal mapped in Figure 3.3.9b. For the signal mapped between scatters produced by a reference signal and another test signal that diverges systematically from the reference at higher signal levels (note signal divergence in a simple way with decreasing slope departing from the 45° slope), the divergence from the fingerprint-line plot is visually and quantitatively a useful diagnostic tool.

The instrumental analysis, data transmission, and interface technologies are the factors that have been optimized, engineered, and now upgraded using SMI methods. The use of these measures also can be exploited in software to help guide the user during a measurement to deliver the best result in the time available for the measurement. However, measurement optimization is possible but can only be delivered in a dynamic on-the-spot guided adjustment and should be part of a nanospectroscopic analysis. There is no need to over-elaborate here and these issues will be simply stated in the reference made to the work.

–   The tensiograph was developed on the basis of maximizing the reciprocal entropy (experience has shown this measure is easier to use visually than entropy) of the various noise components – electronic, optical, thermal, and mechanical – for this instrument. This work led to the development of an optical instrument with unrivalled sub-ppb detection levels [91].

- Noise issues were subsequently minimized using wavelet measures in the signals [92].
- The vital issue of minimizing the measurement error in the limited measurement time has been dealt with to ensure the user decides on the best experimental option that will deliver the minimum error in the work done by Bertho and McMillan [93].
- The data scatter method has been developed for data mining in spectroscopy and other fields. This is a sensitive fingerprinting technique that has two sides: the quantitative, from SMI measures on the goodness of fit, and the qualitative, from what is a data scatter-based pattern that identified the cause of the differences between signals. In addition, the scatter pattern between two measurements shows the noise type (thermal, shot, etc.) for experimentally equivalent signals [94].
- Shannon and Boltzmann entropy methods have been investigated to optimize the transmission of digital optical data in fibers [95]. The resulting Mark 1 tensiograph optimizing the performance has been described and its performance has been quantified [96].
- The tensiograph was tested in an EU project and subsequently in an NRA study showing sub-ppb detection levels for PAH compounds [97].
- These methods have been used in the development of all ANT products and the numerous applications of these technologies [98]. The tensiograph has also been used for quality control of Guinness stouts, identifying differences in stout mixtures in a matter of seconds; a similar analysis with GC-MS takes hours [92].
- The spectra to be analyzed require the following processing: distinguishing the signal from its contextual background; distinguishing the signal from noise arising from the environment or from the detector and measuring signal; and clearly demarcating patterns and trends that, in effect, amount to zooming in on the spectrum. The main method that supports all such processing is the wavelet transform. In wavelet transform space, the spectrum's background is determined, and with a noise model that is well expressed in terms of resolution level for prominent statistical distributions of noise, wavelet-based filtering can be carried out. Zooming in on the spectrum is carried out through the coefficients defined at the resolution scales or resolution levels provided by the wavelet transform. Köküer presents a thorough analysis carried out on tensiographic spectra of beverages. The outcome of the wavelet transform is displayed in [98]. Information content at each resolution level can be determined and expressed as entropy. The overall entropy of the spectrum is calculated as Shannon entropy, summed over the set of resolution scales, and over the set of samples, i. e., discrete elements of the signal, at each resolution level. Furthermore, if the signal is considered as signal plus noise, then as stated in the cited article, the signal's "entropy is proportional to energy divided by noise." A consequence of basing our entropy definition on multiple scales is the following, from Starck et al. [99]: "One consequence is that correlation between pixels is taken into account when measuring information."

Impact

The use of SMI is already well represented in the literature with information measurements to the base 2, spectroscopic measures such as absorbance to the base 10, and chemical reaction measures of Shannon entropy and biodiversity using Shannon–Weiner based on the natural logarithm base of Euler's number "e." Conversion from one scale to the other is a simple matter of a logarithmic ratio of the two respective bases.

The idea of a Shannon depository is one that can use the single ruler of the dimensionless SMI to unify the data representation. For nanospectroscopy, this is very important in allowing a vision of the nanospectroscopy set against that of traditional and microspectroscopies. Let us consider the subset of Raman and SERS. The scatter analysis of these signals shows these distinct and different signal types, with different scatter signatures. The two representations immediately illustrate the importance of the various data probes of the statistical data distributions, data scatter patterns and measures, fingerprint-line structures, and other means of analyzing the data and from there decision making. The recommendation we make is to convert all to base 2, but this is an optional issue.

The final point is to explain why data scatter is so extraordinarily sensitive and able to deal with complex admixed multimeasurand or spectral signals. The tensiograph will deliver ultrasensitive measurements of surface tension, UV-Vis spectra, turbidity, and refractive index, but beyond that it can deliver measures of other dependencies such as reaction rates on solid substrates as explained in the State-of-the-art section. From fingerprinting using data scatter signals such as the tensiotrace it is able to give ultrasensitive measures of any dependencies using calibration for example of spiked samples and using the data scatter measure plotted to give a measure of an unknown. This approach perhaps has enormous importance for forensics.

(iii) Fusion challenges: the development of nanotensiospectroscopy

The integration of nanospectroscopies and their fusion with micro- and traditional spectroscopies requires re-evaluations of many fundamentals. While this millstone is turning, it may just as well address the issue of the fusion of spectroscopies with relevant physical measurements.

This work on the presentation of all spectroscopic data in a universally consistent way has been largely achieved in a broad brushstroke way using SMI methods. The future development of "optical tensiospectroscopy" is arguably best achieved with a nanoresearch focus. Moreover, the proposal is in an absolutely critical area to the new sustainable Europe 4.0 extending the methods and building tensionanospectroscopy.

Lest we forget, spectroscopists are famously conservative, for example resisting the adoption of the desire to remove the unit of wavenumber (unit cm^{-1}) proposed in the *General Conference on Weights and Measures*. The professional resistance eventually prevailed, and this "wavenumber" unit was admitted as what is termed a "coherent derived unit in the SI expressed in terms of base units." Nevertheless, it has to be possible to have the spectroscopic representation of the diverse but familiar and accepted forms for any modality representing spectral information, toggling with a dimensionless normalized SMI representation.

Integrating nanospectroscopies requires deliberations beyond the obvious of ensuring the scales on the axes for all the various nanospectroscopies are identical. The first issue is combining quantitative information from two sets of data. ANT technologies, e. g., allow microvolume and nanovolume spectra to be recorded for various modalities (UV-Vis, NIR, MIR, Raman, fluorescence, or DLS); this comes from the necessity to deal with the issue of how to properly use quantitative measurements from, for example, UV-Vis and NIR in a statistically valid way. There is no space here to go into any detail on specifics, but some elementary thoughts on these complex matters are useful in what, in effect, is simply little more than raising an important question. Clearly, if two wavelengths can both deliver quantitative measurement of an analyte (perhaps in a reagent as a single component or with more than one component) from either modality, then the best quantitative result should be used. There are, however, various considerations here, not just the obvious judgement based on metrological issues (accuracy, reproducibility, etc.), but other quantitatively relevant information such as chemical interferences, specificity, signal drift, etc., that could come into play. To illustrate the problem, given a superior measurement of a component from one method, there are possible reasons for using the other result if the first is compromised, for example, by interference. The usual objective is to minimize the statistical measurement error. These problems are ones that should be as far as possible flagged in nanospectrometers to dynamically feedback with instrumental intelligence. This is an ambition of the multimodality omnispectroscopic systems engineered by the authors. A few comments here explain how this work has already begun.

Qualitative analysis is typically based on reference libraries and requires attention to issues on how to integrate data and maximize the value. These issues are complex in nanospectroscopy with the need to combine quantitative and qualitative data, and connect and fully exploit data from more than one modality. Here the rotating loading system for nanodrops can be extended to allow, for example, double-fiber UV-Vis and NIR transmission measurements on the single droplet sample rotated from one position to the next. These plinths can use the back-read architecture and magnetic concentration for NP reads for SERS or nanofluorescence.

The instrument can learn from past experience via cloud data storage and analysis and in other ways can advance beyond the limited boundaries of microspectroscopy. Integration of spectroscopies has been proposed via surface science but at the microspectroscopic level. At the microvolume level there are various opportunities for quantitative and qualitative statistics that can be represented in a complex analysis on two orthogonal axes, and these maps can be used to deliver new insights into analytical science. Combining NIR (transmission mode) and Raman (epi mode) data can be experimentally investigated in the NanoWand as both modalities are powerful qualitative methods, but the former is also useful as a quantitative method. The complex issues of combining the analysis of micro- and nanospectroscopic systems need serious consideration, which has not begun yet but will be needed with the advance of nanospectroscopy.

The fusion challenge here has at its center the role of software for the instrumental control, ultraleveling systems, environmental conditioning, etc., for experimental decision making integrating the analytical power with micro- and nanospectroscopic and microscope systems, control of 3D raster scanning of the laser, and the integration of measurement systems including magnetic field switching. These are just some of the vital software issues that must be addressed in taking nanotensiospectroscopy forward. The integration of legacy knowledge using spectral libraries and the utilization of all chemical and spectral parameters to drive toward the fusion of optimized measurement with analysis is an objective. Software control is a given when moving toward automation, but increasingly in laboratories there is a need to track workflow and ensure high productivity.

3.3.6 Summary

This chapter is based on a review of nanospectroscopy modalities beginning from the examination of the connectivities among the microspectroscopies from which the nanospectroscopies evolved. The analysis on using AMRA in this review in fact was used in the content design of the Volumes 1 and 2 of "Optical Nanospectroscopy". Here the application study of tensionanospectroscopy begins with analysis of the corresponding nanospectroscopies of this nanodroplet tensioscience. AMRA has indeed moved from there to address the problems of tensionanospectroscopy identifying innovative successful approaches:

- The problem of combining different spectral analyses, e. g., NIR and Raman (this extends to all modalities), can be overcome by using the same rule for all the analyses, which is dimensionless SMI for both axes.
- The micro- and nanospectral data OR spectral and physical and nanospectral OR indeed any type of experimental data on the single-droplet "drop laboratory" must be combined. These integrations are to be done using SMI.
- SMI measures provide a whole range of assay possibilities: (i) in spectroscopy absorbance the Shannon measure base is 10; in natural biological processes and water measurements the Shannon entropy measure is used with base e; in signal

transmission and digital signal processing Shannon measures are base 2. The issue here is not the merit of the base, but rather that information measures are being used. In connecting analysis it is important to use one base and for forensic analysis SMI would be perhaps the obvious choice. (ii) In nanospectroscopy Raman, fluorescence, SERS, TERS, nanofluorescence are all nephelometric methods and the SMI measures are applicable e.g. with laser read/excitation at the same spatial position while using magnetic switching. (iii) Toggling between Raman and SERS with magnetic switching using back-read provides a way of reading micro and then nanospectroscopic signals at the same point in a sample and with the same environmental conditions. This is possible with the MagSERS method.

– AMRA analysis shows the problem of undertaking NP measurements in liquids without the sample drying on a substrate, but in the liquid or a droplet. The "back-read method" offers a practical solution to that problem.

– SERS has a quantitation problem highlighted in the AMRA analysis. The problem led to a proposal for a new methodology of measurement called here the "Aliquant."

– The AMRA method has identified that the shape analysis used in tensiography pioneered by some of the authors has relevance to micro- and nanospectroscopy (nephelometric modalities). The requirement baseline subtractions have been overlooked, but it needs to be flagged. These algorithms change the spectral line shape and error distribution, so they fundamentally impact the statistical analysis. Data scatter methods are the most sensitive shape analysis tool available and now dimensionless scales for spectral representation of the scatter are all SMIs.

– Quantitative and qualitative measurements are the ones that in analytical science are rather distinct, which is an observation from AMRA analysis. The SMI approach brought forward to address this issue is a result.

– There are other issues that come from AMRA, but due to space limitations they cannot be expanded on.

The development of tensiospectroscopy, both micro and nano, has been directly inspired in much of its detail by Irish science and technology research [100]. George Berkeley, philosopher, mathematician, economist, and educational pioneer, can be said in the British-speaking world to be the father of environmental/economic self-sufficiency with the publication of "The Querist" (1735). The application study of carbon sequestration at the core of this chapter is inspired directly by the local genius John Tyndall and extends his work on the issues of climate change to nanospectroscopy. Furthermore, the Shannon measures discussed here come from Berkeley's insistence on graphical methods (measurements) being numerical (dimensionless) and here SMI units are used universally throughout this nano- and microtensiospectroscopic analysis.

Now, we can appreciate the incredible insight of Berkeley's contribution in explaining colors being a product of the animal/human mind that has transformed electromagnetic wavelengths into colors, but the minds that "see" these colors can indeed do optical nanospectroscopy observations connecting the nephelometric scatter sizes of particles or molecules in the medium under study. This nephelometric color series is based on Tyndall's experimental discovery of the reciprocal fourth-power scattering law $(1/\lambda^4)$; this diagram provides an example coming directly from Faraday's method first highlighted by Tyndall [101]. The mind can "size" and "quantify" the particulate of molecular matter at the nanometer scale; the human mind can be said to perform quantitative nanospectroscopy at admittedly a basic level but then we know, more incredibly, birds use quantum tunneling in navigation! It should be remembered that the electromagnetic field concepts were first explained by Faraday and mathematically delivered from Faraday's Royal institution exposition by Maxwell and from there by G.F. FitzGerald, opening the way for Hertz and Lorentz. The beauty of the Tyndall spectral (nephelometric) series should not obscure the fact this is a spectroscopic example of the Faraday method and a special development of physics that characterized explicitly the role of "human imagination" as Berkeley, Hamilton, Tyndall, FitzGerald, Bell, and Tyndall insisted. In short, this comes directly from this philosophical and Irish-based research and most specifically from Tyndall's work [102] and his method which is illustrated in Figure 3.3.10. The beauty of Tyndall colors and also "pseudo-color-coded" data scatter patterns have a utility that nanoscience should utilize.

Tyndall Colour Series

Tyndall (Sky) Blue goes from almost white to a dark blue. The sunset (Tyndall scattering) colours are more intense the lower the light levels and the Tyndall (sunset) Egg is shown here to illustrate the actual colours produced by the Tyndall scattering effect.

Figure 3.3.10: Nephelometric color series now extending into NIR and MIR, a gift from Irish spectroscopy, from Tyndall, Berkeley, Hamilton, FitzGerald, and Bell.

The nanotransformative science today needs a new unifying numerical representation, not just in operating computer systems but also in the practice of optical nanospectroscopy. Optical nanospectroscopy demands new metrological standards, unifying principles, and internationally agreed standards for its procedures. This chapter presented some of the issues that ESRs' work may address to help present a vision of the solution to these enormous challenges. A rapid advance will be possible once the conceptual hurdles here of the need for integrating nanospectroscopic capabilities have been taken. This fusion will be advanced using Big Data, information theory, and software control systems that will combine to be an essential driver in delivering this transformative societal change.

The current inventions discussed briefly above and being specifically flagged for the reader who is assumed to be new to the field of optical nanospectroscopy specifically relate to the intelligent real-time monitoring and control of surface metrology to extend the measurement range and outcomes in droplet tensiometry and tensiospectroscopy. In particular, the practical problems of nanodroplet analysis are addressed to avoid the need for drying NP dispersions or to manipulate pipetted magnetic NP samples using new sample handling innovations. A meta-review approach is highlighted to provide a fundamental approach for a literature review.

The final example briefly touched here reflects the first pilot-scale use of nanotensiospectroscopy for environmental improvements in agriculture. It is fitting to consider that in Ireland, the birthplace of nano and spectroscopic pioneers Tyndall and Synge, a cooperative of scientists, engineers, environmentalists, and farmers known as Green Restoration Ireland (GRI) is applying nanospectroscopic analysis to assess the beneficial effects of remediating degraded peat bogs and managing them back to their original condition. In doing so, these peat bogs can once again function as effective carbon sinks with excellent biodiversity and water quality characteristics.

The Farm Carbon EIP [103] importantly is employing the STRAPS software tool to analyze holistic quantification of the benefits of environmental improvement measures for drainage-based agriculture on peatland soils beginning from the fusion of micro- and nanospectroscopic data. This will integrate multidisciplinary measurements to establish viable sustainable farming options for these soils which account for approximately 25 % of all EU agricultural carbon emissions from just 3 % of agricultural lands.

In rounding up this brief vision on the forensic and agricultural applications of concern to this chapter it is important to mention that tensionanospectroscopy is a green science for a number of reasons: samples are all spatially contained, have the smallest possible practical volume, and are measured usually by intelligent algorithmic methods selecting the measured pipetted volume to optimize the measurement of absorbance, e. g., removing the need to make up solutions, reducing for this and other reasons the volume of sample/solvent/chemicals used and minimizing the disposal issues. It facilitates multimodal measurements on single samples with "drop laboratory" measurements in many applications including protein kinetic dynamics. A fundamental point of this chapter is to underline that the forensic methodologies for traceability and calibration connecting to "chain-of-evidence" methods are ones that translate directly to the environmental applications discussed here. The contention is that this benefit of tensionanospectroscopy is unchallengeable from any other analytical techniques in open lab work where flexibility is a vital consideration and in systems it is only rivaled by lab-on-a-chip.

It is a consensus view that humanity's future food production systems must be resource-efficient, clean, green, safe, and circular, providing fair and healthy toxin-free food to regenerate our degraded soils, biodiversity, and waterways. Ecosystem services must be restored to address the climate and biodiversity crises, water pollution, and other key issues. Quantitation and qualitative analysis are prerequisites, with results delivered immediately so real-time action can be established. This is an impossible ambition without economically priced testing capability and beyond that, a full fusion of information from diverse sources with all the diversity of inputs bringing unique quantifiable information to those requiring decisions. One player of this transformative process is nanotensiospectroscopy, because it is a low-cost economic green technology that connects and fuses with the established methods. The NanoWand [104] has seen the delivery of the first realization of integration of two droplet modalities for a portable fiber system to be used for integrated bimodality spectrometers for field work on bogs and in agricultural research. The design is positioned now to be developed from this initial modular design by the PhotonHub EU Consortium for carbon monitoring of cut faces. This project will also develop nanospectroscopic technologies and other portable technologies such as the Soil Organic Carbon IR tracked (SOCit) Spectrometer for profiling cut faces of bogs and other carbon deposits. Perhaps the greatest social and economical application of optical nanospectroscopy, as outlined here by the example of tensionanospectroscopy and SMI, will come from the democratization of scientific knowledge and research. Within a short period, the ability of everyone to conduct analytical science may increasingly become as mundane and as widespread as a Google search is currently. Until now, the accurate quantification of any chemical, physical, or biological substance was necessarily mostly the preserve of natural scientists using expensive, time-consuming, and laboratory-based equipment and requiring specialist technical expertise. Future nanospectroscopy techniques, such as the green "nanodroplet" chemistry, might deliver measurements and analysis in any farm, village/city, factory school, shop, office, amenity, or region. This could provide continuous monitoring and condition profiling for any defined substance within any area or activity under analysis in our built or natural environment.

As new multidisciplinary perspectives emerge, many positive social and economic outcomes are envisaged. This will occur due to social scientists engaging with natural scientists in research countering poor and unevidenced risk communication and harnessing the existing and widespread goodwill towards nature in all its forms. Individuals of all ages will soon be able to ask, what is in my food, how safe is my drinking water, what is the quality of the air being breathed, how hazardous is my workplace, and what chemicals and pathogens will I be exposed to today and in what quantities? They may receive quick "forensic-quality" answers to these important questions that will be transparent, accurate, reliable, and valid.

Bibliography

[1]   Gorelick R. Combining richness and abundance into a single diversity index using matrix analogues of Shannon's and Simpson's indices. Ecography. 2006;29:525–30.

[2]   Doyle G, McMillan ND, Murtagh F, O'Neill M, Riedel S. An introduction to the generalised theory of data scatter. In: Murtagh FD, editor. SPIE. https://doi.org/10.1117/12.606077.

[3]   McMillan N et al. New procedure for direct measurements of absorbance of thin films of ultra-high absorbance UV blocks. In: Bowe BW, Byrne G, Flanagan AJ, Glynn TJ, Magee J, O'Connor GM, O'Dowd RF, O'Sullivan GD, Sheridan JT, editors. Opto-Ireland 2005: Photonic Engineering. Proc. of SPIE. vol. 5827. Bellingham, WA: SPIE; 2005. p. 428–37. https://doi.org/10.1117/12.629159. 0277-786X/05/$15.

[4]   Type 20-C/Q/0.008 specifically for a sunscreen test conducted according to an Australian TSA standard. This is a demountable cell comprised of a recessed body with an 8 µm path length, and a cover plate window and a cuvette that has been designed that can be produced in path lengths between 10 and 100 microns to 0.5 micron accuracy.

[5]   Perry S, McMillan N, Rodriguez R, Mackowski S, Sheremet E, Fleischer M, Kneipp K, Zhan D, Adam P-M, Madeira T. Amorim in collaboration with the membership of the COST MP1302 Action's Workgroup 4, Novel advanced scoping meta-review methodology for defining a graduate level textbook in an emerging subject area. Liber Quarterly. 2018;27:1–20.

[6]   McMillan N. A micro and nanodrop multimeasure & imaging system, Irish Patent PTIE20220000000193. 12 May 2022.

[7]   Tematys, exploring the photonics market, miniature and micro spectrometers: end-users needs, market and trends. EPIC EU Report, September 2016.

[8]   Hansen HN, Carneiro K, Haitjema H, De Chiffre L. Dimensional micro and nano metrology. CIRP Ann. 2006;55(2). https://doi.org/10.1016/j.cirp.2006.10.005.

[9]   Burgess C, Frost T, editors. Standards and Best Practice in Absorption Spectrometry, in which the crucial and key contribution came from J. Hammond and C. Burgess provided the two linked contributions 'Regulatory Overview' and Recommended Procedures for Standardization', pp. 120–129 and 130–140. Oxford: Blackwell; 1999.

[10]   Smith SRP, McMillan ND, O'Neill M, Riedel SM, Hammond J. Determining the absorption or turbidity coefficient of a liquid EPO 18774059.2. 2015.

[11]   López M, Guarcia-Ruiz C. Infrared and Raman spectroscopy techniques applied to identification of explosives. Trends Anal Chem. 2014;54:36–44.

[12]   Pettinger B, Ren B, Picardi G, Schuster R, Gerhald E. Nanoscale probing of adsorbed species by tip-enhanced Raman spectroscopy. Phys Rev Lett. 2004;92:096101.

[13]   Prater C, Kjoller K, Shetty R. Nanoscale infrared spectroscopy. Mater Today. 2010;13(11):56–60.

[14]   McDowell RS, Patterson CW, Harter WG. The modern revolution in infrared spectroscopy. Los Alamos Science. 38–65 (see https://www.researchgate.net/publication/237373887_the_modern_revolution_in).

[15]   Cantu AA. Nanoparticles in forensic science. Proc SPIE. 2008;7119:71190F-1, 1–8.

[16]   Yang Meng, Li M, Yu A, Zu Y, Yang Minying, Mao C. Fluorescent nanomaterials for the development of latent fingerprints in forensic sciences. Adv Funct Mater. 2017;27(14). https://doi.org/10.1002/adfm.201606243.

[17]   Oroian M, Ropciuc S, Paduret S. Honey adulteration detection using Raman spectroscopy. Food Anal Methods. 2018;11:959–96.

[18]   Vašková H, Tomeček M. Rapid spectroscopic measurement of methanol in water-ethanol-methanol mixtures. MATEC Web Conf. 2018;210:02035. https://doi.org/10.1051/matecconf/201821002035.

[19]   Synge EH. A suggested method for extending the microscopic resolution into the ultramicroscopic region. Philos Mag. 1928;6:356. https://doi.org/10.1080/14786440808564615.

[20]   Synge EH. An application of piezoelectricity to microscopy. Philos Mag. 1932;13:297. https://doi.org/10.1080/14786443209461931.

[21]   Kim JH, Song H-B. Recent progress of nanotechnology with NSOM. Micron. 2007;38:409–26.

[22]   Lopez GA, Estevez M-C, Soler M, Lechuga LM. Recent advances in nanoplasmonic biosensors: applications and lab-on-a-chip integration. Nanophotonics. 2017;6(1):123–36. https://doi.org/10.1515/nanoph-2016-0101.

[23]   Lin P-C, Lin S, Wang PC, Sridhar R. Techniques for physicochemical characterization of nanomaterials. Biotechnol Adv. 2014;32:711–26.

[24]   McMillan ND, Vallely L. John Tyndall and the foundation of the sciences of infra-red spectroscopy and nephelometry. Technol Ireland. 1989;21(4):37–40.

[25]   Tyndall J. Contributions to Molecular Physics in the Domain of Radiant Heat. vol. 8, London; 1872.

[26]   Cao A. Light scattering, recent applications. Anal Lett. 2003;36(15):3185–225.

[27]   Bhattacharjee S. Review article: DLS and zeta potential – What they are and what they are not? J Control Release. 10 August 2016;235:337–51.

[28]   The ultimate guide: absolute macromolecular characterization. Wyatt Technology Corpora. 2016.

[29]   Xu R. Light scattering: a review of particle characterisations. Particuology. 2015;18:11–21.

[30]   Hyvärinen A-P, Vakkari V, Laakso L, Hooda RK, Sharma VP, Panwar TS, Beukes JP, van Zyl PG, Josipovic M, Garland RM, Andreae MO, Pöschl U, Petzold A. Correction for a measurement artefact of the multi-angle absorption photometer (MAAP) At high black carbon mass concentration levels. Atmos Meas Tech. 2012;5:6553–75.

[31]   AZoNano, Uv/visible spectroscopy methodology for particle characterisation, https://www.Azonano.com/atticle.aspx?ArticleID=3588. six pages hosted by Perkin Elmer.

[32]   Petryayeva E, Algar WR, Medintz IL. Quantum dots in bioanalysis: a review of applications across various platforms for fluorescence spectroscopy and Imaging. Appl Spectrosc. 2013;67(3):215–52.

[33]   Hauppauge NY. Quantum Dots: Applications, Synthesis, and Characterization. Nova Science Publishers, Inc.; 2012.

[34]   Ki Y. Application of CdSe quantum dots for the direct detection of TNT. Forensic Sci Int. 2016;259:101–5. https://doi.org/10.1016/j.forsciint.2015.12.028. Epub 2015 Dec 29.

[35]   Yi K-Y, Wei C-S. Electroluminescence of CdTe quantum dots and sensitive detection of haemoglobin. Int J Electochem Sci. 2017;12:3472–3482.

[36]   Kawata SR. Plasmonics for nanoimaging and nanospectrocopy. Appl Spectrosc. 2013;67(2):117–25.

[37]   Rodríguez-Oliveros R, Paniagua-Dominguez R, Sánchez-Gil JA, Macrías D. Plasmonic spectroscopy: theoretical and numerical calculations, and optimization techniques. Nanospectroscopy. 2015;1(1):67–96.

[38]   Kvítek O, Siegel J, Hnatowicz V, Švorlík V. Hindawi Publishing Corporation, J Nanomater. 2013;2013: Article ID 743684, 15 pages. https://doi.org/10.1155/2013/743684.

[39]   Nugroho AFA, Xu C, Hedin N, Langhammer C. UV–Visible and plasmonic nanospectroscopy of the CO_2 adsorption energetics in a microporous polymer. Anal Chem. 2015;87:10161–5. https://doi.org/10.1021/acs.analchem.5b03108.

[40]   McMillan ND, Finlayson O, Fortune F, Fingelton M, Daly D, Townsend D, McMillan DDG, Dalton MJ. The fibre drop analyser: a new multi-measurand analyser with applications in sugar processing and for the analysis of pure liquids. Meas Sci Technol. 1992;3:746–64.

[41]   McMillan ND, O'Neill M, Riedel SM, Smith SRP. A microvolume analysis system, Int. App. No.
         PCT/EP2007/054546 & EPO 07728998.1.
[42]   McMillan ND, Dunne G, Smith SRP, O'Rourke B, Morrin D, McDonnell L, O'Neill M,
         Riedel S, Krägel J, Mitchell CI, Scully P. New tensiographic approach to surface studies
         of protein kinetics showing possible structural rearrangement of protein layers
         on polymer substrates. Colloids Surf A, Physicochem Eng Asp. 2010;365:112–21.
         https://doi.org/10.1016/j.colsurfa.2010.05.020.
[43]   Pringuet P. The construction and validation of a numerical tensiograph simulation model and
         the investigation of a modelling platform for the fingerprinting of wine. IT Carlow. 2011.
[44]   Lambert C. Studies on Fluoranthene and Water Run-off from Irish Motorways. M.Eng Thesis,
         Trinity College Dublin; 2013.
[45]   Javadi A, Fainerman VB, Miller R. Drop volume tensiometry. In: Bubble and Drop Interfaces.
         2011. p. 119–42.
[46]   O'Neill M, McMillan ND, Dunne G, Mitchell CI, O'Rourke B, Morrin D, Brennan F, Miller R,
         McDonnell L, Scully P. New tensiographic studies on protein cleaning of polymer surfaces.
         Colloids Surf A, Physicochem Eng Asp. 2008;323:109–11.
         https://doi.org/10.1016/j.colsurfa.2007.12.047.
[47]   Dunne G, McMillan N, O'Rourke B, Morrin D, O'Neill M, Riedel S, McDonnell L, Scully P.
         Colloids Surf A, Physicochem Eng Asp. 2010;354:364–7.
         https://doi.org/10.1016/j.colsurfa.2009.06.030.
[48]   McMillan N, Smith SRP, O'Neill M, Tiernan K et al. The tensiograph platform for optical
         measurements. In: Miller R, Liggieri L, editors. Bubble and Drop. Progress in Colloid and
         Interface Science Interfaces. vol. 2. Leiden, Boston: Brill; 2011. p. 401–79.
[49]   Konidakis I, Androulidaki M, Zito G, Pissadakis S. Growth of ZnO nanolayers inside the
         capillaries of photonic crystal fibres. Thin Solid Films. 2014;555:76–80.
[50]   Konidakis I, Konstantaki M, Tsibidis GD, Pissadakis S. A light driven optofluidic switch
         developed in a ZnO-overlaid microstructured optical fiber. Opt Express. 2015;23:31496–509.
[51]   Schwartz VB, Thétiot F, Ritz S, Pütz S, Choritz L, Lappas A, Förch R, Landfester K,
         Jonas U. Antibacterial surface coatings from zinc oxide nanoparticles embedded in
         poly(N-isopropylacrylamide) hydrogel surface layers. Adv Funct Mater. 2012;22:2376–86.
         https://doi.org/10.1002/adfm.201102980.
[52]   Comini E, Baratto C, Faglia G, Ferroni M, Sberveglieri G. Single crystal ZnO nanowires as
         optical and conductometric chemical sensor. J Phys D, Appl Phys. 2007;40(23):7255–9.
[53]   Klini A, Pissadakis S, Das RN, Giannelis EP, Anastasiadis SH, Anglos D. ZnO-PDMS
         nanohybrids: a novel optical sensing platform for ethanol vapour detection at room
         temperature. J Phys Chem C. 2015;119:623–31.
[54]   Hanumegowda NM, Stica CJ, Patel BC, White I, Fan X. Refractometric sensors based on
         microsphere resonators. Appl Phys l ett. 2005;87:201107-3.
[55]   Francois A, Rowland KJ, Monro TM. Highly efficient excitation and detection of whispering
         gallery modes in a dye-doped microsphere using a microstructured optical fiber. Appl Phys
         Lett. 2011;99:141111-3.
[56]   Yang S et al. Reconfigurable liquid whispering gallery mode microlasers. Sci Rep.
         2016;6:27200. https://doi.org/10.1038/srep27200.
[57]   Haustein E, Schwille P. Fluorescence correlation spectroscopy: novel variations of
         an established technique. Annu Rev Biophys Biomol Struct. 9 June 2007;36:151–69.
         https:doi.org/10.1146/annurev.biophys.36.040306.132612.
[58]   Alexiev U, Farrens DL. Fluorescence spectroscopy in rhodopsins: insights and approaches.
         Biochim Biophys Acta. 2014;1837:694–709. https://doi.org/10.1016/j.bbabio.2013.10.008.
[59]   Cricenti A et al. Optical nanospectroscopy applications in material science. Appl Surf Sci.
         2004;234:374–86.

[60] Schermelleh L, Heintzmann R, Leonhardt H. A guide to super-resolution fluorescence microscopy. J Cell Biol. 2010;190(2):165–75.

[61] Das S, Agrawal YK. Recent advancement, techniques and applications. Vib Spectrosc. 2011;57:163–76. https://doi.org/10.1016/j.vibspec.2011.08.003.

[62] Lombardi J, Birk RL. A unified view of surface-enhanced Raman scattering. Acc Chem Res. 2009;42(6):734–42.

[63] Itoh T, Yoshida K, Tamaru H, Biju V, Ishikawa M. Experimental demonstration of electromagnetic mechanisms underlying surface enhanced Raman scattering using single nanoparticle spectroscopy. J Photochem Photobiol A, Chem. 2011;219:167–79. https://doi.org/10.1016/j.jphotochem.2011.03.001.

[64] Luo S, Sivashanmugan K, Liao J-D, Yao C-K. Nanofabricated SERS-active substrates for single-molecule to virus detection *I vitro*: review. Biosens Bioelectron. 2014;61:232–40.

[65] Betz JF, Yu WW, Cheng Y, White IM, Rubloff GW. Simple SERS substrates: powerful, portable, and full of potential. Phys Chem Chem Phys. 2014;16:2224–39.

[66] Kahraman M, Mullen ER, Korkmaz A, Wachsmann-Hogiu S. Fundamental and applications of SERS-based bioanalytical sensing. Nanophotonics. 2017:1–22.

[67] Ouyang L, Ren W, Zhu L, Irudayaraj J. Prosperity to challenges: recent approaches in SERS substrate fabrication. Rev Anal Chem. 2017;36(1):1–22.

[68] Çetin AE, Yilmaz C, Ali Yanik A, Somu S, Busnaina A, Altug H. Plasmonic monopole antenna arrays for biosensing, spectroscopy and nm-precision optical trapping. In: OSA, CLEO. 2011.

[69] McMillan ND, O'Neill M, Riedel S, Smith SRP. Apparatus for analysing the optical properties of a sample EPO18774059.2. 2018.

[70] McMillan ND. Device with exchange operator & new digitisation application of statistics PTIE2020000000187. 2022.

[71] Yang S, Dia X, Stogin BB, Wang T. Ultrasensitive SERS scattering detection in common fluids. Proc Natl Acad Sci USA. 2016;113(2):268–73.

[72] McMillan ND. Nephelometric drop analyte up concentration method PTIE2020000000198. 2022.

[73] Deckert-Gaudig T, Taguchi A, Kawata S et al. Chem Soc Rev. 2017;46(13):4077–110.

[74] Sheremet E, Rodriguez RD, Agapov AL, Sokolov AP, Hietschold M, Zahn DRT. Nanoscale imaging and identification of a four-component carbon sample. Carbon. 2016;96:588–93.

[75] Hartschuh A. Tip-enhanced near-field optical microscopy. Angew Chem, Int Ed. 2008;47:8178–91.

[76] Bailo E, Deckert V. Angew Chem, Int Ed. 2008;47:1658.

[77] Kurouski D, Zaleski S, Casadio F, Van Duyne RP, Shah NC. Tip-enhanced Raman spectroscopy (TERS) for in situ identification of indigo and iron gall ink on paper. J Am Chem Soc. 2014;136(24):8677–84.

[78] Kharintsev SS, Hoffmann GG, Fishman AL, Salakhov MK. Plasmonic optical antenna design for performing tip-enhanced Raman spectroscopy and microscopy. J Phys D, Appl Phys. 2013;46(14).

[79] Snitka V, Rodrigues RD, Lendraitis V. Novel gold cantilever for nano-Raman spectroscopy of graphene. Microelectron Eng. 2011;88(8):2759–62.

[80] Horrer A, Krieg K, Freudenberger K, Rau S, Leidner L, Gauglitz G, Kern DP, Fleischer M. Plasmonic vertical dimer arrays as elements for biosensing. Anal Bioanal Chem. 2015;407(27):8225–31.

[81] Johnson PB, Christy RW. Optical constants of the noble metals. Phys Rev B. 1972;6:4370.

[82] McMillan ND. Tensiograph drophead, EPO 06708399.8 International application No. PCT/EP2006/060118. 2006.

[83] Smith SRP, McMillan ND, Riedel S, O'Neill M. Absorption and turbidity constant, EPO 1578076.6. 2015.

[84] Violakis G, Pissadakis S. Final NanoWand report. A comprehensive feasibility study of fiber-optic based illumination and detection architecture for the NanoWand: individual design and tolerancing analysis for the dual modality EPI-fluorescence and transmission NIR probe nanodrop tensiospectrometer. EU ACTPHAST Consortium. 30 May 2021.

[85] McMillan ND. A plinth, Irish Patent 202000000190. 2022. McMillan ND. A Microscope Plinth, Irish Patent 202000000191.

[86] Centrone A. Infrared imaging and spectroscopy beyond the diffraction limit. Annu Rev Anal Chem (Palo Alto Calif). 2015;8:101–26. https://doi.org/10.1146/annurev-anchem-071114-040435. Epub 2015 May 18.

[87] Lahiri B, Holland G, Centrone A. Chemical imaging beyond the diffraction limit; experimental validation of PTIR technique. Small. 2013;9:439–45.

[88] McMillan ND, O'Rourke B, Riedel SM, Skelly DO, O'Neill M, O'Neill AE, Boller D, Bertho AC, Doyle G, Hammond J, O'Neill AT. A new democratic phase coherent data-scatter technique for calibration, measurement, fingerprinting and rapid archival identification of ultraviolet-visible multi-component food spectra. Anal Chim Acta. 2004;V511(1):119–35.

[89] Moore RA, Unnikrishnan S, Perova TS, McMillan ND, Riedel S, O'Neill M, Doyle G. Investigation of correlation between characteristics of Raman spectra and parameters of data-scatter obtained from phase coherence theory. Proc SPIE. 2005;5826:379–86.

[90] McMillan ND, Riedel SM, McDonald J, O'Neill M, Whyte N, Augousti A, Mason J. A Hough transform inspired technique for the rapid fingerprinting and conceptual archiving of multianalyser tensiotraces. In: Irish Machine Vision Conference Proceedings. 1999. p. 330–46.

[91] McMillan ND. Device with exchange operator & new digitisation application of statistics PTIE2020000000187. 2022.

[92] Kököer M, Murtagh F, McMillan ND, Riedel S, O'Rourke B, Beverly K, Augousti AT, Mason J. A wavelet, Fourier, and PCA data analysis pipeline: application to distinguishing mixtures of liquids. J Chem Inf Comput Sci. https://doi.org/10.1021/ci025601j.

[93] Bertho AC. Drop spectroscopy. PhD thesis. IT Carlow; 2012.

[94] Doyle G, McMillan ND, Murtagh F, O'Neill M, Riedel S, Perova TS, Unnikrishnan S, Moore RA. Phase coherence theory for data-mining and analysis: application studies in spectroscopy. In: Murtagh FD, editor. Opto-Ireland 2005: Imaging and Vision. Proceedings of SPIE. vol. 5823. Bellingham, WA: SPIE; 2005. 0277-786X/05/$15, https://doi.org/10.1117/12.606073.

[95] Egan J, McMillan N, Denieffe D. Theoretical enquiry into the use of data-entropy methods for optoelectronic-fiber systems-digital design. In: Proceedings of IT&T Annual Conference. Letterkenny. 2003. p. 249–59.

[96] Tiernan K, Kennedy D, McMillan N. Tensiograph instrument for measuring liquid material properties. Mater Des. 2005;26:197–201. https://doi.org/10.1016/j.matdes.2004.02.019.

[97] McMillan ND, Smith SRP, Bertho AC, Morrin D, O'Neill M, Tiernan K, Hammond J, Barnett N, Pringuet P, O'Mongain E, O'Rourke B, Riedel S, Neill M, Augousti A, Wüstneck N, Wüstneck R, McMillan DDG, Colin F, Hennebert P, Pottecher G, Kennedy D. Quantitative drop spectroscopy using the drop analyser: theoretical and experimental approach for microvolume applications of non-turbid solutions. Meas Sci Technol. 2008;19:055601. (18 pp).

[98] Kököer M, Murtagha F, Augousti AT, Mason J, McMillan ND. Wavelet- and entropy-based feature extraction: application to distinguishing mixtures of beverages. In: Shearer A, Murtagh FD, Mahon J, Whelan PF, editors. Opto-Ireland 2002: Optical Metrology, Imaging, and Machine Vision. Proceedings of SPIE. vol. 4877. 2003. p. 175–82.

[99] Starck JL, Murtagh F, Bonnarel F. Multiscale entropy for semantic description of images and signals. 2000.

[100] McMillan ND. Prometheus's Fire: History of Scientific and Technological Education in Ireland. Carlow: Tyndall Publications; 2000.

[101] Tyndall J. Faraday as a Discoverer. London: Longman, Green Co.; 1868.

[102] Tyndall J, McMillan ND. Dictionnaire des philosophes de France. Paris 1982 and second enlarged entry in 2nd edition, 1993.

[103] EIP Farm Carbon website. https://ec.europa.eu/eip/agriculture/en/find-connect/projects/farm-carbon-eip.

[104] Dutschk V, Karapantsios T, Liggieri L, McMillan N, Miller R, Starov VM. Smart and green interfaces: from single bubble/drops to industrial, environmental and biomedical applications. Adv Colloid Interface Sci. 2014;209:109–26.

Part 4: **Life sciences**

In life sciences, relevant processes take place at the cellular and subcellular/molecular levels, i. e., clearly on the micro- and nanoscale. Life itself depends on solar energy being harvested by proteins in the process of photosynthesis. Therefore, techniques are indispensable that can address and gather information at just these length scales. This concerns both learning more about objects of interest such as bacteria [1], viruses [2], single proteins [3], neurotransmitters [4], nucleic acids [5], lipids [6], peptides [7], and many more and the manipulation of cells in view of medical applications, e. g., by targeted drug delivery [8] or hyperthermal therapies [9]. High-resolution techniques such as fluorescence imaging, superresolution imaging, or scanning near-field optical microscopy (SNOM) as well as chemically specific techniques such as tip-enhanced Raman spectroscopy (TERS) or surface-enhanced Raman spectroscopy (SERS) and its resonant or nonlinear variations offer huge analytical potential. For SERS, signal enhancement is achieved by bringing the substance of interest to the surface of illuminated metallic nanoparticles or vice versa. For this purpose, suitable nanoparticle substrates need to be designed and incorporated for in situ investigations and applications. Here additional questions such as toxicity and biocompatibility issues emerge. SERS offers the advantages of label-free detection, ultrahigh sensitivity, chemical specificity, and the possibility to build SERS labels or probes for targeting specific analytes. Applications of nanospectroscopy where further advances may be expected include such varied directions as the observation of conformational transitions [10], pH probing in living organisms [11], or cancer detection [12]. As shown in the following examples, optical nanospectroscopy plays an increasing role in the life sciences. By exploiting optical forces to aggregate nanoparticles in liquids and using them as SERS probes, label-free sensing of dyes, chemicals, or biomolecules is achieved. The concept of SERS is extended to the analysis of proteins, cells, and tissues with the aim of label-free cancer detection. By incorporating nanoparticles in cells and tissues, it is shown that even complex biological samples may be investigated, with a perspective of targeting whole organisms for diagnostics and therapeutics.

Bibliography

[1] Jarvis RM, Goodacre R. Discrimination of bacteria using surface-enhanced Raman spectroscopy. Anal Chem. 2004;76(1):40–7.
[2] Shanmukh S et al. Rapid and sensitive detection of respiratory virus molecular signatures using a silver nanorod array SERS substrate. Nano Lett. 2006;6(11):2630–6.
[3] Xu HX et al. Spectroscopy of single haemoglobin molecules by surface enhanced Raman scattering. Phys Rev Lett. 1999;83(11):4357–60.
[4] Tang L et al. SERS-active Au@Ag nanorod dimers for ultrasensitive dopamine detection. Biosens Bioelectron. 2015;71:7–12.
[5] Garcia-Rico E et al. Direct surface-enhanced Raman scattering (SERS) spectroscopy of nucleic acids: from fundamental studies to real-life applications. Chem Soc Rev. 2018;47(13):4909–23.

https://doi.org/10.1515/9783110442908-010

[6]   Zivanovic V et al. Molecular structure and interactions of lipids in the outer membrane of living cells based on surface-enhanced Raman scattering and liposome models. Anal Chem. 2021;93(29):10106–13.

[7]   Bantz KC et al. Recent progress in SERS biosensing. Phys Chem Chem Phys. 2011;13(24):11551–67.

[8]   Yilmaz H et al. Pharmaceutical applications of a nanospectroscopic technique: Surface-enhanced Raman spectroscopy. Adv Drug Deliv Rev. 2022;184:114184.

[9]   Kaddi CD et al. Computational nanomedicine: modelling of nanoparticle-mediated hyperthermal cancer therapy. Nanomedicine. 2013;8(8):1323–33.

[10] Seweryn S et al. Plasmonic hot spots reveal local conformational transitions induced by DNA double-strand breaks. Sci Rep. 2022;12(1):12158.

[11] Pan CY et al. Dynamically monitoring pH in living organisms based on a SERS-active optical fiber. Adv Mater Interfaces. 2022;9:2200328.

[12] Vendrell M et al. Surface-enhanced Raman scattering in cancer detection and imaging. Trends Biotechnol. 2013;31(4):249–57.

Antonino Foti, Barbara Fazio, Cristiano D'Andrea,
Maria Grazia Donato, Valentina Villari, Onofrio M. Maragò,
Ramzi Maalej, Sameh Kessentini, and Pietro G. Gucciardi

4.1 Molecular and biomolecular SERS detection in liquid environment

4.1.1 Key messages

- Surface-enhanced Raman spectroscopy (SERS) can be employed for the implementation of high-sensitivity label-free molecular sensors working in a liquid environment.
- Optical forces can be used for manipulating metal nanoparticles (NPs) in liquid and to foster the controlled aggregation of SERS-active plasmonic nanostructures.
- Chemically and optically induced NP aggregates in liquid environment enable SERS detection of dyes, chemicals, and biomolecules at ultralow molecular concentrations.

4.1.2 Pre-knowledge

This chapter gives a review of SERS molecular and biomolecular sensors in liquid environment with metal NPs aggregates. For full comprehension of the chapter the reader is expected to have a good knowledge of SERS, which involves both surface plasmon resonance (SPR) and Raman scattering (see Volume 1, Sections 2 and 3). In addition, notions regarding the synthesis (Volume 2, Section 6) and localized SPR (LSPR) (Volume 1, Section 1) of metal NPs and nanoantennas are required. Optical spectroscopy is largely employed to establish the plasmonic properties of the NPs; the concept can be found in Volume 1, Section 2. Finally notions of optical microscopy, microspectroscopy, and related instrumentation are required to understand the experimental implementation of optical and Raman/SERS tweezers (Volume 1, Section 1 and Volume 2, Section 4).

In the following, some fundamental optical properties of metal NPs are recapitulated.

Optical properties of metal nanoparticles

An introduction to the optical properties of metal NPs and plasmons is given in Volume 1, Chapter 1.6. In the most general case, calculation of the optical properties of NPs is accomplished by applying the full electromagnetic theory [1, 2]. When the scatterer dimensions are much smaller than the wavelength of light, as for NPs, it is also possible, and convenient, to use approximated theories. These can give useful physical insight into complex problems, in spite of the approximated nature of the solution. The size parameter x, defined as

$$x = k_m a = \frac{2\pi n_m}{\lambda_0} a, \tag{4.1.1}$$

is the quantity used to determine the range of validity of these approaches. Here $k_m = n_m \omega/c$ is the wavenumber of the incident field in the medium surrounding the particle, c is the speed of light in vacuum, ω is the angular frequency, a is the characteristic dimension of the particle that for a sphere is

https://doi.org/10.1515/9783110442908-011

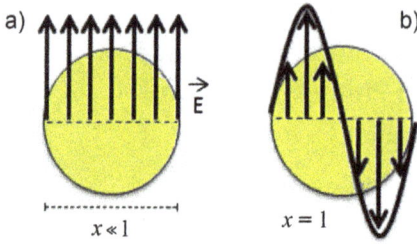

Figure 4.1.1: The electric field inside a particle with a size parameter much smaller than 1 (a) can be considered as homogeneous. Conversely, when the size of the particle approaches the wavelength (b), phase retardation effects cannot be neglected, leading to an inhomogeneous field distribution inside the particle.

the radius, λ_0 is the light wavelength in vacuum, and n_m is the refractive index of the medium in which the particle is located. When $x \ll 1$, the dipole approximation holds. The particles "see" the incident radiation field as a homogeneous field with no phase retardations within its volume (Fig. 4.1.1a). Otherwise, there would be a phase difference between the incident field at the two opposite sides of the particle, leading to interference effects not considered by the approximation (Fig. 4.1.1b) [3, 4]. This is a valid hypothesis when also the relation $|n_p/n_m|x \ll 1$ is true, where n_p is the refractive index of the particle. This condition has to be taken with care, especially for high-refractive index materials or when considering materials with a complex refractive index, e. g., plasmonic NPs [5]. However, in order to extend the validity of this approximation also to larger particles without involving the full electromagnetic theory, some corrections, like the *radiative* or damping *correction*, can be applied [3, 4]. In the following paragraphs we briefly summarize formulas for the polarizability and plasmon resonance of nanospheres and nanorods (NRs).

Polarizability and localized plasmon resonances of nanospheroids
When an electric field of wavevector k_m is applied to a small neutral object placed in a dielectric medium ($k_m a \ll 1$, with a being the typical dimension of the particle), the negative electron clouds of the particle's atoms will be displaced according to the electric field polarization, leading to a separation of the centers of mass of the electron distributions with respect to the position of the positive nuclei. Within the dipole approximation, we can model the response of our scatterer as a dipole moment induced by an external field to retrieve the properties of the scattered field exploiting the radiation properties of an electric dipole in an oscillating electric field (*dipole approximation*). The atomic induced dipole moment $\vec{p}$ and the excitation field $\vec{E}_i$, in the linear regime, are related by [5]:

$$\vec{p} = \alpha \vec{E}_i, \tag{4.1.2}$$

where α is the linear atomic polarizability. In a first approach, α is calculated resolving the electrostatic problem, i. e., considering a homogeneous, isotropic, nonmagnetic spherical particle with radius a, immersed in an electrostatic field embedded in a dielectric medium. For a sphere of volume $V = 4\pi a^3/3$, the polarizability α is given by the well-known Clausius–Mossotti relation [3, 5]:

$$\alpha_{CM} = 3V\varepsilon_0\varepsilon_m \frac{\varepsilon_r - \varepsilon_m}{\varepsilon_r + 2\varepsilon_m}, \tag{4.1.3}$$

where $\varepsilon_0, \varepsilon_m$, and ε_r are the dielectric constants of the vacuum, the medium, and the particle, respectively. The field outside the sphere is the sum of the incident field $\vec{E}_i$ and the electric field due to a dipole $\vec{p} = \alpha_{CM}\vec{E}_i$ placed at the center of the sphere. For particles with a diameter much smaller than the optical wavelength, the static model can be used as a first-order approximation of the radiative case, in which the electromagnetic field oscillates at frequency ω. The rate at which energy is removed from the incident field through scattering and absorption is described by the extinction cross-section $\sigma_{ext} = \sigma_{scat} + \sigma_{abs}$, where σ_{scat} and σ_{abs} are the scattering and the absorption cross-section, respectively. The extinction and scattering cross-sections read as [3][1]

$$\sigma_{ext} = k_m \frac{\text{Im}\{\alpha(\omega)\}}{\varepsilon_0\varepsilon_m} \tag{4.1.4}$$

and

$$\sigma_{sca} = k_m^4 \frac{|\alpha(\omega)|^2}{6\pi(\varepsilon_0\varepsilon_m)^2}. \tag{4.1.5}$$

In the case of small spheres with radius a, we can use the Clausius–Mossotti expression $\alpha_{CM}(\omega)$ for the polarizability. Defining the extinction yield $Q_{ext} = \sigma_{ext}/\sigma_{geom}$, where $\sigma_{geom} = \pi a^2$ is the geometrical cross-section (the NP area intercepted by the incident beam), we find

$$Q_{ext} = 4k_m a \ \text{Im}\left\{\frac{\varepsilon_r(\omega) - \varepsilon_m(\omega)}{\varepsilon_r(\omega) + 2\varepsilon_m(\omega)}\right\}. \tag{4.1.6}$$

Similarly we can obtain the scattering yield as

$$Q_{scat} = \frac{8}{3}(k_m a)^4 \left|\frac{\varepsilon_r(\omega) - \varepsilon_m(\omega)}{\varepsilon_r(\omega) + 2\varepsilon_m(\omega)}\right|^2. \tag{4.1.7}$$

These expressions represent a quasi-static approximation to the scattering problem, where the polarizability takes the analytical form of the Clausius–Mossotti expression $\alpha(\omega) = \alpha_{CM}^0(\omega)$, but with frequency-dependent dielectric constants. Equations (4.1.6) and (4.1.7), however, are only valid for extremely small metal NPs ($\sim$5 nm) [6], for which the scattering efficiency is negligible and the extinction is mainly due to absorption ($Q_{ext} \approx Q_{abs}$). The extinction and scattering properties of metallic NPs are dictated by the dynamics of the free conduction electrons. The free electron cloud can easily move in response to an external electromagnetic field, sustaining collective oscillations in both surface and volume charge densities, referred to as *free-electron plasma*, the quanta of which here are simply referred to as *plasmons* [7]. In the most general case the oscillating electrons in bulk metals are subject to a restoring force due to the positive charges of the nuclei, leading to resonances called plasmon resonance. When the metallic particle dimensions are in the nanometer scale, the plasmon resonance is influenced by the geometry of the particle and is referred to as LSPR. The dynamics of the plasma are described by the dielectric function of the metal $\varepsilon(\omega)$ [3, 7]. The LSPRs of metal NPs are encoded in the extinction cross-section, i. e., LSPR(λ) $\propto Q_{ext}(\lambda)$. For a spheroid, this latter is well approximated by the relation [8, 1]

$$\text{LSPR}(\lambda) \propto \frac{\text{Im}[\varepsilon_r(\lambda)]}{\{\text{Re}[\varepsilon_r(\lambda)] + \kappa\varepsilon_m\}^2 + \{\text{Im}[\varepsilon_r(\lambda)]\}^2}, \tag{4.1.8}$$

---

1 See also Volume 1, Chapter 1.6, and Chapter 3.1 above. Since absorption is approximately linear to the nanoparticle volume and scattering is approximately proportional to the volume squared, scattering can be neglected for very small particles, hence σ_{ext} becomes approximately equal to σ_{abs}. As the particles become larger, both contributions have to be considered.

where κ accounts for the different possible geometries. For spheres $\kappa = 2$, leading to the above cited *Clausius–Mossotti* formula (for negligible losses in the surrounding medium), and it increases for particles with higher aspect ratio [9]. Whenever the real part of the dielectric function matches the value of $-2\varepsilon_m$ and $\mathrm{Im}\{\varepsilon_r(\lambda)\} \approx 0$ (Fröhlich resonance condition), the extinction cross-section experiences a strong enhancement due to the increase of the polarizability. A peak in the LSPR spectrum therefore appears. These conditions are verified in the near-UV, visible, and near-IR ranges for metals like gold, silver, aluminum, or copper, making such kinds of nanosystems good antennas for light engineering [3].

The size of a spherical particle spectrally shifts the LSPR position. This can be accounted for by introducing a radiative correction to the polarizability [3, 4], leading to the resonance relation

$$\mathrm{Re}\{\varepsilon_r(\lambda)\} = -\left(\frac{2 + (k_m a)^2}{1 - (k_m a)^2}\right)\varepsilon_m. \tag{4.1.9}$$

Increasing the radius a of the sphere, the term $k_m a$ is increased, with $\mathrm{Re}\{\varepsilon_r(\lambda)\}$ decreasing. This redshifts the LSPR resonance position (Fig. 4.1.2a). Similarly, when the permittivity of the surrounding medium ε_m increases, $\mathrm{Re}\{\varepsilon_r(\lambda)\}$ decreases, resulting in a red shift of the LSPR position. Moreover, the interaction with the surrounding medium creates an energy loss and damping of the LSPR. The dipole approximation is valid in the visible range for nanosphere diameters lower than 150 nm since the relation $(k_m a)^2 < 1$ holds. For larger spheres calculations must be performed in the Mie theory framework [1]. Regarding ellipsoidal NPs, analytical solutions can be found in the dipole approximation [4], highlighting the effect of particle shape on its polarizability and thus on its optical response. Due to the different interactions of the incident electric field with each principal axis of the spheroid, the particle will show a different polarizability α_j (with $j = a$, b, or c corresponding to the three semiaxes of the ellipsoid) for each axis, which reads as [1]

$$\alpha_j = \varepsilon_0 \varepsilon_m V_e \left(\frac{\varepsilon_m}{\varepsilon_r(\omega) - \varepsilon_m} + L_j\right)^{-1}. \tag{4.1.10}$$

Here $V_e = 4/3\,\pi abc$ is the volume of the ellipsoid and L_j is the *geometrical factor* (or depolarization factor) relative to each axis of the ellipsoid given as

$$0 \leq L_a \leq L_b \leq L_c \leq 1,$$
$$L_a + L_b + L_c = 1. \tag{4.1.11}$$

The plasmon resonance condition is reached when the denominator in equation (4.1.10) is equal to zero, i. e.,

$$\mathrm{Re}\{\varepsilon_r(\omega)\} = \left(1 - \frac{1}{L_j}\right)\varepsilon_m. \tag{4.1.12}$$

When $L_j > 1/3$ the NP is flatter than a sphere, shaped somewhat like a pumpkin, and the corresponding resonance condition is blue-shifted with respect to the case of a sphere. On the other hand, when $L_j < 1/3$, the NP is sharper than a sphere, like a rugby ball, and the corresponding resonance condition is red-shifted with respect to the case of a sphere [4]. Thus considering spheroids we have two resonance conditions for the same system, related to the long and short axes (Fig. 4.1.2b). This offers a further advantage, allowing one to exploit the plasmon resonances parallel to the different antenna axes and tuned to different wavelengths, simply changing the direction of polarization of the incident field [10].

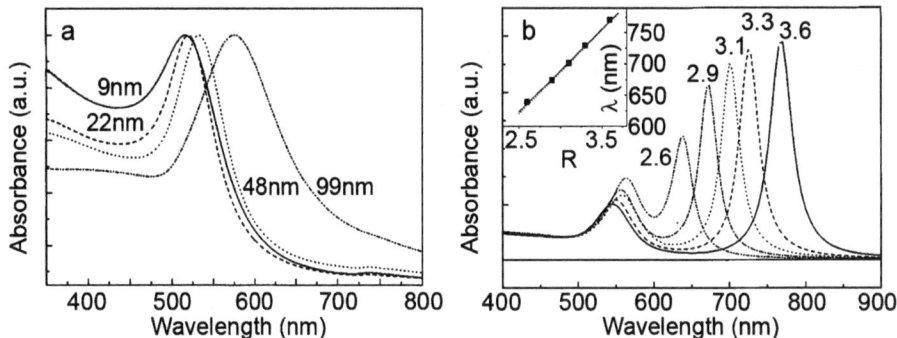

Figure 4.1.2: (a) LSPR of gold nanospheres of increasing diameter (reproduced from Ref. [11]). (b) Calculated extinction spectra of elongated gold ellipsoids as a function of the aspect ratio R (reproduced from Ref. [9]).

Finally, taking into account the radiative correction of the first order, equation (4.1.12) is replaced by the following relation [4]:

$$\text{Re}\{\varepsilon_r(\omega)\} = -\left(\frac{3L_j - 3 + (k_m a)^2}{3L_j - (k_m a)^2}\right)\varepsilon_m. \tag{4.1.13}$$

The relations developed so far show how the LSPR properties can be tuned by modifying the aspect ratio of the ellipsoids (Fig. 4.1.2b).

4.1.3 Introduction

The fields of molecular and biomolecular sensors have become fundamental in biomedical research for the improvement of human health. Early-stage diagnosis of diseases is, in fact, based on the identification and characterization of some specific protein biomarkers [12] that may be present in very low quantities in body fluids. Proteins, in particular, are very important biomarkers. They are crucial for biological systems, as they assume important functions such as enzymatic or hormonal activities, and even carrying small molecules. Altered values of protein concentrations in body fluid can reveal ongoing pathologies. Immunoassays [13] such as enzyme-linked immunosorbent assays (ELISAs), protein arrays, Western blots, immunohisto-chemistry, and immunofluorescence are well-assessed means to detect and quantify biomarkers with sensitivities that in the best cases can reach the subfemtomolar range [14], but are typically in the nM to pM region. These methods, moreover, require complex sample preparation, such as staining with fluorescent probes, and they are time consuming, being therefore unsuitable for rapid (minutes) and ultrasensitive (pM to aM) detection.

Plasmonics introduces new original approaches in the field of molecular detection, offering strategies and solutions with unprecedented sensitivity, among which we find SPR, LSPR, SERS, surface-enhanced infrared absorption/scattering (SEIRA/SEIRS), and metal-enhanced fluorescence (MEF) [15–17, 8].

SERS [3] provides a huge enhancement of the Raman signal (104-109) related to the electromagnetic field amplification due to LSPRs in metallic NPs [18]. This huge signal amplification can bring the sensitivity down to the attomolar range [19], and even to the single-molecule level in particular configurations [20, 21]. SERS has large application potential in label-free detection of biomolecules and pathology biomarkers [22–25]. Different configurations of SERS biosensors have been shown. "Raman reporter" concepts exploit SERS-active labels (particles coated with dyes featuring a high Raman cross-section, which are functionalized with specific receptors against the target molecule) to capture proteins and protein modifications also in vivo [26, 27]. The target molecule is detected through an indirect measurement of the SERS signal of the Raman reporter molecule. Direct, label-free SERS sensors, instead, achieve molecular detection by acquiring the vibrational molecular fingerprint [25, 28]. They are rapid, simple, and rich of information. The vibrational spectrum of proteins, in fact, can give insight in their conformation, structure, and functional state [29, 30], although it must be taken into account that detection of biomolecules in the dry state as well as the interaction with the NP surface can alter their native state [31, 32]. SERS detection of biomolecules in liquid keeps them in their natural habitat. This latter is, however, still a challenge. The SERS amplification provided by individual NPs (10^2 to 10^3), in fact, is not strong enough to render the weak Raman scattering of biomolecules detectable. NPs aggregates offer a viable solution to this problem, with enhancement factors reaching 10^9 in the nanoscale interstices between NPs (the so-called hotspots) [3]. In the last years intense research has been carried out towards the development of effective ways to obtain SERS-active aggregates in liquid solutions containing biomolecules. The idea is to embed the target molecules at the *hotspot locations* during the aggregation process and detect their highly enhanced vibrational fingerprint by Raman spectroscopy.

The aim of this chapter is to review the newest implementations of SERS sensors in liquid environment, and in particular those new routes that exploit optical forces [5, 33]. The chapter is organized as follows. In Section 4.1.4.1 we introduce the concepts of field enhancement effects in metal NPs. Section 4.1.4.2 reviews relevant chemical and thermophoretic methods to aggregate NPs in liquid and perform SERS detection of molecular compounds. Section 4.1.4.3 introduces the concept of optical force and provides calculations of the optical forces on metal NPs. Section 4.1.4.4 describes the use of optical forces for NPs aggregation and SERS detection of molecules in liquid environment.

4.1.4 State-of-the-art

4.1.4.1 Field and SERS enhancement of individual and near-field-coupled nanoparticles

Metal NPs, when excited at resonance, act as nanoantennas, strongly enhancing and confining the electromagnetic field into nanoscale sites, named hotspots. The electromagnetic field exponentially decreases far from the hotspot. Consequently, molecules benefit from the field amplification only when they are located at few nanometers away from the NPs' surface. At the hotspots, molecules take advantage of the twofold amplification of the excitation field and of the Raman field, leading to a SERS electromagnetic enhancement factor that can be written as

$$SERS_{EF} = \Gamma_{exc}(\lambda_{las}) \times \Gamma_{rad}(\lambda_R), \qquad (4.1.14)$$

where $\Gamma_{exc}(\lambda_{las})$ is the enhancement factor of the excitation field calculated at the laser wavelength λ_{las} and $\Gamma_{rad}(\lambda_R)$ is the enhancement factor of the Raman field at the wavelength of the emitted photons [3, 34]. The term $\Gamma_{exc}(\lambda_{las})$ is phenomenologically calculated as the ratio $|\vec{E}_{loc}|^2/|\vec{E}_i|^2$, where $\vec{E}_{loc}$ is the local enhanced field intensity at the hotspot and $\vec{E}_i$ is the incident field intensity.

The position of the hotspots and the field amplification strongly depend on the geometry of the NPs, the field polarization, and the resonance order (dipole, quadrupole, etc.). Figure 4.1.3 displays the field distribution around a 200×60 nm^2 gold NR calculated by discrete dipole approximation [35]. When the long-axis LSPR is excited (a)

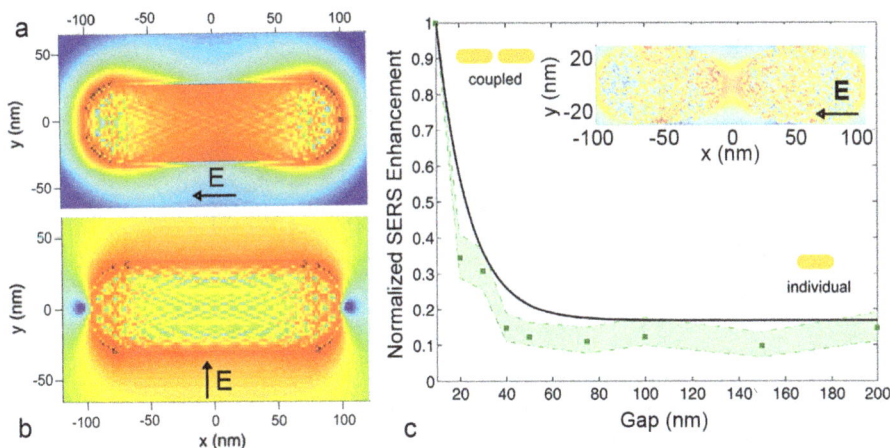

Figure 4.1.3: Map of the electromagnetic field distribution at the surface of an NR when the long axis (a) or the short axis (b) LSPRs are resonantly excited. (c) Experimental (dots) and theoretical (line) values of the SERS enhancement measured on coupled NRs (100×60 nm^2) when the gap is decreased. The inset shows the map of the enhanced field for small gaps (reproduced from Ref. [35]).

two hotspots show up, localized at the NR tips. When the short axis LSPR is excited (Fig. 4.1.3b) the hotspot region is, instead, extended along the NR edges. Notably, the field enhancement provided by a single NR is further amplified in a configuration in which this latter is near-field-coupled with a second NR, e. g., in a tip-to-tip dimer arrangement. This is due to the occurrence of gap resonances, red-shifted with respect to the resonance of the single particle, and which are the result of the near-field interaction among individual NPs [3, 36]. Figure 4.1.3c shows how the SERS enhancement exponentially increases when the gap between two NRs (100 nm long) becomes smaller than 40 nm. In particular, the higher field enhancement gives rise to a net signal gain of a factor of 10 when the gap of the dimer is reduced to 15 nm [35]. As shown by the calculations (inset of Fig. 4.1.3c) in the dimer configuration the hotspot region is located at the gap between the rods.

Silver and gold nanocolloids are suitable for SERS detection in liquids. They provide the building blocks to create multiparticle clusters closely interacting with the analytes in their native environment, leading to very high SERS enhancement factors. Liu et al. [37] studied the aggregation dynamics of silver colloids from the wet to the dry state during the water evaporation process in a microdroplet of Ag sol cast on a hydrophobic surface. Monitoring the structural properties and the SERS enhancement of rhodamine 6G (R6G) molecules dispersed in the solution, they found three different regimes (Fig. 4.1.4): (a) the wet state in which NPs are free in the droplet, (b) the transition state in which NPs are closely packed but the solvent is still in the system, and (c) the dried state in which NPs are in a close-packed arrangement. The SERS intensity increases during the evaporation, reaching a maximum value in the transition state (b), where a signal amplification factor of $\sim 10^4$ is observed. Suddenly after, an

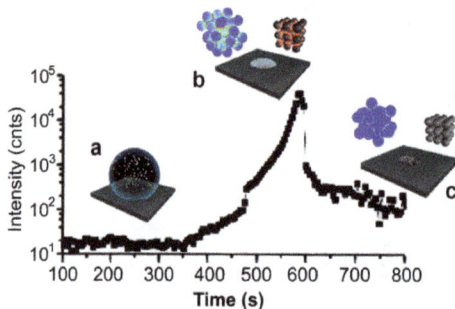

Figure 4.1.4: Time-dependent SERS intensity of the 610 cm^{-1} peak of R6G 50 aM mixed with Ag sol during evaporation of the solvent. Sketches representing the initial wet state (a), the aggregation of particles on the substrate in the presence of some residual solvent (b), and the close-packed arrangement of the dried NPs (c). In (b) and (c) are also displayed the theoretical simulations of the field intensity distributions for a cluster of 35 Ag particles of 50 nm in diameter for gaps $g = 2$ nm (b) and $g = 0$ nm (c) (adapted from Ref. [37]).

abrupt collapse of almost two orders of magnitude was observed, corresponding to the onset of the dried state (c), where the NPs start to get in ohmic contact with each other. As the droplet shrinks the interparticle distance decreases, and at the same time the ionic strength increases, reducing the energetic barrier of the potential responsible for colloidal stabilization [3, 38]. In the presence of ohmic contact, charge transfer phenomena strongly deplete the electromagnetic field. Simulations of the electric field distribution calculated for average gaps of 2 and 0 nm (corresponding to the transition state and the dried state, respectively) highlight the presence of a large number of intense hotspots in the former case (b) and a weaker field in the latter (c). These results highlight that a reduction of the gap size down to a few nanometers can yield an advantage in terms of $SERS_{EF}$ of the order of 10^2 to 10^4.

4.1.4.2 SERS detection in liquid environment with chemically and thermophoretically aggregated nanoparticles

A simple way to aggregate plasmonic nanocolloids is via the addition of chemical agents, including salts [39] or pyridine [40]. Such practices provide structures with large SERS enhancement, but also cause remarkable signal instabilities and low reproducibility. Aggregation is, in fact, a dynamical process based on both the irreversible NP agglomeration and the diffusion properties of the analyte in the solution. If the interaction between the analytes and the NPs is weaker than the particle–particle one, the number of molecules in the hotspots will be low and the SERS signal will be less intense [41]. With statistical treatment of the signals, however, even quantitative SERS detection can be achieved. An example was shown on the detection of uric acid present in human serum, in which the authors demonstrated limit of detection (LOD) values of ~240 μM (40 μg/mL), besides achieving quantitative evaluation [42]. An efficient approach to increase the sensitivity of this method is to induce aggregation via addition of acidified sulfate to the solution containing the protein to be detected (Fig. 4.1.5a) [43]. This yields NP–protein complexes where the biomolecule is located in the NP nanogaps. Quantitative detection of nonresonant proteins could be achieved at concentrations down to 5 μg/mL on Lysozime (Lys) as shown in Figure 4.1.5b. The same concept was applied to optical fiber SERS sensors. By exploiting NP–protein–NP sandwich structures (Fig. 4.1.5c), a sensitivity of 0.2 μg/mL was shown (Fig. 4.1.5d) [44].

The analytes can play an active role in the aggregation process of plasmonic NPs [45], when their net charge is different from zero. Proteins carry a zero net charge when the pH of the system is at the isoelectric point (pI). An overall positive charge is carried if the pH is below the pI of the protein, and a negative charge is brought when the pH is above the pI. Electrostatic interactions can foster or hinder the formation of highly compact plasmonic architectures in which proteins are located in the interstices. Kahraman et al. [46] highlighted the importance of electrostatic interactions in

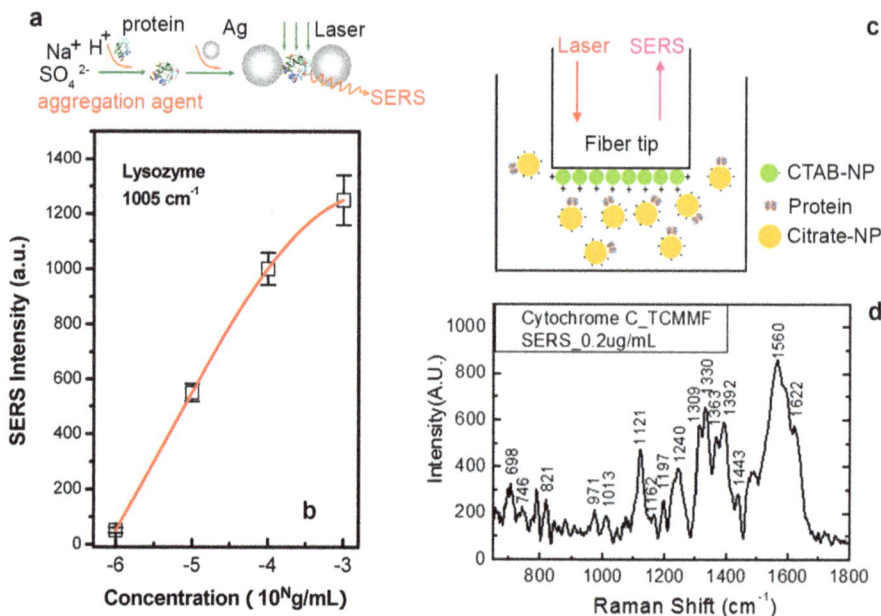

Figure 4.1.5: (a) Aggregation protocol of silver colloids for label-free proteins. (b) Concentration-dependent SERS intensities of Lysozime (1005 cm^{-1}). Reproduced from Ref. [43]. (c) Schematic tip-coated multimode fiber SERS probe in aqueous protein detection. (d) SERS spectra of cytochrome *c* detected by tip-coated multimode fiber SERS probe. Re-adapted from Ref. [44].

thin films made from citrate-reduced silver NPs mixed with two different protein solutions at neutral pH, namely Lys (pI = 11.4) and bovine serum albumin (BSA, pI = 4.7). Adsorption of citrate ions yielded AgNPs with a negative surface charge density. The net charge of Lys was positive, while that of BSA was negative. The aggregation process was driven by capillary forces (Fig. 4.1.6a), but electrostatic interactions between NPs and analytes affected the packing of the NPs. This was proved by the thin film images, showing up as more compact structures when using Lys, whose surface charge is opposite with respect to that of the AgNPs (Fig. 4.1.6b). SERS, however, was not dramatically affected by the packing difference, reaching sensitivity down to 0.5 µg/mL for both proteins.

Chemically induced aggregation in liquid environment was also demonstrated using iodide-modified AgNPs (Ag IMNPs) [47]. The iodide ions formed a one atom thick layer on the NPs' surface without creating an interfering signal, preventing SERS from eventual surface impurities (Fig. 4.1.7a). The aggregation process was triggered by the protein addition, as shown in Figure 4.1.7b (red line). A LOD of 3 µg/mL was achieved for Lys, while the LOD for BSA was only 300 µg/mL. The almost identical features between SERS and normal Raman spectra (Fig. 4.1.7c) evidenced a weak alteration of the protein structure, suggesting this protective layer could also be useful for keeping the native structure of the biomolecules. Other methods for SERS detec-

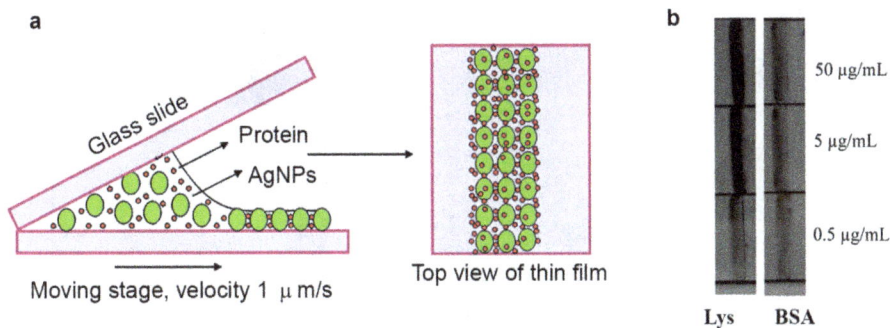

Figure 4.1.6: (a) Schematic of the assembly of protein/AgNP thin films. (b) Picture of films of Lys and BSA for different protein concentrations. Adapted from Ref. [46].

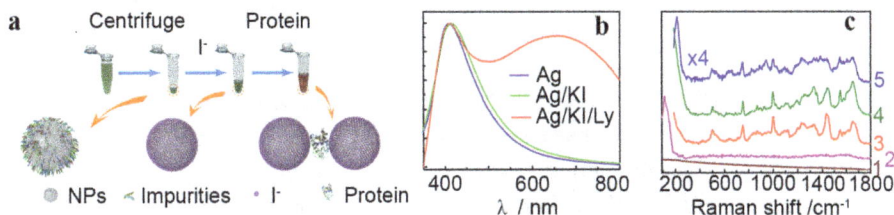

Figure 4.1.7: (a) Preparation of the aggregated system with iodide-modified AgNPs. (b) UV-Vis spectra of silver colloids in AgNPs and AgNPs mixed with Lys. (c) SERS spectra of Ag particles (1), Ag IMNPs (2), 300 µg/mL Lys with Ag IMNPs (3), and 300 µg/mL Lys with AgNPs (5) and Raman spectra of 100 mg/mL Lys solution (4). Adapted from Ref. [47].

tion of biomolecules in liquid can take advantage of hydrophobic interactions (LOD ~ 5 µg/mL for Lys and cytochrome *c*) [48], heat-induced self-assembly (LOD ~ 50 nM in glutathione) [49], or mechanical aggregation of AuNPs in microfluidic channels (LOD ~ 0.1 nM for BSA) [50]. Accumulation of NPs using biocompatible coatings is a further methodology for SERS detection in liquid (LOD ~ 50 nM for cytochrome *c*) [51].

Optical radiation can trigger a series of interesting physicochemical processes yielding the formation of SERS-active structures that can be used for detection of chemical compounds in liquid. A pioneering experiment in this field showed that, by illuminating a solution of Ag ions dispersed in dye molecules, it was possible to locally produce Ag NPs that, once aggregated, permitted the SERS detection of the dye molecules [52]. Ag reduction can also be helped by plasmon-mediated photoemitted electrons, yielding the formation of hybrid Au-Ag sandwich structures that are SERS-active [53]. Remote SERS sensing in liquid based on optical fibers can also take advantage of NP aggregation induced by dipolar and thermophoretic interactions [54, 55]. As shown by Liu et al. [55], clusters of colloidal silver nanocubes (NCs) mixed with *p*-aminothiophenol (*p*-ATP) were produced at the apex of an optical fiber im-

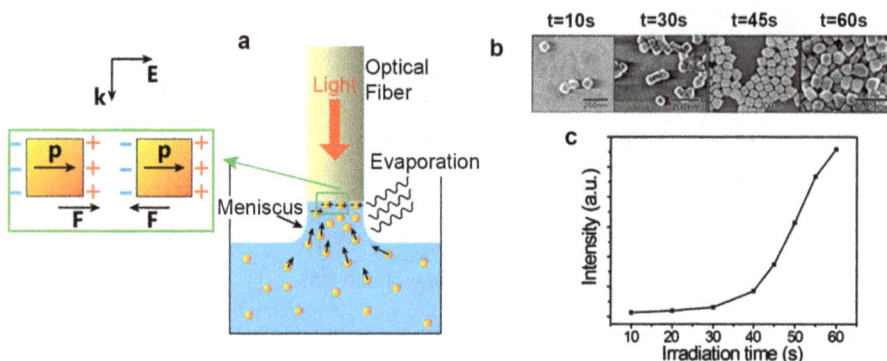

Figure 4.1.8: (a) Schematic of the particle aggregation on a fiber SERS probe by light-induced self-assembly of metal NPs. The meniscus occurs by lifting the fiber facet up from the solution surface. (b) SEM of the center of the fibre facets taken at subsequent laser irradiation intervals. (c) SERS intensity of the 1142 cm^{-1} band of p-ATP versus laser irradiation times. Adapted from Ref. [55].

mersed in liquid (Fig. 4.1.8a) upon light irradiation and consequent heating of the NPs [56]. The fiber was then used as a waveguide to collect the SERS signal of the p-ATP molecules. The aggregation process was fast (only 60 s, Fig. 4.1.8b, and c) and more efficient when the optical fiber formed a meniscus structure, allowing for LODs in the 100 pM range. In this configuration capillarity forces act against gravity, bringing NCs near the cleaved surface of the fibre. Heat transfer from excited plasmonic NCs to the solvent increased the water evaporation, enhancing the capillarity effects and accelerating the aggregation.

Aggregation processes triggered by chemical agents typically lead to irreversible binding of metal NPs. Reversible aggregation is possible exploiting thermophoretic effects [57] or temperature-responsive coil-to-globule transition of polymers [58] to trigger by plasmonic heating [56]. Li and coworkers [57] controlled the aggregation of Au nanotriangles (NTs) through the resonant excitation of a nanostructured gold surface. In their experiments plasmonic heat generated a temperature gradient, which affected the diffusion properties of the colloidal suspension, fostering the accumulation of the gold NTs in the laser spot region (Fig. 4.1.9a). The field enhancement provided by the coupled system caused a further temperature increase, with a consequent acceleration of the aggregation process. Turning off the laser disassembled the aggregate. Different laser beams permitted to pattern the substrate with SERS-active clusters and reach an LOD value of ~1 μM for R6G. Ding et al. [58] exploited the coil-to-globul transition of poly(N-isopropyl-acrylamide) (pNIPAM) used as functionalizing agent to aggregate 60 nm diameter citrate-stabilized gold nanospheres. Below the critical temperature T_c, pNIPAM is hydrophilic and swelled in the gel by the water. When heated above T_c pNIPAM became hydrophobic and expelled the water, collapsing to a much smaller volume. This caused a reduction in the energy barrier that stabilized the colloid, fostering NP assembly (Fig. 4.1.9b). As the system was cooled down, the pNIPAM

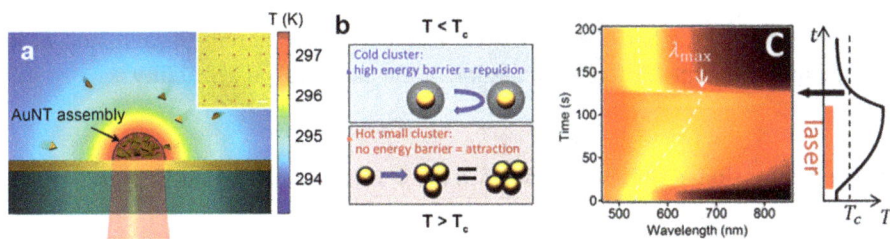

Figure 4.1.9: (a) Calculated temperature at the interface between substrate and particle solution after NT assembly. Inset: Optical image of the NT assemblies. Adapted from Ref. [57]. (b) Schematics showing the repulsive (top panel) and the attractive (bottom panel) regime of individual AuNPs functionalized with pNIPAM for $T < T_c$ and $T > T_c$, respectively. (c) LSPRs maximum position (λ_{max}) versus time, as the laser is switched on and off. The LSPR red shift highlights the light-induced nature of the aggregation. Adapted from Ref. [58].

swelled again, allowing for the re-dispersion of the individual nanostructures. The aggregation was controlled detecting the red shift of the LSPR position (Fig. 4.1.9c) when aggregation occurred. As the laser was switched off the NPs disassembled, and the LSPR maximum position was restored to its initial value.

4.1.4.3 Optical forces

Light exerts forces and torques on matter due to the momentum conservation in light–matter interaction [5, 59, 60]. Optical forces enable the manipulation of plasmonic NPs, providing novel routes for the synthesis of efficient SERS-active aggregates in liquid in a controlled and contactless way [33, 61, 62]. Optical forces offer unique advantages over the other aggregation methods, in terms of chemical-free operation, simplicity of operation, control of the process, reversibility, and the possibility for in vivo application. An example is given by the tunability of the light–particle interaction. We can change from the attractive optical trapping regime to the repulsive optical pushing regime [63] by simply varying the intensity or wavelength of the light field. Generally speaking, in fact, when the energy of the laser field is far off-resonance with respect to the LSPR, the gradient force prevails (see below), attracting the metal NPs into the high-field intensity focus [64–66]. Conversely, when the field is resonant with the LSPR, the radiation pressure prevails (see below), leading to a pushing of the metal NPs along the optical axis of the beam [67–70].

Calculations of the optical forces, in the most general approach, are accomplished through the full electromagnetic theory [5, 71]. If the scatterer dimensions are much smaller or much bigger with respect to the wavelength of light, however, it is possible and convenient to use approximated theories. Again, we can use the size parameter x, defined in equation (4.1.1) to discriminate the different regimes. When $x \gg 1$, the *ray optics approximation* can be used, and the results are retrieved by applying Snell's law. On the contrary, when $x \ll 1$, we use the *dipole approximation* [5] combined with the dipolar approximate for the calculation of the particles' polarizability.

General expression of the optical forces in the dipole approximation

The optical force exerted on an electric dipole $\vec{p}_d$ in the presence of time-varying electric and magnetic fields (with the electric field strength $\vec{E}_i(\vec{r}, t)$ and its complex conjugate $\vec{E}_i^*(\vec{r}, t)$, magnetic field strength $\vec{H}_i(\vec{r}, t)$, and magnetic flux density $\vec{B}_i(\vec{r}, t)$) is given, for nonrelativistic dipoles, by [5]

$$\vec{F}_{DA}(\vec{r}_d, t) = (\vec{p}_d \cdot \vec{\nabla})\vec{E}_i(\vec{r}_d, t) + \vec{p}_d \times (\vec{\nabla} \times \vec{E}_i(\vec{r}_d, t)) + \frac{d}{dt}(\vec{p}_d \times \vec{B}_i(\vec{r}_d, t)), \quad (4.1.15)$$

where $\vec{r}_d$ is the position of the dipole's center of mass. Due to the fact that the electromagnetic field oscillates very fast (optical frequencies $\approx 10^{15}$ Hz), we are interested in the time average of $\vec{F}_{DA}(\vec{r}_d, t)$, obtaining

$$\vec{F}_{DA} = \overline{\vec{F}_{DA}(\vec{r}_d, t)} = \overline{(\vec{p}_d \cdot \vec{\nabla})\vec{E}_i(\vec{r}_d, t)} + \overline{\vec{p}_d \times (\vec{\nabla} \times \vec{E}_i(\vec{r}_d, t))}. \quad (4.1.16)$$

The induced dipole moment depends on the electric field through the relation $\vec{p}_d = \alpha \vec{E}_i$, where α is the complex polarizability. As a result equation (4.1.16) can be written as

$$\vec{F}_{DA} = \frac{1}{2}\alpha' \operatorname{Re}\{(\vec{E}_i \cdot \vec{\nabla})\vec{E}_i^* + \vec{E}_i \times (\vec{\nabla} \times \vec{E}_i^*)\} - \frac{1}{2}\alpha'' \operatorname{Im}\{(\vec{E}_i \cdot \vec{\nabla})\vec{E}_i^* + \vec{E}_i \times (\vec{\nabla} \times \vec{E}_i^*)\}, \quad (4.1.17)$$

where $\alpha' = \operatorname{Re}\{\alpha\}$ and $\alpha'' = \operatorname{Im}\{\alpha\}$. By using some vector identities and the extinction cross-section $\sigma_{\text{ext}} = k_m \alpha'' / (\varepsilon_0 \varepsilon_m)$, we can finally obtain [5]

$$\vec{F}_{DA} = \frac{1}{4}\alpha' \vec{\nabla}|\vec{E}_i|^2 + \frac{n_m \sigma_{\text{ext}}}{c}\vec{S}_i - \frac{1}{2}\frac{\sigma_{\text{ext}}c}{n_m}\vec{\nabla} \times \vec{s}_d, \quad (4.1.18)$$

where c is the speed of light in vacuum, $\vec{S}_i$ is the time-averaged Poynting vector of the incident wave, calculated as

$$\vec{S}_i = \frac{1}{2}\operatorname{Re}\{\vec{E}_i \times \vec{H}_i^*\} = \frac{1}{2}c\varepsilon_0 n_m |\vec{E}_i|^2 \hat{z}, \quad (4.1.19)$$

with $\hat{z}$ representing the unity vector in the propagating field direction, and the time-averaged spin density of the incoming wave is given by [72]

$$\vec{s}_d = i\frac{\varepsilon_0 \varepsilon_m}{2\omega}\vec{E}_i \times \vec{E}_i^*. \quad (4.1.20)$$

The first term in equation (4.1.18) expresses the gradient force, which is responsible for the optical trapping. The gradient force has a conservative nature, deriving from the potential energy of a dipole in an electric field. The intensity of the electric field is $I_i = |\vec{S}_i| = 1/2c\varepsilon_0 n_m |\vec{E}_i|^2$. We can thus re-write the gradient force in terms of the gradient of the field intensity as

$$\vec{F}_{grad} = \frac{1}{2}\frac{\alpha'}{c\varepsilon_0 n_m}\vec{\nabla}I_i. \tag{4.1.21}$$

Particles with positive polarizability, e. g., dielectric particles with refractive index higher than the one of the surrounding medium, are attracted in the high-intensity region of the laser field.

The second term in equation (4.1.18) is the scattering force [5]:

$$\vec{F}_{scat} = \frac{n_m \sigma_{ext}}{c}\vec{S}_i = \frac{n_m \sigma_{ext}}{c}I_i\hat{z}. \tag{4.1.22}$$

The scattering force is dissipative. It is due to the transfer of momentum from the field to the particle as a result of scattering and absorption processes. $\vec{F}_{scat}$ is, in fact, proportional to the extinction cross-section σ_{ext}. This force is directed parallel to the Poynting vector $\vec{S}_i$ and pushes the particle along the field propagation direction $\hat{z}$.

The third term in equation (4.1.18) is the spin-curl force:

$$\vec{F}_{sc} = -\frac{1}{2}\frac{\sigma_{ext}c}{n_m}\vec{\nabla}\times\vec{s}_d. \tag{4.1.23}$$

It originates from polarization gradients in the electromagnetic field, being nonconservative. The spin-curl force is relatively small with respect to the gradient and scattering components and, therefore, is generally negligible in optical trapping experiments [5].

Optical forces on plasmonic nanoparticles

For particles smaller than the laser wavelength, the dipole approximation describes quite well the mechanical interactions between NPs and incident optical fields. Within this hypothesis, neglecting the *spin-curl* contribution, the optical force exerted by a Gaussian beam of power P and waist w_0 propagating along the direction $\hat{z}$ on a small scatterer immersed in a medium with refractive index n_m ($\varepsilon_m = n_m^2$) is given by the sum of equation (4.1.21) and equation (4.1.22):

$$\vec{F}_{rad} = \vec{F}_{grad} + \vec{F}_{scat} = \frac{1}{2}\frac{\alpha'}{c\varepsilon_0 n_m}\vec{\nabla}I_i + \frac{n_m \sigma_{ext}}{c}I_i\hat{z}, \tag{4.1.24}$$

where σ_{ext} is the extinction cross-section, α' is the real part of the polarizability, c is the speed of light, and $I_i = (2P)/(\pi w_0^2)$ is the intensity of the incident laser beam at the focus. $\vec{F}_{grad}$ is conservative and controls the operation of optical tweezers [33, 60].

The gradient force is repulsive when $\alpha' < 0$, e. g., if the incident wavelength is blue-shifted with respect to the plasmon resonance. It is directed towards the laser focus for $\alpha' > 0$. It increases by exploiting high-numerical aperture objectives that create larger intensity gradients. In order to have a stable 3D optical trap, the $\vec{F}_{grad}$ component along the z axis has to overcome $\vec{F}_{scat}$, which is responsible for pushing particles along the light propagation direction [5]. The scattering force is proportional to the extinction cross-section and to the intensity of the incident beam, so in proximity to plasmon resonances, where σ_{ext} can be large, dissipative forces acting on the metal NP can be very strong.

The increased polarizability of metallic particles generates enhanced optical forces with respect to dielectric particles of equal size. It was demonstrated that the longitudinal resonance of gold NRs can be exploited for stable trapping. Using a laser red-detuned with respect to the LSPR, an increased residence time in the laser focus region was observed, together with a suppression of rotational diffusion, indicative of an alignment effect [73]. At wavelengths shorter than the LSPR, the residence time decreased and the NRs were repelled. On the other hand, an increased polarizability may lead also to a larger scattering force that may push NPs away from an optical trap [5]. As an example, we calculate here the optical forces generated by a focused 632.8 nm laser on an NR featuring a major-axis LSPR at 687 nm, i. e., the laser is blue-detuned. The polarizability is calculated from the *Clausius–Mossotti* relation given by equation (4.1.10), with the optical constants measured by Johnson and Christy for gold [74]. The geometrical factors L_j are defined in terms of the particle eccentricity, $e^2 = 1 - (c/a)^2$. For a prolate ellipsoid they are [1]

$$L_a = \frac{1-e^2}{e^2}\left(\frac{1}{2e}\ln\frac{1+e}{1-e} - 1\right),$$
$$L_b = L_c = \frac{1 - L_a}{2}. \tag{4.1.25}$$

Calculations for the two configurations in which the field is polarized along the NR's long (a) and short axis (b = c) as a function of the rod position with respect to the laser focus center (z = 0) are shown in Figure 4.1.10 [70]. When the NR is oriented parallel to the incident field, the gradient force (a, red line) is repulsive. The scattering force (a, blue line) is even more intense, resulting in a net optical pushing effect. When the NR is orthogonal to the field, the gradient force (b, red line) shows an attractive, "trapping" character. However, the scattering force is markedly larger (b, blue line), due to the fact that we are near the plasmonic resonance and σ_{ext} is large, resulting in a pushing of the NR from the laser focus also in this case, as experimentally observed [70].

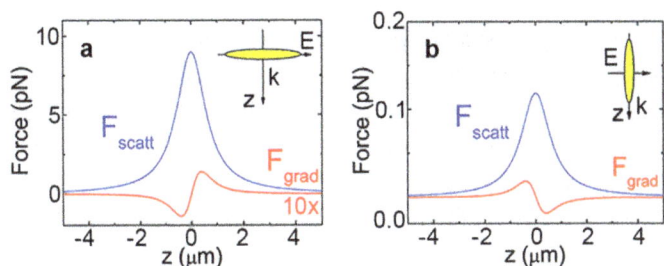

Figure 4.1.10: Optical forces acting on gold NRs. The $z = 0$ position corresponds to the laser spot center. The $\vec{z}$-axis is parallel to the wavevector $\vec{k}$ and points "downward." (a) When the incident field is parallel to the long axis, the gradient force (red line) has a repulsive nature. The scattering force is always oriented parallel to $\vec{z}$ (blue line), and is even more intense. (b) For a field parallel to the short axis, the gradient force (red line) features an attractive nature. Also in this case, however, the scattering force is much higher (blue line), resulting in a net positive optical force acting on the NRs that will expel the rod from the laser focus. Redrawn from Ref. [70].

4.1.4.4 SERS molecular detection in liquid environment mediated by optical forces

Inducing the near-field coupling of plasmonic NPs in a liquid environment is a crucial point to efficiently perform enhanced molecular spectroscopy, due to the more effective field enhancement provided by plasmonic nanocavities with respect to single NPs [3, 37, 75]. Toussaint et al. [76] proved the possibility to stably trap multiple plasmonic NPs simultaneously. To do this they used a laser red-shifted with respect to the LSPR of the NPs that is capable of enhancing the gradient force with respect to the scattering one. When more particles are in the trapping volume, the optical coupling among the particles caused a red shift of the resonance relative to the single particle [77]. The change in the scattering spectrum was explained as due to the resonance of the new coupled dimer (Fig. 4.1.11).

The temperature increase due to the absorbed-light-to-heat conversion was a further obstacle to the stability of the optical trap, leading to NPs escaping from the

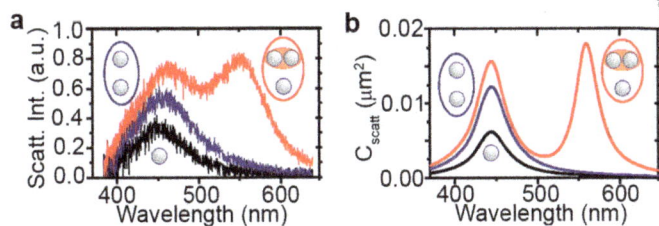

Figure 4.1.11: (a) Scattering spectra (experiment) of one (black), two (blue), and three (red) optically trapped silver NPs. (b) Scattering cross-section spectra (theory) of one (black), two uncoupled (blue), and two coupled ($d = 2$ nm) plus one uncoupled (red) NPs for randomly changing orientation (adapted from Ref. [77]).

trapping region as these coupling effects became more important [77]. The formation of plasmonic dimers can be controlled if a trapped single metal NP is brought
into the near-field of another metal NP fixed on a substrate (Fig. 4.1.12a-b). Svedberg
et al. [78] pioneered this approach, showing it for SERS-active dimers in liquid environment and detecting thiophenol (TP) molecules dissolved therein (10 μM). AgNPs
of 40 nm diameter were used, and two different laser wavelengths were exploited for
trapping (830 nm) and SERS excitation (514.5 nm). A SERS signal from TP was observable when the dimer formation was optically induced (Fig. 4.1.12c). These first experiments paved the way to new aggregation strategies for the formation of SERS-active
complexes in liquid [80] or in lab-on-a-chip architectures [81, 79]. The detection of organic compounds (pseudoisocyanine 10 fM, naphtalenethiol 50 μM) has been shown
by applying these methods in solution. Figure 4.1.12d illustrates the optical trapping
and aggregation as a function of time (see laser spot area) of metal NPs flowing in a
microfluidic channel. The NPs were dispersed in TP or 2-naphtalenethiol (2-NT) solutions. A corresponding increase of the SERS signal of TP and 2-NT (Fig. 4.1.12e,f) was
observed as the aggregate size increased with time. Metalized silica beads have also
been used for trapping and detection of emodin at micromolar concentrations [82].
Efficient optical trapping of metal NPs in a photonic crystal led to SERS detection of
4-aminothiophenol molecules down to a concentration of 10 nM [83]. Alternatively to
optical aggregation of individual NPs, it is also possible to trap NP aggregates already

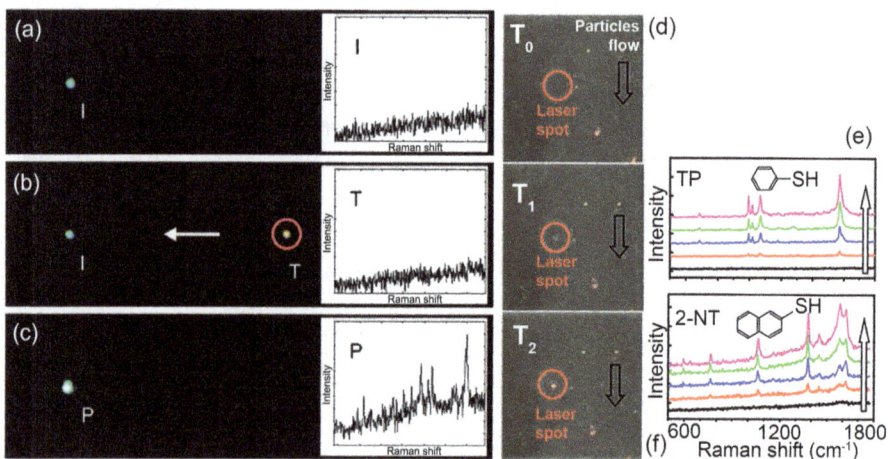

Figure 4.1.12: (a, b) Dark-field images and corresponding SERS signals (white panels) of an immobilized and a trapped Ag NP (labeled with I and T, respectively). (c) When the particles are in near-field
contact, the dimer (labeled with P) shows a strong SERS signal of the thiophenol (TP) molecules in
which the NPs are embedded. (d) Time sequence (T_0, T_1, T_2) of images showing the optical trapping
of AgNPs flowing in a microfluidic channel inside the laser spot. NPs trapped and accumulated induce an increasing SERS signal of the surrounding analytes, namely TP (e) and 2-NT(f). Panels (a, b,
and c) adapted from Ref. [78]; panels (d, e, and f) adapted from Ref. [79].

formed in liquid solution, on which a protein has been absorbed, carrying out SERS detection of biomolecules in liquid [84].

NP aggregation induced with optical forces can be reversible. This offers advantages in terms of manipulation, placement on-demand of the aggregates, and "reusability" of the NPs [85, 86, 72]. One strategy to achieve reversible aggregation is to exploit the radiation pressure [85]. A convergent laser beam (Fig. 4.1.13a) can collect and compress against gravity gold NPs of about 100 nm on the upper window of a glass cell, creating a dense plasmonic aggregate. Moreover, by using independent traps, the plasmonic aggregates can be manipulated independently, brought together, or set apart (Fig. 4.1.13b). Patra et al. [86] showed that reversible optical aggregation can lead to SERS molecular detection with single-molecule sensitivity. They exploited the evanescent waves arising from the excitation of surface plasmon polaritons (SPPs) at the metal–water interface (Fig. 4.1.13c, d) for both trapping and SERS. The authors pointed out how, in addition to optical forces, the temperature gradient induced by plasmonic heating of the NPs created an advection flow, which fostered the accumulation of metal NPs in the focus region where they were stably blocked. Notably, the evanescent field was effective enough to excite the Raman scattering of molecules in the interstices of the aggregate, allowing to reach single-molecule SERS sensitivity (Fig. 4.1.13e) of R6G and Nile Blue (NB). The technique took advantage of silver–gold hybrid core-shell NPs. Laser power is an important parameter for optically induced aggregation, regulating the strength of the optical forces and the rate at which NPs interact. Usual values for aggregating power densities are in the range from mW to tens of mW [53, 68, 81, 85, 87]. Nevertheless, it was observed that this value can be

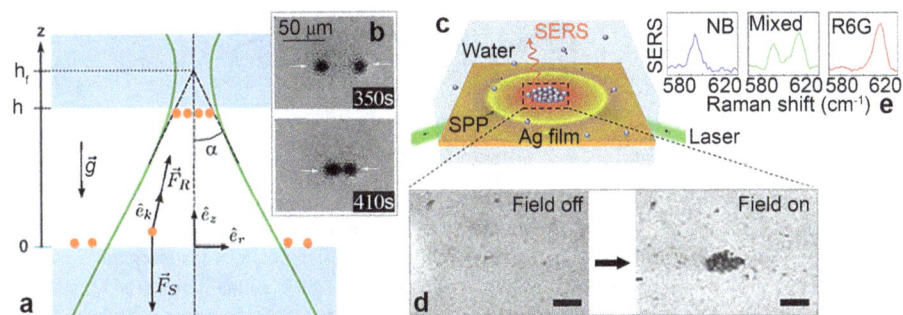

Figure 4.1.13: (a) Reversible aggregation of NPs is achieved in an inverted geometry (light propagates along the z-direction). The focal point is placed just inside the upper glass window of the microchamber where metal NPs are dispersed. The radiation pressure pushes and aggregates the NPs against gravity at the bottom surface of the glass window. (b) Images of aggregated Au colloids in a multiple trap allowing for their manipulation (reproduced from Ref. [85]). (c) Illustration of the optical assembly of NPs using the evanescent waves generated by SPPs in a silver film on a glass prism. The same radiation is used to excite SERS. (d) The aggregate reversibly disappears when the field is switched off and re-appears when the laser (532 nm) is re-applied. (e) SERS spectra of single molecules of NB (blue), R6G (red), and combined NB and R6G (green). Reproduced from Ref. [86].

reduced once the aggregate is formed without affecting dynamics of the system and at the same time stabilizing the trap population [85] and reducing thermal effects, and therefore the NPs' escaping chances [77].

Our group has investigated the potential of optically induced aggregation for spectroscopy in liquid environment. In particular, we focused our attention on the exploitation of the radiation pressure as a means to selectively aggregate metal NPs and perform SERS detection of amino acids and proteins in buffered solutions at nanomolar and even picomolar concentrations [70]. This approach is a step forward for the development of SERS sensors operating in liquid, allowing one to use lasers in a broader range of wavelengths, no more limited by LSPR of the NPs. At the same time, it enables the controlled positioning of metal NPs on surfaces [85, 88] and living cells [69].

This methodology was referred to as LIQUISOR, from LIQUId SERS sensOR. The LIQUISOR, as sketched in Figure 4.1.14, uses commercial gold NRs covered by a sta-

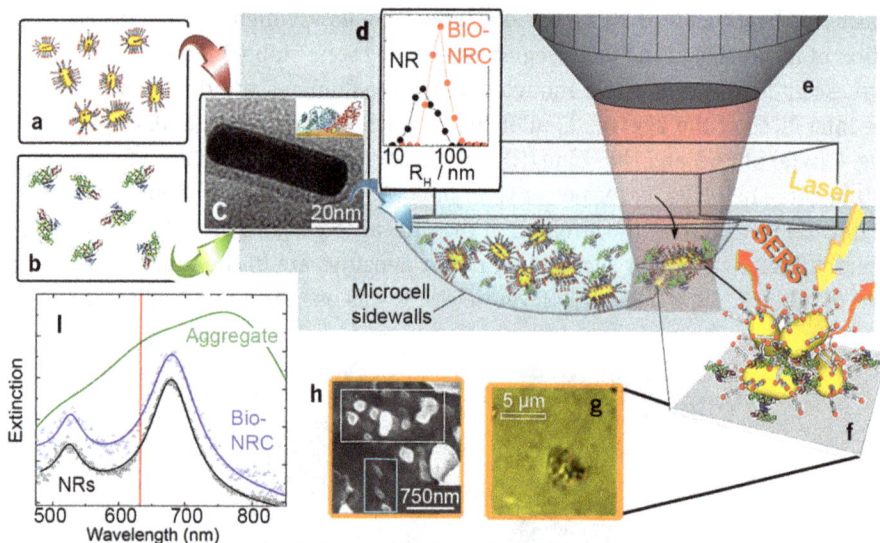

Figure 4.1.14: Sketch of the LIQUISOR. (a) Gold NRs capped with CTAB are mixed with the target molecules in PBS solution (b). The PBS action destabilizes the surfactant layer fostering the binding of the biomolecules on the NRs' surface. (c) TEM image of a BIO-NRCs highlighting the NR surrounded by a BSA corona (5 nm). The inset shows a schematic of the BSA interaction with the gold surface by means of the Au-S coordination (from Ref. [32]). (d) DLS measurements of NRs coated with CTAB (black) and after binding with BSA (red). (e) A small aliquot of the biomolecule–NR solution is transferred into a hemispheric glass microcell, which is sealed with a coverslip and subjected to laser irradiation. To this aim, a laser beam (633 nm) is focused on a near-diffraction-limited spot, inside the liquid and close to the sidewalls of the cell. The NRs are pushed and aggregated at the bottom of the cell, creating SERS-active structures that embed the target molecules (f). (g) Bright-field image of an aggregate reaching micron-scale dimensions. (h) SEM zoom on the aggregate. (i) Extinction spectrum of an optically induced BIO-NRC aggregate (green line), compared to the ones acquired on the NRs as purchased (black dots) and on the BIO-NRCs (blue symbols). Solid lines are fits using a Lorentz model. Adapted from Ref. [70].

bilizing capping layer of cetyltrimethylammonium bromide (CTAB) and dispersed in de-ionized water. The NRs (Fig. 4.1.14a) are mixed with biomolecules in phosphate-buffered saline (PBS, Fig. 4.1.14b) preserving the pH of the solution. Upon mixing, the biomolecules bind to the gold NRs (PBS, Fig. 4.1.14c), due to destabilization of the CTAB bilayer induced by the PBS and intercalation of the amino acid residues of the protein [31, 32, 89]. In the resulting biomolecule–NR complexes (BIO-NRCs) [32] the individual NRs are now stabilized by the protein layer. Dynamic light scattering (DLS) measurements (Fig. 4.1.14d) showed an increase of the mean hydrodynamic radius of the NRs from 35 nm (black symbols) to 65 nm (red symbols) when bound with BSA. The mixture was finally transferred into a microcell under a Raman microspectrometer (Fig. 4.1.14e). Optically induced aggregation was carried out by focusing the laser spot near the bottom sidewall of the microcell. Such a process enabled the dynamic aggregation of the BIO-NRCs present in solution (Fig. 4.1.14f). NR aggregates reached the size of several microns (Fig. 4.1.14g). Inside the aggregate (Fig. 4.1.14h) randomly oriented gold NRs were visible, both individual or clusterized, mixed with the biomolecules. Some complexes showed rods in a side-by-side configuration. The extinction spectrum of the aggregate (Fig. 4.1.14l, green line) showed a broadening and red shift of the plasmon resonance with respect to both the pristine NRs (black line) and BIO-NRCs in solution (blue line), as expected when the NRs aggregate [61].

The growth in size of an optically induced aggregate of NRs in BSA was followed as a function of time (Fig. 4.1.15a,b). The aggregate dynamics are characterized by a steep linear increase in size between 0 and 20 min. After 30 min the aggregate size reached saturation. The growth kinetics follow a $D(t) = A(1-e^{-t/t_0})$ trend, with a final aggregate diameter of $A = (7.2\pm0.2)\,\mu m$ and a typical time scale on which the aggregation occurs

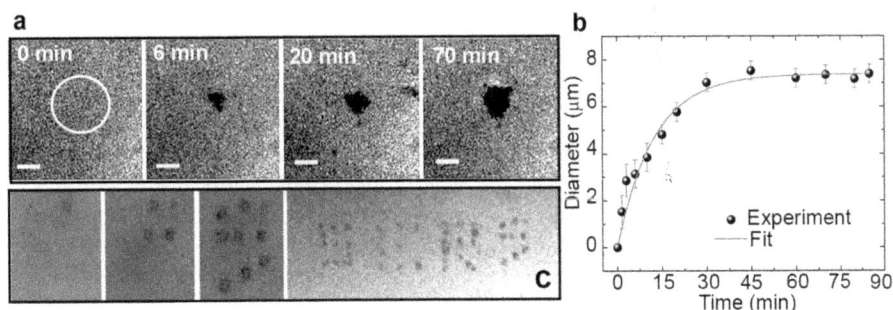

Figure 4.1.15: (a) Time series images showing the formation of an aggregate under laser irradiation of gold NRs in BSA. Scale bar is 2.5 μm. Optical images are acquired monitoring the light transmitted from a white lamp placed underneath the sample after repeated periods of laser irradiation (638 nm, 13 mW). Adapted from Ref. [70]. (b) Variation of the aggregate diameter as a function of time (solid symbols). Data are fitted (solid line) with a function $D(t) = A(1 - e^{-t/t_0})$ indicating an initial phase in which the aggregate diameter increases nearly linearly, followed by a saturation regime. Redrawn from Ref. [70]. (c) Optical images of patterns produced with optically induced spots made from gold NRs ($30 \times 90\ nm^2$) mixed with BSA (100 μM) in PBS (200 mM). Each spot took 10 to 15 min to produce.

of $t_0 = (13 \pm 1)$ min. The stable link of the clusters produced with gold NRs in BSA with the glass substrate paves the way to novel possibilities for patterning in liquid environment. Figure 4.1.15c shows the sequential process of writing of the letter "S" and of the word "SERS" using spots made with optically induced aggregates.

The NRs aggregates thus produced proved to be SERS-active and were applied for high-sensitivity molecular detection. Studies conducted on BSA are shown in Figure 4.1.16. BSA is a model protein rich in sulfur spread over 17 disulfide bridges (S-S), five methionine residues (S-CH$_3$), and one free thiol group (a cysteine residue) [90]. Disulfide bridges give solidity to the α-helix bundles. When BSA interacts with gold NPs, the sulfur affinity causes the disruption of the S-S bridges and the fast creation of the Au-S coordination. A protein corona all-around the surface is thus formed, which stabilizes the colloidal suspension [32]. This is shown at BSA concentrations of 100 μM in PBS (200 mM) [89], where its fast interaction with the gold surface (a stable BSA-Au bond is formed in less than 1 μs) ensures a protein capping layer of a few nanometer thick (Fig. 4.1.4c), as reported in the work of Wang et al. [32]. DLS measurements confirmed these values also in our case [70].

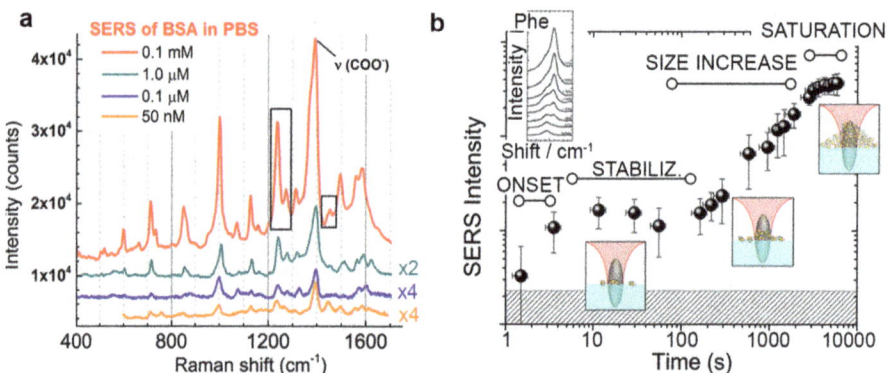

Figure 4.1.16: LIQUISOR detection of BSA. (a) BSA SERS at concentrations of 100 μM (red), down to 1 μM (green), 0.1 μM (blue), and 50 nM (orange). Spectra are offset for clarity. Experiments are carried out at 632.8 nm, with a power of 6.7 mW, focused through a 100× microscope objective. (b) SERS intensity of the Phe ring breathing at 1006 cm^{-1} (inset) increasing with time during the optically induced aggregation of gold NRs in BSA. A continuous signal increase is observed, characterized by several steps related to the NR aggregation dynamics on the different time scales (seconds, minutes, tens of minutes). Adapted from Ref. [70].

Figure 4.1.16a shows the SERS spectra of BSA, acquired by LIQUISOR at decreasing concentrations from 100 μM (red line), down to 1 μM (green line), 100 nM (blue line), and 50 nM (orange line). SERS spectra feature intense peaks in the same spectral range of the BSA vibrations measured in solution phase or the lyophilized state (400–1700 cm^{-1}). These were attributed to the phenylalanine (Phe), tyrosine (Tyr), and tryptophan (Trp) aromatic amino acids [42, 70, 91]. The strongest SERS peaks were attributed to the disulfide bridges (500 cm^{-1} region), the Tyr doublet (850 cm^{-1} region), Phe ring stretching (1004 cm^{-1}), the CH modes (1300 cm^{-1} and 1450 cm^{-1}), COO$^-$ stretching (1395 cm^{-1}), amide III (1239 cm^{-1} and 1274 cm^{-1}), amide I (1650 cm^{-1}), and the CH stretching vibrations (2820–3000 cm^{-1}). Indeed, such mode assignment must be considered as tentative. The interaction of the protein with the gold surface and the residual surfactant molecules could be the reason for the observed spectral shifts and intensity changes. We estimated that saturated aggregates under prolonged laser irradiation could reach temperatures higher than 40 °C. Temperature increments can trigger the formation of β-amyloid structures at the hotspots [92], although in our case the protein showed still some form of tertiary structure, suggesting only partial modification of the global conformation. At 10 nM precipitation of gold NRs was observed. Very likely the quantity of protein at this concentration was not sufficient to completely surround the NRs, yielding unstable BIO-NRCs. The LOD of BSA with the LIQUISOR methodology was found between 10 and 50 nM, a concentration more than 10^4 to 10^5 times lower than what we can detect by conventional Raman spectroscopy (~1 mM).

The LIQUISOR methodology proved to be a quite fast technique. Figure 4.1.16b highlights insight into the dynamics and time scale of the SERS signal increase during BIO-NRC aggregation. The SERS signal of the protein appears over the PBS emission in the first few seconds. The signal grows during the following minutes, after which a saturation plateau is reached. The process is characterized by different time scales: the onset of the aggregate creation occurs after few seconds from when the irradiation starts; the aggregates stick on the cell sidewall and are stabilized in the next few tens of seconds (up to 1 min), leading to a strong SERS signal from BSA; over longer periods (1–30 min) the aggregate repeatedly enlarges, adding up BIO-NRCs, further increasing the SERS signal. Aggregates up to several microns are obtained at saturation, typically after some tens of minutes (30–50 min), when the particles totally fill up the laser focal spot.

Application of the LIQUISOR to different biomolecules is shown in Figure 4.1.17, with SERS spectra of lysozyme (Lys) (a), phenylalanine (Phe) (b), and hemoglobin (Hgb) (c) in buffered solutions. Detection of Lys by LIQUISOR at concentrations of 1 μM is achieved after 2–5 min of irradiation (Fig. 4.1.17a, blue line). A more clear signal is obtained after 60 min (Fig. 4.1.17a, red line) on a saturated aggregate. Strong vibrational modes are observed at ~1000 cm^{-1} (the Phe-Trp doublet, distinctive of Lys) and in the 1100–1650 cm^{-1} range. Lys was still detectable at 100 nM (Fig. 4.1.17a, black line), i. e., at a concentration 10^5 times smaller than by means of confocal Raman (LOD = 10 mM). Phe is an amino acid with a much simpler structure with respect to a protein. Optically induced aggregates provide a clear SERS fingerprint of Phe in PBS (Fig. 4.1.17b) at concentrations of 1 mM, i. e., 100 times smaller than the LOD of normal Raman (100 mM) [93]. The most intense SERS vibrations (red line) are in the in-plane ring stretching region (1595 and 1616 cm^{-1}), in the C-H bending and scissoring range (1400–1500 cm^{-1}), at 1290 cm^{-1} (CH$_2$), and in the range 1160–1220 cm^{-1}, where

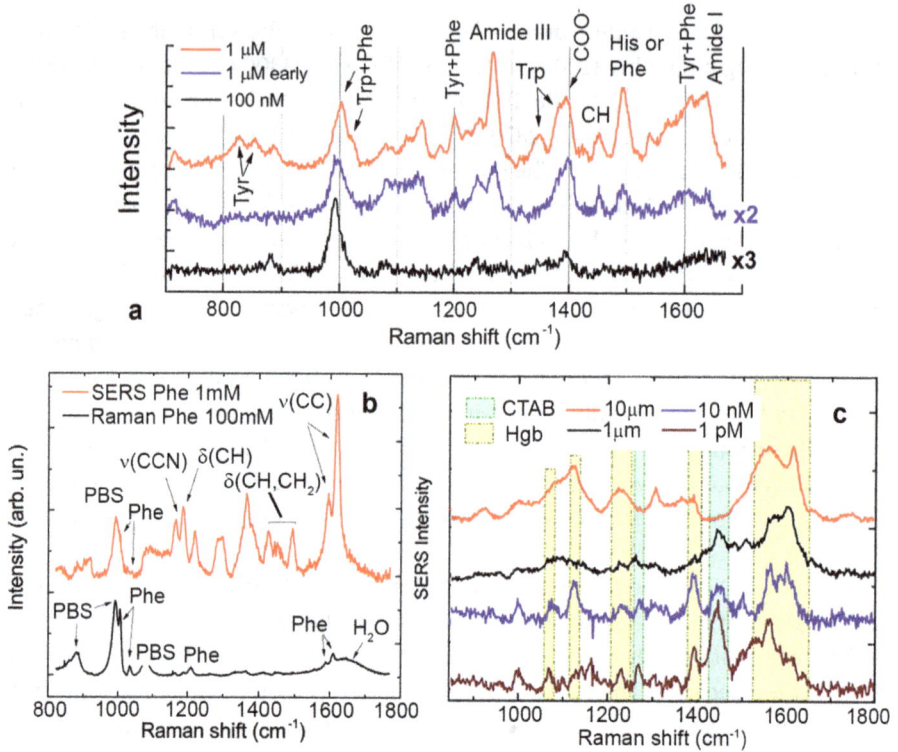

Figure 4.1.17: (a) LIQUSOR detection of Lys. SERS spectra at 100 nM (black line) and 1 μM acquired after a few minutes of irradiation (blue line) and after 60 min of irradiation (red line). Adapted from Ref. [70]. (b) LIQUISOR detection of Phe in PBS at a concentration of 1 mM (red line) compared with the Raman spectrum of Phe at 100 mM (black line). Adapted from Ref. [70]. (c) LIQUISOR detection of Hgb down to 1 pM. Hgb bands are highlighted in yellow. The CTAB signal is highlighted with cyan bands.

the C-CN stretch, CH bending, and the phenyl-C stretch occur. Similar vibrations are found also in the solution-phase Raman spectrum (black line). Even if the SERS effect has allowed the single-molecule detection of Hgb [94], detection in liquid turned out to be much more complicated, but it is crucial for the implementation of new diagnostic techniques in the biomedical field. The best reported sensitivity is of the order of 100 nM, and has been achieved via aggregation of silver NPs inside Hgb [43, 47]. Figure 4.1.17c shows applications of the LIQUISOR to detect Hgb in buffer at decreasing concentrations from 10 μM down to 1 pM. Hgb was dispersed in PBS (10 mM), mixed with 30×50 nm^2 gold NRs (LSPR peaked at 540 nm) in a 4:1 (v/v) ratio, and the solutions were incubated for 3 h. We found reproducible SERS spectra different from CTAB even at 1 pM. The shape of the amide II band (1550 cm^{-1}) and the presence of the amide III band at 1232 cm^{-1} was clear evidence for the protein, together with the 1395 cm^{-1} stretching vibration of the COO$^-$ group [95, 96]. SERS bands at 1268 and

$1442\,\mathrm{cm}^{-1}$ (cyan boxes in Fig. 4.1.17c) were attributed to the CH_2 wagging and scissoring of the CTAB [97]. The LOD of Hgb found with the LIQUISOR was ~1 pM, i. e., ca. 10^7 times smaller than for normal Raman spectroscopy in liquid (LOD between 10 and 100 µM).

4.1.5 Summary and impact

In this chapter we have reviewed different strategies for molecular detection via SERS in a liquid environment. SERS is achieved via aggregation of metal NPs in a solution where the target analyte is dispersed. Several advantages can be listed with respect to conventional "dry" SERS methods, among which are high sensitivity, reproducibility, and ease of use. NP aggregation can be achieved by solvent evaporation or, in a more controlled manner, using chemical agents or light-induced effects. Thermophoretic transport and optical forces are consequences of light irradiation of particles in liquid. Optical forces allow us to manipulate, trap, and aggregate metal NPs in a contactless, chemical-free way. Optical pushing creates in a very simple way 2D and 3D SERS-active aggregates that can be used for high-sensitivity molecular detection in liquids.

SERS allows to achieve single-molecule sensitivity on dye molecules (featuring a large Raman cross-section) and sensitivities in the nM, pM, or even fM range in the detection of biomolecules. We expect a strong impact from the integration of SERS analytical platforms in "lab-on-a-chip" devices. Specificity can potentially be increased by a proper functionalization of the NPs with aptamers or antibodies, in order to capture selectively the target molecules in solution. Important potential applications could be in the field of specific detection of pathology biomarkers in body fluids, such as saliva, urine, or blood. Higher sensitivity can be potentially achieved using efficient core-shell NPs or nanostars [98, 99], gold/silver alloys [100], or aluminum NPs to combine the resonance Raman scattering of biomolecules in the UV spectrum with plasmonic enhancement [101]. The use of lasers in the window of optical transparency of biological tissues could allow one to apply our methodology in combination with the injection of NPs into living organisms for in vivo SERS detection [69].

Bibliography

[1]   Bohren CF, Huffman DR. Absorption and Scattering of Light by Small Particles. Weinheim, Germany: John Wiley & Sons; 1983.
[2]   Borghese F, Denti P, Saija R. Scattering from Model Nonspherical Particles: Theory and Applications to Environmental Physics. Heidelberg, Germany: Springer-Verlag; 2007.
[3]   Le Ru E, Etchegoin P. Principles of Surface Enhanced Raman Spectroscopy. Amsterdam, The Netherlands: Elsevier; 2009.

[4]   Guillot N, De La Chapelle ML. Nanoantenna. In: Wiley Encyclopedia of Electrical and
        Electronics Engineering. 2012. p. 1–13.
[5]   Jones PH, Maragò OM, Volpe G. Optical Tweezers. Cambridge, UK: Cambridge University
        Press; 2016.
[6]   Le Ru EC, Somerville WRC, Auguié B. Radiative correction in approximate treatments of
        electromagnetic scattering by point and body scatterers. Phys Rev A. 2013;87:012504.
[7]   Maier SA. Plasmonics: Fundamentals and applications. New York, NY, USA: Springer-Verlag;
        2007.
[8]   Willets KA, Van Duyne RP. Localized surface enhanced spectroscopy and sensing. Annu Rev
        Phys Chem. 2007;58:267–97.
[9]   Link S, Mohamed MB, El-Sayed MA. Simulation of the optical absorption spectra of gold
        nanorods as a function of their aspect ratio and the effect of the medium dielectric constant. J
        Phys Chem B. 1999;103:3073–7.
[10]   D'Andrea C et al. Optical nanoantennas for multiband surface-enhanced infrared and Raman
        spectroscopy. ACS Nano. 2013;7:3522–31.
[11]   Link S, El-Sayed MA. Size and temperature dependence of the plasmon absorption of colloidal
        gold nanoparticles. J Phys Chem B. 1999;103:4212–7.
[12]   Hanash S. Desease proteomics. Nature. 2003;422:226–32.
[13]   Uhlen M. Mapping the human proteome using antibodies. Mol Cell Proteomics.
        2007;6:1455–6.
[14]   Rissin DM et al. Single-molecule enzyme-linked immunosorbent assay detects serum proteins
        at sub-femtomolar concentrations. Nat Biotechnol. 2010;28:595–9.
[15]   Brolo A. Plasmonics for future biosensors. Nat Photonics. 2012;6:709–13.
[16]   Anker JN et al. Biosensing with plasmonic nanosensors. Nat Mater. 2008;7:442–53.
[17]   Di Fabrizio E et al. Roadmap on biosensing and photonics with advanced nano-optical
        methods. J Opt. 2016;18:063003.
[18]   Le Ru EC, Meyer M, Etchegoin PG. Surface enhanced Raman scattering enhancement factors: a
        comprehensive study. J Phys Chem C. 2007;111:13794–803.
[19]   De Angelis F et al. Breaking the diffusion limit with super-hydrophobic delivery of molecules
        to plasmonic nanofocusing SERS structures. Nat Photonics. 2011;5:682–7.
[20]   Nie SM, Emory SR. Probing single molecules and single nanoparticles by surface-enhanced
        Raman scattering. Science. 1997;275:1102–6.
[21]   Kneipp K et al. Single molecule detection using surface-enhanced Raman scattering (SERS).
        Phys Rev Lett. 1997;78:1667–70.
[22]   Porter MD et al. SERS as a bioassay platform: fundamentals, design, and applications. Chem
        Soc Rev. 2008;37:1001–11.
[23]   Xie W, Schlücker S. Medical applications of surface-enhanced Raman scattering. Phys Chem
        Chem Phys. 2013;15:5329–44.
[24]   Han XX, Zhao B, Ozaki Y. Surface-enhanced Raman scattering for protein detection. Anal
        Bioanal Chem. 2009;394:1719–27.
[25]   Shanmukh S et al. Rapid and sensitive detection of respiratory virus molecular signatures
        using a silver nanorod array SERS substrate. Nano Lett. 2006;6:2630–6.
[26]   Qian XM et al. In vivo tumor targeting and spectroscopic detection with surface-enhanced
        Raman nanoparticle tags. Nat Biotechnol. 2008;26:83–90.
[27]   Shachaf CM et al. A novel method for detection of phosphorylation in single cells by Surface
        Enhanced Raman Scattering (SERS) using composite organic-inorganic nanoparticles (COINs).
        PLoS ONE. 2009;4:e5206.
[28]   Cottat M et al. High sensitivity, high selectivity SERS detection of MnSOD using optical
        nanoantennas functionalized with aptamers. J Phys Chem C. 2015;119:15532–40.

[29] Tuma R. Raman spectroscopy of proteins: from peptides to large assemblies. J Raman Spectrosc. 2005;36:307–19.
[30] Feng M, Tachicawa H. Surface-enhanced resonance Raman spectroscopic characterization of the protein native structure. J Am Chem Soc. 2008;130:7443–8.
[31] Tsai D-H et al. Adsorption and conformation of serum albumin protein on gold nanoparticles investigated using dimensional measurements and in situ spectroscopic methods. Langmuir. 2011;27:2464–77.
[32] Wang L et al. Revealing the binding structure of the protein corona on gold nanorods using synchrotron radiation-based techniques: understanding the reduced damage in cell membranes. J Am Chem Soc. 2013;135:17359–68.
[33] Maragò OM, Jones PH, Gucciardi PG, Volpe G, Ferrari AC. Optical trapping and manipulation of nanostructures. Nat Nanotechnol. 2013;11:807–19.
[34] Fazio B et al. Re-radiation enhancement in polarized surface-enhanced resonant Raman scattering of randomly oriented molecules on self-organized gold nanowires. ACS Nano. 2011;5:5945–56.
[35] Kessentini S et al. Gold dimer nanoantenna with slanted gap for tunable LSPR and improved SERS. J Phys Chem C. 2014;118:3209–19.
[36] Nikoobakt B, El-Sayed MA. Surface-enhanced Raman scattering studies on aggregated nanorods. J Phys Chem A. 2003;107:3372–8.
[37] Liu HL, Yang ZL, Meng LY, Sun YD, Wang J, Yang LB, Liu JH, Tian ZQ. Three-dimensional and time-ordered surface-enhanced Raman scattering hotspot matrix. J Am Chem Soc. 2014;136:5332–41.
[38] Ghosh SK, Pal T. Interparticle coupling effect on the surface plasmon resonance of gold nanoparticles: from theory to applications. Chem Rev. 2007;107:4797–862.
[39] Jana NR. Silver coated gold nanoparticles as new surface enhanced Raman substrate at low analyte concentration. Analyst. 2003;128:954–6.
[40] Creighton JA, Blatchford CG, Albrecht MG. Plasma resonance enhancement of Raman scattering by pyridine adsorbed on silver or gold sol particles of size comparable to the excitation wavelength. J Chem Soc Faraday Trans. 1979;75:790–8.
[41] Bell SEJ, McCourt MR. SERS enhancement by aggregated Au colloids: effect of particle size. Phys Chem Chem Phys. 2009;11:7455–62.
[42] Zakel S et al. Double isotope dilution surface-enhanced Raman scattering as a reference procedure for the quantification of biomarkers in human serum. Analyst. 2011;136:3956–61.
[43] Han XX et al. Label-free highly sensitive detection of proteins in aqueous solutions using surface-enhanced Raman scattering. Anal Chem. 2009;81:3329–33.
[44] Yang X et al. Highly sensitive detection of proteins and bacteria in aqueous solution using surface-enhanced Raman scattering and optical fibers. Anal Chem. 2011;83:5888–94.
[45] Marsich L, Bonifacio A, Mandal S, Krol S, Beleites C, Sergo V. Poly-L-lysine-coated silver nanoparticles as positively charged substrates for surface-enhanced Raman scattering. Langmuir. 2012;28:13166–71.
[46] Kahraman M et al. Label-free detection of proteins from self-assembled protein-silver nanoparticle structures using surface-enhanced Raman scattering. Anal Chem. 2010;82:7596–602.
[47] Xu LJ et al. Label free detection of native proteins by surface enhanced Raman spectroscopy using iodide modified nanoparticles. Anal Chem. 2014;86:2238–45.
[48] Mehmet K, Balz BN, Wachsmann-Hogiu S. Hydrophobicity-driven self-assembly of protein and silver nanoparticles for protein detection using surface-enhanced Raman scattering. Analyst. 2013;138:2906–13.
[49] Huang GG et al. Development of a heat-induced surface-enhanced Raman scattering sensing method for rapid detection of glutathione in aqueous solutions. Anal Chem. 2009;81:5881–8.

[50] Zhou J, Ren K, Zhao Y, Dai W, Wu H. Convenient formation of nanoparticle aggregates on microfluidic chips for highly sensitive SERS detection of biomolecules. Anal Bioanal Chem. 2012;402:1601–9.

[51] Arumugam S et al. Functionalized Ag nanoparticles with tunable optical properties for selective protein analysis. Chem Commun. 2011;47:3553–5.

[52] Bjerneld EJ et al. Laser-induced growth and deposition of noble-metal nanoparticles for surface-enhanced Raman scattering. Nano Lett. 2003;3:593–6.

[53] Lee SJ et al. Photoreduction at a distance: facile, nonlocal photoreduction of Ag ions in solution by plasmon-mediated photoemitted electrons. Nano Lett. 2010;10:1329–34.

[54] Tanaka Y et al. Laser induced self assembly of silver NPs via plasmonic interactions. Opt Express. 2009;17:18761.

[55] Liu Y et al. Highly sensitive fibre surface enhanced Raman scattering probes fabricated using laser induced self assembly in a meniscus. Nanoscale. 2016;8:10607–14.

[56] Baffou G, Quidant R. Thermo-plasmonics: using metallic nanostructures as nano-sources of heat. Laser Photonics Rev. 2013;7:171–87.

[57] Lin L et al. Light-directed reversible assembly of plasmonic nanoparticles using plasmon-enhanced thermophoresis. ACS Nano. 2016;10:9659–68.

[58] Ding T et al. Light-induced actuating nanotransducers. Proc Natl Acad Sci USA. 2016;113:201524209.

[59] Ashkin A. Acceleration and trapping of particles by radiation pressure. Phys Rev Lett. 1970;24:156–9.

[60] Ashkin A et al. Observation of a single-beam gradient force optical trap for dielectric particles. Opt Lett. 1986;11:288–90.

[61] Messina E et al. Plasmon-enhanced optical trapping of gold nanoaggregates with selected optical properties. ACS Nano. 2011;5:905–13.

[62] Lehmuskero A, Johansson P, Rubinsztein-Dunlop H, Tong L, Kall M. Laser trapping of colloidal metal nanoparticles. ACS Nano. 2015;9:3453–69.

[63] Arias-González JR, Nieto-Vesperinas M. Optical forces on small particles: attractive and repulsive nature and plasmon-resonance conditions. J Opt Soc Am A. 2003;20:1201–9.

[64] Selhuber-Unkel C et al. Quantitative optical trapping of single gold nanorods. Nano Lett. 2008;8:2998–3003.

[65] Brzobohatý O et al. Three-dimensional optical trapping of a plasmonic nanoparticle using low numerical aperture optical tweezers. Sci Rep UK. 2015;5:8106.

[66] Dienerowitz M et al. Optical vortex trap for resonant confinement of metal nanoparticles. Opt Express. 2008;16:4991–9.

[67] Nedev S et al. Optical force stamping lithography. Nano Lett. 2011;11:5066–70.

[68] Guffey MJ, Scherer NF. All-optical patterning of Au nanoparticles on surfaces using optical traps. Nano Lett. 2010;10:4302–8.

[69] Li M, Lohmueller T, Feldmann J. Optical injection of gold nanoparticles into living cells. Nano Lett. 2015;15:770–5.

[70] Fazio B et al. SERS detection of biomolecules at physiological pH via aggregation of gold nanorods mediated by optical forces and plasmonic heating. Sci Rep UK. 2016;6:26952.

[71] Saija R, Iatì MA, Giusto A, Denti P, Borghese F. Transverse components of the radiation force on nonspherical particles in the T-matrix formalism. J Quant Spectrosc Radiat Transf. 2005;94:163–79.

[72] Rodrigo JA, Alieva T. Light driven transport of plasmonic nanoparticles on demand. Sci Rep UK. 2016;6:33729.

[73] Pelton M, Liu M, Kim HY, Smith G, Guyot-Sionnest P, Scherer NF. Optical trapping and alignment of single gold nanorods by using plasmon resonances. Opt Lett. 2006;31:2075–7.

[74] Johnson PB, Christy RW. Optical constants of noble metals. Phys Rev B. 1972;16:4370–9.

[75] Juan ML, Righini M, Quidant R. Plasmon nano-optical tweezers. Nat Photonics. 2011;5:349–56.

[76] Toussaint et al. Plasmon resonance based optical trapping of single and multiple Au NPs. Opt Express. 2007;15(19):12018.

[77] Ohlinger A et al. Optothermal escape of plasmonically coupled silver nanoparticles from a three-dimensional optical trap. Nano Lett. 2011;11:1770–4.

[78] Svedberg F et al. Creating hot nanoparticle pairs for surface-enhanced Raman spectroscopy through optical manipulation. Nano Lett. 2006;6:2639–41.

[79] Tong T et al. Optical aggregation of metal nanoparticles in a microfluidic channel for surface-enhanced Raman scattering analysis. Lab Chip. 2009;9:193–5.

[80] Tanaka Y et al. Surface enhanced Raman scattering from pseudoisocyanine on Ag nanoaggregates produced by optical trapping with a linearly polarized laser beam. J Phys Chem C. 2009;113:11856–60.

[81] Liu Z et al. Optical manipulation of plasmonic nanoparticles, bubble formation and patterning of SERS aggregates. Nanotechnology. 2010;21:105304.

[82] Bálint S et al. Simple route for preparing optically trappable probes for surface-enhanced Raman scattering. J Phys Chem C. 2009;113:17724–9.

[83] Lin S et al. Surface-enhanced Raman scattering with Ag nanoparticles optically trapped by a photonic crystal cavity. Nano Lett. 2013;13:559–63.

[84] Messina E et al. Manipulation and Raman spectroscopy with optically trapped metal nanoparticles obtained by pulsed laser ablation in liquids. J Phys Chem C. 2011;115:5115–22.

[85] Königer A, Köhler W. Optical funneling and trapping of gold colloids in convergent laser beams. ACS Nano. 2012;6:4400–9.

[86] Patra PP et al. Plasmofluidic single-molecule surface-enhanced Raman scattering from dynamic assembly of plasmonic nanoparticles. Nat Commun. 2014;5:4357.

[87] Zhang Y, Gu C. Optical trapping and light induced aggregation of gold nanoparticle aggregates. Phys Rev B. 2006;73:165405.

[88] Violi IL et al. Light-induced polarization-directed growth of optically printed gold nanoparticles. Nano Lett. 2016;16:6529–33.

[89] Kah JCY, Zubieta A, Saavedra RA, Hamad-Schierli K. Stability of gold nanorods passivated with amphiphilic ligands. Langmuir. 2012;28:8834–44.

[90] Peters T. All About Albumin: Biochemistry, Genetics, and Medical Applications. San Diego, CA, USA: Academic Press; 1996.

[91] David C et al. Raman and IR spectroscopy of manganese superoxide dismutase, a pathology biomarker. Vib Spectrosc. 2012;62:50–8.

[92] Chakraborty S et al. Contrasting effect of gold nanoparticles and nanorods with different surface modifications on the structure and activity of bovine serum albumin. Langmuir. 2011;27:7722–31.

[93] Zhu G, Zhu X, Fan Q, Wan X. Raman spectra of amino acids and their aqueous solutions. Spectrochim Acta A. 2011;78:1187–95.

[94] Xu H et al. Spectroscopy of single hemoglobin molecules by surface enhanced Raman scattering. Phys Rev Lett. 1999;83(21):4357–60.

[95] Kaminska A et al. Chemically bound gold nanoparticle arrays on silicon: assembly, properties and SERS study of protein interactions. Phys Chem Chem Phys. 2008;10:4172–80.

[96] Kaminska A, Forster RJ, Kayes TE. The impact of adsorption of bovine pancreatic trypsin inhibitor on CTAB-protected goldnanoparticle arrays: a Raman spectroscopic comparison with solution denaturation. J Raman Spectrosc. 2010;41:130–5.

[97] Dendramis AL et al. SERS study of CTAB on copper. Surf Sci. 1983;134:675–88.

[98] Li JF et al. Shell-isolated nanoparticle-enhanced Raman spectroscopy. Nature. 2010;464:392–5.

[99] Indrasekara ASDS et al. Gold nanostar substrates for SERS-based chemical sensing in the femtomolar regime. Nanoscale. 2014;6:8891–9.

[100] Messina E et al. Tuning the structural and optical properties of gold/silver nano-alloys prepared by laser ablation in liquids for optical limiting, ultra-sensitive spectroscopy, and optical trapping. J Quant Spectrosc Radiat Transf. 2012;113:2490–8.

[101] Knight MW et al. Aluminum for plasmonics. ACS Nano. 2014;8:834–40.

Gamze Yesilay, Ertug Avci, Mine Altunbek, Sevda Mert, and
Mustafa Culha

4.2 SERS in label-free detection of cancer from proteins, cells, and tissues

4.2.1 Key messages

- Label-free cancer detection provides an opportunity to bypass possible false positive results of label-based methods.
- Label-free surface-enhanced Raman spectroscopy (SERS) can be utilized as an alternative approach to existing label-based assays by reducing analysis time and increasing diagnostic accuracy.
- SERS substrates with optimized properties are needed to increase spectral reproducibility and reliability to exploit the actual power of the technique in clinical diagnostics.
- Important SERS applications in biosciences are explained for protein, cellular, and tissue applications with special reference to cancer diagnostics.

4.2.2 Pre-knowledge

Vibrational spectroscopies have been increasingly investigated for clinical applications in recent years because they deliver rapid molecular-level information about the status of a clinical sample. In vibrational spectroscopy, bond vibrations in molecules are monitored through infrared (IR) or Raman spectroscopy. Although both types of spectroscopic techniques arise from the same fundamentals, IR spectroscopy is based on the absorption of the electromagnetic radiation in the IR region of the spectrum, while Raman spectroscopy is based on the scattered photons resulting from collisions of photons with the vibrating molecules. Since the recorded spectrum of these techniques is a "fingerprint" of a molecule, both can be used for the qualitative investigation of the composition of a sample. Raman spectroscopy has the advantages of having an improved spatial resolution (<1 μm, as compared to ~5–10 μm for IR) and limited interference from water [1–4]. However, Raman spectroscopy has the disadvantages of lower sensitivity and longer analysis times. Readers can find details on vibrational spectroscopy in Chapter 2.3 in Volume 1 of this book. The discovery of SERS is important, offering a way of overcoming the limitation of the low sensitivity of Raman spectroscopy in important situations. However, in such assays, the use of a noble metal surface introduces additional uncertainties for measurements. The details of SERS and SERS substrates can be found in Chapter 3.3.1 in Volume 1 and Chapter 6.2.3 in Volume 2 of this book, respectively. Combined with multivariate statistical analyses, an improved level of information can be extracted from SERS spectra,

---

Acknowledgement: The authors acknowledge the financial support of The Scientific and Technological Council of Turkey (TUBITAK) (Project No. 113Z554) and Yeditepe University. The authors also acknowledge financial support through the COST Action MP1302 Nanospectroscopy.

https://doi.org/10.1515/9783110442908-012

especially when differentiating healthy from cancerous cells or tissues. More detailed information on the basics of such analyses and the data pre-processing approaches can be found in Chapter 4.6 in Volume 2.

4.2.3 Importance of the application

Today, cancer continues to be one of the deadliest diseases, and its diagnosis and treatment pose major challenges. Figure 4.2.1 schematically presents the stages of cancer formation and diagnosis processes at different stages. At stage 0 (zero), cells are not cancerous, but they have the propensity to become cancerous. Diagnosis in this stage is based only on the detection of biomarkers. However, the concentration of biomarkers is not enough to detect their presence at this early stage. Although it is very difficult, at stages I, II, and III, malformation in tissue can be detected with imaging techniques including magnetic resonance imaging (MRI), computed tomography (CT), endoscopy, colonoscopy, X-ray imaging, and ultrasound with the use of an appropriate contrast agent. At stage IV, the cancerous tissue develops into a tumor, and the cancer cells might spread through the body. It is much easier to locate a tumor or metastasis with the conventional imaging techniques mentioned above. After locating malignant formation, a biopsy is performed for histopathological examinations. It can be efficacious to remove the whole suspicious tissue and examine it. Meanwhile, body fluids are analyzed for biomarker detection. After all these steps, clinicians decide for further interventions. At every step of the current diagnosis approach, several obstacles are hindering the accurate diagnosis. For example, imaging techniques have resolution

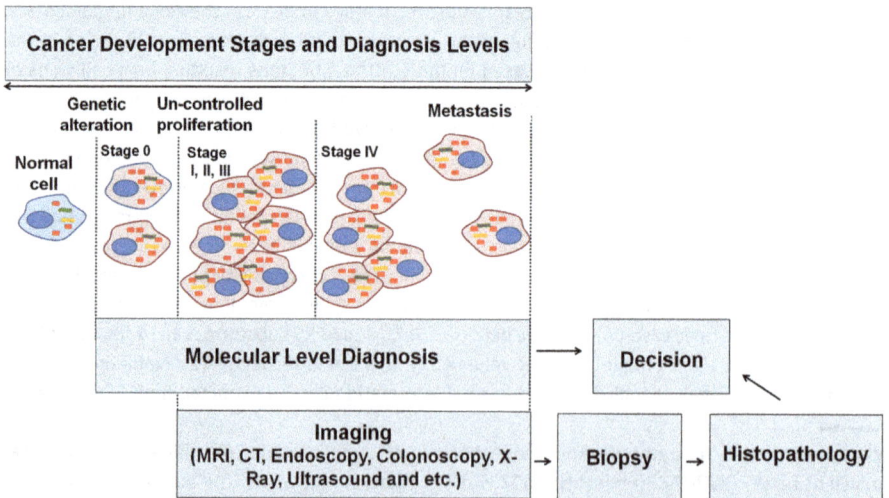

Figure 4.2.1: Cancer development stages and its diagnosis procedures.

limitations and cannot show tumor borders efficiently. Although the histopathological examination is the golden standard for diagnosis, it may not accurately determine the stage of cancer.

In order to save patients' lives and provide a better life quality, it is important to detect diseases in the early stages. Any suspicious biochemical markers or morphological changes are evaluated with a hope of early diagnosis or improved prognosis. For the detection of biomarkers, a very sensitive technique is required. The current techniques used for biomarker detection in clinics are mass spectroscopy, fluorescence spectroscopy, colorimetric immunoassays, and radioimmunoassays. However, each of these techniques has shortcomings. For example, mass spectroscopy-based detection is expensive and requires professional personnel. Fluorescence-based assays require a label that limits the accuracy of the detection of a tumor. Immunoassay-based approaches have sensitivity, specificity, and quantification issues. Vibrational spectroscopic technologies are highly sensitive for the detection of any changes in the composition and structure of biological molecules. It has been suggested that vibrational spectroscopy can be used in the concept of "spectral cytopathology" and "spectral histopathology" [5–7].

Importantly, Raman spectroscopy, one of the vibrational spectroscopic techniques, can be helpful for cancer diagnosis by detection capabilities in body fluids, at the single-cell and tissue levels. With the focusing of laser light on a sample through an objective with less than 1 μm of spatial resolution, molecular-level information from subcellular structures can be obtained. The limitation of low sensitivity due to the weak Raman scattering, on the other hand, can be overcome with the incorporation of noble metals such as gold and silver nanoparticles to enhance Raman scattering exploiting SERS, and the resolution limit can be further lowered to the nanometer scale. The molecular information provided by SERS can be extremely useful for clinicians in a range of decision-making cases. Help with tumor border identification during surgery allowing accurate excision may prevent cancer re-occurrence. In addition, since fingerprint information originates from the molecules near or on the surface of the noble metal particles in SERS, small molecular changes in cells and tissues can be detected, which can provide additional information for decision making by the surgeon. Since decision making is a subjective process with histopathological examinations, the incorporation of fingerprint information provided by SERS can help to improve the accuracy, and is consequently of importance to real-time diagnosis.

SERS has enormous potential to speed up the treatment process with a detection sensitivity at the single-molecule level, fast analysis time, and low cost, and thus can be considered as an alternative technique to the current techniques for cancer biomarker detection. With its translation into clinics, it will not only increase hope in fighting against this devastating disease, but also reduce costs of public health.

4.2.4 State-of-the-art and the application

4.2.4.1 Label-free detection and SERS

Detection of a biomarker in a very complex biological environment is a challenging task due to the similar structures of biomarkers with biomolecules present in the matrix. The use of a label that selectively responds to the respective technique is one approach for the detection. The methods employing labels might use a fluorescent, isotopic, chemiluminescent, electrochemically active molecule or a nanoparticle attached to the molecule to recognize the targeted molecule. However, the labeling process is usually one of considerable difficulty and may disrupt the function of the targeted molecular structure. It unfortunately sometimes introduces additional uncertainties to the measurement. For example, conjugated labels sometimes cause false positive results by nonspecific interaction between the biomarker and labeled molecule.

The label-free approach depends on the detection of intrinsic biophysical properties of biomarkers such as molecular structure, weight, and refractive index. The reproducibility and reliability of these methods are higher than those of the methods employing label-based approaches, yet there are similar challenges in the label-free approaches for cancer diagnosis.

Among the label-free approaches, SERS is emerging as a label-free cancer diagnosis and perhaps prognosis technique. It is a very sensitive technique to detect very low concentrations of molecules with and without biological origin, and can thus be utilized for either labeled or label-free approaches. Figure 4.2.2 schematically presents the comparison of label-based and label-free SERS detection in a sandwich-based immunoassay. In the labeled scheme, a SERS-active reporter molecule is attached either

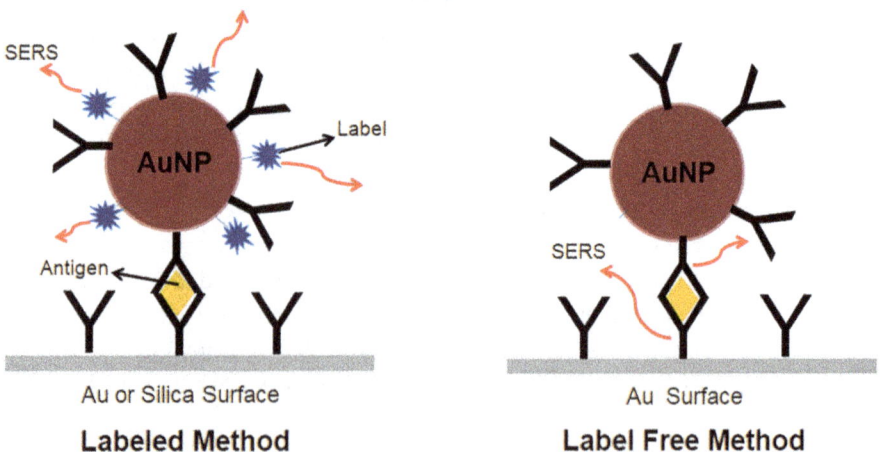

Figure 4.2.2: Representation of labeled and label-free SERS methods.

chemically to a targeting molecule or directly to the surface of a noble metal nanoparticle. In order to complete the assay, the targeting antibody specific for the same antigen of interest is immobilized on a suitable surface (e. g., Au or silica surface). First, the antigens are captured by the antibodies immobilized onto the surface. Then, the SERS label composed of both targeting antibody and reporter molecule is allowed to interact with the target antigen that is captured by the antibody immobilized on the surface. After several washing steps, the SERS signal originating from the reporter molecule is recorded for detection. If there is no SERS signal, no antigen is detected. In the label-free scheme, the changes in the intrinsic SERS spectra before and after binding of an antigen to the antibody chemically attached to a SERS-active substrate are used for detection.

Several examples demonstrate that SERS can be used for label-free detection of various cancers, such as gastric, nasopharyngeal, and colorectal cancer, by simply mixing colloidal suspensions of silver or gold nanoparticles with blood samples of patients [8–12]. The basis of cancer detection relies on the compositional difference of body fluid of healthy and cancer patients. Although the reports demonstrate the proof of concept, there are a number of issues such as choice of substrate and excitation wavelength that should be addressed for reproducible and reliable data for cancer detection [13]. The design of such diagnostic assays requires much practical knowledge about suitable SERS-active nanostructures with high enhancement factor (EF) and spatial reproducibility, which are discussed in Chapter 3.3.1 in Volume 1 and Chapter 6.2.3 in Volume 2, respectively. Otherwise, the obtained spectra will be trivial to be interpreted due to the variances in SERS hotspot formation, or the reproducibility will not be guaranteed in the complex environment of biological systems.

In SERS experiments, either a nanostructured surface or colloidal silver or gold nanoparticles (AgNPs and AuNPs, respectively) are commonly used as SERS substrates. For colloidal nanoparticles, their size, shape, and aggregation status play important roles for the SERS outcome [14]. When a body fluid is used as a sample, the properties of the colloidal suspension should be adjusted to the sample characteristics such as pH and ionic strengths. Otherwise, mixing can significantly change the nature of the interactions of biomacromolecules with the nanoparticles and act as a diagnostic interference [15]. For example, the surety of the analysis can depend on the pH, as blood and serum have a pH of 7.4, while urine has a pH ranging from 4.5 to 8.0 [16, 17].

Both AuNPs and AgNPs are used for cancer biomarker detection. One should also realize the difference in using AuNPs and AgNPs. Although AuNPs are considered to be stable, less toxic, and inert, a higher SERS activity is obtained with AgNPs [18]. This is attributed to the rough surface morphology, optical properties, and high polydispersity of AgNPs compared to AuNPs synthesized through Turkevich's citrate reduction method [19, 20]. The limitation of AgNPs is their rapid oxidation, influencing the SERS activity and resulting in additional uncertainty [21].

The selection of the laser wavelength plays a role in the enhancement of Raman spectra. There are three scenarios in the spectral outcome, depending on the chosen laser wavelength: (i) If the scattered laser wavelength is used at which the substrate has strong absorption, it causes a strong SERS signal from the molecule through electromagnetic enhancement. (ii) If the laser has the same wavelength as the absorption of the molecule to be analyzed, it increases the signals from that molecule while not influencing the signal of other molecules, which is called "resonance Raman scattering (RR)." (iii) The combination of the enhancement from the noble metal substrate and RR produces a greater enhancement, which is called surface-enhanced resonance Raman scattering (SERRS). In the case of biological applications, some biological species have absorbance in the visible region, and here the example of hemoglobin provides a useful illustration of this serious practical issue. Hemoglobin is of course one of the abundant proteins in the blood, but problematically absorbs in the blue-green region of the spectrum. If the laser wavelength used is in the blue-green region, the hemoglobin will dominate the whole spectrum of the blood, which has the direct consequence of suppressing the signals of other molecules that could be spectroscopic targets for the diagnosis. Therefore, the use of a 785 nm laser wavelength for excitation is reported to be appropriate for analysis of the blood, serum, and plasma [22–24], highlighting the importance of laser wavelength selection for designing SERS/SERRS diagnostics.

After summarizing the essentials for obtaining reproducible and reliable data from SERS studies, the focus will be on the label-free detection of cancer using SERS at the protein, cell, and tissue levels in the following sections.

4.2.4.2 Proteins

Having structural and functional roles, proteins are the major biological entities serving for survival and maintenance of cells and higher-order organisms. Detection and identification of proteins thus have critical importance in clinics for diagnosis and can be delivered using several different techniques. From the clinical perspective, immunoassay-based approaches are the golden standard for detection and identification. The information from these approaches is limited and usually as a consequence these methods are performed using labels. Among them, the enzyme-linked immunosorbent assay (ELISA) is the most commonly used technique in medicine and molecular biology. In this assay, a color change due to enzymatic reactions reveals the presence of the protein of interest. Apart from colorimetric labels, radioactive isotopes, fluorogenic reporters, and electrochemiluminescent labels are also used for immunoassays. Although these approaches provide detection and identification of protein structures, they are unable to provide fingerprint information about proteins. The spectroscopic techniques such as IR, Raman spectroscopy, and mass spectrometry (MS) can however here provide structural information. Among them, MS is the most commonly used technique for proteomics, but it has its own issues such as being destructive and difficulties in interpretation of the spectra. Due to these difficulties, the use of MS requires skilled personnel and brings high costs. IR spectroscopy is another technique that can be utilized, but its high sensitivity to water dominating the whole spectral regions and broad spectral bands seriously hamper the protein analysis.

Raman spectroscopy is one of the techniques utilized for protein detection and identification. With its nondestructive nature and low sensitivity to water, it can provide fingerprint information about proteins. However, Raman scattering is an inherently very weak phenomenon, and Raman scattering from proteins is very inefficient due to their chemical structure. Enhancement of Raman scattering of proteins using noble metal surfaces, so-called SERS, drastically lowers the detection limits for proteins and has been used since the 1980s [25–27]. In order to familiarize the readers with typical SERS spectra of proteins, spectra of eight different blood proteins are shown in Figure 4.2.3. The spectra were acquired using an 830 nm near-IR (NIR) laser after mixing proteins with a suspension of AgNPs with an average size of 50 nm and drying of the mixture. Human serum albumin accounts for 55 % of blood plasma proteins. Globulins are the second prevalent proteins with 30–40 % abundance. Fibrinogen, a blood clotting protein, makes up around 5 % of blood plasma proteins. Transferrin is an iron transporter protein and covers a similar percentage of blood proteins as fibrinogen.

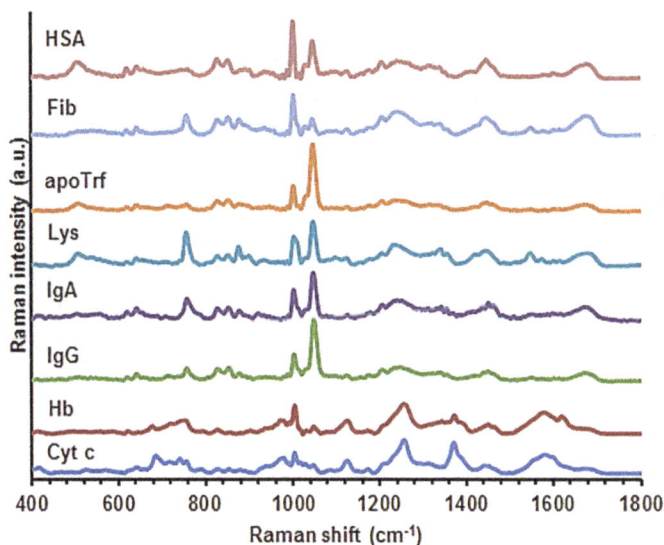

Figure 4.2.3: Label-free SERS spectra of human serum albumin (HSA), fibrinogen (Fib), apotransferrin (apoTrf), lysozyme (Lys), immunoglobulin A (IgA), immunoglobulin G (IgG), hemoglobin (Hb), and cytochrome *c* (Cyt c).

As seen in the figure, proteins have similar SERS bands. This is expected because proteins are composed of different combinations of the same 20 amino acids, and these amino acids bind each other with the same peptide bond, which is an amide bond. Structural and functional differences among proteins stem from having different combinations of these amino acids. In label-free SERS, these differences are directly reflected in their spectra. The numbers of each amino acid and the position of each amino acid in the 3D protein structure define the final detail of the spectrum. At first glance, it is seen that SERS spectra of Hb and cytochrome c are quite different from the spectra of other proteins. This is because of the presence of a heme group in their structure. The heme group is composed of a porphyrin ring having an iron ion in its center and has a distinct Raman and SERS spectrum [28]. Its SERS spectrum suppresses the SERS signals of amino acids. For nonheme proteins, Raman bands between $400 \, cm^{-1}$ and $1200 \, cm^{-1}$ stem from amino acids [29]. Mostly aromatic amino acids (phenylalanine (Phe), tyrosine (Tyr), and tryptophan (Trp)) contribute to the spectrum. Bands between 1200 and $1800 \, cm^{-1}$ are mainly due to the backbone structure of the proteins. The distribution of α-helices and β-sheets in the 3D structure of proteins defines the shape of this part of a protein spectrum.

Since the 1980s, many studies of label-free protein SERS have been reported. Most of these studies were about protein structure and function [30–33]. Interactions of different molecules such as drugs with proteins were also studied [34]. On the other hand, very few descriptions of label-free SERS-based immunoassays have been published up to date [35, 36]. This is due to an inherent difficulty of label-free working with proteins. Their spectra possess similar bands as seen in Figure 4.2.3, and therefore small spectral differences occur upon protein–protein interactions. For example, the human IgG concentration in blood was determined using a Fe_3O_4@Ag nanocomposite SERS substrate [37]. Fe_3O_4@Ag was first functionalized with PEG and streptavidin. Then, attached anti-IgG was used for IgG capture from blood obtained by finger pricking. The limit of detection of the technique was 0.6 ng/L.

Body fluids such as blood and saliva are extremely complex mixtures. Direct analysis of proteins of these body fluids is very challenging; therefore, some separation steps are required before SERS measurements. Notable contributions to this field have been published in recent years. In one of the reports, plasma proteins of gastric cancer patients and healthy volunteers were fractionated by protein electrophoresis on cellulose acetate membranes [38]. Albumin- and globulin-containing parts of the membranes were soaked in acetic acid and then mixed with colloidal AgNP suspension for SERS analysis. The spectra were analyzed using principal component analysis (PCA), and results showed 100 % diagnostic sensitivity and specificity. They performed the same experiments for hepatocellular carcinoma detection [39] and differentiation of digestive system cancers [40], and the results were also satisfactory. The same group also performed label-free diagnosis studies using saliva and achieved detection of nasopharyngeal cancer [41] and differentiation of benign and malignant breast tissue [42].

The label-free SERS technique has a potential to be routinely used in clinics. On the other hand, several issues such as reproducibility of acquired data, analysis of spectra of complex mixtures, and choice of SERS substrate for the experiments must be resolved for protein-based cancer detection. As the number of reports increases, the potential of the technique is hoped to be realized in the near future to become more amenable to the analyst.

4.2.4.3 Cells

Being the building blocks of our bodies, cells carry vital information about our health state. Understanding the cellular processes enables us to deeply understand the connection between health and disease and propose novel solutions to existing health problems. These are particularly important points in cancer research because processes such as drug resistance, cell death, cellular uptake of drugs or other molecules, cell differentiation, or pH changes in the cell upon stimuli are all playing roles in diagnostic, therapeutic, and prognostic stages. However, there are still gaps in our understanding of these processes.

The existing methods to investigate real-time cellular events mainly rely on fluorescence-based markers. Even though an incredible amount of knowledge has been extracted from these studies, fluorescence is prone to photobleaching, which in simple terms is the decomposition of molecules under light exposure and loss of their fluorescence. In Figure 4.2.4, an example of photobleaching is shown in a thin cryosection of mouse intestine. The nuclei are pseudo-colored with green (Sytox Green), the mucus of goblet cells with blue (Alexa Fluor 350 wheat germ agglutinin), and the filamentous actin in the brush border with red (Alexa Fluor 568 phalloidin) [43]. Over the course of 10 min, snapshots were taken at 2-min intervals, as shown in Figure 4.2.4a–f. In this example, at minute 8 (Fig. 4.2.4e), the signals from blue and red stains almost completely faded away, whereas the green stain could resist longer.

Moreover, developing labels for small organic molecules (size < 900 Da) such as metabolites, cell signaling molecules, and other organic compounds is a challenging task with the existing approaches [44].

For the above reasons, using SERS has been seen as a practical alternative. However, before mentioning advantages of SERS in cellular research, it is important to understand what kind of information can be extracted from a SERS spectrum of a cell. Eukaryotic cells consist of DNA, RNA, lipids, carbohydrates, and proteins of various types such as enzymes and structural proteins, as well as other small molecules, all of which are either enclosed within an organelle or free in the cytosol (Fig. 4.2.5). Therefore, from the vibrational spectroscopy point of view, a cell is composed of various types of organic compounds, most of which carry, for instance, carbon–carbon bonds or carbon–hydrogen bonds.

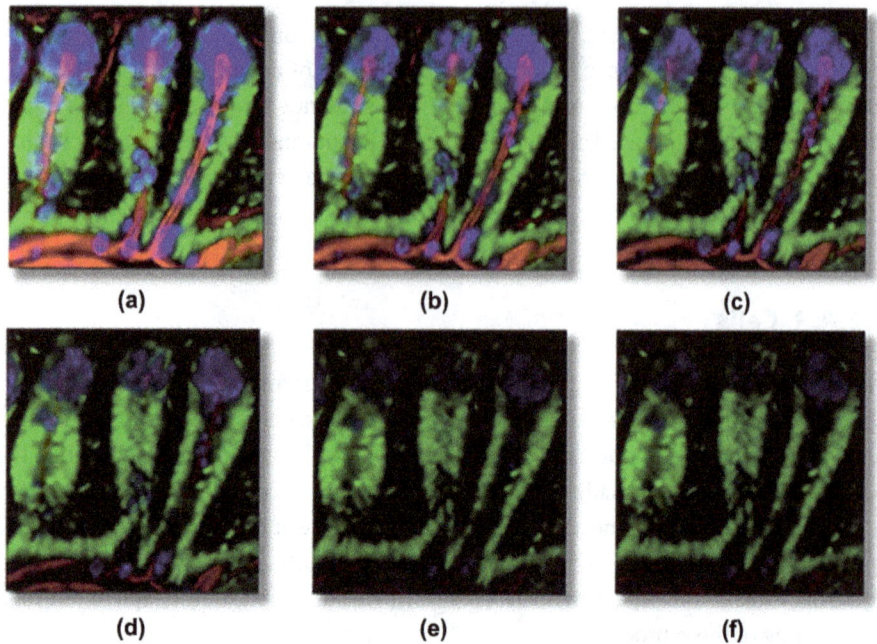

Figure 4.2.4: Differential photobleaching of multiple stained tissues. From (a) to (f), snapshots were taken from the mouse intestine sections at time zero to 10 min with 2-min intervals. Image courtesy of Molecular Expressions at Florida State University (https://micro.magnet.fsu.edu/).

At first sight, the similarities between all cellular molecules might be seen as a problem for a SERS-based approach. However, SERS can be used as a genuine tool for specific targeted cellular identifications. For instance, within the context of the COST Action Nanospectroscopy, the cytotoxicity of nanomaterials (NMs) was investigated by using SERS [45]. The main problem in NM cytotoxicity detection with conventional methods, which mostly rely on colorimetric absorbance or fluorescence-based assays, is obtaining false positive or negative results. NMs might absorb or scatter light, thus altering the absorbance results, or they might interact with the assay component and alter the results. We have shown that SERS can be utilized as an alternative for cytotoxicity detection in living cells. In Figure 4.2.6, the experimental procedure is shown. Cells seeded on calcium fluoride (CaF_2) slides were incubated with NM of interest at increasing concentrations, in this example with single-walled carbon nanotubes (SWCNTs) together with 50 nm diameter-sized spherical AuNPs for 24 h. After incubation, slides were rinsed with phosphate-buffered saline (PBS) and placed on a polydimethylsiloxane (PDMS)-coated polystyrene Petri dish. The coating was applied to prevent strong background polystyrene-related signals in the cell spectra.

The measurements were taken in line-mapping mode with the help of a high-speed encoded stage that enabled one single cell scan to be completed in around 2

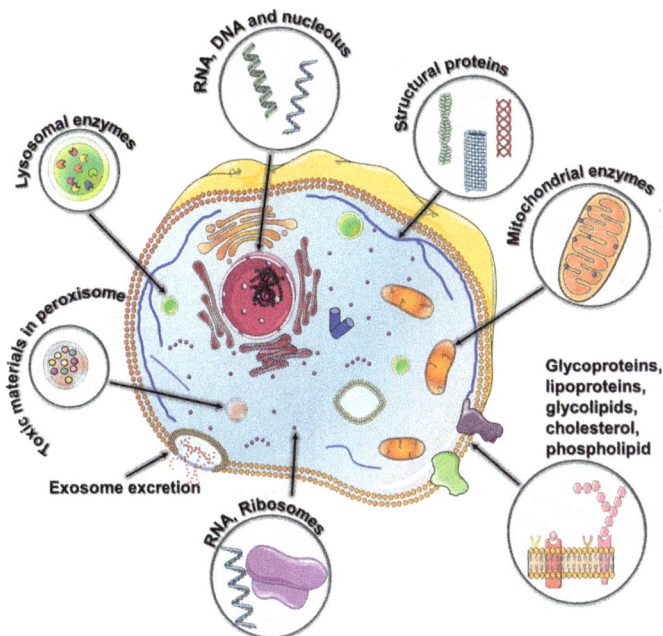

Figure 4.2.5: Cellular compartments and their major components. This figure was created using Servier Medical Art templates, which are licensed under a Creative Commons Attribution 3.0 Unported License; https://smart.servier.com/.

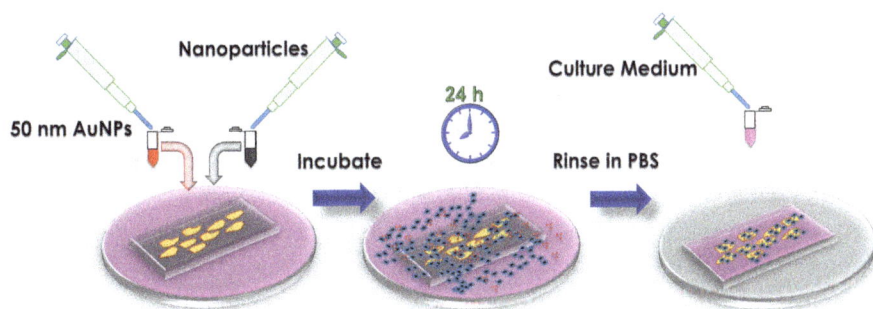

Figure 4.2.6: Experimental procedure of a simple "living cell SERS" analysis of cells exposed to NMs to test nanotoxicity. The NM of interest (blue dots on cells) is introduced into cells simultaneously with the SERS substrate, i. e., AuNPs (red dots on cells). Upon 24 h incubation, the slide seeded with cells is rinsed with cell culture medium, and a drop of fresh culture medium is added on top of the cells to prevent drying throughout the SERS experiment.

to 3 min, depending on the cell size. A 20× long-distance objective was used, with a laser spot diameter of around 2.5 µm. Therefore, spectra were collected at 2- µm intervals. An 830 nm NIR laser was set at 150 mW power with 2 s exposure time. The

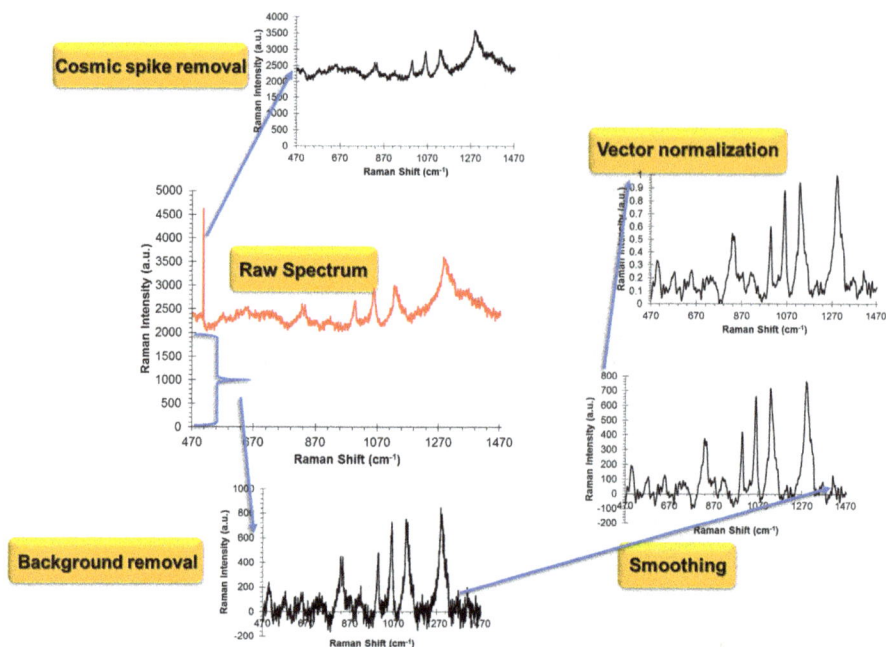

Figure 4.2.7: Pre-processing steps of a cellular SERS spectrum.

collected spectra were background-corrected to remove autofluorescence originating from the cells. Cosmic spikes were removed, and spectra were smoothed and vector-normalized using the instrument's built-in software. The procedure is exemplified on a spectrum in Figure 4.2.7 (see Chapter 4.6 in Volume 2 for more details). To get the representative spectrum of the scanned cell, spectra collected from the cell were averaged. As a side note, it is important to scan at least 20 cells, preferably a minimum of 30 cells, to obtain better sampling from the seeded cell population.

An averaged spectrum obtained from a test group was used to plot the representative spectrum of that test group. In Figure 4.2.8, resultant spectra for cells incubated with increasing concentrations of SWCNTs are shown with the control group where the cells were only incubated with AuNPs. In the literature, there are excellent reports and reviews on the peak assignments of biological molecules [46, 47]. By the help of these, combined with a molecular biology background, it is possible to comment on the cellular conditions. However, there are very important points to be made clear prior to interpretation. One of them is to make sure AuNPs are distributed fairly homogenously in the cells. If AuNPs end up only in the endosomes, it is not wise to comment on the changes in the nucleus. Similarly, if a targeting strategy was applied for the AuNP localization, it should be made clear whether the AuNPs were successfully localized in the desired area. The other point is to consider the SERS or Raman scattering from the material itself that is tested on the cells, such as drugs, nanoparticles, or toxic chem-

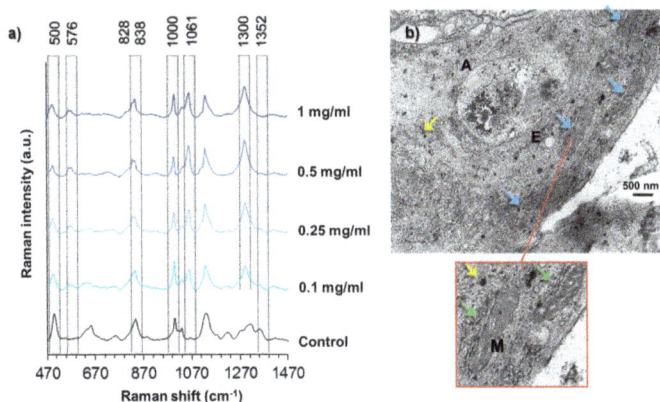

Figure 4.2.8: (a) SERS spectra of A549 cells exposed to increasing concentrations of SWCNTs. (b) TEM image of A549 cells exposed to AuNPs and 0.25 mg/ml SWCNTs. A, autophagosome; E, endosome; M, mitochondrion; yellow arrows show AuNPs, blue arrows show mitochondria, and green arrows point to the ER.

icals. In the case below, SWCNTs are strong Raman scatterers, and they show their characteristic peaks in the spectra, as can also be seen in Figure 4.2.8 at $1300 \, \text{cm}^{-1}$. In the spectra, peaks related to phosphatidyl inositol (PPI) ($576 \, \text{cm}^{-1}$) and ceramide ($1061 \, \text{cm}^{-1}$) increased at increasing concentrations of SWCNTs. It was shown that upon endoplasmic reticulum (ER) stress, mitochondria are colocalized with the ER to initiate Ca^{2+} release that increases PPI and ceramide production [48, 49]. These two then initiate cell death pathways and some lipid kinases. Also, protein denaturation can be traced in the spectra. When proteins denature, they start to lose their tertiary and secondary structures by losing the disulfide bridges and α-helix structures, whereas their hydrophobic amino acid residues such as Trp and Phe are exposed to the aqueous medium, which enables the access of SERS substrates to these residues. From this information, it is expected to see a decrease in the disulfide-related region ($500 \, \text{cm}^{-1}$) and the protein secondary structure-related region ($1200–1300 \, \text{cm}^{-1}$) in contrast to an increase or some changes in peaks related to Phe ($1000 \, \text{cm}^{-1}$) and Trp ($828–838 \, \text{cm}^{-1}$), all of which are visible in the spectra. The colocalization of ER and mitochondria was confirmed by transmission electron microscopy (TEM) imaging (Fig. 4.2.8b).

Similar to this study, it is possible to extract cellular information from the cells in cancer research, for instance on the effect of cancer drugs. Indeed, the first SERS study on cells was reported in 1991 by Nabiev and colleagues where they showed the spectral difference of the nucleus and cytoplasm of doxorubicin-treated, live K562 erythroleukemia cells by using AgNPs as SERS substrates [50]. Many further label-free SERS studies followed this study on cancer cells. As mentioned earlier, biochemical changes upon drug treatment or during metastatic processes or cancer development are now possible to be detected at the single-cell level. Although fixated cells and tissues provide precious spectral information, the in situ investigation of living single cells brings out another dimension to our understanding on cellular dynamics as well as bypassing the fixative-related peaks in the obtained spectra [51]. Moreover, it is possible to further develop an understanding of the cellular events in real-time through monitoring cellular compartments such as the endolysosomal pathway components [52] or targeting the SERS substrates into chosen cellular compartments, such as the nucleus [53]. Furthermore, the stages of mitosis can be monitored in healthy and cancerous cells to compare biochemical differences [54]. The effect of photothermal therapy can be traced in real-time by utilizing AuNRs both for photothermal and SERS purposes [55].

The advantage of SERS beyond all the mentioned points above is that unlike with the conventional methods, it is possible to obtain multidimensional information from the technique with minimum sample preparation, which reduces labor. Especially in the case of living cell analysis, a PBS rinse to get rid of extracellular materials is enough in most cases. However, there are some limitations and points to be considered in living cell, label-free SERS studies. The SERS substrate is one of the most contributing factors to the overall information obtained from the cells, and in the case of colloidal particles, the aggregation state and the cellular localization need to be well understood to prevent any misinterpretation of the data.

4.2.4.4 Tissues

Histopathology is the use of microscopic techniques to analyze cellular morphology from the stained tissue samples of biopsies. Malformations, infectious diseases, and tumor progression have been detected by utilizing histopathology since the nineteenth century [56, 57]. A schematic illustration of the steps used in conventional histopathology is given in Figure 4.2.9. Tissue preparation for pathological examinations starts with fixation of the tissue using formalin and embedding in paraffin. This is then called formalin-fixed paraffin-embedded (FFPE) tissue. After that, FFPE tissues are sliced using a microtome, and unfixed frozen tissue specimens are sliced using a cryostat, generally at 5 μm thickness. The latter is used for the examination of tissue during surgery for rapid decision making, but it has lower quality than FFPE tissue processing. In addition, FFPE tissue processing provides long-term storage of excised tissues; therefore, it is the more widely preferred option. Then, the tissue sections are stained to generate a visual contrast between the cellular components. The hematoxylin/eosin (H&E) combination stain is commonly used in histopathology as the golden standard. Hematoxylin has an affinity for basophilic molecular structures such as DNA and RNA in the nucleus and RNA in ribosomes, while eosin has an affinity for acidophilic substances such as amino acids with positively charged side chains, resulting in dark bluish or purple nuclei and a pinkish cytoplasm [58].

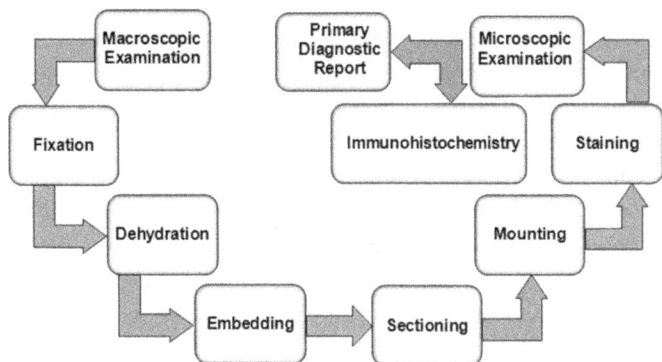

Figure 4.2.9: Steps of tissue processing in clinical settings.

To meet the needs of oncologists for more specific diagnostic markers, immunohistochemistry (IHC) was developed in 1942 [59]. Since then, it has become the most common technique in oncologic pathology. IHC distinguishes normal tissues from diseased tissues with the help of detectable signals produced by a fluorescent label that binds to an antigen. On top of the laborious fixation and paraffin embedding prior to tissue sectioning stages in standard histopathology, IHC has many more procedures taking at least two days, including deparaffinization, rehydration, antigen retrieval, blocking endogenous enzymes, blocking nonspecific sites, incubation with primary and secondary antibodies, washing, color development, counterstaining, and mounting. Several steps used in these processes cause many possible artifacts such as strong background staining, weak target antigen staining, and many others, all of which affect the accuracy of the diagnosis [60]. Moreover, the pathology reports can take a few weeks changing from hospital to hospital, affecting the survival rate of patients. Therefore, more accurate and faster diagnosis approaches are necessary, especially when the cancer becomes invasive, spreading to lymph nodes and other parts of the body. In addition, the pathology report is issued by a subjective decision-making procedure; therefore, interpretation of the data can vary from pathologist to pathologist. Several studies have estimated the error rate in surgical pathology to range from 0.25 % to 43 % [61, 62]. As seen, obtaining a pathology report is not only a time-consuming and laborious process, but the resulting report might also not accurately reflect the actual state of the patient.

For the above reasons, alternative pathological approaches or complementary approaches to conventional clinical pathology methods are needed to provide accurate diagnosis because the accuracy of diagnosis directly affects the treatment type of cancer patients. The diagnostic method is preferred to be rapid and nondestructive with a minimal sample preparation requirement, and the most important features of the diagnostic approach are its objectivity and versatility. All of the mentioned requirements point to SERS as a potential tool for clinics.

In the literature, there are studies related to cancer diagnosis using SERS-labeled assays where Raman reporters have been successfully applied on FFPE tissue sections to detect various cancer biomarkers [63–65]. However, SERS-based label-free detection has gained much more interest for being a fast and low-cost method. Label-free detection of deep tissues has been reported by Stone et al., through injecting citrate-reduced colloidal AgNPs as SERS substrates into mammalian tissue [66]. As a mammography supporting method in early diagnosis of breast cancer, gold nanoparticles that were coated with an ultrathin shell of silica were added onto a breast lesion surface [67]. A cervical cancer detection-based SERS study placed tissue samples on a hot bar and incubated them with colloidal AuNP solution for about 2 h [68]. Czaplicka et al. prepared 100 nm thick Ag layers sputtered on the surface of modified silicon for SERS measurements of tissue specimen [69]. Another sampling method alternative to the conventional methods was carried out producing AgNPs in situ by hydrothermal synthesis assisted by microwave radiation [70]. However, to obtain a highly uniform SERS-active substrate, Yamazoe et al. developed a self-assembly fabrication method using a boehmite template for Au deposition [71]. The silica coating was applied to minimize chemical interaction between the colloids and the adsorbed molecules on them. Zheng et al. fabricated Au@SiO$_2$ shell-isolated nanoparticles and added them onto the surface of the frozen sections [72]. Kowalska et al. performed physical vapor deposition coating to build up a silver platform on which tissue homogenates were placed [73]. The silica coating was applied to minimize chemical interaction between the colloids and the adsorbed molecules on them. Unlike these methods, Çulha et al. developed a simple sampling method by crushing frozen tissue and mixing the homogenized tissue with colloidal AgNP suspension to differentiate rat tissue specimens and classify kidney tumor biopsies in different stages [74, 75]. The steps of this sampling and differentiation method are shown in Figure 4.2.10. Liquid nitrogen was used to avoid the dissolution of the tissue specimen throughout the sampling procedure, which might cause biochemical changes in the biopsy sample. In addition, the frozen tissue sample provided an increased rate of homogenization. The acquired spectral data were used in discrimination algorithms to classify different tissue types.

In another recently developed label-free SERS-based method, hydroxylamine-reduced AgNPs were synthesized in situ in homogenized tissue [76]. Figure 4.2.11 illustrates the application of the in situ method for cancer diagnosis and fabrication of the SERS substrate. The in situ method is aimed to absorb the analyzed sample directly on the nanostructured plasmonic surface to enhance the scattering. Mert in 2018 employed a systematic SERS-based label-free approach called the cryosectioned-PDMS method optimizing each technical and analytical parameter to meet the needs of the pathology clinic [76]. The parameters for the cryosectioned-PDMS method were evaluated including the substrate that the cross-sectioned tissue sample is placed on, tissue thickness, objective lens, laser power, acquisition type (random selection/mapping), mode (high resolution/line mapping), acquired-mapping size, spectral range, SERS substrate concentration, and the droplet position (suspended/sessile).

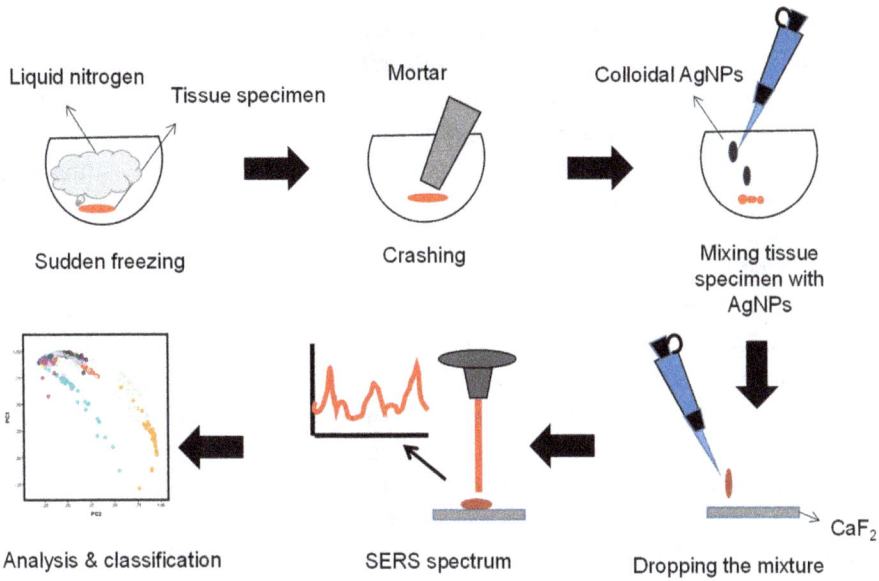

Figure 4.2.10: A schematic illustration of the tissue sampling method for SERS-based cancer diagnosis.

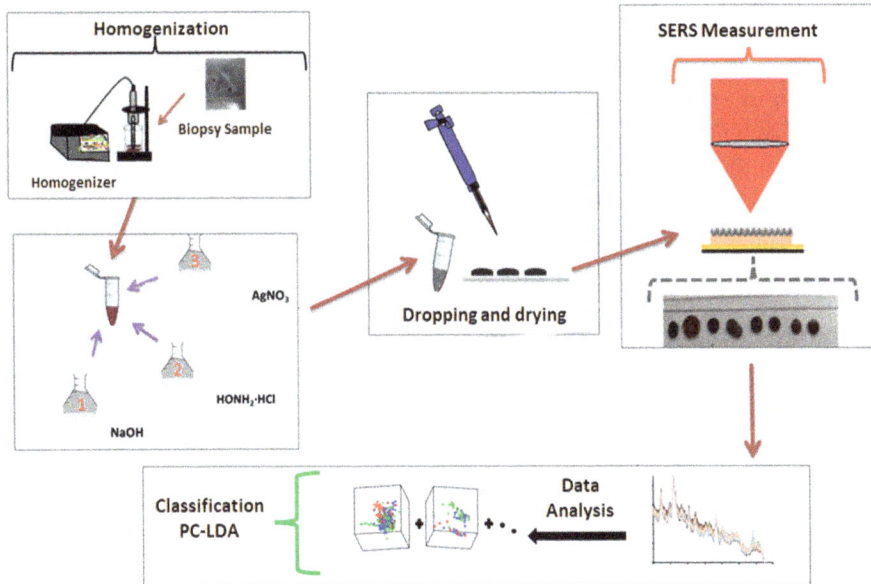

Figure 4.2.11: A schematic illustration of the in situ sampling method for SERS-based cancer diagnosis.

Other label-free approaches in the literature comprise modified SERS substrates, tissue sampling methods, or various types of substrates for biopsy specimens to increase data reliability even more. As an example, a SERS-active titanium dioxide and silver (TiO$_2$-Ag) nanostructured substrate was used for SERS analysis of oral cancer tissue specimens [77]. Here, TiO$_2$ was used in the controlled growth of Ag nanostructures, and the resulting substrate provided highly enhanced SERS spectra and rapid analysis of specimens. Another study used an aluminum plate as SERS substrate to mount diabetic and healthy tissue slices for possible use in cancer tissue-related studies [78, 79]. A different sampling method was developed by incubating NPs with deparaffinized tissue mounted on a silanated glass slide [80]. Another study injected the NPs into tissue mounted on a stainless steel slab [81]. Ma et al. used the method of centrifuging homogenized tissues and mixing tissue samples with colloidal AuNPs [82]. Mert used PDMS-covered glass slides to obtain a background-free platform that is low-cost, easy to produce, and possibly applicable in clinical settings [76].

SERS is an effective technique for cancer diagnosis since biopsy differentiation is based on chemical information rather than morphological information used in conventional methods. However, all parameters influencing a tissue-SERS experiment should be well established and optimized to utilize the technique for cancer diagnosis.

4.2.5 Some challenges and solutions

As demonstrated throughout the chapter, the SERS technique has the potential to become a versatile tool for clinics in the near future. On the other hand, the complexity of biological samples and the lack of a worldwide accepted SERS substrate are the main obstacles for the broader clinical translation of label-free SERS at the moment. There are too many SERS substrates reported for a variety of applications. However, almost all display different performance and characteristics, which affect the reproducibility of results.

Either colloidal nanoparticles or nanostructured surfaces of Au or Ag can be used as SERS substrates. A problem for colloidal substrates, apart from reproducible synthesis difficulties, is stability. Colloidal nanoparticles are stable if their zeta potential is outside the +30 to −30 mV range. Nanoparticles possessing surface charge between these two values are prone to aggregation and the following sedimentation upon prolonged waiting. Even nanoparticles with charges out of this range age, and their SERS activity diminishes in time. Therefore, a universal expiration date for all kinds of colloidal SERS substrates should be defined for their potential use in clinics. During the application of colloidal nanoparticles, they also have a tendency to become aggregates in body fluids due to salinity and in the presence of complex biomolecules, which can neutralize the charge of nanoparticles or change the surface chemistry. This situation brings laser compatibility problems because not all laser wavelengths are suitable for aggregated nanoparticles. This problem can be solved by a systematic study of the interaction of different laser wavelengths with nanoparticle clumps having different degrees of aggregation. On the other hand, not each laboratory has all kinds of lasers and facilities. Collaborations among different laboratories in different countries can be extremely useful to deal with this problem. One of the goals of COST Actions (BM1401) throughout Europe was to foster this collaborative environment, and the COST Action MP1302 Nanospectroscopy paved the road to more fruitful relations among laboratories in different countries.

Nanostructured surface-based SERS substrates are not affected by the presence of salty body fluids; therefore, they can provide more reproducible results than colloidal ones. On the other hand, they have the same abovementioned problems as the colloidal ones. Besides, preparation methods can be more complex and usually require lithographic techniques. More complexity brings cost and uncertainty. In addition, upon application of the substrates, biomolecules stick to the substrate surface and make it of only single use. This is an extremely important problem because it can significantly increase the costs. Therefore, different approaches should be investigated to reduce preparation costs of this type of substrates, and we believe it will be achieved in the near future parallel to advances in many related scientific fields.

Another challenge is how to apply SERS substrates to cells, body fluids, and tissues to obtain label-free information from them. As mentioned in the previous section, there are many ways, but the hard part is to find the optimum approach. For cell studies, colloidal nanoparticles are introduced into cells to acquire spectra, but the distribution of nanoparticles can be chaotic inside of cells. Before the integration of SERS into clinical cell studies, the nature of this distribution and its effects on the final acquired spectra should be well understood. In addition, since these nanoparticles are foreign to the cells, their response to them on the molecular level should also be well understood. Label-free SERS can provide nanometer-scale information; thus, the acquired spectrum should be well analyzed to discriminate the spectra due to the response of the cell from the signals stemming from any disease. For biomarker detection, the interaction of biomolecules with the substrate surface and Raman scattering ability of these molecules should be clearly understood. There are thousands of molecules in the body fluids, and they will compete to interact with the substrate surface. Not all molecules adhering to the substrate surface will contribute to the acquired spectra. Spectra will be composed of the bands of more Raman scattering molecules. To overcome this problem, substrate surfaces can be functionalized with molecules possessing a high affinity to the molecule of interest. Spectral differences before and after target molecule binding will provide information.

Challenges for tissue-level label-free SERS studies are more difficult to solve. Some problems are due to the current technological situation of SERS instrumentation, and some are due to the complexity of tissues and living organisms. With label-free SERS, spectral information from the close vicinity of the SERS substrate can be obtained, and the scale for this is just a few nanometers. Therefore, nanoparticles should be injected into the tissue. At that time, other problems emerge. The first one is controlling the ability of uniform distribution of nanoparticles inside tissue. The second one is the penetration depth of the laser. Tissue has three dimensions, and for thorough application of SERS to clinical operations and surgery, information on the centimeter scale with single spectrum acquisition should be obtained. Moreover, in the case of surgery, information from tumor tissue and its nearby environment should be obtained very quickly. To realize this, injection must be very quick, and the 3D mapping ability of the instrument must be very fast, which is not easy with the current technology of instruments. Faster ones should be produced, and seeing the pace of advances in science, there is no reason not to believe that these will be realized in the near future. Furthermore, another challenge is the accessibility of tissues. The curved and moving structure of the tissues and organs needs flexible fiber optic-based SERS substrates. Again, fast and accurate acquisition with these devices is a must. Contamination of the probe during its injection towards target tissue with various kinds of molecules is the other problem to deal with.

Another problem comes from a statistical perspective. The complexity of organisms causes the presence of many spectral signals, which are not relevant to the aim during an operation. When this problem merges with the problem stemming from the low SERS signal of many significant biomolecules relevant to cancer, a thorough statistical analysis of extremely complex data becomes very crucial. A clinician or a surgeon may not be able to handle these huge datasets and lots of mathematical problems. Therefore, mathematicians and computer engineers should merge their power with the medical experts acquiring the spectra to handle this extensive information obtained using SERS for better diagnosis.

As seen, there are many challenges in the incorporation of label-free SERS techniques in clinical settings. On the other hand, knowing that humans always find solutions for their problems, use of SERS in clinical settings at least as a complementary technique is possible in the near feature.

4.2.6 Summary and impact

Cancer remains to be a challenging disease at both diagnostic and therapeutic levels. An early as well as correct diagnosis is crucial for the patients. Understanding the processes related to drug resistance and metastasis are also key points in developing more precise therapeutics. The conventional methods to diagnose cancer are expensive and time consuming, which also require professional staff. On the other hand, the detection reliability might even drop down to 43 % in the case of FFPE histopathology. Even though many types of labels containing fluorescent, isotopic, chemiluminescent, or electrochemically active molecules or a nanoparticle attached to the molecule that recognizes the targeted molecule increase specificity of diagnosis, the limit of detection is not always sufficient, especially for early diagnosis. False positive results might arise from nonspecific interactions between labels and biological components. Molecular functions of the labeled molecule might also be disrupted due to the physicochemical properties of the label. Therefore, alternative approaches that bypass the mentioned problems are needed in clinical settings.

The utilization of SERS as a label-free cancer diagnostic tool is promising in this sense. It requires a basic sample preparation; often mixing colloidal SERS substrate with the sample and rinsing or washing out the unwanted components, such as erythrocytes from the blood plasma or cell culture medium from the cells, would suffice. Upon proper spectral acquisition and data pre-processing steps, the obtained data might either be interpreted for the spectral changes or be combined with multivariate statistical analyses for sample classification and differentiation. Moreover, the reduced steps in sample preparation and analysis significantly lower the analysis time.

The technique, however, has some issues regarding data reproducibility and discrimination of similar molecules as in the case of proteins. To overcome the existing challenges, ongoing research is mainly focused on the development and optimization of SERS substrates and sampling methods.

Bibliography

[1]   Wu JG, Xu YZ, Sun CW, Soloway RD, Xu DF, Wu QG, Sun KH, Weng SF, Xu GX. Distinguishing malignant from normal oral tissues using FTIR fiber-optic techniques. Biopolymers. 2001;62:185–92.
[2]   Fujioka N, Morimoto Y, Arai T, Kikuchi M. Discrimination between normal and malignant human gastric tissues by Fourier transform infrared spectroscopy. Cancer Detec Prev. 2004;28:32–6.
[3]   Stone N, Kendall C, Shepherd N, Crow P, Barr H. Near-infrared Raman spectroscopy for the classification of epithelial pre-cancers and cancers. J Raman Spectrosc. 2002;33:564–73.
[4]   Teh S, Zheng W, Ho K, Teh M, Yeoh K, Huang Z. Diagnostic potential of near-infrared Raman spectroscopy in the stomach: differentiating dysplasia from normal tissue. Br J Cancer. 2008;98:457–65.
[5]   Baker MJ, Trevisan J, Bassan P, Bhargava R, Butler HJ, Dorling KM, Fielden PR, Fogarty SW, Fullwood NJ, Heys KA. Using Fourier transform IR spectroscopy to analyze biological materials. Nat Protoc. 2014;9:1771–91.
[6]   Baker MJ, Hussain SR, Lovergne L, Untereiner V, Hughes C, Lukaszewski RA, Thiéfin G, Sockalingum GD. Developing and understanding biofluid vibrational spectroscopy: a critical review. Chem Soc Rev. 2016;45:1803–18.
[7]   Nallala J, Diebold M-D, Gobinet C, Bouché O, Sockalingum GD, Piot O, Manfait M. Infrared spectral histopathology for cancer diagnosis: a novel approach for automated pattern recognition of colon adenocarcinoma. Analyst. 2014;139:4005–15.
[8]   Li S, Zhang Y, Zeng Q, Li L, Guo Z, Liu Z, Xiong H, Liu S. Potential of cancer screening with serum surface-enhanced Raman spectroscopy and a support vector machine. Laser Phys Lett. 2014;11:065603.

[9]   Feng S, Chen R, Lin J, Pan J, Wu Y, Li Y, Chen J, Zeng H. Gastric cancer detection based on blood plasma surface-enhanced Raman spectroscopy excited by polarized laser light. Biosens Bioelectron. 2011;26:3167–74.

[10] Ito H, Inoue H, Hasegawa K, Hasegawa Y, Shimizu T, Kimura S, Onimaru M, Ikeda H, Kudo S-e. Use of surface-enhanced Raman scattering for detection of cancer-related serum-constituents in gastrointestinal cancer patients. Nanomedicine. 2014;10:599–608.

[11] Lin D, Pan J, Huang H, Chen G, Qiu S, Shi H, Chen W, Yu Y, Feng S, Chen R. Label-free blood plasma test based on surface-enhanced Raman scattering for tumor stages detection in nasopharyngeal cancer. Sci Rep. 2014;4.

[12] Lin D, Feng S, Pan J, Chen Y, Lin J, Chen G, Xie S, Zeng H, Chen R. Colorectal cancer detection by gold nanoparticle based surface-enhanced Raman spectroscopy of blood serum and statistical analysis. Opt Express. 2011;19:13565–77.

[13] Bonifacio A, Cervo S, Sergo V. Label-free surface-enhanced Raman spectroscopy of biofluids: fundamental aspects and diagnostic applications. Anal Bioanal Chem. 2015;407:8265–77.

[14] Álvarez-Puebla RnA. Effects of the excitation wavelength on the SERS spectrum. J Phys Chem Lett. 2012;3:857–66.

[15] Le Ru E, Meyer S, Artur C, Etchegoin P, Grand J, Lang P, Maurel F. Experimental demonstration of surface selection rules for SERS on flat metallic surfaces. Chem Commun. 2011;47:3903–5.

[16] Tortora GJ, Derrickson BH. Principles of Anatomy and Physiology. John Wiley & Sons; 2008.

[17] Strasinger SK, Di Lorenzo MS. Urinalysis & Body Fluids. FA Davis; 2008.

[18] Le Ru E, Etchegoin P. Principles of Surface-Enhanced Raman Spectroscopy: and Related Plasmonic Effects. Elsevier; 2008.

[19] See KC, Spicer JB, Brupbacher J, Zhang D, Vargo TG. Modeling interband transitions in silver nanoparticle-fluoropolymer composites. J Phys Chem B. 2005;109:2693–8.

[20] West PR, Ishii S, Naik GV, Emani NK, Shalaev VM, Boltasseva A. Searching for better plasmonic materials. Laser Photonics Rev. 2010;4:795–808.

[21] Erol M, Han Y, Stanley SK, Stafford CM, Du H, Sukhishvili S. SERS not to be taken for granted in the presence of oxygen. J Am Chem Soc. 2009;131:7480–1.

[22] Bonifacio A, Dalla Marta S, Spizzo R, Cervo S, Steffan A, Colombatti A, Sergo V. Surface-enhanced Raman spectroscopy of blood plasma and serum using Ag and Au nanoparticles: a systematic study. Anal Bioanal Chem. 2014;406:2355–65.

[23] Li S, Zhang Y, Xu J, Li L, Zeng Q, Lin L, Guo Z, Liu Z, Xiong H, Liu S. Noninvasive prostate cancer screening based on serum surface-enhanced Raman spectroscopy and support vector machine. Appl Phys Lett. 2014;105:091104.

[24] Huang S, Wang L, Chen W, Feng S, Lin J, Huang Z, Chen G, Li B, Chen R. Potential of non-invasive esophagus cancer detection based on urine surface-enhanced Raman spectroscopy. Laser Phys Lett. 2014;11:115604.

[25] Xu H, Bjerneld EJ, Käll M, Börjesson L. Spectroscopy of single hemoglobin molecules by surface enhanced Raman scattering. Phys Rev Lett. 1999;83:4357.

[26] Han XX, Jia HY, Wang YF, Lu ZC, Wang CX, Xu WQ, Zhao B, Ozaki Y. Analytical technique for label-free multi-protein detection based on Western blot and surface-enhanced Raman scattering. Anal Chem. 2008;80:2799–804.

[27] Culha M. Surface-enhanced Raman scattering: an emerging label-free detection and identification technique for proteins. Appl Spectrosc. 2013;67:355–64.

[28] Kitahama Y, Ozaki Y. Surface-enhanced resonance Raman scattering of hemoproteins and those in complicated biological systems. Analyst. 2016;141:5020–36.

[29] Das G, Gentile F, Coluccio M, Perri A, Nicastri A, Mecarini F, Cojoc G, Candeloro P, Liberale C, De Angelis F. Principal component analysis based methodology to distinguish protein SERS spectra. J Mol Struct. 2011;993:500–5.

[30] Arif M, Kumar GP, Narayana C, Kundu TK. Autoacetylation induced specific structural changes in histone acetyltransferase domain of p300: probed by surface enhanced Raman spectroscopy. J Phys Chem B. 2007;111:11877–9.

[31] Abdali S, De Laere B, Poulsen M, Grigorian M, Lukanidin E, Klingelhofer J. Toward methodology for detection of cancer-promoting S100A4 protein conformations in subnanomolar concentrations using Raman and SERS†. J Phys Chem C. 2010;114:7274–9.

[32] Brulé T, Bouhelier A, Dereux A, Finot E. Discrimination between single protein conformations using dynamic SERS. ACS Sensors. 2016.

[33] Paidi SK, Siddhanta S, Strouse R, McGivney JB, Larkin C, Barman I. Rapid identification of biotherapeutics with label-free Raman spectroscopy. Anal Chem. 2016;88:4361–8.

[34] Vicario A, Sergo V, Toffoli G, Bonifacio A. Surface-enhanced Raman spectroscopy of the anti-cancer drug irinotecan in presence of human serum albumin. Colloids Surf B, Biointerfaces. 2015;127:41–6.

[35] Han XX, Chen L, Ji W, Xie Y, Zhao B, Ozaki Y. Label-free indirect immunoassay using an avidin-induced surface-enhanced Raman scattering substrate. Small. 2011;7:316–20.

[36] Hodges MD, Kelly JG, Bentley AJ, Fogarty S, Patel II, Martin FL, Fullwood NJ. Combining immunolabeling and surface-enhanced Raman spectroscopy on cell membranes. ACS Nano. 2011;5:9535–41.

[37] Balzerova A, Fargasova A, Markova Z, Ranc V, Zboril R. Magnetically-assisted surface enhanced Raman spectroscopy (MA-SERS) for label-free determination of human immunoglobulin G (IgG) in blood using Fe3O4@ Ag nanocomposite. Anal Chem. 2014;86:11107–14.

[38] Lin J, Chen R, Feng S, Pan J, Li Y, Chen G, Cheng M, Huang Z, Yu Y, Zeng H. A novel blood plasma analysis technique combining membrane electrophoresis with silver nanoparticle-based SERS spectroscopy for potential applications in noninvasive cancer detection. Nanomedicine. 2011;7:655–63.

[39] Wang J, Feng S, Lin J, Zeng Y, Li L, Huang Z, Li B, Zeng H, Chen R. Serum albumin and globulin analysis for hepatocellular carcinoma detection avoiding false-negative results from alpha-fetoprotein test negative subjects. Appl Phys Lett. 2013;103:204106.

[40] Lin J, Wang J, Xu C, Zeng Y, Chen Y, Li L, Huang Z, Li B, Chen R. Differentiation of digestive system cancers by using serum protein-based surface-enhanced Raman spectroscopy. J Raman Spectrosc. 2016.

[41] Feng S, Lin D, Lin J, Huang Z, Chen G, Li Y, Huang S, Zhao J, Chen R, Zeng H. Saliva analysis combining membrane protein purification with surface-enhanced Raman spectroscopy for nasopharyngeal cancer detection. Appl Phys Lett. 2014;104:073702.

[42] Feng S, Huang S, Lin D, Chen G, Xu Y, Li Y, Huang Z, Pan J, Chen R, Zeng H. Surface-enhanced Raman spectroscopy of saliva proteins for the noninvasive differentiation of benign and malignant breast tumors. Int J Nanomed. 2015;10:537.

[43] Horwitz AV. Distinguishing distress from disorder as psychological outcomes of stressful social arrangements. Health. 2007;11:273–89.

[44] Zhang Y, Li B, Chen X. Simple and sensitive detection of dopamine in the presence of high concentration of ascorbic acid using gold nanoparticles as colorimetric probes. Mikrochim Acta. 2010;168:107–13.

[45] Kuku G, Saricam M, Akhatova F, Danilushkina A, Fakhrullin R, Culha M. Surface-enhanced Raman scattering to evaluate nanomaterial cytotoxicity on living cells. Anal Chem. 2016;88:9813–20.

[46] Talari ACS, Movasaghi Z, Rehman S, Rehman Iu. Raman spectroscopy of biological tissues. Appl Spectrosc Rev. 2015;50:46–111.

[47] De Gelder J, De Gussem K, Vandenabeele P, Moens L. Reference database of Raman spectra of biological molecules. J Raman Spectrosc. 2007;38:1133–47.

[48] Kim T-J, Mitsutake S, Igarashi Y. The interaction between the pleckstrin homology domain of ceramide kinase and phosphatidylinositol 4, 5-bisphosphate regulates the plasma membrane targeting and ceramide 1-phosphate levels. Biochem Biophys Res Commun. 2006;342:611–7.

[49] Mitsutake S, Kim T-J, Inagaki Y, Kato M, Yamashita T, Igarashi Y. Ceramide kinase is a mediator of calcium-dependent degranulation in mast cells. J Biol Chem. 2004;279:17570–7.

[50] Nabiev I, Morjani H, Manfait M. Selective analysis of antitumor drug interaction with living cancer cells as probed by surface-enhanced Raman spectroscopy. Eur Biophys J. 1991;19:311–6.

[51] Draux F, Gobinet C, Sulé-Suso J, Trussardi A, Manfait M, Jeannesson P, Sockalingum GD. Raman spectral imaging of single cancer cells: probing the impact of sample fixation methods. Anal Bioanal Chem. 2010;397:2727–37.

[52] Kneipp J, Kneipp H, McLaughlin M, Brown D, Kneipp K. In vivo molecular probing of cellular compartments with gold nanoparticles and nanoaggregates. Nano Lett. 2006;6:2225–31.

[53] Dreaden EC, Austin LA, Mackey MA, El-Sayed MA. Size matters: gold nanoparticles in targeted cancer drug delivery. Ther Delivery. 2012;3:457–78.

[54] Panikkanvalappil SR, Hira SM, Mahmoud MA, El-Sayed MA. Unraveling the biomolecular snapshots of mitosis in healthy and cancer cells using plasmonically-enhanced Raman spectroscopy. J Am Chem Soc. 2014;136:15961–8.

[55] Ali MR, Wu Y, Han T, Zang X, Xiao H, Tang Y, Wu R, Fernandez FM, El-Sayed MA. Simultaneous time-dependent surface enhanced Raman spectroscopy, metabolomics and proteomics reveal cancer cell death mechanisms associated with Au-nanorod photo-thermal therapy. J Am Chem Soc. 2016.

[56] Frey W, Suter F. Historisches, Nieren und Ableitende Harnwege. Springer; 1951. p. 3–43.

[57] Hajdu SI. Rudolph Virchow, pathologist, armed revolutionist, politician, and anthropologist. Ann Clin Lab Sci. 2005;35:203–5.

[58] Avwioro G. Histochemical uses of haematoxylin—a review. J Phys Conf Ser. 2011;1:24–34.

[59] Coons AH, Creech H, Jones R, Berliner E. The demonstration of pneumococcal antigen in tissues by the use of fluorescent antibody. J Immunol. 1942;45:159–70.

[60] Ramos-Vara J. Technical aspects of immunohistochemistry. Vet Pathol. 2005;42:405–26.

[61] Renshaw AA. Measuring and reporting errors in surgical pathology. Am J Clin Pathol. 2001;115:338–41.

[62] Troxel DB. Medicolegal aspects of error in pathology. Arch Pathol Lab Med. 2006;130:617–9.

[63] Schütz M, Steinigeweg D, Salehi M, Kömpe K, Schlücker S. Hydrophilically stabilized gold nanostars as SERS labels for tissue imaging of the tumor suppressor p63 by immuno-SERS microscopy. Chem Commun. 2011;47:4216–8.

[64] Salehi M, Steinigeweg D, Ströbel P, Marx A, Packeisen J, Schlücker S. Rapid immuno-SERS microscopy for tissue imaging with single-nanoparticle sensitivity. J Biophotonics. 2013;6:785–92.

[65] Lutz B, Dentinger C, Sun L, Nguyen L, Zhang J, Chmura A, Allen A, Chan S, Knudsen B. Raman nanoparticle probes for antibody-based protein detection in tissues. J Histochem Cytochem. 2008;56:371–9.

[66] Stone N, Faulds K, Graham D, Matousek P. Prospects of deep Raman spectroscopy for noninvasive detection of conjugated surface enhanced resonance Raman scattering nanoparticles buried within 25 mm of mammalian tissue. Anal Chem. 2010;82:3969–73.

[67] Liang L, Zheng C, Zhang H, Xu S, Zhang Z, Hu C, Bi L, Fan Z, Han B, Xu W. Exploring type II microcalcifications in benign and premalignant breast lesions by shell-isolated nanoparticle-enhanced Raman spectroscopy (SHINERS). Spectrochim Acta, Part A, Mol Biomol Spectrosc. 2014;132:397–402.

[68] Ceja-Fdez A, Carriles R, González-Yebra AL, Vivero-Escoto J, de la Rosa E, López-Luke T. Imaging and SERS study of the Au nanoparticles interaction with HPV and carcinogenic cervical tissues. Molecules. 2021;26:3758.

[69] Czaplicka M, Kowalska A, Nowicka A, Kurzydłowski D, Gronkiewicz Z, Machulak A, Kukwa W, Kamińska A. Raman spectroscopy and surface-enhanced Raman spectroscopy (SERS) spectra of salivary glands carcinoma, tumor and healthy tissues and their homogenates analyzed by chemometry: Towards development of the novel tool for clinical diagnosis. Anal Chim Acta. 2021;1177:338784.

[70] Ferreira N, Marques A, Águas H, Bandarenka H, Martins R, Bodo C, Costa-Silva B, Fortunato E. Label-free nanosensing platform for breast cancer exosome profiling. ACS Sensors. 2019;4:2073–83.

[71] Yamazoe S, Naya M, Shiota M, Morikawa T, Kubo A, Tani T, Hishiki T, Horiuchi T, Suematsu M, Kajimura M. Large-area surface-enhanced Raman spectroscopy imaging of brain ischemia by gold nanoparticles grown on random nanoarrays of transparent boehmite. ACS Nano. 2014;8:5622–32.

[72] Zheng C, Liang L, Xu S, Zhang H, Hu C, Bi L, Fan Z, Han B, Xu W. The use of Au@ SiO_2 shell-isolated nanoparticle-enhanced Raman spectroscopy for human breast cancer detection. Anal Bioanal Chem. 2014;406:5425–32.

[73] Kowalska AA, Berus S, Szleszkowski Ł, Kamińska A, Kmiecik A, Ratajczak-Wielgomas K, Jurek T, Zadka Ł. Brain tumour homogenates analysed by surface-enhanced Raman spectroscopy: discrimination among healthy and cancer cells. Spectrochim Acta, Part A, Mol Biomol Spectrosc. 2020;231:117769.

[74] Aydin Ö, Kahraman M, Kiliç E, Çulha M. Surface-enhanced Raman scattering of rat tissues. Appl Spectrosc. 2009;63:662–8.

[75] Mert S, Özbek E, Ötünçtemur A, Çulha M. Kidney tumor staging using surface-enhanced Raman scattering. J Biomed Opt. 2015;20:047002.

[76] Mert S. Development of SERS based methods for early cancer detection, Biotechnology, Turkey: Yeditepe University, İstanbul; 2018.

[77] Girish CM, Iyer S, Thankappan K, Rani VD, Gowd GS, Menon D, Nair S, Koyakutty M. Rapid detection of oral cancer using Ag–TiO_2 nanostructured surface-enhanced Raman spectroscopic substrates. J Mater Chem B. 2014;2:989–98.

[78] Pînzaru SC, Andronie L, Domsa I, Cozar O, Astilean S. Bridging biomolecules with nanoparticles: surface-enhanced Raman scattering from colon carcinoma and normal tissue. J Raman Spectrosc. 2008;39:331–4.

[79] Huang H, Shi H, Feng S, Lin J, Chen W, Huang Z, Li Y, Yu Y, Lin D, Xu Q. Silver nanoparticle based surface enhanced Raman scattering spectroscopy of diabetic and normal rat pancreatic tissue under near-infrared laser excitation. Laser Phys Lett. 2013;10:045603.

[80] Jehn C, Küstner B, Adam P, Marx A, Ströbel P, Schmuck C, Schlücker S. Water soluble SERS labels comprising a SAM with dual spacers for controlled bioconjugation. Phys Chem Chem Phys. 2009;11:7499–504.

[81] Dey P, Olds W, Blakey I, Thurecht KJ, Izake EL, Fredericks PM. SERS-based detection of barcoded gold nanoparticle assemblies from within animal tissue. J Raman Spectrosc. 2013;44:1659–65.

[82] Ma J, Zhou H, Gong L, Liu S, Zhou Z, Mao W, Zheng R-e. Distinction of gastric cancer tissue based on surface-enhanced Raman spectroscopy. In: Photonics Asia, International Society for Optics and Photonics. 2012. p. 855328–9.

Janina Kneipp and Daniela Drescher

4.3 Advanced nanospectroscopy in bioapplications

4.3.1 Key messages

- Since surface-enhanced Raman scattering (SERS) delivers fingerprint-like vibrational information, it is ideally suited to probe different kinds of complex biological samples, including cells and tissues.
- SERS probing is restricted to the proximity of plasmonic nanostructures; therefore, often gold or silver nanoparticles are used, since they can be positioned in the region of interest in a sample.
- Based on the beneficial properties of gold and silver nanoparticles, uptake and processing of SERS substrates can be followed inside animal cells from tissues and cell cultures by complementary methods that enable characterization of their distribution in the cell ultrastructure and quantification by, e. g., transmission electron microscopy, cryo X-ray tomography (cryo-XT), and mass spectrometry.
- The particle properties, aggregate geometry, and surface modification play a critical role in the utilization of gold and silver nanoparticles as efficient SERS nanoprobes.

4.3.2 Pre-knowledge

As discussed extensively in Volume 1, Section 3, surface-enhanced Raman scattering (SERS), the strong enhancement of the inelastic scattering (Raman) process (Fig. 4.3.1), is observed when molecules reside in the close proximity of plasmonic (mostly gold or silver) nanostructures [1]. High local optical fields, which occur due to resonances between excitation and scattered field and surface plasmons, provide the main mechanism for the SERS effect (cf. Volume 1, Section 3). Additionally, a second, so-called chemical enhancement mechanism, based on interaction of the molecule with the metal nanostructures, can increase the Raman cross-section of the adsorbed molecule. The plasmonic field enhancement can result in more than 10 orders of magnitude signal enhancement, whereas chemical enhancement is discussed to contribute an enhancement factor of $10-10^3$. If the excitation light is in resonance with an electronic transition in the molecule, additional enhancement is observed due to resonant Raman scattering (SERRS) (Fig. 4.3.1). The strong SER(R)S signal can serve as spectroscopic signature of a label in a similar way to that of a fluorescent dye.

Since SERS delivers fingerprint-like, vibrational information (cf. Volume 1, Section 3), it is ideally suited to probe different kinds of complex, microstructured systems. In this chapter, we will discuss the application of mobile plasmonic nanostructures and hybrid systems for the probing of complex biological samples. After introducing the concepts of SERS labels and SERS probes, we want to provide an overview of how SERS nanoprobes can be characterized in complex biological environments, how their interaction with cells and biomolecules can be understood at the microscopic and nanoscopic levels, and finally how the properties of the nanoprobes can be modified in order to influence their

https://doi.org/10.1515/9783110442908-013

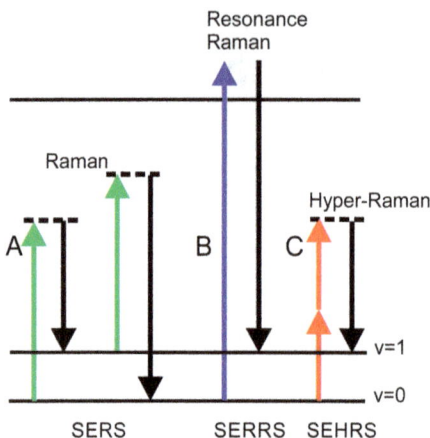

Figure 4.3.1: Spontaneous Raman scattering processes that are used in SERS labels and probes. (a) Nonresonant Raman scattering (Stokes and anti-Stokes). (b) Resonant Raman scattering for SERRS labels. (c) Hyper-Raman scattering for two-photon-excited SEHRS.

interaction with their biological environment. All these aspects are important if one aims for the interpretation of the enhanced vibrational signatures that can be gained from cells, tissues, or other microstructured, heterogeneous materials. Understanding the behavior of SERS substrates in complex biosystems is of great interest for medical diagnostics, for therapeutics, and in bioanalytical approaches.

A SERS (or SERRS) label, sometimes also called a SERS tag, typically consists of plasmonic metal nanoparticles and a reporter molecule that provides a characteristic Raman spectrum (Fig. 4.3.2a). The nanostructure–reporter constructs can be embedded for protection and stability even in sometimes harsh bioenvironments by glass or a polymer [2]. To reach a biological target, such as a cellular compartment or specific positions at the cellular surface or a microscopic region of a tissue, targeting molecules are used, such as antibodies, oligonucleotides, targeting peptides, or small molecules that enable binding of a particular receptor. SERS labels can be taken up by animal cells, in a process called endocytosis. There, the cell membrane engulfs a nanoparticle probe that resides on the outer side of the membrane and forms a vesicle around the particle. As a consequence, a vesicle, termed the endosome, containing the nanoparticle is formed at the inside of the membrane, and the SERS label has successfully entered the cell, where it is located inside the endosome. In Raman experiments with individual cells, high Raman signals are collected from those positions in a cell where the vesicles that contain the SERS labels are located, and mapping of them can be achieved. A significant advantage compared to fluorescence labels are the great multiplexing capabilities of SER(R)S labels, resulting from the many narrow lines in the vibrational spectra, as they yield good separation of the spectral fingerprints, e. g., by multivariate statistical tools or machine learning [3].

4.3.3 State-of-the-art

4.3.3.1 The concept of SERS labels and probes

SERS labels have been used for almost two decades now [4], and they were first proposed for multiplex detection in bioanalytical applications, such as microarrays [5, 6] or immunoassays [7]. Later, they were used for histopathological applications, such as excised tumor tissues [8], or also of living tumor cells in culture [9] and in tissues [10] and even in live animals [11]. Since the encapsulated SER(R)S labels are very stable, their spectra can be used for quantification [12, 13], and in the case of a defined targeting also for a quantification of the targeted molecule or structure, e. g., in the case of a particular antigen that is bound by SERS label directed by a respective antibody. Also nonencapsulated SERS labels can be very stable, as shown for gold nanoparticles coated with 4-aminothiophenol and folate [14]. When tumor cells express a receptor for folate, SERS labels carrying folate can bind to the receptor. On the other hand, also cells without the receptor may bind SERS labels nonspecifically in small amounts. Therefore, although the SER(R)S signals give high sensitivity, there must be a targeting unit (folate) to add sufficient specificity to the system.

A SERRS label provides very high sensitivity, similar to or higher than the fluorescence signal obtained with the same number of reporter molecules [9]. Nanoparticle–reporter constructs can be optimized to a great extent to obtain high signal strength and hence sensitivity, by both a very well-performing plasmonic nanostructure and a suitable resonant molecule. SERS labels, moreover, have the advantage that the number of Raman photons emitted by the Raman process of the reporter molecule in a certain time can be higher than the number of photons coming from a fluorescence process of the same molecule. This is the case because of the shorter vibrational relaxation times compared to electronic relaxation times, so a molecule can go through more "Raman cycles" than "fluorescence cycles" per time interval (cf. Volume 1, Section 2). It must be noted as well that a molecule when at the surface of a metal structure can have higher photostability compared to a free molecule [15], which also supports a higher amount of SERRS photons compared to fluorescence photons.

Although much was stated regarding resonant SERS labels (SERRS labels), nonresonant excitation (Fig. 4.3.1) is particularly desirable in bioprobing. If resonance conditions can be disregarded, many reporter molecules can be used with one and the same excitation wavelength. In such an experiment, the excitation energy would be chosen particularly in the near-infrared (NIR) wavelength range, where photodecomposition of the biological tissue can be neglected. Often, the nonresonant labels are much more photostable over time.

Nevertheless, a SERS label has more or less the same function as a fluorescence tag: it can be used to visualize and to image biological structures. Despite their improved stability this may not be fully convincing of the advantages of such labels. However, also other molecules residing in the plasmonic near-field, that is, intrinsic biomolecules, not just the reporters, undergo SERS, and therefore can be studied. This means that a SERS label is much more than just a bright optical marker, but in fact a SERS *probe*.

4.3.3.2 Probing the chemical environment beyond SERS tagging: SERS probes

The reporter molecules in a SERS probe can interact with the bio-organic molecules in the cellular or tissue surroundings, and a SERS label can at the same time act as a sensitive nanoprobe. Such SERS nanosensors infer information using the relative signals of spectrally narrow pairs of Raman lines in the same reporter spectrum and therefore do not need intensity calibration. This allows for quantitative measurements also before complex background signals in a cell.

Reporter molecules that show a pH-sensitive Raman spectrum can be used to monitor pH distributions in cell populations at subendosomal resolution [16]. When the reporter molecule undergoes protonation/deprotonation in the subcellular compartment, the structure and interaction of the reporter molecule with the metal nanostructure changes, and the spectrum changes [17]. Approaches that can monitor this interaction, such as surface-enhanced hyper-Raman scattering (SEHRS) probes (Fig. 4.3.1), improve the performance of the pH sensors even further and extend their operation range far into acidic values [18]. Learning about acidification in cells, that is, the pH of certain subcellular structures, has consequences for therapy of certain diseases, e. g., pH-responsive drug targeting, and also viral entry into cells. In tissues, pH regulation in the space between the cells (the interstitium) plays a major role, not only in many critical illnesses or in tumor microenvironments, but also during normal physiological operation, e. g., exercise-induced acidosis, brain function, or embryogenesis. SERS probes that are targeted to the interstitium of cells (Fig. 4.3.2b, red nanoprobes) can have such a function.

Other important parameters can be monitored as well by SERS reporters, such as H_2O_2, a reactive oxygen species (ROS) and second messenger in all animal cells. It can be quantified using 4-mercaptophenylboronic ester as reporter molecule [19]. Also hypochlorite (ClO), a ROS that is responsible for killing of pathogens in immune cells, and glutathione (GSH), an important antioxidant related to oxidative stress in cells and tissues, can be monitored with SERS probes, in this case by the reporter 4-mercaptophenol [20].

The anti-Stokes-to-Stokes signal ratio in a Raman spectrum is determined by the Boltzmann distribution, and therefore influenced by temperature (cf. Volume 1, Section 2). Nevertheless, in SERS, also processes other than temperature determine the anti-Stokes-to-Stokes signal ratio and can give rise to deviations from the expected Boltzmann distribution! Provided that those processes that disturb the temperature-dependent anti-Stokes-to-Stokes ratios are understood for the reporter molecule and

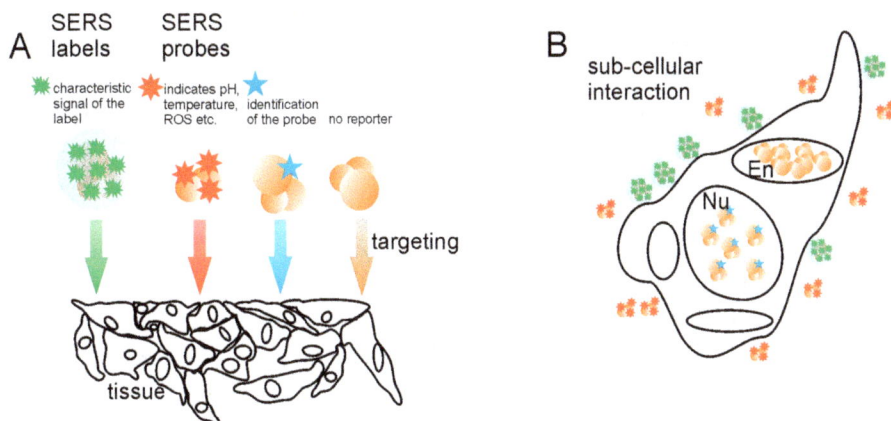

Figure 4.3.2: Molecular imaging in tissues using SERS labels and probes. (a) Schematic of SERS labels and different types of SERS probes that have been reported. Different reporter functions are indicated with star-shaped symbols of different colors. Some SERS probes do not carry a reporter. Targeting can be achieved using different kinds of targeting units, such as antibodies, nuclear localization sequences, oligonucleotides, or substrates of specific receptors. (b) Interaction in a tissue occurs at the microscopic, subcellular level, and the compartments of the cell, its surface, and the interstitium can be probed. In principle, due to the multiplexing capabilities, several different probes with different targeting can be applied simultaneously. Abbreviations: Nu, nucleus; En, endosome.

the plasmonic nanostructure of a particular SERS probe, temperature sensing with SERS probes is possible as well [21, 22].

Direct probing of the structure and composition of the intrinsic molecules in cells, *independent of any reporter*, provides completely new information on the structure and composition of cells and tissues. Of course, it requires an interaction of the bioenvironment with the plasmonic part of the nanoprobe. Probing of cellular biochemistry can be done with labels that carry very few reporter molecules [23] or that do not carry reporter molecules at all [24] (see Fig. 4.3.2 for schematics of both types of SERS probes). Images of the samples obtained by the "label" as discussed in the above paragraphs show the distribution of the SERS probes, but the spectra reveal the actual biochemical composition in the environment and the type of their interaction of the probes with the cell. For example, drugs that are transported as cargo by gold nanoparticles into cells can be followed using the drugs' SERS spectrum [25, 26]. Likewise, the biomolecules that reside on the surface of the nanostructures (termed corona) inside the living cell can be characterized over time [27], and the interaction of cultured cells with nanomaterials that have different surface properties can be investigated. As an example, the surface of silica nanostructures, important in many delivery applications, can be mimicked by core-shell structures with a plasmonic core that give SERS spectra of the different molecules interacting with their surface depending on the cell type and the associated route of uptake [28]. The cell cycle has been characterized by SERS probes targeted to the nuclei [25], and cellular transport pathways can be followed by following endosomes that carry SERS probes [29]. Furthermore, SERS nanoprobes can be used to image metabolically important species in living cells, such as adenosine monophosphate in macrophages [24], and even in whole microscopic animals, e. g., lipid storage granules in the intestine of the microscopic nematode *Caenorhabditis elegans* [30], or trehalose, a disaccharide that enables cryptobiosis in tardigrades [31]. When reporter molecules are present in a SERS probe as well, they can be helpful to identify different types of probes when present at the same time, for example in two different cellular compartments (Fig. 4.3.2, blue stars) [3]. In such experiments, multivariate data analysis helps to identify different probes and different cellular signatures.

4.3.3.3 Multimethod approach to understand SERS nanoprobes in cells

Relating SERS signals to ultrastructural information

SERS provides detailed information about the nano–bio interaction at the surface of the plasmonic nanoparticles in live cells. To elucidate the interactions of gold and silver nanoparticles with biomolecules and cells and to draw conclusions about the origin of the SERS signals, the localization of the nanoparticles inside the cell and the cellular structures they interact with must be analyzed. The combination of the SERS information on the molecular composition in the immediate environment of the nanostructures, e. g., of gold [32] and silver [27] nanoparticles, with information about the distribution and arrangement of the metal nanoparticles in cellular substructures allows for the deduction of transport mechanisms and molecular changes within the cell throughout the cell lifetime.

The standard method for the visualization of single nanoparticles in cellular substructures is transmission electron microscopy (TEM, cf. Volume 2, Chapter 6.5.1) due to its nanoscale resolution [33]. The contrast is based on the interaction of the biological material with the electron beam. Since electron transparent samples are required, ultramicrotome cuts with a few nanometer thickness of the several micrometers thick cells have to be prepared. Consequently, by TEM, 2D resolution can be obtained from animal cells after extensive sample preparation. Chemical fixation and staining of the cells are required and can lead to artifacts of the cell structure and the nanoparticles' localization [34]. While nanoparticles with a high electron density, such as gold and silver nanoparticles [33, 35], can be displayed in high contrast in the cellular ultrastructure [36], particles with a low electron density, for example, organic nanoparticles or carbon nanotubes, are often not visible in TEM images [37]. The uptake and localization of nanoparticles in cells [33, 36] or in tissue [35, 38] have been described in numerous studies. TEM images of gold nanoparticles in cells probe their size-dependent internalization by endocytosis and the localization of the nanomaterials in endosomal structures after incubation under certain cell culture conditions [33, 39]. Additional information can be obtained by combining TEM with energy dispersive X-ray spectroscopy (EDX) or electron energy loss spectroscopy (EELS), which reflect the distribution of elements in the biological sample. For example, in zebrafish embryos, the distribution of silver nanoparticles was detected in various organs with TEM and EDX [35].

In studies of Peckys et al. gold nanoparticles were visualized in living cells by liquid-scanning TEM (STEM) enabling single-nanoparticle resolution on fully hydrated, nonfixed, intact cells (Fig. 4.3.3a and b) [40]. Pristine cells were grown in a microfluidic chamber and incubated with 30-nm gold nanoparticles for 24 h. STEM images show particle aggregates distributed inside the cell (Fig. 4.3.3a). Blurred structures are located outside the focal plane, suggesting a 3D character of the aggregates. The cellular ultrastructure of whole animal cells cannot be visualized by liquid STEM due to a low contrast of the biological material in water, but the STEM images can be

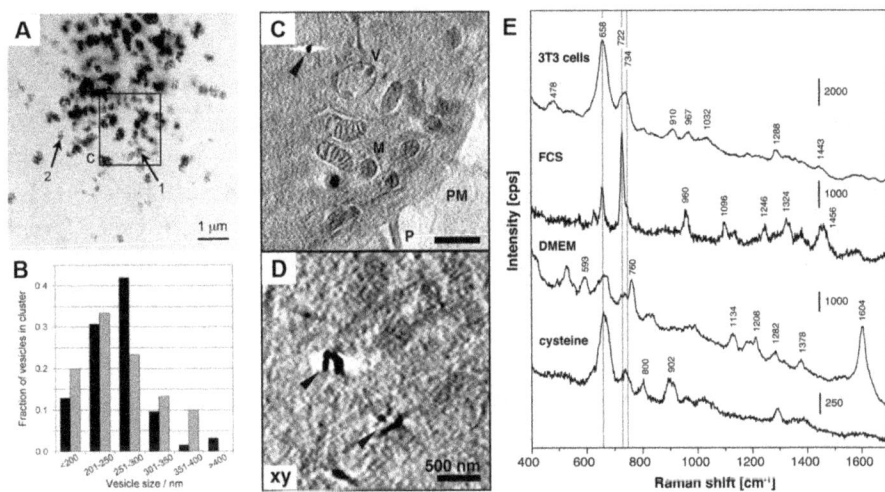

Figure 4.3.3: (a, b) Liquid-scanning transmission electron microscopy of live cells in a microfluidic chamber after 24 h incubation with gold nanoparticles. (a) Intracellular gold aggregates inside the cell. (b) Normalized size distribution histogram of vesicles containing gold particles for two different cells (gray and black bars). Reprinted with permission from [40]. Copyright © 2011, American Chemical Society. (c, d) Slices of a tomographic reconstruction of J774 macrophages after 3 h exposure to 2 pM silver nanoparticles in cell culture medium. (c) The particles are located in vesicular structures in the proximity of cell compartments. (d) The projection image in the xy-plane reveals the 2D linear arrangement in nanorings of the silver particles. XRM images were acquired with a 25-nm zone plate (9.8 nm pixel size). PM, plasma membrane; P, pseudopod; M, mitochondrion; V, vesicle. (e) SERS spectra of a 3T3 fibroblast and of cell culture relevant media (excitation wavelength 785 nm; acquisition time 1 s; laser intensity 8×10^{4} W cm^{-2}). Reprinted with permission from [27]. Copyright © 2013, The Royal Society of Chemistry.

used for quantitative analysis of particle uptake and spatial distribution (Fig. 4.3.3b). A total of 117 ± 9 and 164 ± 4 vesicles were counted in two individual cells with 57 ± 26 particles/vesicle.

In a combined SERS/TEM study of gold nanoprobes in an epithelial cell line we have shown that after endocytosis the aggregates inside the cell change their morphology over time, and also the plasmonic conditions should be altered because of this [24]. These changes can be explained by the varying biochemical surroundings and decreasing pH values in the vesicles during the aging (termed maturation) of the endosomes. Due to the importance of nanoaggregates for the electromagnetic enhancement [41, 42], their formation in the course of endosomal maturation should be monitored in the SERS experiments and complementary TEM studies. In addition to aggregation also adsorption of molecules on the particle surface, for example, reporter species, can change their properties, which may result in different uptake behavior of the cells [43, 44].

An alternative method to TEM that can visualize cellular substructures, in particular membranes as boundaries of cellular compartments, is nanoscale cryo soft X-ray tomography (cryo-SXT). Cryo-SXT is characterized by a high spatial resolution of a few nanometers and high penetration depth of X-rays, and is therefore suitable for the study of cells in their physiological environment. In cryo-SXT, whole snap-frozen cells of up to 10 μm thickness are visualized in the so-called X-ray water window at 510 eV photon energy employing the natural contrast due to the stronger absorption of soft X-rays in organic material or by nanoparticles as compared to the surrounding water [27, 45–47]. The acquisition of images at different tilt angles of the sample allows for 3D imaging of the ultrastructure of the intact frozen cells. Cryo-SXT has been utilized in numerous studies for the imaging of cells with subcellular resolution [27, 46, 47]. Due to the small number of synchrotron radiation sources, this method has so far been used only rarely compared to TEM or fluorescence microscopy for biomedical or toxicological problems worldwide. In addition to 3D tomography, X-ray microscopy (XM) allows for the quantification of the absorption and the chemical identification by X-ray fluorescence (XRF) and X-ray absorption [48, 49]. Graf et al. proved the presence of gold nanoparticle-based core-shell nanoparticles in skin tissue sections [50].

The distribution of individual nanoparticles or small aggregates, e. g., SERS probes, and the identification of the 3D aggregate morphology in subcellular structures were investigated by cryo-SXT and combined with SERS information (Fig. 4.3.3c–e) [27]. We showed that uptake and processing in complex biological systems differ for different nanomaterials due to different physicochemical properties. For silver nanoparticles, the X-ray tomograms prove the presence of 2D ring-like nanostructures in the 3D cellular ultrastructure (Fig. 4.3.3c and d). This can be correlated with the formation of a specific biomolecule corona in the culture medium suspension as investigated by SERS (Fig. 4.3.3e) [27]. From the comparison of the intracellular SERS signals with spectra of important components of the cell incubation medium (Fig. 4.3.3e), it becomes obvious that cysteine from DMEM or cysteine-rich proteins from fetal calf serum interact with the particles. This is in contrast to SERS spectra of fibroblasts containing gold nanoparticles, which show considerable variation in surface composition [24], as well as in the intracellular arrangement of the particles. These studies demonstrate that ultrastructural information about nanoparticle distribution in the cellular architecture from nanoscale cryo-SXT or TEM can be efficiently combined with the chemical characterization of the nanoparticle surface obtained by SERS to characterize nano–bio interactions.

Combining SERS characteristics with spatially resolved nanoaggregate quantification

To understand the influence of the physicochemical properties of nanomaterials on the local distribution and the number of nanoparticles, the uptake of a nanomaterial must be quantified and the intracellular processing has to be determined. There are very few methods that allow spatially resolved localization of inorganic nanomaterials in a cell and give quantitative information. Optical methods such as dark-field microscopy [51], SERS [23, 29], and photothermal microscopy [52, 53] show only limited use for the quantification of nanoparticles in individual cells, as they are based on the plasmonic properties of the nanoparticles. These may change in nanoparticle uptake and processing in the biological system and vary for different cell lines [24].

Inductively coupled plasma mass spectrometry (ICP-MS) is a well-established method to determine the metal content per cell [54, 55]. Inorganic nanoparticles can be quantified by elemental analysis in particular by ICP-MS or ICP optical emission spectrometry (ICP-OES) after extraction or digestion of a cell suspension [33, 56]. In recent years, alternative analysis techniques have been developed which enable the spatial information for bioimaging, mainly to visualize elemental and molecular distributions in tissue sections and cells [57]. In addition to mass spectrometric techniques such as secondary ion mass spectrometry (SIMS) [58] and matrix-assisted laser desorption–ionization mass spectrometry (MALDI-MS) [59], methods based on X-ray radiation such as proton-induced X-ray emission (PIXE) [60], XRF [61], and scanning electron microscopy with energy dispersive X-ray analysis (SEM-EDX) [62] are employed.

A technique that offers several advantages for the imaging of elements in tissue sections and single cells is laser ablation with ICP-MS (LA-ICP-MS), as it combines a simple sample preparation, multielement detection with high sensitivity and high spatial resolution [54, 63–65]. This method is mainly applied in medical diagnostics [66, 67], but it was also utilized for spatially resolved bioimaging of the distribution of silver and gold nanoparticles and their quantification at the single-cell level [63]. Subcellular mapping enables visualization of nanoparticles in the cellular ultrastructure, showing accumulation in the perinuclear region with increasing incubation time.

The combination of LA-ICP-MS and SERS data can provide valuable insights into the interaction of metal nanoparticles with single cells to associate certain factors such as the number of nanoparticles per laser spot and the intracellular particle localization. Büchner et al. showed in pulse-chase experiments that the nanoparticle pathway from endocytic uptake through intracellular processing to cell division can be monitored for gold nanoparticles by combining SERS data with the quantitative information from LA-ICP-MS (Fig. 4.3.4) [32]. In a series of pulse-chase experiments, fibroblasts were incubated with gold nanoparticles for 30 min, followed by different chase times ranging from 15 min to 48 h in the absence of nanoparticles. The distribution of gold nanoparticles and their aggregates is followed in individual cells by LA-ICP-MS micromapping (Fig. 4.3.4a–c). As evidenced by these data, the nanoaggregates accumulate close to the nucleus of the cells (Fig. 4.3.4a and b) and re-organize during mitosis (Fig. 4.3.4c). To relate the information about their localization at the micron scale with the molecular composition of the nanoparticle corona, SERS spectra were recorded after different chase times. They indicate that the interaction with

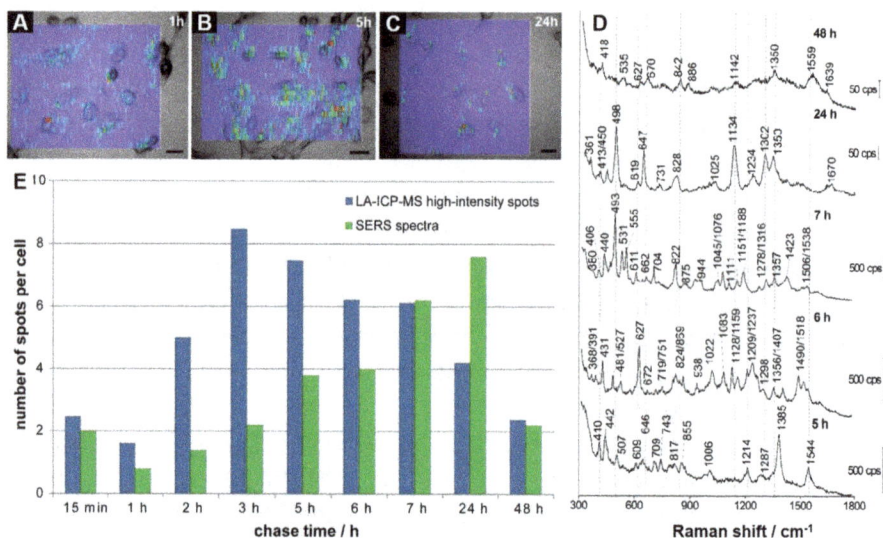

Figure 4.3.4: Combination of SERS information with LA-ICP-MS data of fibroblasts exposed to 1 nM gold nanoparticles with a size of 14 nm in a pulse-chase regime. An incubation pulse of 30 min and chase times ranging from 15 min to 48 h in cell culture medium in the absence of nanoparticles were applied. (a–c) Bright-field micrographs of fixed cells superimposed with the corresponding LA-ICP-MS images of the $^{197}Au^+$ intensity distribution. Scale bars represent 25 microns. Parameters: Laser spot size 8 μm, scan speed 8 μm/s, repetition rate 5 Hz, pixel size 1.5 × 7 μm, fluence 0.4 J cm^{-2}. (d) SERS spectra obtained from fibroblasts after incubation with gold nanoparticles for 30 min followed by different chase times (5 to 48 h). Excitation wavelength 785 nm, excitation intensity 1.9 × 10^5 W cm^{-2}, acquisition time 1 s. (e) Comparison of the number of high-intensity spots (>50,000 cps) determined by LA-ICP-MS and the number of SERS spectra per fibroblast. Reprinted with permission from [32]. Copyright © 2014, Springer.

some biomolecules, mainly proteins, persists throughout the endosomal matura-tion process (Fig. 4.3.4d). By combination of this complementary information, that is, of the SERS characteristics with spatially resolved nanoaggregate quantification (Fig. 4.3.4e), we are able to assess the efficiency of the gold nanoaggregates to act as SERS nanoprobes. The results suggest that the high SERS signals are neither directly related to the number of particles nor to high local nanoparticle densities (compare green and blue bars in Fig. 4.3.4e). Rather, we can conclude from the data that aggre-gate geometry and interparticle distances in the cells must change during intracellular processing. Combining information from complementary methods, such as SERS and LA-ICP-MS, is helpful to evaluate different plasmonic nanostructures in order to act as efficient SERS nanoprobes for bioanalytical applications.

4.3.3.4 Multifunctional plasmonic probes to understand nanoparticle–bio interactions

Silica-coated plasmonic nanoparticles

Silica nanoparticles with a plasmonic core are interesting systems for many applica-tions as they combine physical and chemical properties of two different materials. The plasmonic core can serve as an optical sensor, and the coating with silica changes the surface properties and enables the environment to interact with the silica material. By adding a gold core to silica nanoparticles, spectral information of the interaction of silica surfaces with their surrounding can be analyzed by SERS. These hybrid probes open a new field of spectral methods to study metal oxide–bio interactions.

Understanding the interactions between silica nanoparticles and animal cells is very important, especially in the context of nanosafety and nanomedicine. We have shown that the molecular interactions at the nano–bio interface of silica nanopar-ticles can be investigated by SERS (Fig. 4.3.5), employing silica nanoparticles with gold or silver core, termed BrightSilica [28]. To ensure that the cells interact exclu-sively with the silica surface, a complete and homogeneous coating of the gold and silver nanoparticles is required. Based on the beneficial properties of the plasmonic metal core, BrightSilica nanoparticles can be investigated inside animal cells by three complementary methods (Fig. 4.3.6): (i) By SERS, intrinsic information about the molecules interacting with the silica surface and penetrating into the silica shell can be obtained. Using BrightSilica as SERS substrate requires a silica shell thickness of very few nanometers, since the electromagnetic field decreases greatly with the distance to the metal surface. (ii) LA-ICP-MS is used to analyze the quantitative up-take and distribution of the silica-coated nanoparticles in single cells. The $^{107}Ag^+$ and $^{197}Au^+$ intensities can be correlated with the number of nanoparticles per cell. (iii) Cryo-SXT gives 3D information about the arrangement of BrightSilica nanoparticles in

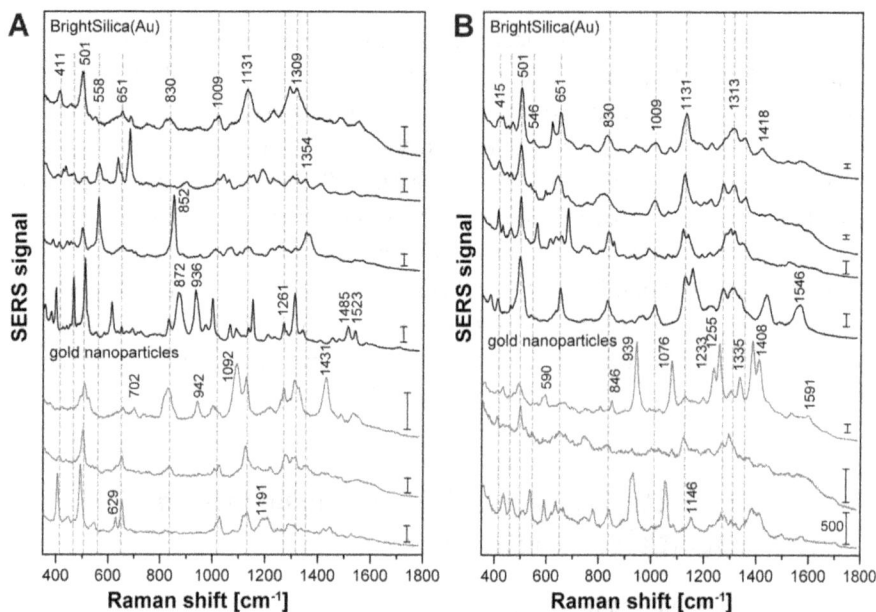

Figure 4.3.5: SERS spectra of 3T3 fibroblasts (a) and J774 macrophages (b) exposed to BrightSilica(Au) (black) and gold nanoparticles (gray) for 24 h. Representative spectra of three individual cells are shown for each incubation condition. Scale bars: 1000 cps. Excitation wavelength 785 nm, accumulation time 1 s, intensity 1.9×10^5 W cm^{-2}. Adapted with permission from Drescher et al. [28].

the cellular ultrastructure due to the high contrast of the metal cores in the biological environment.

The SERS data show that the hybrid silica particles interact with amino acid side chains of proteins adsorbed to their surface (Fig. 4.3.5). Based on synchrotron cryo-XT (Fig. 4.3.6d and h) the high SERS signal intensities can be explained by agglomeration of the plasmonic BrightSilica nanoparticles in the endosomal system. Inside the endosomes, the homogeneous silica layer enables the coupling of the localized surface plasmons of the metal cores. BrightSilica(Ag) is less suitable for in situ studies than BrightSilica(Au), since extracellular molecules from culture medium irreversibly bind to the silver surface, which has implications for systems that use silica-coated plasmonic nanostructures as SERS sensors. The 3D reconstruction of the X-ray microscopic images proves the stability of the silica layers under physiological conditions. The quantification of the BrightSilica nanoparticles using LA-ICP-MS indicates a different interaction of silica-coated nanoparticles compared to gold nanoparticles under the same experimental conditions (Fig. 4.3.6b and f). The metal core enables the investigation of the chemical environment, precise 3D ultrastructural imaging, and a quantitative comparison regarding uptake rates and accumulation behavior for different materials at the particle surface. These findings have implications for understand-

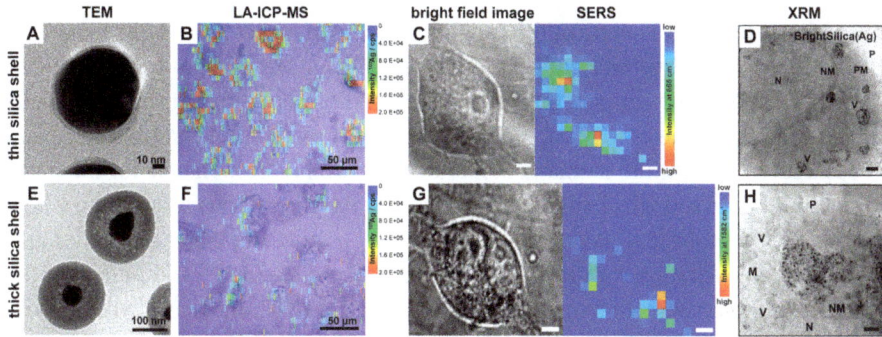

Figure 4.3.6: In situ characterization of particle–cell interaction using silica-coated silver nanoparticles. Transmission electron micrographs (TEM) reveal a thin (a) and thick silica shell (e) around silver nanoparticles. (b, f) Bright-field micrographs superimposed with LA-ICP-MS images of an $^{107}Ag^+$ intensity distribution of fixed fibroblasts after exposure to core-shell nanoparticles. Parameters: laser spot size 8 μm, scan speed 8 μm/s, repetition rate 5 Hz, fluence 0.6 J cm^{-2}. (c, g) SERS chemical maps and the corresponding bright-field images of fibroblasts incubated with silica-coated silver nanoparticles (scale bars represent 4 μm). The intensity distributions of the SERS signal at 665 cm^{-1} (ν(C-S) of cysteine) and at 1582 cm^{-1} (ν(C-C) of *para*-aminothiophenol) are displayed. Excitation wavelength 785 nm, accumulation time 1 s, intensity 1.9×10^5 W cm^{-2}. (d, h) Nanoscale X-ray microscopic imaging of vitrified macrophages after incubation with silica-coated silver nanoparticles for 24 h. Scale bars: 1 μm. Abbreviations: N, nucleus; NM, nuclear membrane; M, mitochondrion; V, vesicle; PM, plasma membrane; P, pseudopod. Adapted with permission from Drescher et al. [28].

ing silica particles and other nanomaterials in the biological context and improve our knowledge of particle–cell interactions.

Magnetic SERS nanoprobes

SERS spectra of reporter or label molecules can be obtained from composite plasmonic nanostructures with magnetic properties [68–73]. Spectra of magnetic SERS labels were also employed to visualize their interaction with cells, e. g., when particles with plasmonic and magnetic properties bind to cell membranes [74–76], or when they enter a cell [77, 78]. In bacteria, plasmonic-magnetic nanostructures were shown to provide SERS spectral information on the molecules in the cell walls of bacterial cells [79–81].

The interaction of composite magnetic-plasmonic nanoprobes (Ag-Magnetite and Au-Magnetite) inside animal cells, especially the molecular composition in the proximity of the composite structures, can be observed by SERS. The nanostructures serve as multifunctional probes of their endosomal environment in cells that can be manipulated, e. g., in microfluidic structures. Ag-Magnetite and Au-Magnetite can, for example, be prepared by linking magnetite particles with Ag and Au nanoparticles, respectively, using (3-aminopropyl) triethoxysilane (APTES) [82]. As was shown by our group, the absorbance spectra of the composite nanostructures show typical features of their respective silver and gold nanoparticles. The silver and gold nanoparticles are stabilized by their connection to the magnetite nanoparticles in the composite nanostructures. In the case of Au-Magnetite, a slight red shift of the plasmon band is observed. In cell culture media, a biomolecular corona – responsible for the particle stability in physiological media during cellular uptake – forms around the whole composite nanoprobe [83], in accord with observations made for magnetite nanoparticles [84]. At an excitation wavelength of 785 nm, both Ag-Magnetite and Au-Magnetite show SERS enhancement factors on the order of 10^3 to 10^4, in accord with the observation that many of the metal nanoparticles do not form aggregates, but are kept separate from one another within their respective nanocomposite particles.

An XTT cytotoxicity test demonstrated no toxicity effects of the composite nanoparticles in a 3T3 cell line [83] similar to gold and silver nanostructures [3, 27]. In contrast, similar Au-Magnetite composite particles were observed to be toxic for bacteria growth [85]. Based on their biocompatibility in cell culture experiments it was concluded that Ag-Magnetite and Au-Magnetite can be used as SERS nanoprobes in animal cells. Due to the presence of the magnetic nanostructures in fibroblasts incubated with Ag-Magnetite and Au-Magnetite for 24 h it was possible to manipulate the cells in an external magnetic field. For both composite structures, displacement in the magnetic field was mainly observed for single cells and also for groups of cells. Such magnet-induced motion can be applied in microfluidic channels or for applications in magnetic cell separations, as well as in microfluidics.

As was revealed by cryo soft-XRT data, the composite nanoparticles were contained in endosomes inside the cells, indicating their endocytic uptake. The nanostructures were stable even in the harsh environment of the late endosomes and lysosomes, suggesting their application as versatile optical and magnetic probes in the characterization of live cells. The high stability was specifically supported by the SERS data. The spectra provided information about the composition of the biomolecules at the nanoparticle surface. From the SERS spectra it could be inferred that the surface composition of the Ag-Magnetite and Au-Magnetite was different from that of pure gold or silver nanoparticles and was influenced by the interaction of cellular biomolecules with the magnetite parts of the nanoprobes [83]. LA-ICP-MS micromapping of intact cells and ICP-MS experiments on cellular extracts suggested that the endosomal uptake is determined by the magnetite component of the composite nanoprobes.

SERS substrates to probe the nucleus

Through specific modification of the particle surface, SERS probes can be developed that address specific cell compartments, such as the nucleus or the cytoplasm. For this purpose, the binding of specific peptide sequences (nuclear targeting sequences, NLSs) [86, 87] to the surface of the nanoparticles for the endosomal escape or the passage of the core pore complex is of great interest. Only the presence of plasmonic nanoparticles in the immediate vicinity of the molecules to be examined makes it possible to investigate the composition and processes within the nucleus or cytoplasm by SERS. Gold nanostructures that serve as probes for nanospectroscopic analysis of the nucleus can be obtained either by modification of the particle surface with certain recognition patterns or by the in situ reduction of tetrachloroauric acid ($HAuCl_4$). The latter approach will be discussed in the following paragraph.

A possibility to introduce SERS-active nanostructures into the cytoplasm or the nucleus is the in situ formation of gold nanoparticles inside the cells by incubation with $HAuCl_4$ [88, 89]. In a recent study it was shown that by applying a multimethod approach the formation process of in situ generated gold nanoparticles as well as specific biomolecule interactions can be followed [90]. Also here, the combination of SERS, TEM, and LA-ICP-MS provides complementary perspectives on plasmonic nanoparticles and nonplasmonic compounds inside the cells (Fig. 4.3.7). The SERS spectra allow on the one hand to draw conclusions about the biomolecular species involved in the in situ nanoparticle formation in fibroblasts and on the other hand to observe the interaction of the in situ formed structures with the surrounding molecules. In the case of in situ generated nanoparticles, SERS can also be used as a probe of particle generation, as long as nanoparticles with significant enhancement factors are formed.

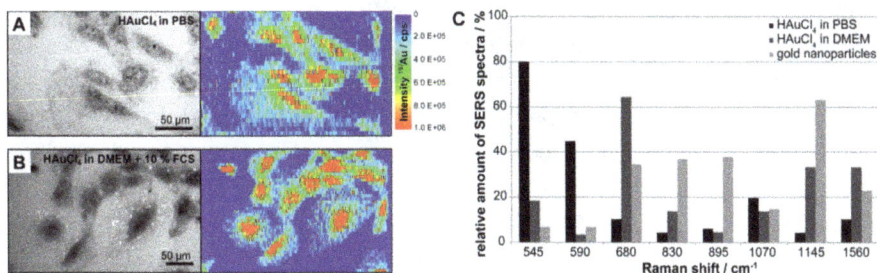

Figure 4.3.7: (a, b) Bright field images and the corresponding intensity distribution of $^{197}Au^+$ in single fibroblasts after 24 h exposure to 1 mM $HAuCl_4$ (a) in PBS and (b) in DMEM with 10 % fetal calf serum determined by LA-ICP-MS. Parameters: laser spot size 4 µm, line distance 6 µm, scan speed 5 µm/s, frequency 10 Hz, pixel size $6 \times 1 µm^2$, fluence 0.7 J/cm^2. (c) Relative amount of SERS spectra with a signal at the given spectral positions as determined from fibroblasts after 24 h exposure to 1 mM $HAuCl_4$ in PBS (black bars) and in DMEM (dark gray bars) and with synthesized gold nanoparticles (light gray bars). Adapted with permission from Drescher et al. [90].

TEM images essentially prove the formation of particles inside the cell and their localization within the cellular ultrastructure (vesicles, cytosol, nucleus). The variation in particle size is correlated with the respective biomolecules that act as reducing agent and naturally vary within the surrounding cellular environment during cellular processing. The SERS and TEM data prove the influence of the biosynthesis conditions on the surface composition, optical properties, size, and subcellular localization of the in situ generated gold nanostructures.

Before application of gold nanoparticles that are generated in situ as optical probes inside animal cells, it is crucial to identify the biomolecular species linked to their formation, as the biomolecular corona will determine the processing of the nanostructures inside the cells and their physical properties. This effect has been investigated in numerous studies in the area of nanotoxicology carried out with gold nanoparticles synthesized in the absence of biomaterials and later incubation of cells [56, 91]. Comparison with data obtained from ready-made gold nanoparticles suggests complementary application of in situ and ex situ generated nanostructures for optical probing. Both strategies proved feasible for probing cellular systems and address different cellular substructures for bioanalytical and nanobiophotonic investigations [24, 27, 28].

4.3.3.5 Multiphoton vibrational probing in complex biological systems

SEHRS (Fig. 4.3.1c) is the two-photon-excited analog of SERS that was proposed very early after the discovery of SERS [92, 93]. The nonlinear process of hyper-Raman scattering (HRS) is enhanced much more than linear Raman spectroscopy is enhanced in SERS. Effective SEHRS cross-sections up to the order of 10^{-45} cm^4 s that were reported [94] are similar or higher than those of two-photon-excited fluorescence. SEHRS can address excellently the need of spectroscopic applications regarding molecular structural sensitivity, detection sensitivity, and spectroscopic imaging, making it ideal for probing complex microstructured bio-organic materials [95]. SEHRS is excited by two photons from the same laser (Fig. 4.3.1c) and gives information about the entire vibrational spectrum. SEHRS presents several advantages, mostly relying on the multiphoton excitation, which other vibrational spectroscopies do not have. Specifically, HRS has other selection rules than Raman scattering, and therefore offers complementary vibrational information, including infrared (IR)-active modes or even additional so-called silent modes, which are seen neither in Raman nor in IR spectra. In a SEHRS experiment, the excitation with light in the NIR, typically convenient for biological samples, is combined with the desirable detection in the visible spectral range. Besides, two-photon excitation is favorable for microscopic applications due to the increased penetration depth and limited probed volume [96], resulting in an improved resolution for imaging. Last but not least, SEHRS has the potential to provide better insight into the structure and interaction of molecules on surfaces, as it is much more sensitive than SERS with respect to surface environmental changes [18, 97–101]. It is therefore a very good complementary tool to characterize SERS nanoprobes and to construct hybrid probes that can be used for both one-photon and two-photon vibrational probing also in cells and tissues [94, 102].

After the first SEHRS spectra from cells were reported by Kneipp et al. [94] and the potential of SEHRS for pH probing inside cells using the SEHRS spectrum of a pMBA reporter molecule was proposed [16], only recently SEHRS-based hyperspectral imaging in live macrophages was demonstrated [102]. In these experiments, macrophages were incubated with nonresonant SEHRS labels that carried the reporter molecules 2-NAT and pMBA, respectively. The spectra of both reporter molecules were obtained in the cell culture medium and inside intracellular vesicles after phagocytosis in live cells. Chemical mapping of both reporter signatures indicated an accumulation of the SEHRS labels in the region close to the nucleus, suggesting their accumulation in phagosomes and lysosomes (Fig. 4.3.8). The SEHRS spectra of pMBA that were used as pH meter verified the localization in lysosomes close to the nuclei in chemical pH maps. The SEHRS spectra of pMBA from the cellular interior resemble those in solution and are in agreement with pH calibration in simpler systems [18]. The two-photon-excited SEHRS spectrum of 2-NAT obtained inside and outside of cells strongly resembles the one-photon-excited SERS spectrum of the molecule.

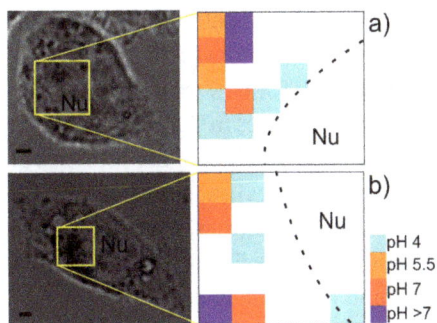

Figure 4.3.8: pH maps (examples) of two J774 macrophages containing pMBA SEHRS labels that act as intraendosomal pH nanosensors. Insets in the bright-field micrographs (left) show the regions of SEHRS mapping. The pH values in each pixel are average values, obtained from calibration curves that use the relative intensity of the two pH-dependent SEHRS bands at 1076 and 1585 cm^{-1}. Pixels without signal or too poor signal-to-noise ratio are white. No probes were localized in the nucleus (marked with dotted lines). Step width 2 microns, scale bars 4 microns. Nu, nucleus. Reprinted with permission from Heiner et al. [102].

Imaging approaches based on SEHRS have similar or higher sensitivity than two-photon fluorescence [94]. The SEHRS spectra that were used for hyperspectral imaging in the example shown in Fig. 4.3.8 were excited off-resonance with electronic transitions in the reporter molecules. This puts SEHRS hyperspectral mapping of cellular biomolecules within reach, as the SEHRS cross-sections of the reporters used here

[18] are very similar to those of typical biomolecules such as nucleobases [94, 100] or amino acids [101].

As a first application, monitoring of drug molecules in cells could in principle be envisioned. Tricyclic antidepressant (TCA) molecules are drugs that have different modes of action and that can be used for therapy of a variety of diseases including tumors. They act on enzymes in the endolysosomal system, including enzymes that are involved in the metabolism of lipids. The SEHRS and SERS spectra of several TCA molecules during their interaction with biocompatible gold nanostructures and with silver nanostructures were reported [103]. Many bands in the SEHRS spectra of the TCA molecules have been known from IR spectra of the molecules, which is in good agreement with the selection rules that govern the hyper-Raman process. Therefore, it is useful to combine SEHRS with SERS in order to get a full picture of the interaction of the drug molecules with silver and gold nanostructures. Due to the local information from the nanostructures' surface, the spectra can also be obtained in cell culture media. Specifically with the use of gold nanoparticles, this is an attractive path to study them in live cells as well. From the SEHRS spectra excited at 1064 nm and SERS spectra obtained at two different excitation wavelengths (532 nm and 785 nm), it could be interpreted that the molecules interact with the silver nanostructures mainly via their ring moiety and less intensely with the alkyl chain. Thereby the SEHRS data corroborated previous SERS findings [104]. The interaction with biocompatible gold, more important in drug delivery, is different. There, the methyl-aminopropyl side chain plays a very important role, together with parts of the ring system. The NIR-excited SERS spectra of the TCA-gold are greatly invariant with respect to changes in TCA concentration and size of the gold nanoparticles. They show remarkable stability in the presence of cell culture media and upon decrease of the pH in the typical ranges of pH values, so they would be stable in late endosomal and lysosomal structures, a region where a typical cell would move them, and where drug action happens. These findings suggest that both SEHRS and SERS should be combined and can thereby help to design and optimize new drug delivery platforms.

4.3.4 Outlook and challenges: from cultured cells to live animals

SERS probes and labels offer several advantages over other optical probes with respect to versatility, sensitivity and selectivity, and biocompatibility. Most exciting current practical applications of SERS nanoprobing certainly seem to occur for advanced sensing and imaging in the biomedical field.

Future challenges include the development of multifunctional SERS probes with optimized plasmonic nanostructures as basic building blocks, and their targeting in whole organisms. Plasmonic structures as key components of SERS probes can offer high local optical fields for both sensitive diagnostic probing with SERS and efficient therapeutic tools, e. g., SERS in combination with plasmon-supported, improved light-based therapies [105].

Particularly for applications in biological objects, probes based on two-photon excitation benefit from excitation at longer wavelengths in the NIR. Several two-photon optical probe methods are used in biomedical imaging, among them second harmonic generation and two-photon fluorescence. As discussed above, the concept of SERS probes and labels can also be extended to two-photon excitation using SEHRS [94]. With effective cross-sections that can be higher than typical two-photon cross-sections of fluorophores [94], SEHRS labels and probes have generated spectra from cultured cells [94, 102] and are well suited for applications in tissues. Due to selection rules different from those of SERS, they deliver additional chemical information and display improved selectivity and specificity.

Inside living cells or in tissues, the spectra measured with SERS probes give information on biological structures and processes from the molecular perspective, with extremely high lateral resolution from nanometer volumes. Since the nanoscopic SERS probes and labels are used to interrogate microscopic (cells and tissues) or macroscopic structures (whole organs or animals), depending on their dose and distribution and the detection modality, they can function as either individual nanosensors or ensembles that provide an "average signal." Especially in macroscopic imaging, where heterogeneity at the microscopic level due to histological or cellular substructures is not evident, the performance of a SERS label will be determined by the selectivity of its targeting unit. Understanding the interaction of SERS probes and labels with their environment at the cellular level will remain indispensable for interpreting spectral signatures and understanding their dynamics. This is also true for microscopic sensing and imaging: Interpreting SERS images or spectra that come from SERS probes inside individual cells requires an understanding of their interaction at the level of the cellular ultrastructure.

Microscopic examination of tissues parallel to in vivo probing is crucial for validation, but even more for understanding the interaction of the probes at the tissue and cellular level and for optimization of probe targeting. Combinations of Raman imaging with other microscopies, e. g., dark-field microscopy [25], or spatially resolved LA-ICP-MS imaging, the latter capable of absolute quantification of plasmonic nanostructures in cells [28], have been proposed to achieve this. Furthermore, high-resolution X-ray nanotomography can give precise information about the localization of SERS nanoprobes with respect to cellular organelles (Fig. 4.3.6) [27].

For basic investigations of SERS probes and labels in vivo, very small nonvertebrate model organisms such as the small nematode *Caenorhabditis elegans* could serve as interesting, dynamic samples, providing different tissue types, as well as uptake and processing pathways [30]. The penetration depth of the excitation and the scattering light can include the whole animal. The Raman microscopic experiments enable retrieval of both histological information and data on subcellular localization in the same experiment and provide good insight into the possible routing of nanoprobes and labels.

For studies of larger animals, for example small rodents in the laboratory, Raman imaging of the whole organism requires several different types of experiments. While confocal approaches can be applied for subcutaneous regions [10] or in ex vivo measurements [106], spatially offset Raman scattering (SORS) is very promising for probing deeper layers of tissue in the body [107] and has been combined with SERS in a number of experiments (SESORS) already, ranging from the recovery of SERRS signals from samples on the order of 45–50 mm thickness [108] and transcutaneous in vivo glucose detection by implanted sensors in living rats [109] to the quantification of diluted SERS labels through bone [110]. As demonstrated recently for porcine skin samples, SESORS in principle enables temperature measurements by subsurface SERS probes in turbid media [21].

Utilizing the progress that has been made in the design of SE(R)RS labels and probes, the application of SERS probes is about to revolutionize bioanalysis and future theranostic applications. The first proofs-of-principle are very promising and will pave the way for subsurface, noninvasive SERS monitoring of cell and tissue parameters with exciting opportunities for in vivo molecular imaging.

Bibliography

[1] Kneipp J, Kneipp H, Kneipp K. SERS-a single-molecule and nanoscale tool for bioanalytics. Chem Soc Rev. 2008;37:1052–60.

[2] Doering WE, Piotti ME, Natan MJ, Freeman RG. SERS as a foundation for nanoscale, optically detected biological labels. Adv Mater. 2007;19:3100–8.

[3] Matschulat A, Drescher D, Kneipp J. Surface-enhanced Raman scattering hybrid nanoprobe multiplexing and imaging in biological systems. ACS Nano. 2010;4:3259–69.

[4] Ni J, Lipert RJ, Dawson GB, Porter MD. Anal Chem. 1999;71:4903–8.

[5] Cao YWC, Jin RC, Mirkin CA. Nanoparticles with Raman spectroscopic fingerprints for DNA and RNA detection. Science. 2002;297:1536–40.

[6] Cao YC, Jin RC, Nam JM, Thaxton CS, Mirkin CA. Raman dye-labeled nanoparticle probes for proteins. J Am Chem Soc. 2003;125:14676–7.

[7] Grubisha DS, Lipert RJ, Park HY, Driskell J, Porter MD. Femtomolar detection of prostate-specific antigen: an immunoassay based on surface-enhanced Raman scattering and immunogold labels. Anal Chem. 2003;75:5936–43.

[8] Schlücker S, Küstner B, Punge A, Bonfig R, Marx A, Ströbel P. Immuno-Raman microspectroscopy: in situ detection of antigens in tissue specimens by surface-enhanced Raman scattering. J Raman Spectrosc. 2006;37:719–21.

[9] Pallaoro A, Braun GB, Moskovits M. Quantitative ratiometric discrimination between noncancerous and cancerous prostate cells based on neuropilin-1 overexpression. Proc Natl Acad Sci USA. 2011;108:16559–64.

[10] Qian XM, Peng XH, Ansari DO, Yin-Goen Q, Chen GZ, Shin DM, Yang L, Young AN, Wang MD, Nie SM. In vivo tumor targeting and spectroscopic detection with surface-enhanced Raman nanoparticle tags. Nat Biotechnol. 2008;26:83–90.

[11] Keren S, Zavaleta C, Cheng Z, de la Zerda A, Gheysens O, Gambhir SS. Noninvasive molecular imaging of small living subjects using Raman spectroscopy. Proc Natl Acad Sci USA. 2008;105:5844–9.

[12] Shaw CP, Fan MK, Lane C, Barry G, Jirasek AI, Brolo AG. Statistical correlation between SERS intensity and nanoparticle cluster size. J Phys Chem C. 2013;117:16596–605.

[13] Kang SY, Wang Y, Reder NP, Liu JTC. Multiplexed molecular imaging of biomarker-targeted SERS nanoparticles on fresh tissue specimens with channel-compressed spectrometry. PLoS ONE. 2016;11.

[14] Fasolato C, Giantulli S, Silvestri I, Mazzarda F, Toumia Y, Ripanti F, Mura F, Luongo F, Costantini F, Bordi F, Postorino P, Domenici F. Folate-based single cell screening using surface enhanced Raman microimaging. Nanoscale. 2016;8:17304–13.

[15] Kneipp K, Wang Y, Dasari RR, Feld MS. Approach to single molecule detection using surface-enhanced resonance Raman scattering (SERRS)- A study using rhodamine 6G on colloidal silver. Appl Spectrosc. 1995;49:780–4.

[16] Kneipp J, Kneipp H, Wittig B, Kneipp K. One- and two-photon excited optical pH probing for cells using surface-enhanced Raman and hyper-Raman nanosensors. Nano Lett. 2007;7:2819–23.

[17] Michota A, Bukowska J. Surface-enhanced Raman scattering (SERS) of 4-mercaptobenzoic acid on silver and gold substrates. J Raman Spectrosc. 2003;34:21–5.

[18] Guhlke M, Heiner Z, Kneipp J. Combined near-infrared excited SEHRS and SERS spectra of pH sensors using silver nanostructures. Phys Chem Chem Phys. 2015;17:26093–100.

[19] Peng RY, Si YM, Deng T, Zheng J, Li JS, Yang RH, Tan WH. A novel SERS nanoprobe for the ratiometric imaging of hydrogen peroxide in living cells. Chem Commun. 2016;52:8553–6.

[20] Wang WK, Zhang LM, Li L, Tian Y. A single nanoprobe for ratiometric imaging and biosensing of hypochlorite and glutathione in live cells using surface-enhanced Raman scattering. Anal Chem. 2016;88:9518–23.

[21] Gardner B, Stone N, Matousek P. Non-invasive chemically specific measurement of subsurface temperature in biological tissues using surface-enhanced spatially offset Raman spectroscopy. Faraday Discuss. 2016;187:329–39.

[22] Pozzi EA, Zrimsek AB, Lethiec CM, Schatz GC, Hersam MC, Van Duyne RP. Evaluating single-molecule Stokes and anti-Stokes SERS for nanoscale thermometry. J Phys Chem C. 2015;119:21116–24.

[23] Kneipp J, Kneipp H, Rice WL, Kneipp K. Optical probes for biological applications based on surface-enhanced Raman scattering from indocyanine green on gold nanoparticles. Anal Chem. 2005;77:2381–5.

[24] Kneipp J, Kneipp H, McLaughlin M, Brown D, Kneipp K. In vivo molecular probing of cellular compartments with gold nanoparticles and nanoaggregates. Nano Lett. 2006;6:2225–31.

[25] Austin LA, Kang B, El-Sayed MA. Probing molecular cell event dynamics at the single-cell level with targeted plasmonic gold nanoparticles: A review. Nano Today. 2015;10:542–58.

[26] Meister K, Niesel J, Schatzschneider U, Metzler-Nolte N, Schmidt DA, Havenith M. Label-free imaging of metal-carbonyl complexes in live cells by Raman microspectroscopy. Angew Chem, Int Ed. 2010;49:3310–2.

[27] Drescher D, Guttmann P, Buchner T, Werner S, Laube G, Hornemann A, Tarek B, Schneider G, Kneipp J. Specific biomolecule corona is associated with ring-shaped organization of silver nanoparticles in cells. Nanoscale. 2013;5:9193–8.

[28] Drescher D, Zeise I, Traub H, Guttmann P, Seifert S, Buchner T, Jakubowski N, Schneider G, Kneipp J. In situ characterization of SiO_2 nanoparticle biointeractions using BrightSilica. Adv Funct Mater. 2014;24:3765–75.

[29] Ando J, Fujita K, Smith NI, Kawata S. Dynamic SERS imaging of cellular transport pathways with endocytosed gold nanoparticles. Nano Lett. 2011;11:5344–8.

[30]   Charan S, Chien F-C, Singh N, Kuo C-W, Chen P. Development of lipid targeting Raman probes for in vivo imaging of caenorhabditis elegans. Chem Eur J. 2011;17:5165–70.

[31]   Kneipp H, Mobjerg N, Jorgensen A, Bohr HG, Helix-Nielsen C, Kneipp J, Kneipp K. Surface enhanced Raman scattering on Tardigrada towards monitoring and imaging molecular structures in live cryptobiotic organisms. J Biophotonics. 2013;6:759–64.

[32]   Büchner T, Drescher D, Traub H, Schrade P, Bachmann S, Jakubowski N, Kneipp J. Relating surface-enhanced Raman scattering signals of cells to gold nanoparticle aggregation as determined by LA-ICP-MS micromapping. Anal Bioanal Chem. 2014;406:7003–14.

[33]   Chithrani BD, Ghazani AA, Chan WCW. Determining the size and shape dependence of gold nanoparticle uptake into mammalian cells. Nano Lett. 2006;6:662–8.

[34]   Chen S, Goode AE, Skepper JN, Thorley AJ, Seiffert JM, Chung KF, Tetley TD, Shaffer MSP, Ryan MP, Porter AE. Avoiding artefacts during electron microscopy of silver nanomaterials exposed to biological environments. J Microsc. 2016;261:157–66.

[35]   Asharani PV, Wu YL, Gong ZY, Valiyaveettil S. Toxicity of silver nanoparticles in zebrafish models. Nanotechnology. 2008;19:255102. 8pp.

[36]   Heller DA, Baik S, Eurell TE, Strano MS. Single-walled carbon nanotube spectroscopy in live cells: towards long-term labels and optical sensors. Adv Mater. 2005;17:2793–9.

[37]   Kempen PJ, Thakor AS, Zavaleta C, Gambhir SS, Sinclair R. A scanning transmission electron microscopy approach to analyzing large volumes of tissue to detect nanoparticles. Microsc Microanal. 2013;19:1290–7.

[38]   Jeong SH, Kim JH, Yi SM, Lee JP, Kim JH, Sohn KH, Park KL, Kim MK, Son SW. Assessment of penetration of quantum dots through in vitro and in vivo human skin using the human skin equivalent model and the tape stripping method. Biochem Biophys Res Commun. 2010;394:612–5.

[39]   Albanese A, Chan WCW. Effect of gold nanoparticle aggregation on cell uptake and toxicity. ACS Nano. 2011;5:5478–89.

[40]   Peckys DB, de Jonge N. Visualizing gold nanoparticle uptake in live cells with liquid scanning transmission electron microscopy. Nano Lett. 2011;11:1733–8.

[41]   Kneipp K, Wang Y, Kneipp H, Perelman LT, Itzkan I, Dasari R, Feld MS. Single molecule detection using surface-enhanced Raman scattering (SERS). Phys Rev Lett. 1997;78:1667–70.

[42]   Moskovits M. Surface-enhanced Raman spectroscopy: a brief retrospective. J Raman Spectrosc. 2005;36:485–96.

[43]   Gregas MK, Yan F, Scaffidi J, Wang HN, Vo-Dinh T. Characterization of nanoprobe uptake in single cells: spatial and temporal tracking via SERS labeling and modulation of surface charge. Nanomedicine. 2011;7:115–22.

[44]   Sirimuthu NMS, Syme CD, Cooper JM. Monitoring the uptake and redistribution of metal nanoparticles during cell culture using surface-enhanced Raman scattering spectroscopy. Anal Chem. 2010;82:7369–73.

[45]   McDermott G, Le Gros MA, Larabell CA. Visualizing cell architecture and molecular location using soft X-ray tomography and correlated cryo-light microscopy. Annu Rev Phys Chem. 2012;63(63):225–39.

[46]   Müller WG, Heymann JB, Nagashima K, Guttmann P, Werner S, Rehbein S, Schneider G, McNally JG. Towards an atlas of mammalian cell ultrastructure by cryo soft X-ray tomography. J Struct Biol. 2012;177:179–92.

[47]   Schneider G, Guttmann P, Heim S, Rehbein S, Mueller F, Nagashima K, Heymann JB, Müller WG, McNally JG. Three-dimensional cellular ultrastructure resolved by X-ray microscopy. Nat Methods. 2010;7:985–7.

[48]   Andrews JC, Meirer F, Liu YJ, Mester Z, Pianetta P. Transmission X-ray microscopy for full-field nano imaging of biomaterials. Microsc Res Tech. 2011;74:671–81.

[49]   Gilbert B, Fakra SC, Xia T, Pokhrel S, Mädler L, Nel AE. The fate of ZnO nanoparticles
        administered to human bronchial epithelial cells. ACS Nano. 2012;6:4921–30.
[50]   Graf C, Meinke M, Gao Q, Hadam S, Raabe J, Sterry W, Blume-Peytavi U, Lademann J, Rühl
        E, Vogt A. Qualitative detection of single submicron and nanoparticles in human skin by
        scanning transmission x-ray microscopy. J Biomed Opt. 2009;14:021015.
[51]   Aaron J, Travis K, Harrison N, Sokolov K. Dynamic imaging of molecular assemblies in live cells
        based on nanoparticle plasmon resonance coupling. Nano Lett. 2009;9:3612–8.
[52]   Huang X, Jain P, El-Sayed I, El-Sayed M. Plasmonic photothermal therapy (PPTT) using gold
        nanoparticles. Lasers Med Sci. 2008;23:217–28.
[53]   Nam J, Won N, Jin H, Chung H, Kim S. pH-induced aggregation of gold nanoparticles for
        photothermal cancer therapy. J Am Chem Soc. 2009;131:13639–45.
[54]   Becker JS, Matusch A, Palm C, Salber D, Morton KA, Becker S. Bioimaging of metals in brain
        tissue by laser ablation inductively coupled plasma mass spectrometry (LA-ICP-MS) and
        metallomics. Metallomics. 2010;2:104–11.
[55]   Krystek P, Ulrich A, Garcia CC, Manohar S, Ritsema R. Application of plasma spectrometry for
        the analysis of engineered nanoparticles in suspensions and products. J Anal At Spectrom.
        2011;26:1701–21.
[56]   Alkilany AM, Murphy CJ. Toxicity and cellular uptake of gold nanoparticles: what we have
        learned so far? J Nanopart Res. 2010;12:2313–33.
[57]   Qin ZY, Caruso JA, Lai B, Matusch A, Becker JS. Trace metal imaging with high spatial
        resolution: applications in biomedicine. Metallomics. 2011;3:28–37.
[58]   Hagenhoff B, Breitenstein D, Tallarek E, Mollers R, Niehuis E, Sperber M, Goricnik B, Wegener
        J. Detection of micro- and nano-particles in animal cells by ToF-SIMS 3D analysis. Surf
        Interface Anal. 2013;45:315–9.
[59]   Stoeckli M, Chaurand P, Hallahan DE, Caprioli RM. Imaging mass spectrometry: a new
        technology for the analysis of protein expression in mammalian tissues. Nat Med.
        2001;7:493–6.
[60]   Tkalec ŽP, Drobne D, Vogel-Mikuš K, Pongrac P, Regvar M, Štrus J, Pelicon P, Vavpetič P, Grlj N,
        Remškar M. Micro-PIXE study of Ag in digestive glands of a nano-Ag fed arthropod (Porcellio
        scaber, Isopoda, Crustacea). Nucl Instrum Methods Phys Res, Sect B, Beam Interact Mater
        Atoms. 2011;269:2286–91.
[61]   James SA, Feltis BN, de Jonge MD, Sridhar M, Kimpton JA, Altissimo M, Mayo S, Zheng
        CX, Hastings A, Howard DL, Paterson DJ, Wright PFA, Moorhead GF, Turney TW, Fu J.
        Quantification of ZnO nanoparticle uptake, distribution, and dissolution within individual
        human macrophages. ACS Nano. 2013;7:10621–35.
[62]   Zvyagin AV, Zhao X, Gierden A, Sanchez W, Ross JA, Roberts MS. Imaging of zinc oxide
        nanoparticle penetration in human skin in vitro and in vivo. J Biomed Opt. 2008;13.
[63]   Drescher D, Giesen C, Traub H, Panne U, Kneipp J, Jakubowski N. Quantitative imaging of
        gold and silver nanoparticles in single eukaryotic cells by laser ablation ICP-MS. Anal Chem.
        2012;84:9684–8.
[64]   Giesen C, Waentig L, Mairinger T, Drescher D, Kneipp J, Roos PH, Panne U, Jakubowski N.
        Iodine as an elemental marker for imaging of single cells and tissue sections by laser ablation
        inductively coupled plasma mass spectrometry. J Anal At Spectrom. 2011;26:2160–5.
[65]   Trouillon R, Passarelli MK, Wang J, Kurczy ME, Ewing AG. Chemical analysis of single cells.
        Anal Chem. 2013;85:522–42.
[66]   Becker JS, Zoriy M, Matusch A, Wu B, Salber D, Palm C, Becker JS. Bioimaging of metals by
        laser ablation inductively coupled plasma mass spectrometry (La-Icp-Ms). Mass Spectrom
        Rev. 2010;29:156–75.
[67]   Konz I, Fernandez B, Fernandez ML, Pereiro R, Sanz-Medel A. Laser ablation ICP-MS for
        quantitative biomedical applications. Anal Bioanal Chem. 2012;403:2113–25.

[68] Donnelly T, Smith WE, Faulds K, Graham D. Silver and magnetic nanoparticles for sensitive DNA detection by SERS. Chem Commun. 2014;50:12907–10.

[69] Gühlke M, Selve S, Kneipp J. Magnetic separation and SERS observation of analyte molecules on bifunctional silver/iron oxide composite nanostructures. J Raman Spectrosc. 2012;43:1204–7.

[70] Han XX, Schmidt AM, Marten G, Fischer A, Weidinger IM, Hildebrandt P. Magnetic silver hybrid nanoparticles for surface-enhanced resonance Raman spectroscopic detection and decontamination of small toxic molecules. ACS Nano. 2013;7:3212–20.

[71] La Porta A, Sanchez-Iglesias A, Altantzis T, Bals S, Grzelczak M, Liz-Marzan LM. Multifunctional self-assembled composite colloids and their application to SERS detection. Nanoscale. 2015;7:10377–81.

[72] Lim JK, Kim Y, Lee SY, Joo SW. Spectroscopic analysis of L-histidine adsorbed on gold and silver nanoparticle surfaces investigated by surface-enhanced Raman scattering. Spectrochim Acta, Mol Biomol Spectrosc. 2008;69:286–9.

[73] Spuch-Calvar M, Rodríguez-Lorenzo L, Morales MP, Álvarez-Puebla RA, Liz-Marzán LM. Bifunctional nanocomposites with long-term stability as SERS optical accumulators for ultrasensitive analysis. J Phys Chem C. 2009;113:3373–7.

[74] Jun B-H, Noh MS, Kim J, Kim G, Kang H, Kim M-S, Seo Y-T, Baek J, Kim J-H, Park J, Kim S, Kim Y-K, Hyeon T, Cho M-H, Jeong DH, Lee Y-S. Multifunctional silver-embedded magnetic nanoparticles as SERS nanoprobes and their applications. Small. 2010;6:119–25.

[75] Liu Y, Chang Z, Yuan H, Fales AM, Vo-Dinh T. Quintuple-modality (SERS-MRI-CT-TPL-PTT) plasmonic nanoprobe for theranostics. Nanoscale. 2013;5:12126–31.

[76] Noh MS, Jun B-H, Kim S, Kang H, Woo M-A, Minai-Tehrani A, Kim J-E, Kim J, Park J, Lim H-T. Magnetic surface-enhanced Raman spectroscopic (M-SERS) dots for the identification of bronchioalveolar stem cells in normal and lung cancer mice. Biomaterials. 2009;30:3915–25.

[77] Bertorelle F, Ceccarello M, Pinto M, Fracasso G, Badocco D, Amendola V, Pastore P, Colombatti M, Meneghetti M. Efficient AuFeOx nanoclusters of laser-ablated nanoparticles in water for cells guiding and surface-enhanced resonance Raman scattering imaging. J Phys Chem C. 2014;118:14534–41.

[78] Charan S, Kuo CW, Kuo Y-W, Singh N, Drake P, Lin Y-J, Tay L, Chen P. Synthesis of surface enhanced Raman scattering active magnetic nanoparticles for cell labeling and sorting. J Appl Phys. 2009;105:07B310.

[79] Fan Z, Senapati D, Khan SA, Singh AK, Hamme A, Yust B, Sardar D, Ray PC. Popcorn-shaped magnetic core–plasmonic shell multifunctional nanoparticles for the targeted magnetic separation and enrichment, label-free SERS imaging, and photothermal destruction of multidrug-resistant bacteria. Chem Eur J. 2013;19:2839–47.

[80] Tamer U, Cetin D, Suludere Z, Boyaci IH, Temiz HT, Yegenoglu H, Daniel P, Dincer I, Elerman Y. Gold-coated iron composite nanospheres targeted the detection of Escherichia coli. Int J Mol Sci. 2013;14:6223–40.

[81] Zhang L, Xu J, Mi L, Gong H, Jiang S, Yu Q. Multifunctional magnetic–plasmonic nanoparticles for fast concentration and sensitive detection of bacteria using SERS. Biosens Bioelectron. 2012;31:130–6.

[82] Liang R-P, Yao G-H, Fan L-X, Qiu J-D. Magnetic Fe3O4@Au composite-enhanced surface plasmon resonance for ultrasensitive detection of magnetic nanoparticle-enriched α-fetoprotein. Anal Chim Acta. 2012;737:22–8.

[83] Buchner T, Drescher D, Merk V, Traub H, Guttmann P, Werner S, Jakubowski N, Schneider G, Kneipp J. Biomolecular environment, quantification, and intracellular interaction of multifunctional magnetic SERS nanoprobes. Analyst. 2016;141:5096–106.

[84] Mbeh DA, Javanbakht T, Tabet L, Merhi Y, Maghni K, Sacher E, Yahia LH. Protein corona formation on magnetite nanoparticles: effects of culture medium composition, and its consequences on superparamagnetic nanoparticle cytotoxicity. J Biomed Nanotechnol. 2015;11:828–40.

[85] Niemirowicz K, Swiecicka I, Wilczewska AZ, Misztalewska I, Kalska-Szostko B, Bienias K, Bucki R, Car H. Gold-functionalized magnetic nanoparticles restrict growth of Pseudomonas aeruginosa. Int J Nanomed. 2014;9:2217–24.

[86] Feldherr CM, Akin D, Cohen RJ. Regulation of functional nuclear pore size in fibroblasts. J Cell Sci. 2001;114:4621–7.

[87] Tkachenko AG, Xie H, Coleman D, Glomm W, Ryan J, Anderson MF, Franzen S, Feldheim DL. Multifunctional gold nanoparticle-peptide complexes for nuclear targeting. J Am Chem Soc. 2003;125:4700–1.

[88] Anshup, Venkataraman JS, Subramaniam C, Kumar RR, Priya S, Kumar TRS, Omkumar RV, John A, Pradeep T. Growth of gold nanoparticles in human cells. Langmuir. 2005;21:11562–7.

[89] Liu H, Dong CQ, Huang XY, Ren JC. Spatially resolved scattering correlation spectroscopy using a total internal reflection configuration. Anal Chem. 2012;84:3561–7.

[90] Drescher D, Traub H, Büchner T, Jakubowski N, Kneipp J. Properties of in situ generated gold nanoparticles in the cellular context. Nanoscale. 2017;9:11647–56.

[91] Khlebtsov N, Dykman L. Biodistribution and toxicity of engineered gold nanoparticles: a review of in vitro and in vivo studies. Chem Soc Rev. 2011;40:1647–71.

[92] Baranov AV, Bobovich YS. Super-enhanced hyper-Raman scattering from dyes adsorbed on colloidal silver particles. JETP Lett. 1982;36:339–43.

[93] Murphy DV, Vonraben KU, Chang RK, Dorain PB. Surface-enhanced hyper-Raman scattering from SO32- adsorbed on Ag powder. Chem Phys Lett. 1982;85:43–7.

[94] Kneipp J, Kneipp H, Kneipp K. Two-photon vibrational spectroscopy for biosciences based on surface-enhanced hyper-Raman scattering. Proc Natl Acad Sci USA. 2006;103:17149–53.

[95] Madzharova F, Heiner Z, Kneipp J. Surface enhanced hyper Raman scattering (SEHRS) and its applications. Chem Soc Rev. 2017;46:3980–99.

[96] Zipfel WR, Williams RM, Webb WW. Nonlinear magic: multiphoton microscopy in the biosciences. Nat Biotechnol. 2003;21:1369–77.

[97] Golab JT, Sprague JR, Carron KT, Schatz GC, Duyne RPV. A surface enhanced hyper-Raman scattering study of pyridine adsorbed onto silver: Experiment and theory. J Chem Phys. 1988;88:7942–51.

[98] Hulteen JC, Young MA, Van Duyne RP. Surface-enhanced hyper-Raman scattering (SEHRS) on Ag film over nanosphere (FON) electrodes: surface symmetry of centrosymmetric adsorbates. Langmuir. 2006;22:10354–64.

[99] Valley N, Jensen L, Autschbach J, Schatz GC. Theoretical studies of surface enhanced hyper-Raman spectroscopy: The chemical enhancement mechanism. J Chem Phys. 2010;133:054103.

[100] Madzharova F, Heiner Z, Guhlke M, Kneipp J. Surface-enhanced hyper-Raman spectra of adenine, guanine, cytosine, thymine, and uracil. J Phys Chem C. 2016;120:15415–23.

[101] Madzharova F, Heiner Z, Kneipp J. Surface enhanced hyper-Raman scattering of the amino acids tryptophan, histidine, phenylalanine, and tyrosine. J Phys Chem C. 2017;121:1235–42.

[102] Heiner Z, Guhlke M, Zivanovic V, Madzharova F, Kneipp J. Surface-enhanced hyper Raman hyperspectral imaging and probing in animal cells. Nanoscale. 2017;9:8024–32.

[103] Zivanovic V, Madzharova F, Heiner Z, Arenz C, Kneipp J. Specific interaction of tricyclic antidepressants with gold and silver nanostructures as revealed by combined one and two-photon vibrational spectroscopy. J Phys Chem C. 2017.

[104] Jaworska A, Malek K. A comparison between adsorption mechanism of tricyclic antidepressants on silver nanoparticles and binding modes on receptors. Surface-enhanced Raman spectroscopy studies. J Colloid Interface Sci. 2014;431:117–24.

[105] Ali MRK, Panikkanvalappil SR, El-Sayed MA. Enhancing the efficiency of gold nanoparticles treatment of cancer by increasing their rate of endocytosis and cell accumulation using rifampicin. J Am Chem Soc. 2014;136:4464–7.

[106] Oseledchyk A, Andreou C, Wall MA, Kircher MF. Folate-targeted surface-enhanced resonance Raman scattering nanoprobe ratiometry for detection of microscopic ovarian cancer. ACS Nano. 2016.

[107] Matousek P. Deep non-invasive Raman spectroscopy of living tissue and powders. Chem Soc Rev. 2007;36:1292–304.

[108] Stone N, Kerssens M, Lloyd GR, Faulds K, Graham D, Matousek P. Surface enhanced spatially offset Raman spectroscopic (SESORS) imaging – the next dimension. Chem Sci. 2011;2:776–80.

[109] Ma K, Yuen JM, Shah NC, Walsh JT, Glucksberg MR, Van Duyne RP. In vivo, transcutaneous glucose sensing using surface-enhanced spatially offset Raman spectroscopy: multiple rats, improved hypoglycemic accuracy, low incident power, and continuous monitoring for greater than 17 days. Anal Chem. 2011;83:9146–52.

[110] Sharma B, Ma K, Glucksberg MR, Van Duyne RP. Seeing through bone with surface-enhanced spatially offset Raman spectroscopy. J Am Chem Soc. 2013;135:17290–3.

The development of novel materials as well as the improvement and control of existing materials requires in-depth knowledge of all relevant material parameters. This includes their mechanical and chemical composition at the nanoscale, which becomes especially apparent in blended materials. Material science constantly strives to develop lighter, stronger, less expensive materials for improved energy efficiency, especially in today's increasingly energy-intensive world [1]. This can concern light-weight yet strong materials for transportation, energy-saving building materials, materials for energy storage, protective materials for electronic devices or the food sector, etc. [2]. Multifunctional materials are developed whose properties may be changed by external stimuli [3] or that perform several functionalities in one [4]. One direction that has emerged over the last two decades concerns 2D materials such as graphene, hexagonal boron nitride (h-BN), or transition metal dichalcogenides (TMDCs) that are taking lightness and strength to the limit, are extremely flexible and highly transparent, and can offer a broad variety of electronic and optical band-gaps and properties [5]. They have, e. g., been shown to qualify as advanced building blocks for extremely miniaturized optoelectronic devices [6]. The properties of these 2D sheets are strongly influenced by defects and impurities, which directly impact device properties, or can be expressly introduced to add functionality [7]. Again, high-resolution spectroscopic techniques that allow characterization all the way to single atomic defects or dopants are indispensable [8]. The same holds for analytical science to probe material characteristics. Suitable analytical methods are highly relevant for probing the authenticity of a product's provenance or of cultural goods, as well as for contamination detection [9, 10]. The lower the required sample volume, the more advantageous the approach proves for production chain monitoring or for probing sensitive unique materials such as artworks. In the present volume, the use of optical nanospectroscopy in the field of food science, packaging, provenance, and agriculture is illuminated. Spectroscopy on food, e. g., allows finding signatures of fermentation processes or spoilage or of contamination by pesticides, fungicides, or bacteria. In another example, applications of surface-enhanced Raman spectroscopy (SERS) for determining the age and provenance of paintings, textiles, or historical artifacts, e. g., based on pigment and dye evaluation, are described. The self-organization of cyanine dyes, which perform important functions, e. g., as biomolecular labels, is elucidated by correlative spectroscopy. Lastly, the properties and potential of the increasingly widespread 2D materials are discussed together with some optical nanospectroscopy techniques that are particularly well suited for their nanoscale characterization.

Bibliography

[1]   Next Generation Materials, Office of Energy Efficiency & Renewable Energy, U.S. Department of Energy, https://www.energy.gov/eere/amo/next-generation-materials.

https://doi.org/10.1515/9783110442908-014

[2] Materials Science and Engineering in Europe: Challenges and Opportunities. Materials Science and Engineering Expert Committee of the European Science Foundation. 2013. http://archives.esf.org/publications/materials-science-and-engineering.html.

[3] Special Issue: From responsive materials to interactive materials. Adv Mater. 2020;32(20).

[4] Nemat-Nasser S et al. Mulifunctional Materials. In: Bar-Cohen Y, editor. Biomimetics: Biologically Inspired Technologies. CRC Press; 2005.

[5] Avouris P, Heinz TF, Low T, editors. 2D Materials: Properties and Devices. Cambridge University Press; 2017.

[6] Tan T et al. 2D material optoelectronics for information functional device applications: status and challenges. Adv Sci. 2020;7(11):2000058.

[7] Liang Q et al. Defect engineering of two-dimensional transition-metal dichalcogenides: Applications, challenges, and opportunities. ACS Nano. 2021;15(2):2165–81.

[8] Kato R et al. Ultrastable tip-enhanced hyperspectral optical nanoimaging for defect analysis of large-sized WS_2 layers. Sci Adv. 2022;8(28):eabo4021.

[9] Hassoun A et al. Fraud in animal origin food products: advances in emerging spectroscopic detection methods over the past five years. Foods. 2020;9(8):1069.

[10] Zalaffi MS et al. Electrochemical and SERS sensors for cultural heritage diagnostics and conservation: recent advances and prospects. J Electrochem Soc. 2020;167(3):037548.

Enisa Omanović-Mikličanin and Amina Stambolić

5.1 Application of nanospectroscopy in food science and agriculture

5.1.1 Key messages

- Nanotechnology is an intriguing new topic of study that has the potential to meet a number of the urgent requirements in the sectors of food and agriculture by revealing new directions for the production, manufacturing, and packaging of food, the cultivation of crops, and nutrition of animals. Nanotechnology offers enormous promise in the food industry to improve both functionality and quality of food [1].
- One of the most important applications of nanotechnology in the sectors of food and agriculture is the development of detection systems on the nanoscale for the improvement of food safety.
- Nanospectroscopy has a great potential to overcome analytical problems and disadvantages of other analytical techniques and to become a very powerful alternative to other methods used in nano-based systems for the evaluation of food safety.

5.1.2 Pre-knowledge

The current chapter focuses on applying optical nanospectroscopic techniques in the field of food sciences and agriculture. Therefore, the reader needs to be familiar with fundamental processes in optical spectroscopy (Volume 1, Chapter 2.1), photoluminescence and fluorescence (Volume 1, Chapter 2.2), vibrational spectroscopies (Volume 1, Chapter 2.3), near-field-enhanced spectroscopies (Volume 1, Chapter 3.3), and nanomaterials for nanospectroscopy (Volume 2, Section 6).

Food quality control is performed using highly chemically specific methods such as vibrational spectroscopy, which is an energy-dependent method. It is based on either periodically changing the dipole moment (in **infrared (IR) spectroscopy**) or polarizability (in **Raman scattering**). The oscillations result from vibrations of molecular bonds or groups of atoms. In the corresponding spectra, the discrete energy transitions, respectively the corresponding frequency changes, during the absorption (IR) or scattering (Raman) of electromagnetic radiation are recorded [2].

In surface-enhanced Raman spectroscopy (SERS), molecules are adsorbed on rough metal surfaces or nanostructures, rendering it a highly surface-sensitive technique [3]. Enhancement factors of the Raman intensity on the order of 10^4 to 10^6 have been recorded and can be as high as 10^8 or even 10^{14} depending on the system [4]. Consequently, SERS is highly sensitive and surface-selective whereas Raman is not.

https://doi.org/10.1515/9783110442908-015

5.1.3 Importance of the application

Applications of nanotechnology in agriculture include the use of nanoagrochemicals (nanofertilizer, nanopesticides, and herbicides), nanobiosensors, and nanomaterials for processing, storing, packaging, and labeling food, enhancing food quality (e. g., by nutritional supplements or in nutritional drinks), preserving fruit, promoting plant growth, or improving crops (nanoparticle-mediated gene transfer). They can likewise be used as nanocarriers for delivering nitric oxide, for nano-enhanced agricultural wastewater biotreatment, etc. [5].

Globalization of food production in combination with consumer concerns with regard to food quality and safety gave rise to interconnected global systems handling the production and distribution of food [1]. These were followed by a significant increase in food standards. To implement such standards, it became necessary to abandon the practice of end-of-line product inspection and establish a new system in which quality control is implemented at every step along the chain of food production such that food safety is ensured and the requirements of both regulations and customers are met [6]. Such an environment ensures standardization of primary processes to mitigate risks and ideally reach zero-defect solutions. In consequence, upcoming assessment procedures for food quality and safety will need to feature additional characteristics. In particular they will require low limits of detection, high specificity, high sensitivity, and simple procedures for sample preparation, and allow for portable use by miniaturizing instrumentation [3, 5, 6].

Food safety raises the issue of detrimental biological and chemical factors. At the laboratory level, it is possible to quantitatively assess food safety using various customized techniques. These analytical methods require evaluation times that range from several hours to days. The preparation typically includes different pre-treatment steps [1]. Common analytical methods that are used to detect biological and chemical risk factors in the food and the agriculture sector with their main advantages and disadvantages are, e. g., the following [7, 3]:

- *Colorimetry* yields sensitive information, but can be difficult to handle.
- *Electrochemical impedance spectroscopy* (*EIS*) is an accurate and simple approach, for which however homogenous samples are required.
- The *enzyme-linked immunosorbent assay* (*ELISA*) offers high sensitivity, but is limited in its detection range.
- *Fluorescence* likewise allows for highly sensitive measurements, whereas signal degradation by photobleaching limits the reproducibility and measurement times.
- *Microcantilevers* offer very low limits of detection; however, due to damping the repeatability in liquid samples is low.
- *Polymerase chain reaction* (*PCR*) approaches have the advantages of high sensitivity and linearity; however, they also lead to nontargeted DNA amplification.

- *Quartz crystal microbalance* (*QCM*) experiments can be strongly miniaturized, requiring low sample volumes, but as a trade-off the sensitivity is limited.
- *Surface acoustic wave* (*SAW*) devices likewise allow for miniaturization, with similar limitations in sensitivity as well as repeatability issues.
- *Surface plasmon resonance* (*SPR*) provides label-free, highly sensitive analyses, yet requires elaborate instrumentation.

EIS has been extensively applied in biosensors for identification of food pathogens. It is usually combined with disposable, screen-printed chips. The method records the changes in the impedance of the electrical sensor that are induced by bacterial metabolism and cell growth as a result of the release of ionic metabolites from the living cells (carbon dioxide and organic acids produced by catabolism and ion exchange through cell membranes) [8]. Recently, it was shown that impedimetric biosensors can be sufficiently miniaturized to be integrated on a chip (lab-on-a-chip (LOC) technology) by making use of advances in microfabrication technology. This format offers new opportunities in the field of bacterial detection in food security monitoring [9].

ELISA is an immunological technique that uses an enzyme to detect whether a certain antibody or antigen is present in a sample. Two possible variations of ELISA methods are used in food analysis and food authentication. The indirect ELISA utilizes two antibodies, one of which is specific to the antigen and the other is coupled to an enzyme. The second antibody gives the assay its "enzyme-linked" name and will elicit a signal from a chromogenic or fluorogenic substrate. In a sandwich ELISA the antigen is bound between two antibodies, the capture and the detection antibody. The detection antibody can be coupled to an enzyme or can bind the conjugate that will produce the biochemical reaction [10, 11].

A PCR constitutes a fast and low-cost technique for creating multiple copies of a DNA fragment at a specific band. PCR acts as a DNA detector with the aim of amplifying and tracing pathogenic strands in processed food.

Fluorescence spectroscopy makes it possible to detect the presence or concentration of food pathogens or pollutants by utilizing the natural fluorophores in food products. Fluorescence spectra are made up of broad, overlapping emission bands that carry data on the sample's components. Fluorescence analysis may seek for peaks that have changed in intensity, wavelength, or bandwidth. Fluorescence can be used in either a right-angle or a front-face setup to allow fast, sensitive, nondestructive analysis of a wide range of food kinds [6].

SPR determines kinetic parameters of the molecular interactions and concentration of an analyte. SPR biosensors measure the binding of analytes to immobilized biomolecules without using labels. SPR systems combined with miniaturized flow systems for efficient sample and buffer delivery to the sensor surfaces permit continuous monitoring of formation of complexes and the dissociation of analyte from immobilized ligand(s) [12, 13].

Microcantilevers are microscale beams anchored at one end. Molecular adsorption on a cantilever surface generates bending due to adsorbate-induced changes in surface stress. Mass loading due to molecular adsorption causes changes in the resonance frequency on the cantilever. Chemical or biological selectivity can be achieved via immobilized receptors on the cantilever surface [14].

QCM provides an analytical technique that offers in situ investigation of surface interactions. A QCM calculates mass per unit area by measuring the frequency shift of a quartz crystal resonator. The addition or removal of a tiny mass at the surface of the acoustic resonator changes the resonance. QCM can be employed in vacuum, gas phase, and liquid conditions [15].

SAW devices belong to the group of acoustic wave devices. Acoustic wave devices are mass sensors which have mechanical acoustic waves as their transduction mechanism. The acoustic wave propagates, guided or unguided, along a single surface of the substrate in SAW devices [16].

Demand for rapid and nondestructive analytical performances makes spectroscopic and nanospectroscopic methods very important in the evaluation of food quality and safety. The vibrational methods, **IR** (Randall 1927) or **Raman** spectroscopy, are the most important spectroscopic methods, while **SERS** (Raman 1928, Raman and Krishnan 1928) is the most important nanospectroscopic method applied in food science and agriculture. Since the application of SERS is still very limited in this field and spectroscopic methods employ nanoparticles in experimental setups, we evaluate the application of both methods in the following text.

5.1.4 State-of-the-art

5.1.4.1 Vibrational spectroscopy

Methods based on vibrational spectroscopy were introduced in the early twentieth century and have quickly developed as rapid and nondestructive tools in different applications. Compared to IR spectroscopy, conventional Raman spectroscopy has seen less use in food analysis. While its disadvantages include comparatively weak signals, fluorescence interference, and more expensive equipment, it has the advantage over IR of less interference from water signals. More widespread and broader applications of Raman spectroscopy have been developed since the discovery of SERS in 1974, where it was shown that the weak Raman scattering intensity is greatly enhanced in combination with noble metal nanostructures, while fluorescence is suppressed [17].

Vibrational IR spectroscopy is based on the irradiation of the sample with infrared light. In the same sample, different chemical bonds absorb at different wavelengths depending on the atoms connected and the surrounding molecules.

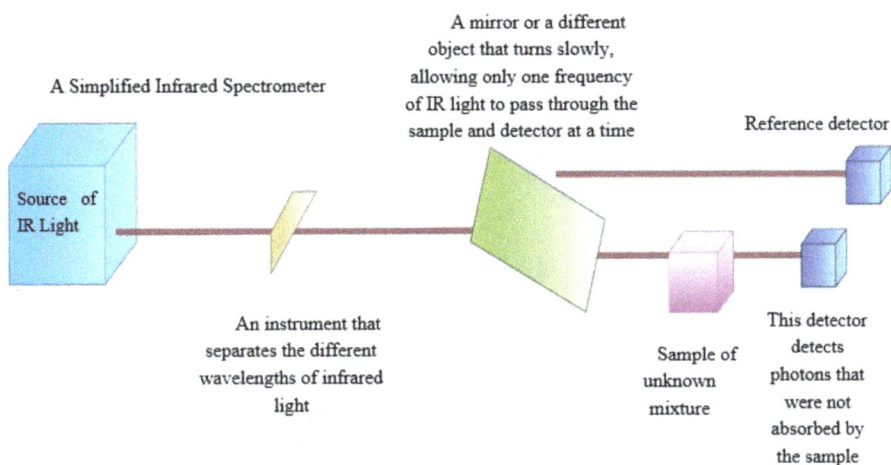

A mirror or a different
object that turns slowly,
allowing only one frequency
of IR light to pass through the
sample and detector at a time

A Simplified Infrared Spectrometer

Reference detector

Source of
IR Light

An instrument that
separates the different
wavelengths of infrared
light

Sample of
unknown
mixture

This detector
detects
photons that
were not
absorbed by
the sample

Figure 5.1.1: The principle of infrared spectrometry (after Harding, www.chem.ucla.edu).

The purpose of IR spectroscopy is to identify the functional groups present in a molecule. IR spectroscopy analyzes the number of IR photons and the amount of energy found in IR photons absorbed by the molecule [18] (Fig. 5.1.1).

IR spectra of a sample can be recorded by illuminating the sample with an IR beam and detecting the transmitted light. If the frequency of the IR light coincides with the vibrational frequency of a bond or a collection of bonds, absorption takes place. By measuring the intensity of the transmitted light it can be determined how much energy was absorbed at each frequency (or wavelength). The information can be gained by either scanning the wavelength range by means of a monochromator, or by measuring the entire wavelength range by means of a Fourier transform instrument. As a result a transmittance or absorbance spectrum is recorded [19].

In Raman spectroscopy, a sample is illuminated with a monochromatic light source, and the scattered light is detected. The resulting Raman spectra provide a signature or "fingerprint" of the sample's molecular vibrations, which can be used for qualitative and quantitative analysis. The majority of the scattered light consists of Rayleigh or elastic scattering that has the same frequency as the excitation source. Only a minute amount of the scattered light (ca. 10^{-5}% of the incident light intensity) exhibits an energy shift relative to the energy of the laser because of interactions between the incident electromagnetic waves and the vibrational energy levels of the molecules in the sample. If the intensity of the "shifted" light is plotted versus the frequency, where the laser frequency and therefore the Rayleigh band marks the origin $0 \, \text{cm}^{-1}$, a Raman spectrum of the sample is obtained. In this representation, the Raman bands will appear at frequencies that indicate the energy levels of the vibrations of different functional groups. The interpretation of the Raman spectrum is therefore analogous to that of the IR absorption spectrum [20] (Fig. 5.1.2).

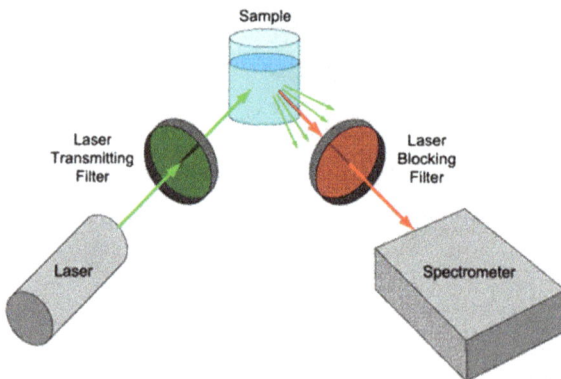

Figure 5.1.2: The principle of Raman spectroscopy. Source: IDEX Health & Science, printed with permission.

Raman spectroscopy applied in food analysis has several advantages over other analytical methods (it is nondestructive, without any pre-treatment of samples, and requires small sample volumes). Raman spectroscopy simultaneously provides qualitative and quantitative information about different food compounds. It also allows structural analysis of food [21].

SERS is a further development of the Raman technique, in which the intensity resulting from the vibrational modes of molecules that are adsorbed on or in the vicinity of noble metal nanostructures is strongly enhanced [6, 22]. The amplification of Raman signals in the presence of noble metal nanostructures has been explained by two mechanisms: electromagnetic enhancement and chemical enhancement. SERS research has recently shifted from metal surfaces to the use of well-defined metallic nanoparticles in order to better understand and control signal enhancement [23] (Fig. 5.1.3).

Application of Raman spectroscopy

Raman spectroscopy has a lot of potential in the food industry (monitoring and quality control in industrial food processing, food safety in agricultural plant production, etc.). Raman spectroscopy has several key advantages over other analytical methods used in food analysis, including a faster analysis time and the capacity to probe water-rich samples. Poor signal strength and a very strong fluorescence signal that can overlap with the Raman signal are disadvantages of Raman spectroscopy in food analysis.

The types of Raman spectroscopy applications in food industry may be classified as follows:

SURFACE ENHANCED RAMAN SPECTROSCOPY

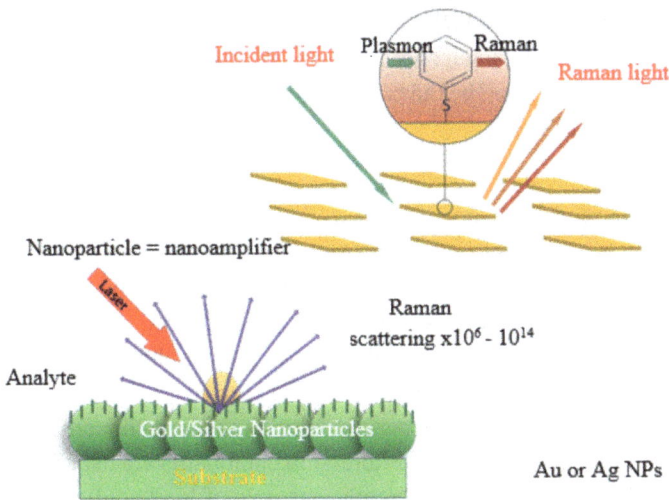

The enhancement mechanisms are roughly divided into chemical enhancement and electromagnetic enhancement

Figure 5.1.3: Surface-enhanced Raman spectroscopy (SERS) (after sciencefacts.net).

a) Raman spectroscopy in monitoring of food processes

– *Fermentation.* Raman spectroscopy is a valuable tool for monitoring fermentation during different food production processes; however, the primary disadvantage is the resulting fluorescence signal. The most common solution is to use a laser light source that emits light in the relatively far ultraviolet or IR range where fluorescence does not interfere with Raman signals [24].

– *Heating and oxidation of edible oils.* Edible oils present a topic in food sciences that is frequently discussed in the modern world. Many types of cooked or processed foods use oil as one of the ingredients. Raman spectroscopy has been used on edible oil as a method for oil characterization and authentication. Most reports on Raman monitoring of the oxidation of edible oil are relatively recent [25].

– *Processing and preservation of meat.* Raman spectroscopy is a relatively new method for monitoring meat processing and preservation. The major disadvantage of the method is overlapping Raman signals in meat samples [24].

b) Raman spectroscopy in food contamination

A number of reviews have been published that focus on the use of Raman spectroscopy in food safety by monitoring food contamination [3, 26].

– *Pesticides and fungicides in food.* Fungal pathogens are the leading cause of economic loss during the post-harvest handling of fruit. Raman spectroscopy has been used to study the antifungal activity of ZnO nanoparticles against *Bortrytis cinerea* and *Penicillium expansum* [27].

- **Bacteria in food.** Traditional Raman spectroscopy can be used to investigate the spoilage trajectory of milk caused by the bacteria *Staphylococcus aureus* and *Lactococcus lactis* ssp. *cremoris* [28].

c) Raman spectroscopy in food forensics

Food forensics is the investigation of food origin, adulteration, and contamination. Raman spectroscopy is ideally suited for this analysis, largely due to the specificity of the method and the diversity of the analytes which can be probed, ranging from the macrofood constituents, lipids, proteins, and carbohydrates, to the minor components, dyes, pigments, or preservatives [5, 30].

- **Adulteration (food authenticity)** is a major concern for the food industry. Adulterants are often difficult to detect since adulterant components are usually similar to the authentic product. The most notable example of food adulteration is in the production of oil, particularly extravirgin oils, where cheaper oil of similar chemical structure is added to reduce manufacturing costs [31, 32, 29] (Fig. 5.1.4). Chromatographic methods are unable to detect adulterations at relatively low concentrations (5–20 %). On the other hand, Raman and IR spectroscopy have been

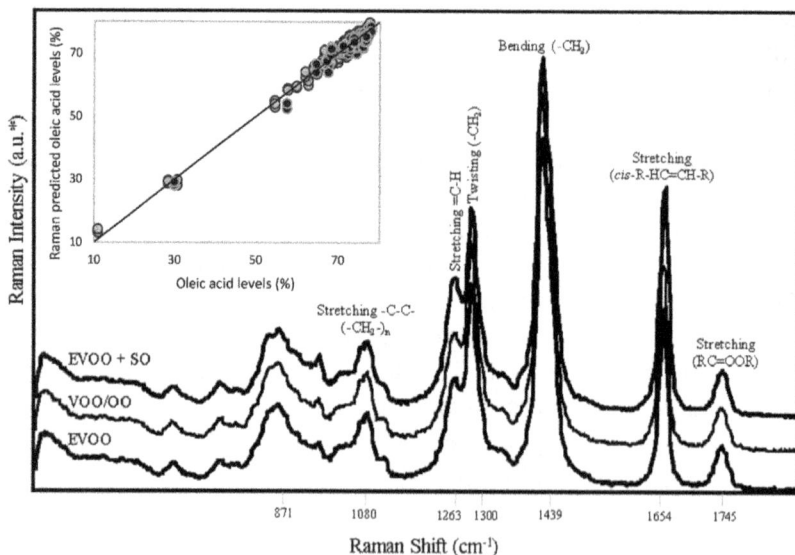

Figure 5.1.4: Raman spectra of olive oils of different quality under 1064 nm excitation (range: 200–1850 cm^{-1}). EVOO, extravirgin olive oil; VOO/OO, blend of virgin olive oil and olive oil; EVOO+SO, extravirgin olive oil + sunflower oil. Inset: Partial least squares regression calibration and external validation plots showing correlation for oleic acid levels in olive oil samples (gray circles, calibration set; black circles, external validation set) [29].

exceptionally useful in the identification of olive oils adulterated with hazelnut oil, soybean oil, corn oil, sunflower oil, and rapeseed oil. Use of Raman spectroscopy in olive oil authentication has also been reported [32–36, 29, 37].

A review of applications for quality assessment in meat and fish was presented by Herero et al. [38].

Another area for food authentication is the production of fruit juices where labeling laws demand that 95 % of the juice must come from the fruit stated on the label. Raman spectroscopy has been employed for the analysis of apple, pomegranate, and bayberry juices [31, 37].

Many consumers put a high premium on the geographical area of food production, or **provenance of a foodstuff**, along with the purity of the product. European laws lay down requirements for obtaining a Protected Designation of Origin (PDO) quality registration. This registration is shown by a label on a product that ensures that the product genuinely originates from a particular geographic region [31, 37]. Raman spectroscopy has been used to confirm the botanical and geographical origins of European honey [39].

Microbial food spoilage and contamination is another key area of food forensics. Over the last decade, there has been a resurgence of concerns about microbiological food safety, particularly the prevalence of *Salmonella*, *Campylobacter*, and *Escherichia coli*. This fear led food industry's attentions to the development of rapid and accurate means of detecting spoiled foodstuffs along all stages of the industrial process. These methods could play a valuable role within the *Hazard Analysis Critical Control Point* (HACCP) process, which is a preventive approach to food safety [31, 37]. Compared to IR spectroscopy, there is not too much published literature on the use of Raman spectroscopy to monitor food spoilage [40, 41].

The **identification of microbes** that facilitate food spoilage is a key point for understanding and detecting the microbial degradation of food. Raman spectroscopy allows analysis of single cells and effective strain-level identification [42, 43]. Strain-level identification is a microbiology term for evaluating tools for strain-level variants within microbial species. A strain is a genetic variant or subtype of a micro-organism.

Application of surface-enhanced Raman spectroscopy

SERS techniques have been used in food analysis since the late 1990s, primarily as a fast and sensitive tool for detecting food contaminants and as a diagnostic tool for foodborne pathogens. SERS is still a new approach in food analysis when compared to traditional procedures (high-performance liquid chromatography (HPLC), gas chromatography, etc.). The increased availability of suitable nanostructured substrates has contributed to the advancement of SERS technologies in food applications [4, 44].

SERS substrates for food chemical analysis

The quality of the SERS substrate is essential for the success of the SERS technique. This is especially true for food and environmental samples. The two most relevant parameters for the chemical analysis are the sensitivity and the reproducibility. A uniform substrate with a homogeneous enhancement factor provides a consistent signal in quantitative analysis. On the other hand, the maximized enhancement is the main prerequisite for a sensitive determination in screening and trace analysis. Compromises between these two parameters are required very often.

Colloid-based substrates

Au and Ag nanoparticles ranging from 10 to 200 nm in diameter are the most traditional SERS-active substrates. Their main advantages are simplicity and low costs of fabrication. Au and Ag colloids are used in SERS analysis of food additives (sodium benzoate, N-methylglycocyamine, monosodium glutamate butylated hydroxyanisole, β-hydroxy-β-methylbutanoic acid (HMB)) and chemical contaminants (tricyclazole, methamidophos, omethoate, dimethoate, nitrofurans, melamine) [9] with limits of detection between 2 and 10 ng/mg, depending on the composition of the colloids, chemical structure of analytes, sample preparation, and analytical procedure.

The primary disadvantage of colloid-based substrates is low signal reproducibility and nanoparticle agglomeration during substrate drying. Many SERS substrates manufactured using the "bottom-up" approach exhibit strong variations in their performance when comparing the signal between different spots on one substrate or between different substrates. The fundamental reason for this is a lack of structural consistency and integrity across the entire substrate region [15].

Solid surface-based substrates

Klarite™ (Renishaw Diagnostic Ltd., Glasgow, UK) and Q-SERS (Nanova Inc., Columbia, Mo., USA) are two commercially available solid surface-based substrates that have been used in food analysis. Their main advantages are simple sample preparation and relatively consistent analysis results, whereas their main disadvantage is a high price on the market.

These substrates have been used for detection of melamine in both pure solvents and in various food matrices, malachite green in fish fillets, and the pesticides carbaryl, phosmet, and anziphos-methyl on apples and tomato surfaces [45–48]. Satisfactory sensitivity was reached for most studies, but it was observed that the responses on the same analytes differed between these two substrates. The reason for these results is a different composition and structure of the substrates.

Other examples of home-made solid surfaces are Au-coated zinc oxide (ZnO) composite nanoarrays, or hexadecyltrimethylammonium bromide (CTAB)-coated Au nanorod arrays which are tested in chemical analysis of melamine [49]. Electropolished Al and ZnO/Ag nanoarrays were used for detection of Sudan dyes [50, 51]. The pesticide thiram could be detected in commercial grape juice by means of graphene oxide nanosheets and Au/Ag nanoparticles [52].

Applications of SERS in the chemical analysis of food

Recent food safety incidents (e. g., the melamine incidences in 2007 and 2008) and public health concerns about synthetic food additives and chemical residues in food have driven the need to develop rapid, sensitive, and reliable methods to detect those food hazards. This increases SERS applications in food analysis [1].

From a compositional perspective, all analytes are divided into food additives and chemical contaminants. Thus, examples of SERS applications are listed based on the proposed classification.

Food additives

Food additives are substances which become part of a food when added during the processing of food. They include preservatives, colorants, flavors and flavor enhancers, texturants, nutrients, and others. With the increased use of processed foods over the last century, the use of food additives has increased significantly. This caused public concern over synthetic food additives [53].

SERS studies applied on food additives are still at the early stage of characterization. Most of the studies were focused on the adsorption behavior of food additives on the surface of metallic nanoparticles by comparing the characteristics of enhanced and normal Raman spectra. The appearance of additional bands, as well as shifts observed in SERS compared to Raman signals, provide structural information important for elucidating the metal–adsorbate interaction. The target analytes for SERS detection include antimicrobial agents, like benzoic acid derivatives [54], antioxidants like butylated hydroxyanisole [55], sweeteners like aspartame [56], flavor enhancers like monosodium glutamate [57], and nutrients like flavones [58].

Chemical contaminants

Chemical contaminants in food include agricultural contaminants, environmental contaminants, chemical adulterants, mycotoxins, and foreign food components. The most common targets in food that have been analyzed with SERS are pesticides, antibiotics, illegal drugs, melamine, illegal food colorants, and foreign allergenic and toxic proteins [15].

Pesticides are widely used in modern agriculture. Most pesticides can be grouped based on their chemical structures: organophosphates, organochlorine, carbamates, etc. Recent developments of SERS methods for pesticide detection progressed from detection in simple solvents to complex food matrices. Detection of pesticides in real-food matrices required studies on sample extraction and optimization for SERS analysis. Different methods were explored, depending on the type of the matrix. To recover pesticides from the surface of solid matrices like apples, homogenization of the peel or surface swab procedures were employed [15, 24, 59]. Although the detection limit was not as good as the detection limit for pesticides in simple solvents, most of these detection limits did meet the regulatory level. The surface-swab method is a straightforward method for recovering pesticides from the surface of fruits [15, 60]. Solvent extraction and solid-phase extraction could be methods of choice for sample preparation for samples from liquid food matrices for detection of pesticides [61].

In farm animals and dairy products, residues of concern are *antibiotics* and *illegal drugs*. Most of the SERS studies have focused on the characterization of SERS spectra of antibiotics and illegal drugs in simple solvents. Reported studies include analysis of amoxicillin [62], tetracycline [63], ciprofloxacin [64], enrofloxacin [65], chloramphenicol [37], and others. Many are used as illegal antifungal agents in fishery. SERS detection of dye molecules is much more sensitive compared with antibiotics studies, because dye molecules are very Raman-active and they can be easily adsorbed onto the metal surface. Because of that, SERS is an ultrasensitive method for studying dye molecules. A few studies focused on detection of prohibited colorants in food [39].

Intentional adulteration with melamine in food caused severe public health issues and huge economic losses in 2007 and 2008. Melamine was deliberately added to food in order to artificially elevate the measured protein content. Thus, FDA set up a tolerance level for melamine of 1 mg/kg for infant formula and 2.5 mg/kg for milk and other food products [66]. At the same time, monitoring the level of residual melamine has become an important issue for industry and government agencies. Many methods have been developed for melamine determination in food such as HPLC-MS, liquid chromatography, or ELISA. SERS has a great advantage over all these methods for melamine determination due to its ultrahigh sensitivity and "fingerprint-like" property. Melamine was also analyzed in gluten, chicken feed, and processed foods using SERS. The results showed that food could be screened for melamine presence by SERS. Melamine content in positive samples was confirmed by HPLC [67].

Due to its characteristic sharp peaks and rising safety issue, melamine is clearly one of the most commonly used chemicals for the evaluation of SERS nanosubstrates. SERS development and melamine detection both considerably benefit from these investigations [68, 69].

Illegal food colorants may cause toxicity in humans [51, 70]. SERS was recognized as a practical screening technique to differentiate samples that were contaminated with unauthorized food colorants due to its ultrasensitivity and the distinct SERS patterns of the dye molecules. Several studies demonstrated the capacity of detection of these chemicals [51, 46].

Figure 5.1.5 presents a schematic illustration of graphene/Ag composites. This composite is used as a substrate for SERS detection of prohibited colorants [70].

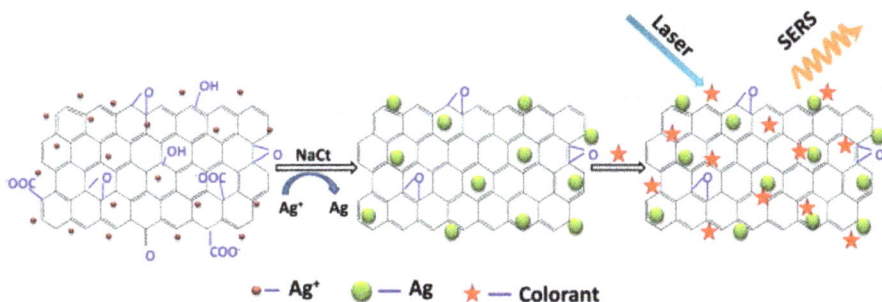

Figure 5.1.5: Schematic illustration of the synthesis of graphene/Ag composites and SERS detection of prohibited colorants using this substrate. Reprinted from [46] with permission from Elsevier.

Mycotoxin residues and other *small chemical toxins* produced by bacteria and other living organisms in foods are of great safety concern. Detection of aflatoxins and ochratoxins with SERS was investigated [71, 72]. SERS has also been successfully employed to detect several marine-derived small-molecule poisons such as saxitoxin, microcystin, and tetrodotoxin [73–75].

Although most of the studies could achieve very sensitive detection limits in simple solvents, they have not been applied on real food samples.

Foreign allergenic and *protein toxins* are a category of toxins which provoke toxicity through cross-contact or intentional adulteration. Many SERS methods have been developed for protein detection. One of the methods uses a unique, sensitive, highly specific, and photobleaching-resistant immunoassay system utilizing gold nanoparticles and SERS for the detection of protein A, which is a specific surface antigen of *S. aureus*. This new system is made by chemisorption of antibody immunoglobulin G (IgG) on gold nanoparticles (AuNPs), followed by coupling the Raman-active reporter molecule 5,5'-dithiobis(2-nitrobenzoic acid) (DTNB) to the surface of IgG–AuNP. The adsorbed DTNB molecules exhibit strong Raman signals via both electromagnetic and chemical enhancement. The narrow spectral widths and high photostability ensure the system to be an excellent detection label. This SERS-based immunoassay was applied to the detection of protein A [56]. A highly sensitive immunoassay based on SERS has been developed with a novel immune marker named Raman reporter-labeled immuno-Au aggregates on a SERS-active immune substrate [76]. The immunoassay is applied for the detection of protein (human IgG) at a level of 100 fg/mL.

These methods are generally very sensitive, but the chemical signature of the target molecule can be lost as the final signal is derived from dye molecules. In addition, secondary conjugation can significantly increase sample processing time. Secondary conjugation is binding of secondary antibodies to primary antibodies which directly bind to the target antigen(s).

5.1.4.2 Nanosensors

Sensors and biosensors based on nanomaterials are called nanosensors. Another definition states that every sensor with a sensing element smaller than 1000 nm is considered to be a nanosensor (Fig. 5.1.6).

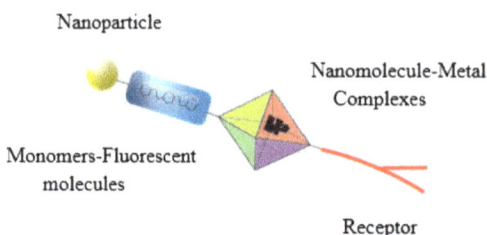

Figure 5.1.6: Example of a nanosensor for molecular recognition (after nanobiotech.wsu.edu).

Aptasensors are aptamer-based biosensors with excellent recognition capability towards a wide range of targets. Specially, there has been ever-growing interest in the development of aptasensors for the detection of small molecules [77].

Nanoparticle-based aptasensors consist of aptamers (the target recognition element) and nanomaterials (the signal transducers and/or signal enhancers) [78].

Aptamers are single-stranded nucleic acid or peptide molecules of less than 25 kDa with natural or synthetic origin. They are highly specific and selective towards their target compound (ions, proteins, toxins, microbes, viruses) due to their precise and well-defined 3D structures. Aptamers have superior physical and chemical attributes to antibodies. Aptamers are named synthetic antibodies due to their selection and generation through an in vitro combinatorial molecular technique called SELEX. Dissociation constants of aptamers are in the nanomolar or picomolar range [1, 79, 80]. Aptasensors containing aptamers with a dissociation constant in the nanomolar and picomolar range will bind even with very low concentrations of targeted compound, which is very important in the determination of food contaminants and constituents.

Aptamers are extensively used as recognition elements in the fabrication of aptasensors [81].

There is a wide variety of nanomaterials which can be used in aptasensors. The list includes metal nanoparticles and nanoclusters, semiconductor nanoparticles, carbon nanoparticles, magnetic nanoparticles, etc. [1].

Based on the signal transduction mode, aptamers can be divided into electrochemical, optical, and mass-sensitive aptamers. Among them, optical aptasensors have been widely developed because of their high sensitivity, quick response, and simple operation. Coupling aptamers as the recognition component with various optical analytical techniques provides special opportunities for the analysis of targets [77].

Nanospectroscopic techniques such as SERS and surface-enhanced resonance Raman scattering (SERRS) have important applications in the construction of aptasensors applied in food packaging.

Application of nanosensors in food packaging

Food is packaged to protect the product from the influence of external environmental conditions like heat, light, presence or absence of moisture, pressure, microorganisms, etc. It provides the consumer with greater ease of use and time-saving convenience. It may provide various sizes and shapes for the final product. The key safety goal for traditional packaging materials which come into contact with food is to be as inert as possible.

Technology innovations move the packaging market from conventional packaging to interactive, aware, and intelligent systems. These "smart" packaging concepts are based on the useful interaction between the packaging environment and the food ingredients to provide active protection to the food. "Smart" packaging utilizes chemical sensors or biosensors to monitor the food quality and safety from the farm to the costumers. It may result in a variety of sensor designs that are suitable for monitoring of food quality and safety (freshness, pathogens, leakage, carbon dioxide, oxygen, pH, time, or temperature). Smart packaging is needed as online quality control for food safety in view of consumers, authorities, and food producers [82]. It has great potential in the development of new sensing systems integrated in the food packaging, which are beyond the existing conventional technologies [83].

Packaging materials can be divided into several groups:
1) packaging materials with improved barrier properties,
2) active packaging materials,
3) intelligent packaging materials,
4) edible coatings,
5) biodegradable packaging materials.

One special category of food packaging design is the application of nanotechnology. The use of nanoparticles in food packaging systems may increase qualities such as barrier properties to various gases, antibacterial properties, biodegradability, or sensor properties that can provide information on food quality, among others. Food packaging materials are currently the most common use of nanotechnology in the food industry [2].

Nanomaterials in food packaging

Nanotechnology can be used in food packaging to make it stronger, lighter, or perform better. Antimicrobials such as nanoparticles of silver or titanium dioxide can be used in food packaging to prevent spoilage of foods. Introduction of clay nanoparticles into packaging is used for blocking oxygen, carbon dioxide, and moisture from reaching the food, and also aids in preventing spoilage [84]. For example, industrial giant Bayer produces a transparent plastic film called Durethan® which contains nanoparticles of clay. This film is an engineering plastic composed of polyamide 6 and polyamide 66 that has been mixed with clay nanoparticles to produce a good combination of qualities such as high strength and toughness, abrasion resistance, chemical resistance, and crack resistance. Durethan® is utilized in a variety of industrial applications, including medical packaging film and food packaging. The nanoparticles are distributed throughout the plastic and can prevent oxygen, carbon dioxide, and moisture from reaching fresh meats or other goods. The use of nanoclay makes the plastic lighter, tougher, and more heat-resistant, which are its key benefits. At the same time, Durethan® film is a low-cost but airtight packaging solution. The incorporated nanoparticles prohibit gases and moisture from permeating the coating [37].

Nanosensors in food packaging

Nanosensors are being added in plastic packaging to detect gases given off by food when it spoils. The packaging itself changes its color to alert users that the food in the package has gone bad. Plastic films are created to allow food to stay fresh longer. These films are packed with silica nanoparticles to reduce the flow of oxygen into the package and the leaking of moisture out of the package.

Nanosensors are also being engineered to detect bacteria and other contaminants such as *Salmonella* on the surface of food and at a packaging plant. Nanosensors will allow more frequent food package testing at a much lower cost than sending the samples to a lab to be analyzed. This packaging testing, if conducted properly, has the potential to reduce the chance of contaminated food reaching grocery store shelves. Another nanosensor being developed is to detect pesticides on fruit and vegetables. This is very useful in a packaging plant because now you can check the quality of your products [85].

a) Smart packaging and food tracking
Nanotechnology in packaging may help to extend the shelf-life of food. Mars Inc. is a company that has patented invisible edible nanowrappers which will act as an envelope for foods to prevent gas and moisture exchange.

Smart packages which contain nanosensors and antimicrobial activators are being engineered to be capable of detecting food spoilage and releasing nanoantimicrobes. Such systems will extend the shelf-life of food. This will enable supermarkets to keep food for greater periods of time before its selling date. Food tracking devices such as the nanosensors embedded into food products as tiny chips that are invisible to the human eye could also act as electronic barcodes. These sensors would emit a signal that would allow food, including fresh food, to be tracked from paddock to factory to supermarket and beyond [85].

b) Interactive "smart" food

Food companies Kraft and Nestle currently work on designing "smart" foods that are intended to interact with the consumers so they can personalize their food, by changing color, flavor, and nutrients on demand. Kraft is developing a clear tasteless drink that contains hundreds of flavors in latent nanocapsules. By using a microwave you would be able to trigger the release of the color, flavor, concentration, and texture of your choice. The technique of nanoencapsulation, or creating nanocapsules, involves coating a nanoparticle so that its contents are released in a controlled way [85].

c) Aptasensors

Aptasensors have broad applications in food and agriculture (Fig. 5.1.7).

Aptasensors for food quality testing are using a wide range of transducing mechanisms. Their respective principles are based on the properties of the nanoparticle

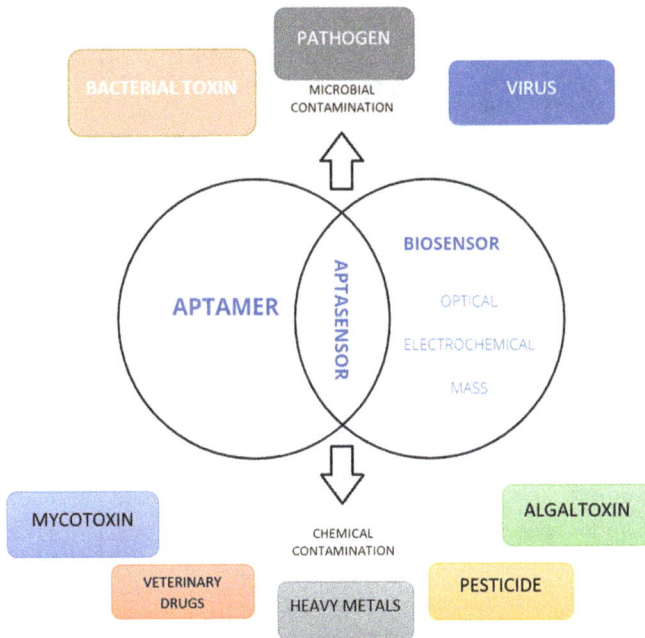

Figure 5.1.7: The formation and applications of aptasensors in food and agriculture.

being employed. Aptamers are divided into optical and electrochemical detection systems depending on their detection systems [1].

Optical aptasensors have transducers that can capture signals in the form of UV, visible, and IR radiation from a chemical/biological/physical reaction or event and can transform them into different data formats. Colorimetric, fluorometric, bioluminescence-based, chemiluminescence-based, and SPR-based optical aptasensors constitute the main types of optical aptasensors [75, 86]. The main advantage of optical aptasensors over other types of aptasensors is their ease of use, analytical speed, and sensitivity.

SPR and Rayleigh scattering aptasensors are a type of refractometric sensor that measures the properties of electromagnetic waves by measuring changes in refractive index caused by any process (chemical, biological, or physical) on the sensor surface. The signal is created by surface plasmon propagation at a noble metal-coated surface and is converted by a dedicated SPR device. In this case, Rayleigh scattering (scattered photons have the same amount of energy as absorbed incident photons) is also found to be a maximum. The phenomenon is called resonance Rayleigh scattering (RRS). The mechanism of RRS assays is the same as the SPR mechanism in the absence of protecting aptamers [75].

SERS- and SERRS-based aptasensors can be used for detection of different components in food analytes. For example, He et al. demonstrated SERS-based aptamers for the qualitative detection of ricin ("bioterror agent") in liquid foods [81].

5.1.5 Some challenges and solutions

Vibrational spectroscopy

Raman spectroscopy, as a fast, noninvasive method, is a powerful tool for analysis of food and agricultural products. It has many advantages over chromatographic and other methods which have been used for food. In order to improve the performance of Raman spectroscopy, certain improvements of the method are necessary. These improvements are, e. g., the development of portable Raman spectrometers, increasing the intensity of the Raman signal, elimination of fluorescence signals, and reduction of the high costs of Raman spectrometers.

SERS is a promising tool for the rapid and sensitive detection of food additives and contaminants. More studies are needed to explore other targets in food since the SERS research of chemical analysis in food is still in the early stage. Most of the studies have focused on substrate development and analysis in simple solvents. There are still several hurdles for the future development of SERS in real-word applications, such as (1) the lack of cost-effective commercial SERS substrates, (2) background interference from food components for SERS analysis of trace amounts of food analytes, and (3) the high cost of the Raman spectroscopy instrumentation [15].

In order to successfully advance the SERS technique in real applications of food analysis, more research is needed in:
1) commercialization and standardization of SERS substrates,
2) technology integration,
3) substrate functionalization and optimization,
4) on-site SERS detection using portable devices,
5) exploration of other applications in food analysis [15].

Nanosensors

The main obstacles in applications of aptasensors are connected to the sample conditions such as pH, ionic strength, nonspecific interactions of aptamers with the sample matrix, and difficulties in selection and development of aptamers against small molecules. Thus, real sample detection in most sensors is achieved either by sample pre-treatment or separation of aptamer–analyte conjugates. Current efforts are directed towards improvement of selection techniques to obtain functional and high-affinity aptamers for several low-molecular weight analytes. The extremely high cost and difficulties in aptamer functionalization have to be reduced for on-field monitoring. Solutions to these problems and development of new nanomaterials will boost the design of affordable and easily operating aptamer-based sensing systems [75].

5.1.6 Summary and impact

Vibrational spectroscopy

Vibrational spectroscopy is an extremely useful tool for food industry and agriculture. Food forensics is a very important and modern application of vibrational spectroscopy nowadays, particularly in determining the origin of food and food adulteration. The quick, noninvasive nature of the techniques, as well as the extensive array of analytes that can be examined, increase the adaptability of Raman and SERS spectroscopies in food analysis. The spectrum of applications in this domain includes the entire industrial process, from plant production and animal feeding, to food manufacturing, monitoring, and packing, and potentially beyond, thereby spanning the entire farm-to-fork chain [37].

This chapter has highlighted the main characteristics of Raman spectroscopy and SERS, discussing their main applications in the sector of food production and agriculture.

Nanosensors

Successful utilization of aptasensors in several detection formats, including food and water analysis, is confirmed by a great amount of available literature related to the application of aptamers. The excellent specificity of the aptamers allows a wide variety of analytes, including toxins, pathogens, heavy metal ions, nucleic acids, and proteins, to be detected. Nanoparticles provide larger surface area for aptamer immobilization as well as benefiting from their own optophysical and electrochemical properties to the sensors. However, for researchers to employ nanoaptasensing in food analysis, it is necessary to prioritize the ultimate aim of the aptasensor application [75, 86].

Bibliography

[1]   Omanovic-Miklicanin E, Maksimovic M. Nanosensors applications in agriculture and food industry. Bull Chem Technol Bosnia and Herzegovina. 2016;47:59–70.
[2]   Windom BC, Hahn DW. Raman Spectroscopy. Encyclopedia of Tribology. Springer; 2013.
[3]   Rai M, Ribeiro C, Mattoso L, Duran N. Nanotechnologies in Food and Agriculture. Springer International Publishing; 2015.
[4]   Smith E, Dent G. Modern Raman Spectroscopy: A Practical Approach. Hoboken, NJ, USA: Wiley; 2005.
[5]   Omanovic-Miklicanin E, Maksimovic M. Applications of nanotechnology in agriculture and food production – nanofood and nanoagriculture. In: Proceedings of IcETRAN. 2018.
[6]   Craig AP, Franca AS, Irudayaraj J. Surface-enhanced Raman spectroscopy applied to food safety. Annu Rev Food Sci Technol. 2013;4:369–80.
[7]   Cho Y-J, Kang S, editors. Emerging Technologies for Food Quality and Food Safety Evaluation. Boca Raton: CRC Press; 2011.
[8]   Poltronieri P, Mezzolla V, Primiceri E, Maruccio G. Biosensors for the detection of food pathogens. Foods. 2014;3(3):511–26.
[9]   Peng B, Li G, Li D, Dodson S, Zhang Q, Zhang J, Lee YH, Demir HV, Ling X, Xiong Q. Vertically aligned gold nanorod monolayer on arbitrary substrates: self-assenbly and femtomolar detection of food contaminants. ACS Nano. 2013;7:5993–6000.
[10] Hu H, Wang Z, Wang S, Zhang F, Zhao S, Zhu S. ZnO/Ag heterogeneous structure nanoarrays: photocatalytic synthesis and used as substrate for surface-enhanced Raman scattering detection. J Alloys Compd. 2011;509:2016–20.
[11] Alhajj M, Farhana A. Enzyme linked immunosorbent assay. In: StatPearls. StatPearls. 2022.
[12] Lopez MI, Ruisanchez I, Callao MP. Figures of merit of a SERS method for Sudan I determination at trace levels. Spectrochim Acta A. 2013;111:237–41.
[13] Retra K, Geitmann M, Kool J, Smit AB, de Esch IJ, Danielson UH, Irth H. Development of surface plasmon resonance biosensor assays for primary and secondary screening of acetylcholine binding protein ligands. Anal Biochem. 2010;407(1):58–64.
[14] Di-Anibal CV, Marsal LF, Callao MP, Ruissanchez I. Surface-enhanced Raman spectroscopy (SERS) and multivariatet analysis as a screening tool for detecting Sudan I dye in culinary spices. Spectrochim Acta A. 2012;87:135–41.
[15] Zheng J, He L. Surface- nhanced Raman spectroscopy for the chemical analysis of food. Compr Rev Food Sci Food Saf. 2014;13:317–28.

[16] Rocha-Gaso M-I, March-Iborra C, Montoya-Baides A, Arnau-Vives A. Surface generated acoustic wave biosensor for the detection of pathogens: a review. Sensors (Basel). 2009;9(7):5740–69.

[17] Zamborini FP, Bao L, Dasari R. Nanoparticles in measurement science. Anal Chem. 2012;84:541–76.

[18] Infrared spectroscopy 2. (2020, September 13). https://chem.libretexts.org/@go/page/15739.

[19] Hsu S. Chapter 15: Infrared spectroscopy. In: Settle FA, editor. Handbook of Instrumental Techniques for Analytical Chemistry. Upper Saddle River, NJ: Prentice Hall PTR; 1997.

[20] Zhang L, Jiang C, Zhang Z. Graphene oxide embedded sandwich nanostructures for enhanced Raman readout and their applications in pesticide monitoring. Nanoscale. 2013;3773–9.

[21] Herrero AM. Raman spectroscopy a promising technique for quality assessment of meat and fish: a review. Food Chem. 2008;107(4).

[22] Perumal J, Wang Y, Attia AB, Dinish US, Olivo M. Towards point-of-care SERS sensors for biomedical and agri-food analysis application: a review of recent advancements. Nanoscale. 2021;13.

[23] He L, Liu Y, Lin M, Awika J, Ledoux DR, Li H, Mustapha A. A new approach to measure melamine, cyanuric acid, and melamine cyanurate using surface enhanced Raman spectroscopy coupled with gold nanosubstrates. Sens Instrumen Food Qual. 2008;2:66–71.

[24] Lin M, He L, Awika J, Yang L, Ledoux DR, Li H, Mustapha A. Detection of melamine in gluten, chicken feed, and processed foods using surface enhance Raman spectroscopy and HPLC. J Food Sci. 2008;73:T129–34.

[25] Kwofie F, Lavine BK, Ottaway J, Booksh K. Differentiation of edible oils by type using Raman spectroscopy and pattern recognition methods. Appl Spectrosc. 2020;74(6).

[26] Zhang Y, Lai K, Zhou J, Wang X, Rasco B, Huang Y. A novel approach to determine leucomalachite green and malachite green in fish fillets with surface-enhanced Raman spectroscopy (SERS) and multivariate analyses. J Raman Spectrosc. 2012;43:1208–13.

[27] Liu B, Zhou P, Liu X, Sun X, Li H, Lin MS. Detection of pesticides in fruits by surface-enhanced Raman spectroscopy coupled with gold nanostructures. Food Bioprocess Technol. 2013;6:710–8.

[28] Gao J, Hu Y, Li S, Zhang Y, Chen X. Adsorption of benzoic acid, phthalic acid on gold substrates studied by surface-enhanced Raman scattering spectroscopy and density functional theory calculations. Spectrochim Acta A. 2013;104:41–7.

[29] Aykas DP, Karaman AD, Keser B, Rodriguez-Saona L. Non-targeted authentication approach for extra virgin olive oil. Foods. 2020;9(2):221.

[30] Omanovic-Miklicanin E. Food forensics. In: Forensic Genetics. Sarajevo: International Burch University; 2018.

[31] Yao W, Sun Y, Xie Y, Wang S, Ji L, Wang H, Qian H. Development and evaluation of a surface enhanced Raman scattering (SERS) method for the detection of the antioxidant butylated hydroxyanisole. Eur Food Res Technol. 2011;233:835–40.

[32] Downey G. Vibrational spectroscopy in studies of food origin. In: New Analytical Approaches for Verifying the Origin of Food. Woodhead Publishing Series in Food Sciences, Technology and Nutrition. 2013.

[33] Peica N, Lehene C, Leopold N, Schlucker S, Kiefer W. Monosodium glutamate in its anhydrous and monohydrate form: differentiation by Raman spectroscopies and density functional calculations. Spectrochim Acta A. 2007;66:604–15.

[34] Peica N. Identification and characterization of the E951 artificial food sweetener by vibrational spectroscopy and theoretical modelling. J Raman Spectrosc. 2009;40:2144–54.

[35] Corredor C, Teslova T, Canamares MV, Chen ZG, Zhang J, Lombardi JR, Leona M. Raman and surface-enhanced Raman spectra of chrysin, apigenin and luteolin. Vib Spectrosc. 2009;49:190–5.

[36] Arlorio M, Coisson JD, Bordiga M, Travaglia F, Garino C, Zuidmeer L, Van Ree R, Giuffrida MG, Conti A, Martelli A. Olive oil adulterated with hazelnut oils: simulation to identify possible risks to allergic consumers. Food Addit Contam Part A Chem Anal Control Expo Risk Assess. 2009;27:11–8. https://doi.org/10.1080/02652030903225799.

[37] Brewster VL, Goodacre R. Vibrational spectroscopy. In: Infrared and Raman Spectroscopy in Forensic Science. John Wiley & Sons Ltd.; 2012.

[38] Liu B, Zhou P, Liu X, Sun X, Li H, Lin MS. Detection of pesticides in fruits by surface-enhanced Raman spectroscopy coupled with gold nanostructures. Food Bioprocess Technol. 2013;6:710–8.

[39] Goodacre R, Radovic B, Anklam E. Progress toward the rapid nondestructive assessment of the floral origin of European honey using dispersive Raman spectroscopy. Appl Spectrosc. 2002;56:7.

[40] Shende S, Inscore F, Sengupta A, Stuart J, Farquharson S. Rapid extraction and detection of trace chlorpyrifos-methyl in orange juice by surface-enhanced Raman spectroscopy. Sens Instrum Food Qual. 2010;4:101–7.

[41] Ji W, Wang L, Qian H, Yao W. Quantitative analysis of amoxicillin residues in foods by surface-enhanced Raman spectroscopy. Spectrosc Lett. 2014;47:451–7.

[42] Li R, Zhang H, Chen Q, Yan N, Wang H. Improved surface-enhanced raman scattering on micro-scale Au hollow spheres: synthesis and application in detecting tetracycline. Analyst. 2011;136:2527–32.

[43] He L, Lin M, Li H, Kim NJ. Surface-enhanced Raman spectroscopy coupled with dendritic silver nanosubstrates for detection of restricted antibiotics. J Raman Spectrosc. 2010;41:739–44.

[44] Zhang Y, Huang Y, Zhai F, Du R, Liu Y, Lai K. Analysis of enrofloxacin, furazolidone and malachite green in fish products with surface-enhanced Raman spectroscopy. Food Chem. 2012;135:845–50.

[45] Ji W, Yao W. Rapid surface-enhanced Raman scattering detection method for cloramphenicol residues. Spectrochim Acta, Part A. 2015;144:125–30.

[46] Xie Y, Li Y, Niu L, Wang H, Qian H, Yao W. A novel surface-enhanced scattering sensor to detect prohibited colorants in food by graphene/silver nanocomposite. Talanta. 2012;100:32–7.

[47] Mecker LC, Tyner KM, Kauffman JF, Arzhantsev S, Mans DJ, Gryniewicz-Ruzicka CM. Selective melamine detection in multiple samples matrices with a portable Raman instrument using surface enhanced spectroscopy-active gold nanoparticles. Anal Chim Acta. 2012;733:48–55.

[48] Lin M, He L, Awika J, Yang L, Ledoux DR, Li H, Mustapha A. Detection of melamine in gluten, chicken feed and processed foods using surface enhanced Raman spectroscopy and HPLC. J Food Sci. 2008;73(8):T129–34.

[49] Lin M. A review of traditional and novel detection techniques for melamine and its analogues in foods and animal feed. Front Chem Eng China. 2009;3(4):427–35.

[50] Bunaciu AA, Aboul-Enein HY, Hoang VD. Vibrational spectroscopy used in milk products analysis: a review. Food Chem. 2016;196:877–84.

[51] Wu X, Gao S, Wang J, Wang H, Huang Y, Zhao Y. The surface-enhanced Raman spectra of aflatoxins: spectral analysis, density, functional theory calculation, detection and differtiation. Analyst. 2012;137:4226–34.

[52] Galarreta BC, Tabatabaei M, Guieu V, Peyrin E, Lagugne-Labarthet F. Microfluidic channel with embedded SERS 2D platform for the aptamer detection of ochratoxin. Anal Bioanal Chem. 2013;405:1613–21.

[53] https://www.who.int/news-room/fact-sheets/detail/food-additives?fbclid= IwAR0gjrbMKUIAXwLMcKz-sVE1kYzO0fyXBs7D2QXp_KljzLFvnX1L6ENuQeY. Accessed: April 22nd 2022.

[54] Zhu Y, Kuang H, Xu I, Ma W, Peng C, Hua Y, Wang L, Xu C. Gold nanorod assemley based approach to toxin detection by SERS. J Mater Chem. 2012;22:2387–91.

[55] Olson TY, Schwartzberg AM, Liu J, Zhang J. Raman and surface-enhanced Raman detection of domoic acid and saxitoxin. Appl Spectrosc. 2012;65:159–64.
[56] Lin WC, Jen HC, Chen CL, Hwang DF, Chang R, Hwang JS, Chaing HP. SERS study of tetrodotoxin (TTX) by using silver nanoparticle arrays. Plasmonics. 2009;4:187–92.
[57] Lin C, Yang Y, Chen Y, Yang T, Chang H. A new protein A assay based on Raman reporter labeled immunogold nanoparticles. Biosens Bioelectron. 2008;24:178–83.
[58] Song C, Wang Z, Zhang R, Yang J, Tan X, Cui Y. Highly sensitive immunoassay based on Raman reporter-labeled immuno-Au aggregates and SERS-active immuno substrate. Biosens Bioelectron. 2009;25:826–31.
[59] He L, Chen T, Labuza TP. Recovery and quantitative detection of thiabendazole on apples using a surface swab capture method followed by surface-enhanced Raman spectroscopy. Food Chem. 2014;148:42–6.
[60] Jin H, Lu Q, Chen X, Ding H, Gao H, Jin S. The use of Raman spectroscopy in food processes: a review. Appl Spectrosc Rev. 2016;51(1):12–22.
[61] Yang D, Ying Y. Applications of Raman spectroscopy in agricultural products and food analysis: a review. Appl Spectrosc Rev. 2011;46:539–60.
[62] He L, Liu Y, Mustapha A. Antifungal activity of zinc oxide nanoparticles against Botrytis cinerea and Penicillium expansum. Microbiol Res. 2011;166:207–15.
[63] Nicolaou N, Xu Y, Goodacre R. Fourier transform infrared and Raman spectroscopies for the rapid detection, enumeration, and growth interaction of the bacteria Staphylococcus aureus and Lactococcus lactis ssp cremoris in milk. Anal Chem. 2013;83:5681–7.
[64] Herrero M. Raman spectroscopy as promising technique for quality assessment of meat and fish. A review. Food Chem. 2008;107.
[65] Zou MQ, Zhang XF, QI XH, Ma HL, Dong Y, Liu CW, Guo X, Wang H. Rapid authentication of olive oil adulteration by Raman spectroscopy. J Agric Food Chem. 2009;30:933–6.
[66] Gossner CM, Schlundt J, Ben Embarek P, Hird S, Lo-Fo-Wong D, Beltran JJ, Teoh KN, Tritscher A. The melamine incident: implications for international food and feed safety. Environ Health Perspect. 2009;117(12):1803–8.
[67] Lopez-Diez EC, Bianchi G, Goodacre R. Rapid quantitative assessmet of the adulteration of virgin olive oils with hazelnut oils using Raman spectroscopy and chemomemtrics. J Agric Food Chem. 2003;51:6145–50.
[68] El-Abassy RM, Donfack P, Materny A. Visible Raman spectroscopy for the discrimination of olive oils from the different vegetable oils and the detection of adulteration. J Raman Spectrosc. 2009;67:77–84.
[69] Schmidt H, Sowoidnich K, Kronfeldt HD. A prototype hand-held Raman sensor for the in situ characterization of meat quality. Appl Spectrosc. 2007;64:888–94.
[70] Fehrmann A, Franz M, Hoffmann A, Rudzik L, Wust E. Dairy product analysis: identification of microoragnisms by mid-infrared spectroscopy and determination of constituents by Raman spectroscopy. J AOAC Int. 1995;78:1537–42.
[71] Rosch P, Harz M, Schmitt M, Peschke KD, Ronneberger O, Burkhardt H, Motzkus HW, Lankers M, Hofer S, Thiele H, Popp. Chemotaxonomic identification of single bacteria by micro-Raman spectroscopy: Application to clean-room-relevant biological contaminations. J Appl Environ Biol. 2005;71:1626–37.
[72] Schmid U, Rosch P, Krause M, Harz M, Popp J, Baumann K. Gaussian mixture discriminant analysis for the single-cell differentiation of bacteria using micro-raman spectroscopy. Chemom Intell Lab Syst. 2009;96:159–71.
[73] Maksimovic M, Vujovic V, Omanovic-Miklicanin E. Application of Internet of things in food packaging and transportation. Int J Sustain Agric Manag Inf. 2015;1(4):333–50.
[74] Lam D. Packaging application in food packaging. Packaging applications using nanotechnology [Online]. 2010;10(2). http://www.iopp.org/files/public/LamDerekSanJose.pdf.

[75] Sharma R, Ragavan KV, Thakur MS, Raghavarao KSMS. In: Chalmers JM, Edwards HGM, Hargreaves MD, editors. Infrared and Raman Spectroscopy in Forensic Science. 2015. Biosensors and Bioelectronics, 74, 612–627.

[76] Yun W, Li H, Chen S, Tu D, Xie W, Huang Y. Aptamer-based rapid visual biosensing of melamine in whole milk. Eur Food Res Technol. 2014;238:989–95.

[77] Feng C, Dai S, Wang L. Optical aptasensors for quantitative detection of small biomolecules: a review. Biosens Bioelectron. 2014;59:64–74.

[78] Maksimovic M, Omanovic-Miklicanin E, Badnjevic A. Nanofood and Internet of Nano-Things. Springer International Publishig; 2019.

[79] Byun J. Recent progress and opportunities for nucleic acid aptamers. Life (Basel). 2021;11(3):193.

[80] Hosseinzadeh L, Mazloum-Ardakani M. Chapter Six – Advances in aptasensor technology. In: Makowski GS, editor. Advances in Clinical Chemistry, vol. 99. Elsevier; 2020.

[81] He L, Lamont E, Veeregowda B, Sreevatsan S, Haynes CL, Diez-Gonzaleza F, Labuza TP. Aptamer-based surface-enhanced Raman scattering detection of ricin in liquid foods. Chem Sci. 2011;2:1579–82.

[82] Maksimovic M, Vujovic V, Omanovic-Miklicanin E. Application of internet of things in food packaging and transportation. Int J Sustain Agric Manag Inf. 2015;1(4).

[83] Kuswandi B, Wicaksono Y, Jayus AA, Yook Heng L, Ahmad M. Smart packaging: sensors for monitoring of food quality and safety. Sens Instrum Food Qual Saf. 2011;5:137–46.

[84] Ashfaq A, Khursheed N, Fatima S, Anjum Z, Younis K. Application of nanotechnology in food packaging: Pros and Cons. J Agric Food Res. 2022;7.

[85] Lam D. Packaging applications using nanotechnology [online]. 2010. http://www.iopp.org/files/public/LamDerekSanJose.pdf.

[86] Sharma R, Ragavan KV, Thakur MS, Raghavarao KSMS. Recent advances in nanoparticle based aptasensors for food contaminants. Biosens Bioelectron. 2015;74:612–27.

Elena Shabunya-Klyachkovskaya

5.2 Surface-enhanced Raman scattering in cultural heritage studies

5.2.1 Key messages

- Surface-enhanced Raman scattering (SERS) allows the identification of organic and inorganic pigments with high molecular specificity and unparalleled sensitivity.
- Natural pigments could be distinguished from their synthetic analogs by means of SERS.
- The choice of SERS-active nanostructures as well as sample treatment procedure depends on the nature of the material (fiber, glass, paper, paint etc.) and its chemical properties.
- Detailed databases of inorganic chromophores have been accumulated in the process of art materials study and SERS technique optimization and are vital tools in the field.

5.2.2 Pre-knowledge

The current chapter focuses on the application of SERS for materials characterization and identification of different objects of cultural heritage. The reader should be familiar with theoretical aspects of Raman scattering and SERS (Volume 1, Chapters 2.3 and 3.3, respectively) as well as the optical properties of metal nanoparticles and the methods for their synthesis and characterization (Volume 1, Chapter 1.6 and Volume 2, Chapter 6).

Raman spectroscopy provides information about the molecular structure of the substance under investigation. Due to the high specificity of the technique, materials with a similar chemical composition can be easily distinguished. However, the application of Raman spectroscopy for cultural heritage studies has been rather limited for a long time because of the very low yield of Raman scattering. Therefore, a large amount of sample was needed for examination. Since Raman scattering is weak, it could be masked by uncontrollable intrinsic fluorescence of the analyzed probe.

The plasmonic effects that take place on the nanotexturized surface of noble metals offer enhancement of light–matter interaction, resulting in an increase of the Raman signal by several orders of magnitude [1]. Thus, the application of SERS techniques leads to the increase of the sensitivity of the method and at the same time reduces the amount of sample needed for the analyses.

The enhancement factor (EF) is highly dependent on the type of substrate. Nonspherical nanoparticles are of particular interest for SERS applications due to the huge local electric field concentration at the sharp edges of these nanoparticles. The nature of the metal of the nanoparticles, the wavelength

Acknowledgement: The author is very grateful to Professor S. Gaponenko and Dr. O. Kulakovich for fruitful discussions as well as to L. Trotsiuk, E. Korza, and A. Matsukovich for their help in the experiments.

https://doi.org/10.1515/9783110442908-016

of the Raman excitation, the mutual metal–analyte position (on the plasmonic film, under the film, or between two plasmonic films), and even optical properties of analytes have a significant impact on developing efficacious SERS applications.

5.2.3 Importance of the application

The study of cultural heritage helps us to learn about the history of our ancestors, their traditions, and their ideology. However, historical artifacts are subjected to significant changes over time. In addition, a number of copies and fakes can distort our understanding of history.

It should be noted that the artistic traditions of different historic periods (Romanesque art, Gothic, Renaissance, Baroque, etc.) are characterized by unique technological features. The use of methods from natural sciences for the comprehensive study of objects of cultural heritage helps to recover the technology of their fabrication as well as to clarify their attribution, provenance, and dating. In some cases cultural heritage studies allow us to understand the commercial communication between different countries in the past.

The main challenge facing researchers is the identification of materials used for the creation of a particular work of art without affecting its esthetic and artistic value. Nowadays, the most common analytical techniques used for cultural heritage studies are X-ray fluorescence (XRF), proton-induced X-ray/gamma-ray emission (PIXE/PIGE), laser-induced background spectroscopy (LIBS), and Fourier transform infrared spectroscopy (FTIR). All techniques mentioned above meet a number of requirements, being nondestructive, rapid, and sensitive.

The application of SERS for art materials identification represents a particular interest. Firstly, the bands in SERS spectra as well as Raman spectra are conditioned by the chemical structure of the substance. Secondly, plasmonic effects provide huge enhancement of the signal obtained and quenching of the luminescence background. An enormous increase in the technique's sensitivity of several orders of magnitude is achieved. Moreover, the amount of the sample to be taken for the testing could be essentially reduced. Thus, SERS offers a universal ultrasensitive method for identification of both organic and inorganic art materials with high spatial resolution.

However, the SERS application in practice is not a trivial task for many reasons. There are some requirements that should be satisfied to provide the interaction between the substance under study and surface plasmons of nanostructures. To summarize, a number of scientific problems should be explored for the successful implementation of SERS in the daily practice of cultural heritage studies. The efficiency of different SERS-active substrates can be engineered, and with sample treatment procedures optimization this is a flexible and increasingly powerful method. The investigation of optical properties of different art materials and the creation of databases makes the technique especially attractive for forensic work for cultural heritage applications.

5.2.4 State-of-the-art

5.2.4.1 Historical perspective

The first case of SERS application for the study of cultural heritage was demonstrated in 1987 [2]. The essentially enhanced Raman spectra of a textile sample were obtained with the help of a silver electrode, and madder was identified in the fibers as a colorant. However, back then this technique could not be widely used because of the lack of reproducibility and sensitivity of the instrumentation available at the time. The number of publications reporting later practical applications of SERS for art materials characterization has increased along with the development of SERS methodology.

The majority of the research to date has been carried out on reference organic materials. The spectral properties of different organic colorants have been well investigated, and the efficiency of various metal substrates for their characterization has been compared [3]. As a result, high-quality and detailed SERS spectra databases of organic dyes have been published [4–7].

To make the transition from the laboratory studies on microscope slides to direct measurements on museum artifacts, SERS-active nanostructures were prepared on different objects of cultural heritage (mural and canvas paintings, manuscripts and artwork on paper, textile and archeological artifacts). Since polychrome objects of cultural heritage are extremely complex systems for analytical study, smart sample treatment protocols (extraction, gas-solid hydrolysis, peelable gels, inkjet nanoparticle deposition, and laser ablation [LA]-SERS) have been designed and evaluated.

There are many successful applications of SERS where case studies of paintings, textiles, manuscripts, artworks from glass and ceramics, etc., have been reported [8–10].

It should be noted that using SERS for inorganic pigments identification was a huge challenge for scientists for a long time. Some successful results have been obtained in manuscripts and watercolors studies [11, 12]. Painting studies are more complex. In the work [13] the attempt of vermilion identification with the help of Ag nanoparticles was performed. However, the vermilion peaks did not appear in the presence of Ag colloids. Ultramarine and lead white have been identified using SERS in [14]. However, this result could be questionable because of a weak signal and high luminescence background. The systematic studies of SERS by reference inorganic art pigments have been demonstrated in recent articles [15–19]. Analytical protocols for colorful pastes identification have been proposed [19, 20].

During the last decades a significant practical experience in SERS application to art materials characterization and identification has been built up in the world's leading investigation centers such as The Metropolitan Museum of Art, the Getty Conservation Institute, The Art Institute of Chicago, the National Centre for Research and Restoration of Museums of France, etc.

5.2.4.2 Art pigments

Organic pigments

Historical organic dyes were produced from plants and insects and widely used in textiles, paintings, and other polychrome works of art [21, 22]. Here examples are the plants *Isatis tinctoria* L. and *Indigofera tinctoria* L., which are used for the preparation of indigo. This blue pigment was quite often used in European easel painting since the Middle Ages. Kermes (wingless insects living on certain species of European oaks) were the source for producing red pigments called carmine. Indian yellow pigment was derived from urine of cows that had been fed mango leaves. This pigment was favored for its great body and depth of tone. It was used by European artists in both oil paints and watercolors from the fifteenth to the nineteenth century. The chromophores of these pigments are molecules of indigotin, carminic acid, and euxanthone, respectively.

The development of chemistry in the nineteenth century has caused the occurrence of synthetic analogs of organic pigments as well as a number of new pigments. The modern classification of these pigments on chemical structure distinguishes several molecular classes such as anthraquinoids, flavonoids, alkaloids, curcuminoids, indigoids, etc. Within the classes dyes can be distinguished from each other by chromophores. Table 5.2.1 represents the most popular organic pigments in Europe and the time periods of their use.

Extensive SERS databases of anthraquinones [4–6, 23–25], flavonoids [4, 6, 26], indigoids [5–7], and other molecular classes [4–6, 27–29] have been published. In the works [30, 31] the correlation between molecular structures of chromophores and lines in SERS spectra have been traced for different classes of dyes.

Inorganic pigments

Inorganic pigments represent the silicates, salts, oxides, and hydroxides of different metals, and initially had been obtained from minerals. The chemical composition of natural inorganic pigments could vary depending on the deposit source of raw mineral.

Table 5.2.1: The most popular organic pigments in Europe and the time periods of their use.

Color	Molecular class	Dyes	Main chromophores	Natural source	Time period of use
Red	Anthraquinoid	Madder	Alizarin (R = H) Purpurin (R = OH)	Plant (root): *Rubia tinctoria*	Natural – from antiquity to the nineteenth century. Synthetic – from the end of the nineteenth century
		Kermes	Kermesic acid	Insects: Kermes	From antiquity to the end of the fifteenth century. Then replaced by cochineal
		Cochineal	Carminic acid	Insects: *Dactylopius coccus* (from Central America)	Natural – from the sixteenth century to the nineteenth century. Synthetic – from the end of the nineteenth century.
		Lac dye	Laccaic acid A	Insects (excretion): *Laccifer (Coccus) lacca*	In Europe – eighteenth/nineteenth century
Yellow	Flavonoid	Weld	Luteolin (R = OH) Apigenin (R = H)	Plant: *Reseda luteola*	From antiquity to the nineteenth century
	Alkaloid	Berberine	Berberine	Plants: *Hydrastis canadensis, Coptis chinensis, Berberis vulgaris, Berberis aristata, Berberis aquifolium*	From antiquity to the nineteenth century
	Carotenoid	Saffron	α-Crocin	Plant: *Crocus sativus*	From antiquity to the twentieth century
Blue	Indigoid	Indigo	Indigotin (R = H)	Plants: *Isatis tinctoria* L., *Indigofera tinctoria* L.	Natural – from antiquity to the nineteenth century. Synthetic – from 1878
		Tyrian purple	6,6'-Dibromindigotin (R = Br)	Mollusks or whelks: Murex brandaris, Purpura haemostoma, Purpura lapillus, Carpillus purpura	From antiquity to the fifteenth/sixteenth century

Silicates are pigments derived from kaolinite, glauconite, dioptase, and lapis lazuli. The salts are sulfides, sulfates, and carbonates of various metals. Mercury sulfide (HgS) is a cinnabar, copper carbonates could be in the form of malachite or azurite, copper sulfate is posnjakite, and lead carbonate is cerussite. Lead oxides could be in the form of massicot or minium.

The red and yellow ochers, siennas, and umbers form a broad class of hydroxide pigments. The chromophores of these pigments are iron hydroxides. Their color could vary from light yellow to dark red or brown depending on the grain size as well as the content of impurities.

According to Pliny [32], the ancient Greek painters used only four colors: white, yellow, red, and black. The first one was melinum, which is a white pigment from the island Melos in the Aegean Sea. The second one was Attic yellow. It is a natural pigment, probably yellow ocher. Sinopia was used as the red pigment. It could be found in many places, but Sinopia from the island Lemnos was especially appreciated. The last one was a black paint called atramentum. This paint was obtained via burning of grape bones and ivory.

In the time of Pliny (first century AD), the number of colors amounted to about 20. In the beginning of the Renaissance the artists used approximately the same number of pigments. It should be noted that the artists of the fourth and third centuries BC worked also with white lead, cinnabar, orpiment, azurite, verdigris, malachite, lapis lazuli, and some others.

Synthetic pigments are varied, namely silicates, salts, oxides, and hydroxides of different metals that are being produced artificially. The most frequent ancient synthetic pigments were lead white and verdigris.

The assortment of artificial pigments began to expand in the eighteenth century. For example, the Prussian blue pigment was synthesized by the colormaker Diesbach of Berlin in about 1704. Three green pigments were created at the end of the eighteenth century: Scheele's green (1778), cobalt green (made by Rinmann in 1780), and chrome green (invented by Vauquelin in 1797). The number of inorganic pigments exceeded 300 in the nineteenth/twentieth century. Table 5.2.2 represents the most popular inorganic pigments in Europe and the time periods of their use.

A systematic SERS study of reference inorganic pigments has not been performed. The state-of-play is that there are only few publications reporting the use of SERS for cinnabar (vermilion), Prussian blue, ultramarine blue (lazurite), cerulean, malachite, chrome green, chrome yellow, emerald green, realgar, orpiment, and cobalt green [11, 12, 15–20]. As has been explained, the plasmonic effects lead only to a simple enhancement of the incident and emitted light intensity. Enhancement then equally occurs for all originally existing Raman lines. Thus, the databases of Raman spectra of inorganic pigments [5, 33, 34] could be used for SERS spectra interpretation.

Table 5.2.2: The most popular inorganic pigments in Europe and the time periods of their use.

Names (natural/synthetic)	Chemical formula	Color	Mineral	Time period of use
Silicates				
Glauconite	$(K, Na, Ca) \times (Fe^{3+}, Mg, Fe^{2+}, Al)_2[(Al,Si)Si_3O_{10}](OH)_2 \times H_2O$	Green	Glauconite	From antiquity
Lapis lazuli/ultramarine	$Na_6Ca_2[AlSiO_4]_6([SO_4]1_{1,4})S_{0,6}$	Blue/ blue, green, red	Lapis lazuli	Natural – from antiquity. Synthetic – from 1818, Guimet
Salts				
Cinnabar/vermilion	HgS	Red	Cinnabar	Natural – from antiquity. Synthetic – from the thirteenth to the nineteenth century
Orpiment	As_2S_3	Yellow	Orpiment	From antiquity to the nineteenth century
Malachite	$CuCO_3 \cdot Cu(OH)_2$	Green	Malachite	Natural – from antiquity to the sixteenth century. Synthetic – from the Middle Ages
Azurite	$2Cu[CO_3] \times Cu[OH]_2$	Blue	Azurite	Natural – from antiquity. Synthetic – from the seventeenth century
Cerussite/lead white	$Pb[CO_3]/2PbCO_3 \cdot Pb(OH)_2$	White	Cerussite	From antiquity
Posnjakite	$Cu_4SO_4(OH)_5H_2O$	Green	posnjakite	Fifteenth/sixteenth century
Lead-tin yellow, Type I	Pb_2SnO_4	Yellow		Natural – from the thirteenth to the eighteenth century. Synthetic – from 1940
Lead-tin yellow, Type II	$Pb(Sn,Si)O_3$	Yellow		From the eighteenth century
Prussian blue	$Fe_4[Fe(CN)_6]_3$	Blue		From 1704, Diesbach
Scheele's green	$Cu_2As_2O_5$	Green		From 1778, Scheele
Chrome yellow	$PbCrO_4$	Yellow		From 1809, Vauquelin
Cadmium yellow	CdS	Yellow		From 1817, Stromeyer
Cadmium red	$CdSe$	Red		From 1919
Cerulean	$CoO \cdot n SnO_2$	Blue		From 1860, Rowney

Table 5.2.2 (continued)

Names (natural/synthetic)	Chemical formula	Color	Mineral	Time period of use
Oxides				
Massicot	PbO	Yellow	Massicot	From antiquity to the seventeenth century
Red lead	Pb_3O_4	Red	Minium	From antiquity to the nineteenth century
Cobalt green	$ZnO \cdot nCoO$	Green		From 1780, Rinmann
Chrome green	Cr_2O_3	Green		From 1797, Vauquelin
Cobalt blue	$CoO \cdot Al_2O_3$	Blue		From 1804, Thénard
Zinc white	ZnO	White		In watercolors – from 1834. In oil – from 1844, LeClaire
Titanium white	TiO_2	White		From 1920
Hydroxides				
Ocher	$Fe_2O_3 \cdot nH_2O$ + clay minerals	Red, yellow, brown	Goethite, hematite, magnetite	From antiquity
Umber	$Fe_2O_3 \cdot nH_2O + MnO_2$ + clay minerals	Brown	Goethite, hematite, magnetite, manganese oxide	From antiquity
Sienna	$Fe_2O_3 + MnO_2$ + clay minerals	Brown	Goethite, hematite, magnetite, manganese oxide	From antiquity
Mars	$Fe_2O_3 \cdot nH_2O + CaCO_3$	Red, yellow, brown, black		From the nineteenth century

5.2.4.3 SERS-active substrates for cultural heritage studies

Colloidal nanoparticles and nanoparticle-based plasmonic films

The colloidal nanoparticles that have been mentioned in connection with the study of art materials were prepared by means of chemical reduction [35–39], LA [40, 41], and photoreduction [42–45]. The main advantage of LA and photoreduction techniques is the absence of any reducing or capping agents that may result in an observation of spurious bands in the SERS spectra [8, 46–48]. The chemical reduction methods offer better control of the synthesis process. Working with the protocols of wet chemistry it is possible to obtain nanoparticles with different shapes. The use of nonspherical colloids (triangles, cubes, flowers, stars) for SERS applications represents a particular interest due to the huge local electric field concentration on the sharp edges [49–53].

Gold nanorods (NRs) offer good prospects as SERS-active substrates due to their tunable longitudinal plasmon bands and the "lightning rod" effect on the surface enhancement [54]. In Figure 5.2.1a–c scanning electron microscopy images of gold NRs with different aspect ratios (ARs) as well as their optical density spectra are presented. The optical density spectra here are characterized by transverse and longitudinal localized surface plasmon resonance (LSPR) bands at 525 nm and 750–820 nm, respectively (Fig. 5.2.1d).

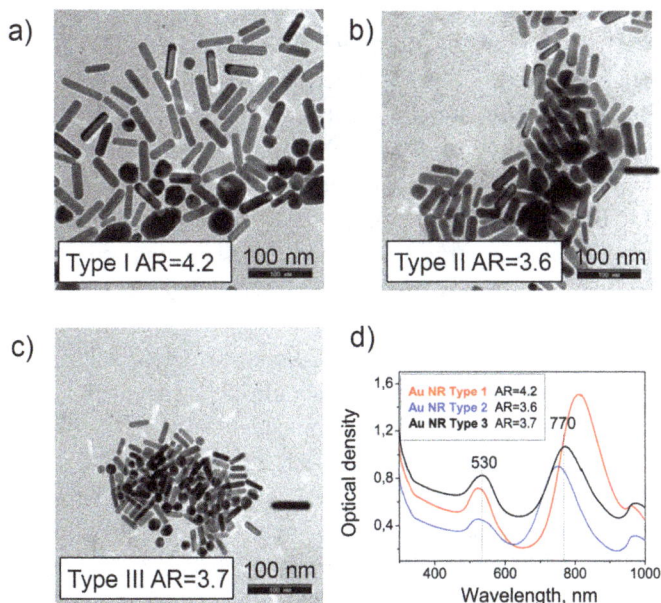

Figure 5.2.1: Transmission electron microscopy (TEM) images of Au nanorods of Type I (AR = 4.2) (a), Type II (AR = 3.6) (b), and Type III (AR = 3.7) (c) and their optical density spectra (d).

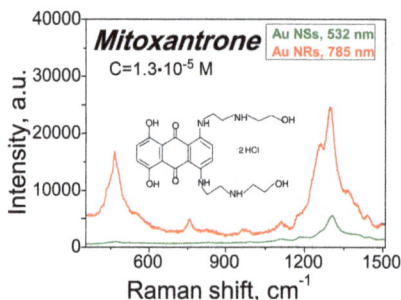

Figure 5.2.2: SERS spectra of mitoxantrone obtained with the help of Au nanospheres at λ = 532 nm and Au nanorods at λ = 785 nm.

The position of the longitudinal LSPR band depends on the AR of gold NRs and can be tuned over a wide range. Obviously, the maximal efficiency of these nanostructures will be achieved using near-infrared (NIR) radiation for Raman scattering excitation. The comparison of SERS efficiency for mitoxantrone detection in the cases of using gold NRs and gold nanospheres (NSs) at the resonance excitation wavelengths (785 nm and 532 nm, respectively) (Fig. 5.2.2) shows a higher sensitivity in the case of using gold NRs. Similar results have been obtained by Nikoobakht and El-Sayed [55].

Moreover, since the majority of materials are characterized by transparency in the infrared region, the use of NIR radiation for SERS excitation addresses the major problem of the high luminescence background. The luminescence quenching must be solved in order to maximize the EF.

Until today these nanostructures have been applied only in model experiments on organic and inorganic pigments detection. It should be noted that the initial hexade-cyltrimethylammonium bromide (CTAB)-stabilized NRs provide more effective SERS for organic molecules (perhaps, due to partially replacing CTAB, providing closer "metal–analyte" contact, and giving an additional contribution to the chemical SERS enhancement mechanism). Since inorganic microcrystals are not able to replace the CTAB, the modification of the gold NRs surface with small L-cysteine molecules was proposed to obtain huge enhancement of Raman scattering by inorganic molecules (see Fig. 5.2.3 [56]).

Plasmonic films are usually fabricated from colloidal nanoparticles by layer-by-layer electrostatic deposition or laser photoreduction. It should be noted that nanoparticle aggregation occurs in plasmonic films during the process of immobilization. As a result, the LSPR band in optical density spectra of such nanostructures becomes broader and shifts into the long-wave region (Fig. 5.2.4a). Localized regions of intense electromagnetic fields could occur in the nanoparticles aggregation. As was previously shown by Shalaev [57] the local enhancement in the hotspots of nanoparticle aggregates could be several orders of magnitude larger than the average.

In some cases the redistribution of the electromagnetic field between two or more metallic nanoparticles located at a distance of a few nanometers from each other leads

a)

b)

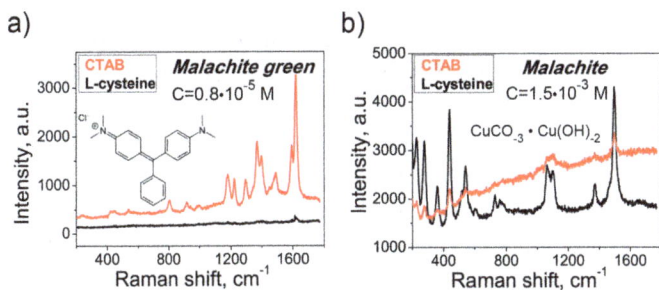

Figure 5.2.3: SERS spectra of the organic colorant malachite green (a) and the inorganic pigment malachite (b) obtained with the help of CTAB-stabilized and L-cysteine-modified Au nanorods at $\lambda = 531\,nm$.

a)

b)

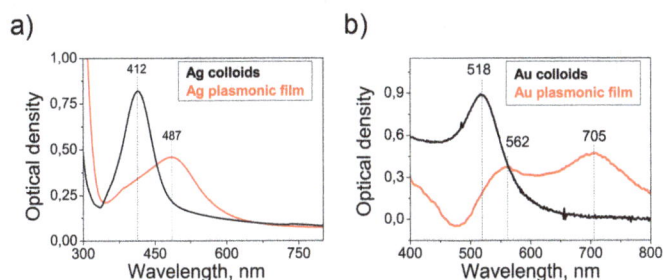

Figure 5.2.4: The optical density spectra of Ag (a) and Au (b) colloids and plasmonic films prepared on their basis.

to the appearance of a second LSPR band in the long-wave region of optical density spectra [58] (Fig. 5.2.4b), although this process is rather incidental and cannot be predefined by synthesis conditions (compare Fig. 5.2.5a and b).

Solid-state substrates

Silver and gold island films (IFs) have been demonstrated as SERS-active substrates in model experiments on the characterization of several organic molecules [23, 59, 60]. A variation of such kinds of substrates is a SERS-active Ag-Al$_2$O$_3$ support that has been initially tested on reference solutions of anthraquinoid dyes, and then was successfully applied for alizarin identification in silk textile [61].

Silver or gold films over NSs (AgFONs and AuFONs) are considered as very promising SERS-active substrates due to their highly ordered, uniform surface. These nanostructures are fabricated by metal vapor deposition on top of silica or polystyrene microspheres obtained on glass slides as the results of a self-assembly process [62]. Also, crystallographic surfaces of colloidal crystals could be used as the templates [63]. These supports are characterized by excellent stability and the possibility of tuning their LSPR band by adjusting the diameter of the spheres and the amount of metal deposited onto them [8]. These substrates have been successfully applied to obtain the high-resolution SERRS spectra of organic colorants such as alizarin, purpurin, carminic acid, laccaic acid, etc. [64, 65]. However, AgFONs are prone to carbon contamination, especially during the silver deposition phase. In this context several strategies for cleaning substrates have been reviewed in [66].

Ge/Si nanostructures covered with silver or gold nanoparticles of 10–20 nm in size are now considered as very good prospective SERS-active substrates due to their nanotextured surfaces. The Ge/Si nanostructures could be obtained by chemical vapor deposition [67] or molecular-beam epitaxy [68]. The surfaces of these nanostructures through manufacture form islets in the forms of pyramids or raspberry-shaped structures, with base diameters of 200 nm and heights of up to 50 nm, respectively (Fig. 5.2.5a and b). The surface topography of these substrates is well reproducible, and their fabrication techniques are now well established in connection with a wide range of applications of Si/Ge nanostructures in optoelectronics [69].

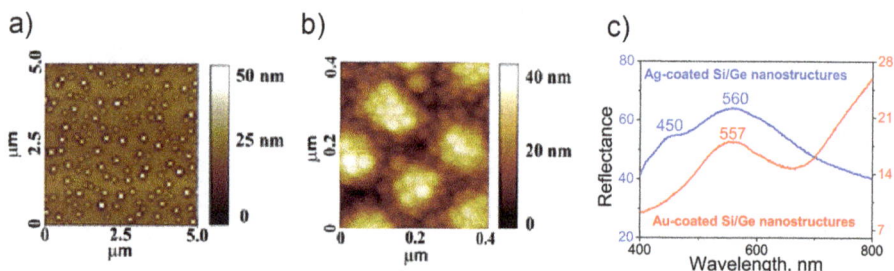

Figure 5.2.5: Atomic force microscopy (AFM) images of Au-coated (a) and Ag-coated Si/Ge nanostructures (b) as well as their reflectance spectra (c).

The nanostructures are characterized by excellent plasmonic properties (see reflectance spectra in Fig. 5.2.5c) and have been initially proposed for the systematic study of ultramarine and cinnabar microcrystals by SERS [15, 17, 18]. The local EF was estimated to as high as 10^5–10^7. In the case of using a pyramid-shaped topography the huge enhancement probably could be explained by the electric field concentration on the sharp edges. In the case of using a raspberry-shaped topography the huge enhancement was apparently caused by "hotspots" in the nanoparticle aggregates.

The nature of the metal nanostructures

Silver and gold nanoparticles have plasmon resonance bands in the visible and NIR spectral ranges and can thus provide the best enhancement to excite Raman modes in these wavelength ranges. Therefore, they are commonly used metals for SERS.

While the majority of the experiments on organic molecules was carried out using silver-based SERS-active substrates, the Au-based substrates appeared more suitable for inorganic pigments study. The experimental results obtained on reference inorganic pigments show that the use of Au-based nanostructures provided the higher signal-to-noise ratio in comparison with the results obtained with the use of Ag-based ones [19]. Moreover, the use of Ag-based nanostructures provides enhancement not only of chromophore peaks but also of the C-C bands at 1375–1600 cm^{-1} caused by the presence of organic components (sodium citrate, polycations PDADMAC) in the samples.

This could be explained with the help of theoretical calculations performed for silver and gold nanoparticles, taking into account that the spatial redistribution of an electromagnetic field occurs both at the frequency of the incident radiation and at the frequency of scattered radiation [70]. Calculations were made under the assumption that the excitation wavenumber, emission wavenumber, and plasmon resonance wavenumber coincide.

As was demonstrated, the maximal enhancement provided by silver and gold nanoparticles occurs at a physically reasonable distance of about 2–2.5 Å because of the core polarization effect [71] and exceeds 10 and five orders of magnitude, respectively (Fig. 5.2.6). However, the enhancement decays rapidly with particle–probe distance [19]. Since the EFs provided by Ag nanoparticles are five orders of magnitude higher than EFs provided by Au nanoparticles, Ag nanoparticles have a larger area of efficient interactions. The consequence is that more substances could appear in the area of efficient interactions with the Ag nanoparticles. In contrast, the smaller EF of the gold nanoparticles provides an increase of selectivity in the SERS spectra.

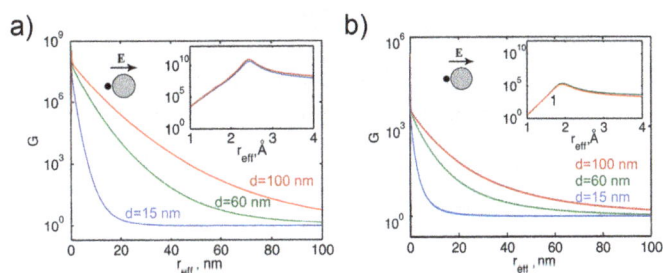

Figure 5.2.6: Calculated Raman scattering enhancement factor G versus distance r_{eff} between a point-like probe and the surface of silver (a) and gold (b) particles with diameters 15 nm, 60 nm, and 100 nm for normal orientation of the induced dipole moment of a probe. The insets show G (r_{eff}) dependences in the close vicinity of the nanoparticles up to 4 Å [19].

5.2.4.4 Sample treatment

The study of objects of cultural heritage offers a great challenge for the investigators because of the heterogeneity of art materials along with the extremely small amount of the sample that could be taken for the analysis. Since huge enhancement could be observed in the close vicinity of noble metal nanoparticles (see Fig. 5.2.6), the main task of the sample treatment steps must aim at providing close interaction between noble metal nanoparticles and organic dyes or the inorganic pigment chromophores.

To date, a number of nondestructive as well as microdestructive approaches for sample treatment have been developed. Microdestructive approaches require a microscopic probe detached from the artwork. Nondestructive approaches allow the study of the artwork without any sampling procedures.

Microdestructive approaches

a) Direct application

Direct application techniques are based on dripping a few microliters of colloids [72, 73] or *colloidal pastes* [11, 65, 74, 75] on top of untreated samples. *Colloidal pastes* produced by colloids centrifuging and concentration provide superior surface coverage to regular colloids and may better preserve the integrity of the substrate. These approaches have been successfully applied in a number of case studies. Different kinds of anthraquinoid dyes were identified in painted objects of French decorative art of the eighteenth century [72], historical textiles [74, 76], Mary Cassatt's pastele [75], mural and canvas paintings [77–79], and wooden statues [77, 79]. Several types of organic as well as numerous inorganic pigments were detected in Islamic manuscripts and Winslow Homer's watercolor [11, 12]. Unfortunately, the application of colloids or colloidal pastes directly on the untreated samples does not provide close interaction between probe molecules and nanoparticles, especially in the case of easel paintings. Moreover, the coatings obtained are characterized by extreme unevenness and consist of sparse islands of various sizes and thicknesses. This results in a lack of reproducibility of experimental data [76, 78].

One more variant of direct application is in situ photoreduction of nanoparticles performed for dyed fibers. For this a fiber fragment was fixed to a cover slide. The central well of a glass slide was filled with $AgNO_3$ solution and covered with the slide supporting the fixed fiber. Then, a laser beam was focused on the surface of the fiber to induce the formation of immobilized Ag nanoparticles on the fiber [43, 44].

b) Extraction techniques

These techniques involve the of pigment from the object (textile fiber, paint layer, etc.), followed by its deposition on SERS-active substrates. The solid-state substrates such as those described above as well as the colloids and plasmonic films on their base are both commonly used to enhance Raman scattering from art materials.

A problem associated with the use of strong acids and alkali such as HCl and NaOH, which were initially used as the reagents for organic dyes extraction, is that they resulted in damage to the host material and the inhibition of the dye–nanoparticles interaction [4, 48]. Thus, milder extracting reagents (pyridine, mixtures of formic acid and methanol, as well as methanol and HCl) have been proposed for colorants extraction [6, 80].

For the extraction of inorganic pigments from the paint layers techniques based on dissolving samples in alcohol and chloroform solutions under slight heating have been proposed [19]. The experimental data obtained with the use of Au colloids as the SERS-active substrates demonstrate the efficiency of the proposed technique for identification of both organic dyes and inorganic pigments (Fig. 5.2.7).

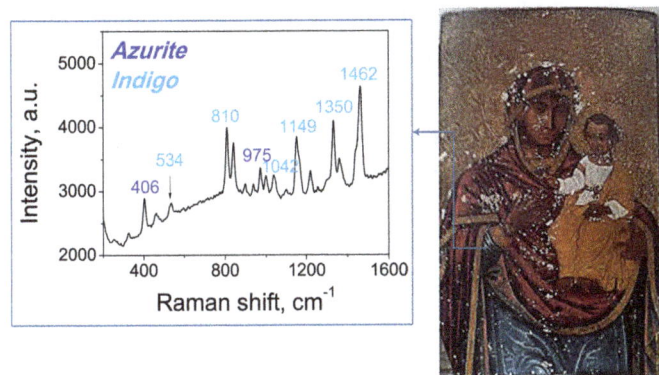

Figure 5.2.7: SERS spectrum of blue paint from the Belorussian icon "Madonna Hodegetria" (seventeenth century) obtained with the help of Au colloids at λ = 531 nm [19].

A variation of this methodology is the so-called "sandwich" configuration, where analyte is placed between two plasmonic films. Theoretical calculations performed for two silver nanoparticles placed close to each other predicted that in the case of normal orientation of the dipole moment the photon density of states between two nanoparticles is five times higher than in the close vicinity of a single nanoparticle [81].

The practical confirmation of the theory was recently presented in [82]. To prepare a "sandwich" configuration, analytes were dripped on the gold plasmon film and dried at room temperature. Then 5 µl of gold sol was dripped on the analytes. In Figure 5.2.8 the SERS spectra of malachite green obtained for different configurations (on plasmonic film, under a plasmonic film, and between films) are presented. The use of the configuration where colorant is placed under plasmonic film provides an enhancement on the order of one magnitude higher than the configuration where colorant is placed on the film. This observation could perhaps be explained by the access reduction of oxygen as well as the changing of the dielectric constant of the medium in contact with the specimens. The use of the sandwich configuration provides an additional five-fold enhancement. This result is in good agreement with the theoretical calculations mentioned above.

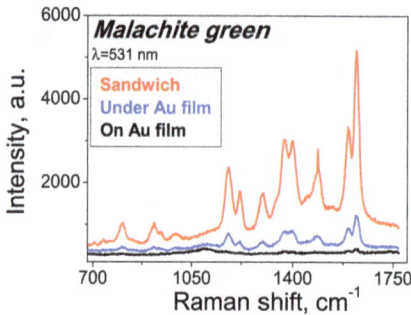

Figure 5.2.8: SERS spectra of malachite green obtained for the different configurations (on the plasmonic film, under the plasmonic film, and between plasmonic films) at $\lambda = 531$ nm [82].

c) Gas-solid hydrolysis

A *gas-solid hydrolysis* of the dye–metal complex is carried out by exposing the sample to a hydrofluoric acid vapor in a closed polyethylene microchamber [4, 30]. Following HF hydrolysis, samples are covered with colloids and analyzed by SERS. This method has become a standard approach for identification of natural dyes in a great variety of polychrome historical objects and works of art [83–85] due to the possibility to detect the colorants with ultrahigh sensitivity.

Nondestructive approaches

a) Peelable gels

Peelable gels have been developed to obtain detachable SERS-active substrates which provide high-sensitivity detection of organic dyes while preserving the esthetic and artistic value of the artwork as well as its physical integrity. Such substrates represent a mixture of silver nanoparticles with hydrophilic polymer gels with small Raman cross-section (for example, polycarbophil polymer [86], methylcellulose matrix [87], gelatin [88]). These gels are applied directly onto the sample surface and used for the SERS measurements after drying.

The efficiency of the method has been verified on anthraquinoid dyes in mock-tempera panel paintings [87] as well as on paper substrates [88].

An alternative matrix transfer procedure is based on using a polymer hydrogel loaded with a solution containing water, dimethylformamide, and disodium ethylene-diaminetetraacetic acid (EDTA) [8, 89]. This gel combines the action of the organic solvent and chelating agent and allows the extraction of small amounts of colorants from the target surface to be analyzed with SERS by focusing the laser beam onto the gel surface.

A variation of the method mentioned above involves the application of a cheap and environmentally friendly Ag-agar hydrogel [8, 90]. The efficiency of dye extraction by nanocomposite Ag-agar gel initially has been tested on three anthraquinoid dyes (alizarin, purpurin, and carminic acid) as well as on mock-up textile dyed with the same colorants, then applied for the study of pre-Columbian textile and Medici's sixteenth century tapestry [91]. The initial formulation was modified by the addition of EDTA, which plays a key role as a matrix stabilizer and improves the gel's microextractive performances [92]. It should be noted that gel materials and solvent systems are not universal and could be different for different analytes.

b) Inkjet nanoparticle deposition

Inkjet technology based on commercially available thermal and piezoelectric inkjet heads allows to deliver silver colloids of 50–150 μm in size onto surfaces for analysis with great precision [93]. The piezoelectric technology uses piezoelements to propel ink onto the substrate, while the thermal method uses heat to create air bubbles, causing the ink to be ejected. Piezoelements are much more amenable to the different kinds of solvents used for printing. The main advantage of this approach is the high spatial resolution that provides highly selective investigation of single features as well as the parts of a feature of complex objects. The method has been successfully applied for the analysis of textile fibers, gel pen ink writing on paper, and a Japanese woodblock print of the nineteenth century [93]. It should be noted that microscopic traces of silver nanoparticles stay on the surface of the object under study, which is considered as the main drawback of this technique, especially in the case of investigating invaluable works of art.

c) LA-SERS

This approach represents the combination of LA microsampling, which could be provided with high spatial resolution, and SERS [94]. The methodology consists of two steps. First, laser radiation is used to ablate the target dye molecules onto the optimized SERS-active surface, i. e., a silver nanoisland film; then, the ablated analytes are excited with a continuous laser to produce SERS spectra [95]. The LA-SERS approach should be applied to detect and identify water-insoluble compounds that would be hard to probe with regular silver colloids and other standard substrates. The possibility here is to focus the analysis on single particles or on thin-layer stratigraphy, which extends the applications of this approach to the study of complex, multilayered, and heterogeneous samples from historical paintings. The technique was successfully applied to resolving multicompound mixtures, dyes in ancient leather samples, and paintings [96].

The treatment methods are not universal. The choice of a suitable approach depends on the class of material and the experimental conditions. SERS substrates that have been mentioned in connection with art materials characterization as well as main treatment procedures for sample preparation are summarized in Table 5.2.3.

Table 5.2.3: Treatment procedures for art pigments characterization by SERS.

Methodology	SERS substrates		Application	
			Model	Real
Direct application				
	Colloids and plasmonic films on their base	Wet chemistry [35–39]	Ultramarine [15, 19] Cinnabar [18] Cerulean [19] Cobalt green [19] Chrome green [19] Malachite [19] Sepia [28]	Saffron, vermilion, carmine, azurite, lazurite, realgar/para-realgar, emerald green, orpiment, and indigo in ancient Islamic manuscripts of Morocco [11] Madder and Cochineal lake in Weaving shuttle (inv13377, Musée des Arts Décoratifs, Paris), Commode "Adélaïde" (inv1965, Musée National du Château de Versailles), Silver case "Carnet de bal" (inv11854, Musée des arts Décoratifs, Paris), and Glass Holder (invPR.2013.2.515, Musée des Arts Décoratifs, Paris) [72] Purple pigment in archeological artifacts from ancient Minoan city of Akrotiri, on the Santorini Island in Greece (1650 BC) and Greco–Roman pink cosmetics [73] Kermesic acid in red glaze in canvas painting "John the Baptist Bearing Witness" (Francesco Granacci, Florence (ca. 1510)) [79] Lac dye in red glaze in the Morgan Madonna. Virgin and Child in Majesty, France [79]
		Laser ablation [40, 41]	Alizarin [3]	
		Photoreduction [42–45]	Alizarin [42–44] Weld [44] Carminic acid [43] Purpurine [43] Madder [44]	

Table 5.2.3 (continued)

Methodology	SERS substrates	Application Model	Real
	Colloidal pastes	Purpurin, carminic acid, madder and cape jasmine in mock-up fibers [65]	Madder, Indian purple, chrome yellow, vermilion, cochineal, carmine in watercolor "For to Be a Farmer's Boy" by Winslow Homer, 1887 [12]
			Lac dye in fiber from carpet from Bursa/Turkey/Istanbul [65]
			Madder in pre-Columbian textile [74]
			Madder in the pastel "Sketch of Margaret Sloane, Looking Right" [75]
			Madder in the painted wooden statue "Virgin with the Child" (Chapel of the Nativity of the Virgin in La Salle, Italy) [77]
			Kermes lake in mural painting "Sant'Anna Metterza" (Santi Pietro e Orso church, Italy) [77]
			Carmine lake in canvas paintings "Portrait of William Nelson" by Robert Feke (1748–1750) and "Portrait of Isaac Bare" by Sir Joshua Reynolds (1766) [78]
Solid-state substrates	Island films	Alizarin, purpurin, kermesic acid, cochineal lake, lac dye, eosin Y, and brazilwood [23]	
		Carminic acid, lac dye [59]	
		Mitoxantrone [60]	
	Films over nanospheres	Eosin Y [29]	
		Mitoxantrone [63]	
		Alizarin and crocin [65]	
		Purpurin, carminic acid, madder, and cape jasmine in mock-up fibers [65]	
	Si/Ge nanostructures	Ultramarine blue [16, 17]	
		Cinnabar [18]	

Table 5.2.3 (continued)

Methodology	SERS substrates	Application	
		Model	Real
Extraction			
	Colloids and Plasmonic films	Anthraquinones, flavonoids, naftoquinones, tannins, orchil dyes and redwoods [4]	Yellow dye in ancient wool threads from Royal Tumulus of In Aghelachem, Libyan Sahara (Garamantian period, second/third century AD) [6]
			Azurite and indigo in Belarusian icon, seventeenth century [19]
			Tyrian purple and madder in ancient textile [48]
			Indigo, madder, weld in Kaitag textiles, seventeenth/eighteenth century [80]
	Sandwich	Malachite green, Mitoxantrone [82]	
Gas-solid hydrolysis			
	Colloids	Anthraquinones, flavonoids, naftoquinones, tannins, orchil dyes and redwoods [4]	Berberine in Tibetan textile of the eighteenth century [27]
		Berberine [27]	Madder in leather from an ancient Egyptian chariot [79]
		Cochineal-based, madder lake-based, and alizarin-based pigments [30]	Madder lake in pink pigment from Corinth, Greece (second century BC) [83]
			Carminic acid in red glaze from canvas painting "The young sailor" by Henri Matisse, 1906. The Metropolitan Museum of Art [83]
			Cochineal lake in canvas painting "Woman Reading" by Édouard Manet, 1879/1880. The Art Institute of Chicago [84]
			Lac dye in nine illuminated manuscripts from medieval Portuguese monasteries (St. Mamede of Lorvão, Holy Cross of Coimbra, and St. Mary of Alcobaça) [85]

Table 5.2.3 (continued)

Methodology	SERS substrates		Application	
			Model	Real
Peelable gels				
	Colloids in	Polycarbophil [86]		
		Methylcellulose	Mock-painting with madder lake and carmine lake in egg tempera [87]	
		Gelatin	Crystal violet and madder [88]	
		EDTA		Crystal violet in Woodblock print on paper "Sekigahara Homare no Gaika" by Toyoharu Kunichika, 1892 [89] Madder in tapestry from "The hunt of the unicorn" series (South Netherlandish, 1495–1505) [89]
		Agar	Alizarin, purpurin, carminic acid in mock-up fibers [90] Turian purple in mock-up fibers [91] Carmine lake and Lac dye in mock-up panel [92] Madder, cochineal, and lac dye in mock-up textile [92]	Madder in pre-Columbian textile [90] Indigo in Italian tapestry of the sixteenth century [91]
Inkjet deposition				
	Colloids		Alizarin in mock-up fiber [93] Methyl violet ink [93]	Methyl violet in a Japanese print "Sekigahara Homare no Gaika" by Toyoharu Kunichika, 1892 [93]
LA-SERS			Crystal violet [95] Lac lake mock-up sample [96]	Madder in decorated dish of the sixteenth century and wooden "Cassone" of the twelfth to the fourteenth century [96] Kermesic lake in "The incredulity of San Thomas," by L. Signorelli (Loreto Basilica, Italy, fifteenth century) and "The adoration of the Shepherds," by Giorgione (fifteenth century) [96]

5.2.4.5 Case studies

SERS has well been proven as a microanalytical technique suitable for ultrasensitive detection of art materials in samples of textiles, archeological objects, pastels, paintings, etc. In a number of studies SERS results both provide important information for the artwork ascription and help in providing other significant historical data. For example, the detection of madder in the 4000-year-old Middle Kingdom leather fragment represents the earliest evidence of obtaining a dye from a plant source and manufacturing a lake pigment from it [9, 79]. SERS has offered the explanation of the fading processes that take place in madder- and cochineal-based purples and reds in watercolors and canvas paintings [11, 84]. The identification of faded colorants in artworks will provide useful insights into how the artifacts under study might have looked like right after completion.

The results obtained with SERS can play a crucial role during the ascription of the artworks. For example, the results obtained during the study of the Belarusian icon of the seventeenth century "Virgin Eleusa" (Fig. 5.2.9) provided the story of existence of the icon [97]. It was known from the inscription on the back side of the icon that the steward of Brest County Adam Spytak donated this icon to the temple in the village Semekhovichi (Pinsk district of the Brest region, Belarus) in 1656. The temple has been "...re-built in the unity with the Roman Catholic church..." after its destruction in a fire. Until today, there is only one church in the village Semekhovichi. It is the Orthodox Church of Nativity that was built in 1830 on the means of Belarusian philanthropist Wojciech Puslovsky. Obviously, the seventeenth century church has been also destroyed.

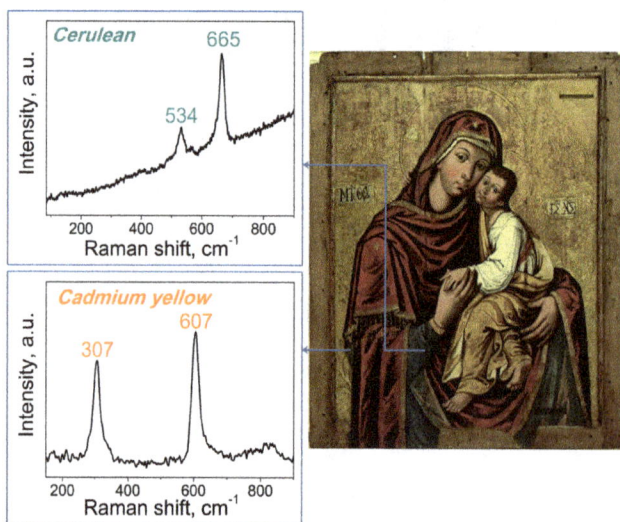

Figure 5.2.9: The Belarusian icon "Virgin Eleusa" (1656 r.) and the SERS spectra of renovation layers obtained with the help of Au colloids at λ = 488 nm [97].

The original art materials used for creating the icon as well as some features of local art traditions have been revealed in the result of examining the icon with the highly sensitive analytical techniques. Also several cases of interventions have been revealed. One of them was performed on a significant part of the icon despite good preservation of the original painting. Cadmium yellow and cerulean were identified by SERS in the colorful pastes of intervention layers. The concentrations of pigments were too low to be detected by other analytical techniques (LIBS, FTIR). It is well known that these pigments came into use after 1830. Therefore, the essential renovation most likely was carried out in this period in order to fit the icon in the new altar.

Sometimes the information about the origin of the identified pigments could be important in clarifying many questions during the ascription of artworks. For example, natural or synthetic pigments have been used for the creation of artwork. Natural inorganic pigments usually contain impurities caused by the joint growth of minerals in the rocks. Besides, the chemical composition of impurities often depends on the geographical location of rocks depositions.

For example, the wollastonite impurity detected in the blue paint of the canvas painting "The removing from the Cross" (Fig. 5.2.10a) is evidence of Chilean lazurite according to [98, 99] while the absence of any impurities in the blue paint of the Be-

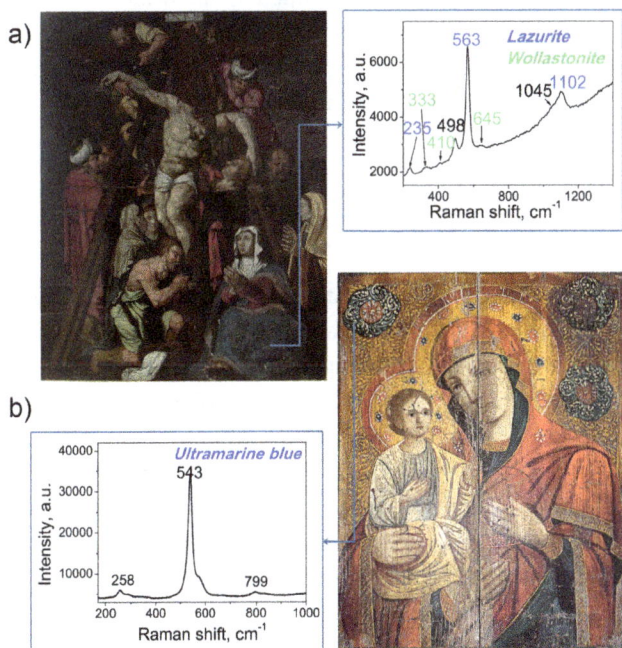

Figure 5.2.10: SERS spectrum of natural lazurite detected in the canvas painting "The removing from the Cross" (seventeenth century) (a) and synthetic ultramarine blue detected in the Belarusian folk icon "Three-handed Theotokos" (nineteenth century) (b).

larusian folk icon "Three-handed Theotokos" (Fig. 5.2.10b) allows it to be deduced that artificial ultramarine blue has been used for its creation. Ultramarine blue was initially synthesized in 1827. Thus, the Belarusian folk icon was created no earlier than the nineteenth century.

General comments

Today SERS is a unique analytical tool that has been successfully applied for the ultrasensitive identification of art materials in a number of case studies including textiles, polychrome objects, works of art on paper, paintings, etc. The considerable efforts and ingenuity of scientists have been applied to create noninvasive or minimally destructive approaches. As a result, a lot of smart sample treatment techniques along with SERS-active substrates have been developed in the last decades. Some of them have already been implemented in the daily practice of cultural heritage study. The results obtained played a crucial role during the clarification of several questions about the origin, history, and preservation of artworks.

Recently such high-resolution techniques like TERS and LA-SERS have been proposed. Although they are in their infancy as analytical techniques for routine application, the further developments of these approaches will open up many new opportunities, for example, in situ layer-by-layer materials identifications by SERS.

The combination of SERS with other analytical techniques is strongly recommended for the comprehensive study of objects of cultural heritage.

5.2.5 Some challenges and solutions

Despite the fact that SERS is successfully applied for cultural heritage studies, a number of questions still need to be addressed to improve and extend its utility. The main challenge is the development of truly nondestructive approaches for painting studies.

Recently, organic sols of gold nanoparticles synthesized in toluene and chloroform were proposed as prospective SERS-active substrates [100]. The organic phase by dissolving binders creates pores in the paint layer. Gold nanoparticles penetrate through the pores in the paint layer. As a result close interaction between pigments and gold nanoparticles occurs, leading to a huge increase in the SERS spectra intensity along with the decrease of luminescence background caused by the presence of the binding media.

In Figure 5.2.11, a comparison of the efficiency of the Au hydro- and organic sols is presented. Both SERS spectra obtained demonstrate the presence of ultramarine blue in the paint layer; however, in the case of using Au hydrosol the chromophore peaks appear on a huge luminescence background while the use of organic Au sol provides

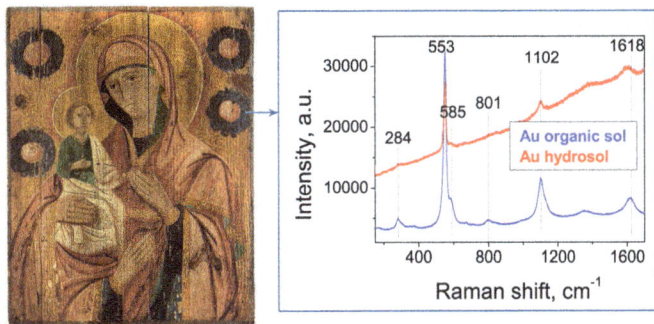

Figure 5.2.11: Comparative SERS spectra of blue paint from the Belarusian icon "Three-handed Theotokos" (nineteenth century) obtained with the help of Au hydro- and organic sols at $\lambda = 488$ nm [101].

an efficient enhancement of Raman lines along with luminescence quenching. Until today, this approach has been tested on microfragments of paint layers taken from different objects of Belarussian icons. However, the further development of this approach will be focused on the possibility to apply the substrates directly to the work of art, without need of removing a paint specimen [101].

Using organic sols seems the most promising nondestructive approach due to the possibility of direct application avoiding sampling as well as time-consuming treatment procedures. Moreover, this approach is suitable for studying both organic and inorganic pigments. Thus, it could also be useful for the identification of pigment mixtures.

5.2.6 Summary and impact

SERS-active substrates that have been mentioned so far in connection with the study of art materials are summarized in this chapter. Nanoparticles obtained via protocols of wet chemistry are most widely used in practical applications due to the possibility of fine-tuning their optical properties during the synthesis process. Such alternative methods as LA and photoreduction provide methods delivering nanoparticles without any organic agents. The experimental results demonstrate the high efficiency of these nanostructures both on reference materials and real objects of cultural heritage. However, they are not yet commonly used in the daily practice of the analysis of art work.

Different approaches aimed at increasing the SERS sensitivity have also been reviewed. One of these is the plasmonic film fabrication where localized regions of intense electromagnetic fields (hotspots) occur as a result of nanoparticle aggregation. The use of a so-called sandwich configuration where analyte is placed between two plasmonic films provides an additional enhancement of up to five times. The development of SERS-active substrates with a sharp-edged topography is also considered as a potentially valuable approach. However, such substrates have been applied only for reference materials study.

Since the preservation of the integrity of cultural heritage objects during the analytical study is strictly necessary a number of sample treatment approaches have been developed in the last decades. Some of them require microscopic samples to be detached from the artwork under investigation, while others could be applied directly to the object without any sampling procedures. The descriptions of these approaches as well as the results of their application are reviewed in this chapter. It should be noted that there is no universal approach. Some techniques are better matched for the characterization of textile fibers while others are more effective for investigating oil paints or watercolors. Hence, the choice of treatment method depends on the class of material and the experimental conditions. This is surely a case of a developing scientific art coming to assist in the preservation of art and important cultural items.

The results of case studies demonstrate that SERS applications play a crucial role for the examination of artworks due to the possibility to detect extremely low concentrations of the pigments or to distinguish natural from synthetic materials. The interpretation of SERS spectra was enabled by the detailed Raman and SERS databases that are continuously developed by accumulation from the results of different scientific groups working in art materials studies.

Bibliography

[1]   Kneipp K, Moskovits M, Kneipp H, editors. Surface-Enhanced Raman Scattering. Berlin, GE: Springer-Verlag; 2006.
[2]   Guineau B, Guichard V. Identification des colorants organiques naturels par microspectrometrie Raman de resonance et par effet Raman exalte de surface (SERS). Sydney: The Getty Conservation Institute; 1987. p. 659–66.
[3]   Canamares MV, Garcia-Ramos JV, Sanchez-Cortes S, Castillejo M, Oujja M. Comparative SERS effectiveness of silver nanoparticles prepared by different methods: a study of the enhancement factor and the interfacial properties. J Colloid Interface Sci. 2008;326:103–9.
[4]   Leona M, Steger J, Ferloni E. Application of surface-enhanced Raman scattering techniques to the ultra-sensitive identification of natural dyes in works of art. J Raman Spectrosc. 2006;37:981–92.
[5]   Burgio L, Clark RJH. Library of FT-Raman spectra of pigments, minerals, pigment media and varnishes, and supplement to existing library of Raman spectra of pigments with visible excitation. Spectrochim Acta, Part A. 2001;57:1491–521.
[6]   Bruni S, Guglielmi V, Pozzi F, Mercuri AM. Surface-enhanced Raman spectroscopy (SERS) on silver colloids for the identification of ancient textile dyes. Part II: pomegranate and sumac. J Raman Spectrosc. 2011;42:465–73.
[7]   Whitnall R, Shadi IT, Chowdry BZ. Case study: the analysis of dyes by SERRS. In: Edwards HGM, Chalmers JM, editors. Raman Spectroscopy in Archaeology and Art History. London, UK: Royal Society of Chemistry; 2005. p. 152–65.

[8]   Pozzi F, Leona M. Surface-enhanced Raman spectroscopy in art and archaeology. J Raman
      Spectrosc. 2016;47:67–77.
[9]   Casadio F, Leona M, Lombardi JR, Van Duyne R. Identification of organic colorants in
      fibers, paints, and glazes by surface enhanced Raman spectroscopy. Acc Chem Res.
      2010;43(6):782–91.
[10] Pozzi F, Zaleski S, Casadio F, Leona M, Lombardi JR, Van Duyne RP. Surface-enhanced Raman
      spectroscopy: using nanoparticles to detect trace amounts of colorants in works of art. In:
      Dillmann P, Bellot-Gurlet L, Nenner I, editors. Nanoscience and Cultural Heritage. Atlantis
      Press; 2016.
[11] El Bakkali A, Lamhasni T, Haddad M, AitLyazidi S, Sanchez-Cortes S, del Puerto Nevado E.
      Non-invasive micro-Raman, SERS and visible reflectance analyses of coloring materials in
      ancient Moroccan Islamic manuscripts. J Raman Spectrosc. 2013;44:114–20.
[12] Brosseau CL, Casadio F, Van Duyne RP. Revealing the invisible: using surface-enhanced Raman
      spectroscopy to identify minute remnants of color in Winslow Homer's colorless skies. J
      Raman Spectrosc. 2011;42:1305–10.
[13] Frano KA, Mayhew HE, Svoboda SA, Wustholz KL. Combined SERS and Raman analysis for
      the identification of red pigments in cross-sections from historic oil paintings. Analyst.
      2014;139:6450–5.
[14] Gui OM, Falamaş A, Barbu-Tudoran L, Aluaş M, Giambra B, Cinta Pinzaru S. Surface-enhanced
      Raman scattering (SERS) and complementary techniques applied for the investigation of an
      Italian cultural heritage canvas. J Raman Spectrosc. 2013;44:277–82.
[15] Klyachkovskaya EV, Guzatov DV, Strekal ND, Vaschenko SV, Harbachova AN, Belkov MV,
      Gaponenko SV. Enhancement of Raman scattering of light by ultramarine microcrystals in
      presence of silver nanoparticles. J Raman Spectrosc. 2012;43:741–4.
[16] Klyachkovskaya E, Strekal N, Motevich I, Vaschenko S, Harbachova A, Belkov M, Gaponenko
      S, Dais Ch, Sigg H, Stoica T, Grützmacher D. Enhanced Raman scattering of ultramarine on
      Au-coated Ge/Si-nanostructures. Plasmonics. 2011;6:413–8.
[17] Klyachkovskaya EV, Strekal ND, Motevich IG, Vashchenko SV, Valakh MY, Gorbacheva AN,
      Belkov MV, Gaponenko SV. Enhancement of Raman scattering by ultramarine using silver films
      on surface of germanium quantum dots on silicon. Opt Spectrosc. 2011;110:48–54.
[18] Shabunya-Klyachkovskaya EV, Gaponenko SV, Vaschenko SV, Stankevich VV, Stepina
      NP, Matsukovich AS. Plasmon enhancement of Raman scattering by mercury sulfide
      microcrystals. J Appl Spectrosc. 2014;81(3):399–403.
[19] Shabunya-Klyachkovskaya E, Kulakovich O, Vaschenko S, Guzatov D, Gaponenko S. Surface
      enhanced Raman spectroscopy application for art materials identification. Eur J Sci Theol.
      2016;12(3):211–20.
[20] Oakley LH, Fabian DM, Mayhew HE, Svoboda SA, Wustholz KL. Pretreatment strategies
      for SERS analysis of indigo and Prussian blue in aged painted surfaces. Anal Chem.
      2012;84:8006–12.
[21] Hofenk de Graaff JH. The Colourful Past – Origins, Chemistry and Identification of Natural
      Dyestuffs. London, UK: Archetype Publications; 2004.
[22] Cardon D. Natural Dyes: Sources, Tradition, Technology and Science. London, UK: Archetype
      Publications; 2007.
[23] Whitney AV, Van Duyne RP, Casadio F. An innovative surface-enhanced Raman spectroscopy
      (SERS) method for the identification of six traditional red lakes and dyestuffs. J Raman
      Spectrosc. 2006;37:993–1002.
[24] Canamares MV, Garcia-Ramos JV, Domingo C, Sanchez-Cortes S. Surface-enhanced Raman
      scattering study of the adsorption of the anthraquinone pigment alizarin on Ag nanoparticles.
      J Raman Spectrosc. 2004;35:921–7.

[25] Shadi IT, Chowdhry BZ, Snowden MJ, Withnall R. Semi-quantitative analysis of alizarin and purpurin by surface-enhanced resonance Raman spectroscopy (SERRS) using silver colloids. J Raman Spectrosc. 2004;35:800–7.

[26] Canamares MV, Lombardi JR, Leona M. Surface-enhanced Raman scattering of protoberberine alkaloids. J Raman Spectrosc. 2008;39:1907–14.

[27] Leona M, Lombardi JR. Identification of berberine in archaeological textiles by surface enhanced Raman spectroscopy. J Raman Spectrosc. 2007;38:853–8.

[28] Centeno SA, Shamir J. Surface enhanced Raman scattering (SERS) and FTIR characterization of the sepia melanin pigment used in works of art. J Mol Struct. 2008;873:149–59.

[29] Greeneltch NG, Davis AS, Valley NA, Casadio F, Schatz GC, Van Duyne RP, Shah NC. Near-infrared surface-enhanced Raman spectroscopy (NIR-SERS) for the identification of eosin Y: theoretical calculations and evaluation of two different nanoplasmonicsubstrates. J Phys Chem A. 2012;116:11863–9.

[30] Pozzi F, Lombardi JR, Leona M. Winsor & Newton original handbooks: a surface-enhanced Raman scattering (SERS) and Raman spectral database of dyes from modern watercolor pigments. Heritage Sci. 2013;1:23. 8p.

[31] Bruni S, Guglielmi V, Pozzi F. Historical organic dyes: a surface-enhanced Raman scattering (SERS) spectral database on Ag Lee–Meisel colloids aggregated by NaClO4. J Raman Spectrosc. 2011;42:1267–81.

[32] Rackham H. Pliny Natural History with an English Translation in Ten Volumes. vol. 9, LIBRI XXXIII–XXXV. London: W. Heinemann; 1965.

[33] Bell IM, Clark RJH, Gibbs PJ. Raman spectroscopic library of natural and synthetic pigments (pre- ≈ 1850 AD). Spectrochim Acta, Part A. 1997;53:2159–79.

[34] Castro K, Pérez-Alonso M, Rodríguez-Laso MD, Fernández LA, Madariaga JM. On-line FT-Raman and dispersive Raman spectra database of artists' materials (e-VISART database). Anal Bioanal Chem. 2005;382:248–58.

[35] Turkevich J, Stevenson PC, Hillier J. A study of the nucleation and growth processes in the synthesis of colloidal gold. Discuss Faraday Soc. 1951;11:55–75.

[36] Lee PC, Meisel D. Adsorption and surface-enhanced Raman of dyes on silver and gold sols. J Phys Chem. 1982;86:3391–5.

[37] Leopold N, Lendl B. A new method for fast preparation of highly surface-enhanced Raman scattering (SERS) active silver colloids at room temperature by reduction of silver nitrate with hydroxylamine hydrochloride. J Phys Chem B. 2003;107:5723–7.

[38] Creighton JA, Blatchford CG, Albrecht MG. Plasma resonance enhancement of Raman scattering by pyridine adsorbed on silver or gold sol particles of size comparable to the excitation wavelength. J Chem Soc Faraday Trans. 1979;II(75):790–8.

[39] Raveendran P, Fu J, Wallen SL. A simple and "green" method for the synthesis of Au, Ag, and Au–Ag alloynanoparticles. Green Chem. 2006;8:34–8.

[40] Fojtik A, Henglein A. Laser ablation of films and suspended particles in a solvent and colloid solutions. Ber Bunsenges Phys Chem. 1993;97:252–4.

[41] Amendola V, Meneghetti M. What controls the composition and the structure of nanomaterials generated by laser ablation in liquid solution? Phys Chem Chem Phys. 2013;15:3027–46.

[42] Canamares MV, Garcia-Ramos JV, Gómez-Varga JD, Domingo C, Sanchez-Cortes S. Ag nanoparticles prepared by laser photoreduction as substrates for in situ surface-enhanced Raman scattering analysis of dyes. Langmuir. 2007;23:5210–5.

[43] Jurasekova Z, del Puerto E, Bruno G, Garcia-Ramos JV, Sanchez-Cortes S, Domingo C. Extractionless non-hydrolysis surface-enhanced Raman spectroscopic detection of historical mordant dyes on textile fibers. J Raman Spectrosc. 2010;41:1455–61.

[44] Jurasekova Z, Domingo C, Garcia-Ramos JV, Sanchez-Cortes S. In situ detection of flavonoids in weld-dyed wool and silk textiles by surface-enhanced Raman scattering. J Raman Spectrosc. 2008;39:1309–12.

[45] Retko K, Ropreta P, Cerc Korošec R. Surface-enhanced Raman spectroscopy (SERS) analysis of organic colourants utilising a new UV-photoreduced substrate. J Raman Spectrosc. 2014;45:1140–6.

[46] Sánchez-Cortés S, Garcia-Ramos JV. Anomalous bands appearing in Raman surface-enhanced spectra. J Raman Spectrosc. 1998;29:365–71.

[47] Bell SEJ, Sirimuthu NMS. Surface-enhanced Raman spectroscopy as a probe of competitive binding by anions to citrate-reduced silver colloids. J Phys Chem A. 2005;109:7405–10.

[48] Bruni S, Guglielmi V, Pozzi F. Surface-enhanced Raman spectroscopy (SERS) on silver colloids for the identificationof ancient textile dyes: tyrian purple andmadder. J Raman Spectrosc. 2010;41:175–80.

[49] Sajanlal PR, Sreeprasad TS, Samal AK, Pradeep T. Anisotropic nanomaterials: structure, growth, assembly, and functions. Nano Reviews. 2011;2:5883. (62 p.).

[50] Rodrıguez-Lorenzo L, Alvarez-Puebla RA, Pastoriza-Santos I, Mazzucco S, Stephan O, Kociak M, Liz-Marzan LM, Garcıa de Abajo FJ. Zeptomol detection through controlled ultrasensitive surface-enhanced Raman scattering. J Am Chem Soc. 2009;131:4616–8.

[51] Nalbant Esenturk E, Hight Walker AR. Surface-enhanced Raman scattering spectroscopy via gold nanostars. J Raman Spectrosc. 2009;40:86–91.

[52] Mathew A, Sajanlal PR, Pradeep T. Molecular precursor-mediated tuning of gold mesostructures: synthesis and SERRS studies. J Cryst Growth. 2010;312:587–94.

[53] Tian F, Bonnier F, Casey A, Shanahan AE, Byrne HJ. Surface enhanced Raman scattering with gold nanoparticles: effect of particle shape. Anal Methods. 2014;6:9116–23.

[54] Gersten JI. The effect of surface-roughness on surface enhanced Raman scattering. J Chem Phys. 1980;72:5779–80.

[55] Nikoobakht B, El-Sayed MA. Surface-enhanced Raman scattering studies on aggregated gold nanorods. J Phys Chem A. 2003;107:3372–8.

[56] Shabunya-Klyachkovskaya E, Trotsiuk L, Matsukovich A, Vaschenko S, Kulakovich O. Electrostatic layer-by-layer deposited Au nanorod films for surface enhanced Raman scattering. In: The 2nd Iran-Belarus International Conference on Modern Applications of Nanotechnology, May 6–8, 2015, Minsk, Belarus. p. P083-1–P083-3.

[57] Shalaev VM, Sarychev AK. Nonlinear optics of random metal-dielectric films. Phys Rev B. 1998;57(20):13265–88.

[58] Jain PK, Huang W, El-Sayed MA. On the universal scaling behavior of the distance decay of plasmon coupling in metal nanoparticle pairs: a plasmon ruler equation. Nano Lett. 2007;7(7):2080–8.

[59] Whitney AV, Van Duyne RP, Casadio F. Silver Island films as substrate for surface-enhanced Raman spectroscopy (SERS): a methodological study on their application to artists' red dyestuffs. Proc SPIE. 2005;117–26.

[60] Strekal N, Oskirko V, Maskevich A, Maskevich S, Jardillier J-C, Nabiev I. Selective enhancement of Raman or fluorescence spectra of biomolecules using specifically annealed thick gold films. Biopolymers (Biospectroscopy). 2000;57:325.

[61] Chen K, Leona M, Vo-Dinh KC, Yan F, Wabuyele MB, Vo-Dinh T. Application of surface-enhanced Raman scattering (SERS) for the identification of anthraquinone dyes used in works of art. J Raman Spectrosc. 2006;37:520–7.

[62] Stiles PL, Dieringer JA, Shah NC, Van Duyne RP. Surface-enhanced Raman spectroscopy. Annu Rev Anal Chem. 2008;1:601–26.

[63] Gaponenko SV, Gaiduk AA, Kulakovich OS, Maskevich SA, Strekal ND, Prohorov OA, Shelekhina VM. Raman scattering enhancement using crystallographic surface of a colloidal crystal. JETP Lett. 2001;74(6):309–11.

[64] Whitney AV, Casadio F, Van Duyne RP. Identification and characterization of artists' red dyes and their mixtures by surface-enhanced Raman spectroscopy. Appl Spectrosc. 2007;61:994–1000.

[65] Brosseau CL, Gambardella A, Casadio F, Van Duyne RP, Grzywacz C, Wouters J. Ad-hoc SERS methodologies for the detection of artist dyestuffs: thin layer chromatography-surface enhanced Raman spectroscopy (TLCSERS) and in situ on the fiber analysis. Anal Chem. 2009;81:3056–62.

[66] Lin XM, Cui Y, Xu YH, Ren B, Tian ZQ. Surface-enhanced Raman spectroscopy: substrate-related issues. Anal Bioanal Chem. 2009;394:1729–45.

[67] Stoica T, Shushunova V, Dais C, Solak H, Grutzmacher D. Two-dimensional arrays of self-organized Ge islands obtained by chemical vapor deposition on pre-patterned silicon substrates. Nanotechnology. 2007;18:455307. 7p.

[68] Lobanov DN, Novikov AV, Vostokov NV, Drozdov YN, Yablonskiy AN, Krasilnik ZF, Stoffel M, Denker U, Schmidt OG. Growth and photoluminescence of self-assembled islands obtained during the deposition of Ge on a strained SiGe layer. Opt Mater. 2005;27:818–21.

[69] Krasilnik ZF, Novikov AV, Lobanov DN, Kudryavtsev KE, Antonov AV, Obolenskiy SV, Zakharov ND, Werner P. SiGe nanostructures with self-assembled islands for Si-based optoelectronics. Semicond Sci Technol. 2011;26:014029. 5p.

[70] Gaponenko S. Introduction to Nanophotonics. New York, NY, USA: Cambridge University Press; 2010.

[71] Ford GW, Weber WH. Electromagnetic interactions of molecules with metal surfaces. Phys Rep. 1984;113(4):195–287.

[72] Daher C, Drieu L, Bellot-Gurlet L, Percot A, Paris C, Le Hô AS. Combined approach of FT-Raman, SERS and IR micro-ATR spectroscopies to enlighten ancient technologies of painted and varnished works of art. J Raman Spectrosc. 2014;45:1207–14.

[73] Van Elslande E, Lecomte S, Le Hô AS. Micro-Raman spectroscopy (MRS) and surface-enhanced Raman scattering (SERS) on organic colourants in archaeological pigments. J Raman Spectrosc. 2008;39:1001–6.

[74] Wustholz KL, Brosseau CL, Casadio F, Van Duyne RP. Surface-enhanced Raman spectroscopy of dyes: from single molecules to the artists' canvas. Phys Chem Chem Phys. 2009;11:7350–9.

[75] Brosseau CL, Rayner K, Casadio F, Van Duyne RP, Grzywacz CM. Surface-enhanced Raman spectroscopy: an in-situ method to identify colorants in various artist media. Anal Chem. 2009;81:7443–7.

[76] Idone A, Gulmini M, Henry AI, Casadio F, Chang L, Appolonia L, Van Duyne RP, Shah NC. Silver colloidal pastes for dye analysis of reference and historical textile fibers using direct, extractionless, non-hydrolysis surface-enhanced Raman spectroscopy. Analyst. 2013;138:5895–903.

[77] Idone A, Aceto M, Diana E, Appolonia L, Gulmini M. Surface-enhanced Raman scattering for the analysis of red lake pigments in painting layers mounted in cross sections. J Raman Spectrosc. 2014;45:1127–32.

[78] Oakley LH, Dinehart SA, Svoboda SA, Wustholz KL. Identification of organic materials in historic oil paintings using correlated extractionless surface-enhanced Raman scattering and fluorescence microscopy. Anal Chem. 2011;83:3986–9.

[79] Leona M. Microanalysis of organic pigments and glazes in polychrome works of art by surface-enhanced resonance Raman scattering. Proc Natl Acad Sci USA. 2009;106:14757–62.

[80] Pozzi F, Poldi G, Bruni S, De Luca E, Guglielmi V. Multi-technique characterization of dyes in ancient kaitag textiles from Caucasus. Anthropol Sci. 2012;4(3):185–97.

[81] Guzatov DV, Klimov VV. Properties of spontaneous radiation of an atom located near a cluster of two spherical nanoparticles. Quantum Electron. 2005;35(10):891–900.

[82] Kulakovich O, Shabunya-Klyachkovskaya E, Matsukovich A, Trotsiuk L, Gaponenko S. Plasmonic enhancement of Raman scattering for sandwich metal–analyte configuration. J Appl Spectrosc. 2016;83(5):860–3.

[83]   Pozzi F, Lombardi JR, Bruni S, Leona M. Sample treatment considerations in the analysis of organic colorants by surface-enhanced Raman scattering. Anal Chem. 2012;84:3751–7.

[84]   Pozzi F, van den Berg KJ, Fiedler I, Casadio F. A systematic analysis of red lake pigments in French impressionist and post-impressionist paintings by surface-enhanced Raman spectroscopy (SERS). J Raman Spectrosc. 2014;45:1119–26.

[85]   Castro R, Pozzi F, Leona M, Melo MJ. Combining SERS and microspectrofluorimetry with historically accurate reconstructions for the characterization of lac dye paints in medieval manuscript illuminations. J Raman Spectrosc. 2014;45:1172–9.

[86]   Bell SJ, Spence SJ. Disposable, stable media for reproducible surface-enhanced Raman spectroscopy. Analyst. 2001;126:1–3.

[87]   Doherty B, Brunetti BG, Sgamellotti A, Miliani C. A detachable SERS active cellulose film: a minimally invasive approach to the study of painting lakes. J Raman Spectrosc. 2011;42:1932–8.

[88]   Doherty B, Presciutti F, Sgamellotti A, Brunetti BG, Miliani C. Monitoring of optimized SERS active gel substrates for painting and paper substrates by unilateral NMR profilometry. J Raman Spectrosc. 2014;45:1153–9.

[89]   Leona M, Decuzzi P, Kubic TA, Gates G, Lombardi JR. Nondestructive identification of natural and synthetic organic colorants in works of art by surface enhanced Raman scattering. Anal Chem. 2011;83:3990–3.

[90]   Lofrumento C, Ricci M, Platania E, Becuccia M, Castellucci E. SERS detection of red organic dyes in Ag-agar gel. J Raman Spectrosc. 2013;44:47–54.

[91]   Ricci M, Lofrumento C, Castellucci E, Becucci M. Microanalysis of organic pigments in ancient textiles by surface-enhanced Raman scattering on agar gel matrices. J Spectrosc. 2016;2016:1380105. 10p.

[92]   Platania E, Lombardi JR, Leona M, Shibayama N, Lofrumento C, Ricci M, Becucci M, Castellucci E. Suitability of Ag-agar gel for the micro-extraction of organic dyes on different substrates: the case study of wool, silk, printed cotton and a panel painting mock-up. J Raman Spectrosc. 2014;45:1133–9.

[93]   Benedetti DP, Zhang J, Tague TJ Jr, Lombardi JR, Leona M. In situ microanalysis of organic colorants by inkjet colloid deposition surface-enhanced Raman scattering. J Raman Spectrosc. 2014;45:123–7.

[94]   Londero PS, Lombardi JR, Leona M. Laser ablation surface-enhanced Raman microspectroscopy. Anal Chem. 2013;85:5463–7.

[95]   Londero P, Lombardi JR, Leona M. A compact optical parametric oscillator Raman microscope for wavelength-tunable multianalytic microanalysis. J Raman Spectrosc. 2013;44:131–5.

[96]   Cesaratto A, Leona M, Lombardi JR, Comelli D, Nevin A, Londero P. Detection of organic colorants in historical painting layers using UV laser ablation surface-enhanced Raman microspectroscopy. Angew Chem, Int Ed. 2014;53:14373–7.

[97]   Shabunya-Klyachkovskaya EV, Kulakovich OS, Mitskevich AG, Moiseev YF, Kiris VV, Matsukovich AS, Karoza AG, Belkov MV. A comprehensive study of the Belarusian icon "Madonna Eleusa" (XVII cent). J Cult Heritage. 2016; submitted.

[98]   De Torres AR, Ruiz-Moreno S, López-Gil A, Ferrer P, Chillónl MC. Differentiation with Raman spectroscopy among several natural ultramarine blues and the synthetic ultramarine blue used by the Catalonian modernist painter Ramon Casas i Carbó. J Raman Spectrosc. 2014;45:1279–84.

[99]   Osticioli I, Mendes NFC, Nevin A, Gil FPSC, Becucci M, Castellucci E. Analysis of natural and artificial ultramarine blue pigments using laser induced breakdown and pulsed Raman spectroscopy, statistical analysis and light microscopy. Spectrochim Acta, Part A. 2008;73(3):525–31.

[100] Trotsiuk L, Korza Ya, Matsukovich A. SERS technique optimization for cultural heritage research. Open Readings 2016. In: 59th Scientific Conference for Students of Physics and Natural Sciences. March 15–18, 2016. Vilnius, Lithuania. p. 162.

[101] Kulakovich O, Shabunya-Klyachkovskaya E, Trotsiuk L, Matsukovich A, Korza Ya. Gold organosols for multicomponent surface-enhanced Raman spectroscopy analysis of paintings. J Raman Spectrosc. 2019;7:936–44.

Bojana Laban, Dragana Vasić-Anićijević, and Vesna Vodnik

5.3 Application of nanospectroscopy methods to study cyanine dyes – J-aggregation on the surface of noble metal nanoparticles

5.3.1 Key messages

- Thiacyanine dyes are synthetic compounds containing a polymethine group. It is the base of artificial light-harvesting systems suitable for use in various spectroscopy detection techniques in many fields of science and technology.
- J- and H-aggregation of cyanine dyes is the spontaneous self-organization of the dye molecules in a parallel way (plane-to-plane stacking). H-aggregates are organized in the form of a sandwich-type arrangement. On the contrary, J-aggregates form a head-to-tail arrangement (end-to-end stacking). This self-organization of dye molecules leads to a blue- or red-shifting of the absorption band in the absorption spectrum concerning the monomer absorption.
- Metal nanoparticles (NPs) are small objects between 1 and 100 nanometers in size with unique optical properties (size, shape, and surface chemistry-dependent localized surface plasmon resonance (LSPR)) and provide in conjunction with cyanine dyes powerful spectroscopic tools.
- Silver and gold NPs induce the formation of thiacyanine dye J-aggregates on their surface. The spectral properties of dyes are strongly influenced by the dye structure, NP size, and surface capping.
- Spectroscopy (UV-Vis, Raman, and fluorescence spectroscopy, Fourier transform infrared spectroscopy (FTIR), dynamic light scattering (DLS)) and microscopy (transmission electron microscopy (TEM), atomic force microscopy (AFM)) techniques in combination with density functional theory (DFT) calculations are widely used for characterization of the mechanisms of dye–NP interactions.
- The self-organized thiacyanine dye–NP assemblies are important for application in many scientific fields, such as nanoelectronics, medical diagnostics, drug delivery, chemical sensing, and catalysis.

Acknowledgement: The authors would like to thank the Ministry of Education, Science and Technological Development of the Republic of Serbia for their financial support (contracts no. 451-03-68/2022-14/200017; 451-03-9/2022-14/200017; 451-03-9/2022-14/200123). Furthermore, financial support by the European Cooperation in Science and Technology through COST Action MP1302 Nanospectroscopy is gratefully acknowledged. The authors are also grateful to Dr. Vesna Vasić, who initiated this work and supported it in all phases.

https://doi.org/10.1515/9783110442908-017

5.3.2 Pre-knowledge

The following chapters in Volume 1 and Volume 2 give the reader support for this application study:

Volume 1, "Optical Nanospectroscopy: Fundamentals & Methods", Section 2. Optical Spectroscopies, Chapters 2.2.1, 2.2.2, and 2.3.1 to 2.3.3, and Section 3. Nanoscopy and Nanospectroscopy, Chapter 3.1.1.

Volume 2, "Optical Nanospectroscopy: Instrumentation, Simulation & Materials", Section 4, Instrumentation for optical microscopy and nanospectroscopy (all chapters); Section 5, Simulations and modeling for nano-optical spectroscopy; Section 6, Nanomaterials for nanospectroscopy, Chapters 6.1 Nanostructures used in nanospectroscopy, 6.2 (Metallic) nanostructures for enhanced spectroscopy, and 6.5 Characterization of nanostructures.

Cyanine dyes are synthesized organic compounds with many applications as photosensitizers in the photographic industry. The positively charged nitrogen center is linked by a conjugated chain of an odd number of carbon atoms to the other nitrogen in the dye structure. The dye molecule is regarded as a point dipole, capable of forming various aggregates due to the self-organization of the molecules in the solution. The optical properties of the dye change because the excitonic state of the dye aggregate splits into two levels through the interaction of transition dipoles.

J-aggregates of cyanine dyes are of interest for many photonic applications. They often possess specific optical spectra, e. g., narrow spectral bands in the visible spectral range and a high fluorescence. The absorption or fluorescence spectra intensity versus dye concentration increase follows the Beer–Lambert law.

J-aggregation is often supported by NPs. The application of various nanospectroscopy techniques described in Volume 1 and Volume 2 enables us to elucidate the nature and mechanism of J-aggregate formation. These are, for example, particle size and surface capping, stability in suspensions, optical properties (UV-Vis, Raman spectra, FTIR), zeta potential, electrophoretic mobility, and conductivity.

5.3.3 Importance of the application

The self-organization of thiacyanine dyes in solution or at a solid–liquid interface is a very interesting behavior of these systems. In dye chemistry, it is usually due to the strong intermolecular van der Waals attractive forces between the molecules, which induce distinct changes in the absorption spectra compared to the monomeric species. Some efforts were made to build artificial light-harvesting systems by adsorbing dye molecules on solid substrates. The solid substrate serves as a matrix for stable fixation and precise stacking of the dye molecules. Their aggregation in solution is analogous to the aggregation of some molecules important in biology, such as those in photosynthetic systems. In addition, aggregation of cyanine dye molecules in solution in the vicinity of metal NPs has captured recent interest due to coupling between dye aggregates and metal nanocrystals. The aggregation or sorption of the dye to the metal surfaces can shift the LSPR of the metal NPs, influence emission in surface-enhanced Raman scattering (SERS), or enhance/quench fluorescence of the dye [1–4].

The combination of emissive labels such as fluorescent dyes and metal NPs to prepare inorganic–organic hybrid materials is a strategy to enhance their photophysical properties, useful for different applications. Such hybrid systems can exhibit advanced properties resulting in a material with intermediate or unique characteristics distinct from the NPs or dye properties alone [5]. Furthermore, the plasmonic coupling between NPs enhances the local electrical field that determines the optical or spectroscopic properties. Consequently, understanding the interparticle molecular interactions and reactivities is still a challenge.

A lot of efforts for the fabrication and characterization of new NP–dye structures have been made. The existing nanospectroscopy tools enable the study of these materials and address some questions in this field, which will be useful for better exploration of novel functionalities useful for the application to molecular plasmonic devices. The size and shape of metal NPs capped by dye aggregates are usually measured by analytical techniques such as TEM, AFM, or scanning electron microscopy (SEM), DLS, or analytical disc centrifugation.

This chapter provides the findings of an investigation of AgNPs and AuNPs (with charged surfaces) and cyanine dyes' interactions with the surface plasmons of these metal nanostructures. The structure and self-organization of cyanine dyes and the term exciton are discussed below. The chapter provides an overview of the application of nanospectroscopic and related analytical techniques to provide an understanding of the specific type of self-organization and insights into J-aggregation of dyes on the surface of AgNPs and AuNPs of different sizes, shapes, and surface chemistry and the relevance of these studies for improving their real-life applications. J-aggregation of cyanine dyes supported by AgNPs and AuNPs has wide applications in photonic devices as materials for spectral sensitization, optical storage, photoelectric cells, and ultrafast optical switching. Besides, this phenomenon is widely used to label biomolecules, such as proteins, antibodies, or fluorescent probes. Dye–peptide conjugates are also useful for diagnostic imaging and therapy. The small sizes of the compounds allow for more favorable delivery to tumor cells. They can be used in endoscopic applications to detect tumors and other abnormalities, provide localized therapy, and perform sonofluorescence tumor imaging.

5.3.4 State-of-the-art

5.3.4.1 Self-organization of cyanine dyes

Cyanine dyes belong to the polymethine group of compounds, representing artificial light-harvesting systems. More than 160 years ago, the chemist Williams was probably the first to observe a cyanine dye, a compound with "a blue of great beauty and intensity" [6], by reacting crude quinoline with ethyl/amyl iodides treated by silver oxide. Similar cyanine dyes with pronounced blue color (*cyanos* = blue) were discovered later, together with their structures [7, 8]. Konig introduced the term polymethine dyes. The color of these dyes is determined by the length of the polymethine chain [8].

The dyes consist of two nitrogen centers. One is positively charged, linked by a conjugated chain of an odd number of carbon atoms to the other nitrogen [9]. Two nitrogens are joined by a polymethine chain. Besides, they are an independent part of a heteroaromatic moiety. These heterocyclic compounds are pyrrole, imidazole, thiazole, pyridine, quinoline, indole, or benzothiazole. The cyanine dyes have high ground-state polarizability of the π-electrons along with the polymethine group in the ground state. This feature gives rise to strong dispersion forces (van der Waals forces) between two cyanine molecules in solution [10]. Cyanine dyes can be cationic, anionic, or neutral polymethines. An example of the general structure, one of the various polymethine dyes with the trivial name thiacyanine dyes, which represents a cationic dye, is shown in Figure 5.3.1.

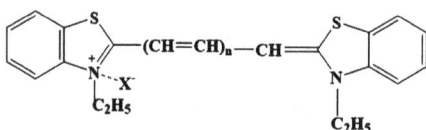

Figure 5.3.1: The structure of polymethine thiacyanine dyes. n, the number of methine groups; X^- = Br^-, Cl^-, ClO_4^-, and various anions.

The noncovalent molecular π-stacking interaction determines the physical and chemical properties, which constitutes the basis of cyanine. The advantages of their photochemical, photoluminescent, electroluminescent, and nonlinear optical properties [11, 12] are responsible for labeling proteins, antibodies, and other biomolecules, for spectral sensitization in the photographic industry or fluorescence labeling as fluorescent probes [13–16].

It is worth pointing out that polymethine thiacyanine dyes can spontaneously self-organize into highly ordered aggregates of various structures and morphologies [10]. Compared to single or individual molecules, aggregates consist of a parallel packing of molecules, causing an interaction between the excited states of each molecule. They form a sandwich-type arrangement (H-dimer) or a head-to-tail arrangement (J-dimer). This organization enables the high polarizability of the π-electrons along with the polymethine group in the ground state. The efficient exciton coupling enables the fast exciton energy migration over thousands of molecules within a few picoseconds. J- and H-aggregates can exist as a 1D assembly of molecules in the solution that could be in a brickwork, ladder, or staircase type of arrangement [17] (Fig. 5.3.2).

The excitonic state of the dye aggregate, regarded as a point dipole, splits into two levels through the interaction of transition dipoles. An electron transition to the upper state in the parallel aggregates and a lower state in a head-to-tail arrangement leads to blue and red shifts of the absorption band concerning the monomer band, respectively [17] (Fig. 5.3.3).

Figure 5.3.2: The possible arrangements of thiacyanine dyes on the solid surface and in solution: ladder arrangement (left), staircase arrangement (middle), and brickwork arrangement (right).

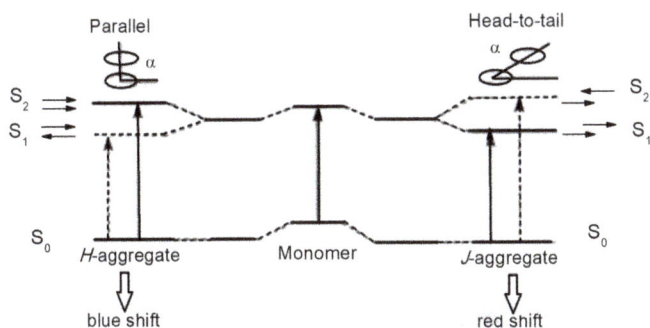

Figure 5.3.3: The relationship between chromophore arrangement and spectral shift. Reprinted with permission from Ref. [18].

These types of aggregation are strongly dependent on dye concentration, ionic strength, pH, solvent polarity, and dye structure [19]. For example, the high dielectric constant of water induces the self-organization of dyes in an aqueous solution due to the reduction of the repulsive forces between the charged molecules [20, 21]. Moreover, the well-organized aggregation of dyes can be enhanced by the addition of inorganic salts, metal ions, or metal NPs [22–26]. The screening factor or effective dielectric constant increases when the salt concentration increases, promoting aggregation [27].

The self-organization of the cyanine dyes in dependence on their concentration is illustrated in Figure 5.3.4, which represents the change of the absorption spectra of thiacarbocyanine (TCC) dye under various conditions. The blue shift of the absorption band due to H-aggregate formation is observed due to the increasing concentration of the dye from 1×10^{-3} mM to 1 mM. Conversely, the rising concentration up to 10 mM induces a red shift of the absorption band due to the J-aggregation.

The absorption spectra suggest that J-aggregates exhibit a narrow red-shifted J band, while H-aggregates exhibit a blue-shifted band concerning the monomer absorption. The distinct changes of the optical properties of aggregates in the solution can be explained in the coupling of transition moments of the constituent dye molecules [29].

Classical J-aggregates of cyanine dyes are formed by aggregation in aqueous solutions in high concentrations of salts [19, 30]. However, the conditions for the desired J-type self-assembly with the specific characteristics (spectral properties) for many ap-

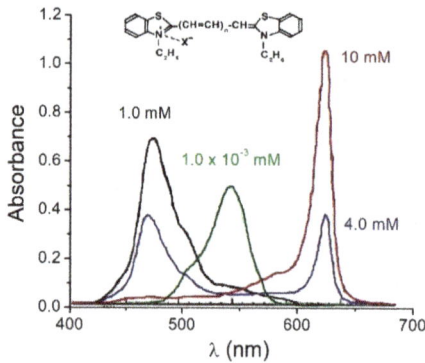

Figure 5.3.4: Absorption spectra of TCC at various concentrations. 1×10^{-3} mM, monomer; 1 mM, H-aggregate; 10 mM, J-aggregate. Adapted with permission from Ref. [28].

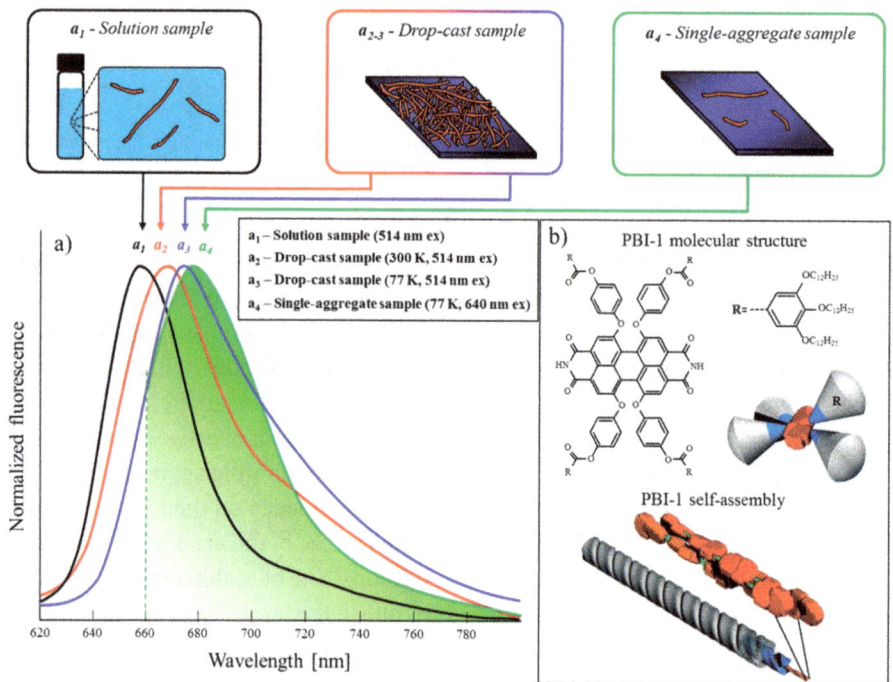

Figure 5.3.5: Respective fluorescence spectra (a_1–a_4) of PBI (b) aggregates formed under various conditions. Reprinted with permission from Ref. [31].

plications can be created separately in other media (e. g., organic solvents) under defined environmental conditions [31] (Fig. 5.3.5).

However, the experimental design to form such dye aggregates requires extremely high concentrations of chromophores [21]. In addition, the high fluorescence yield of

J-aggregates and a very large transition dipole moment offer extremely efficient energy migration [22, 28]. Therefore, they are of special interest for many photonic applications in optoelectronic and electroluminescence devices and as luminescent probes in biology and medicine.

5.3.4.2 Optical properties of AuNPs and AgNPs

Noble metal NPs have unique optical properties that arise from the collective oscillations of their free electrons induced by incident light, which produces an electric field close to the metal surface. External electromagnetic waves excite these oscillations at the metal surface known as LSPR. Their strength and frequency depend on the density of electrons, the effective electron mass, NP size and shape, interparticle distance, and dielectric constant of the medium [32]. The spectral response of spherical AuNPs and AgNPs in solution as a function of particle diameter is shown in Figure 5.3.6. The LSPR band shifts to longer wavelengths and broadens as the diameter increases.

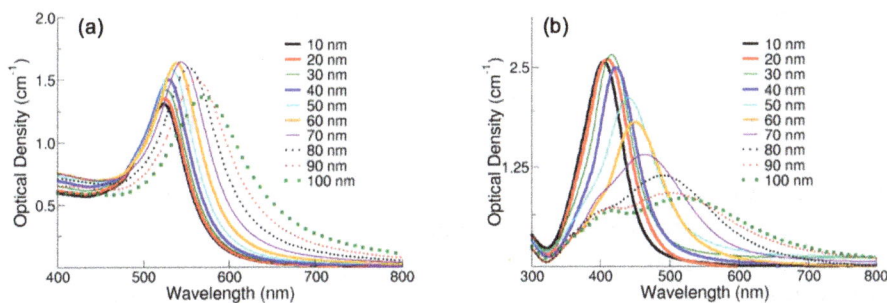

Figure 5.3.6: Dependence of extinction (scattering + absorption) spectra of (a) gold and (b) silver NPs on diameters (used with permission by nanoComposix [33]).

AgNPs and AuNPs are among the most frequently used nanomaterials in basic research [34–44]. Besides, they are the most desirable materials for strong electric field enhancements at their surfaces, which are important for applications in photovoltaic devices, sensors, SERS, and biomedicine [45–47]. Electromagnetic enhancements of 10^6-fold were reported for systems incorporating isolated AgNPs [48]. The photochemical processes on their surface also lead to changes in absorption and luminescence/fluorescence spectra of the sorbed organic dyes.

In general, the aggregation of thiacyanine dyes in the presence of metal NPs, particularly gold and silver, has gained great interest due to the coupling between H- and J-aggregates and metal nanocrystals [23–26, 49–51]. The intriguing spectral properties of J-aggregates and the possibility of technological application have attracted the attention of many researchers. Studies of spectroscopic properties of systems composed of dyes and metal NPs indicated that the optical absorption and fluorescence properties of dye–NP assemblies provide important insights into the NP–chromophore interactions. Due to their very good stability and photophysical properties, AuNPs are used more frequently than AgNPs for these investigations because of the higher desorption rates of the ligands from their surface. However, AgNPs are more interesting from the aspect of the change in the optical properties of the dye–NP composites compared to AuNPs.

5.3.4.3 Application of nanospectroscopy methods in dye aggregation studies

Modifying NP surfaces by thiacyanine dyes strongly changes their physicochemical properties [23, 24, 49, 51]. Metal NPs can induce the desired modulation of the chromophores' optical characteristics due to the electronic coupling of the dye exciton to the polarization of the metal NPs. However, J-aggregation of thiacyanine dye on the NP surfaces strongly depends on the particle surface chemistry and the dye's structure. The self-organization of dye molecules mediated by NPs is especially important for applying dye–NP assemblies in nanoelectronics, medical diagnostics, drug delivery, chemical sensing, and catalysis.

This section deals primarily with applying a set of nanospectroscopy methods to elucidate the mechanism of J-aggregation of one specific dye, 3,3'-disulfopropyl-5,5'-dichlorothiacyanine (TC), on the surface of AgNPs and AuNPs. This dye is chosen because of its appropriate solubility in water, high extinction coefficient, and fluorescence properties. Moreover, it easily undergoes self-organization, forming J-aggregates in the presence of metal ions and NPs. The structure of the dye is presented in Figure 5.3.7.

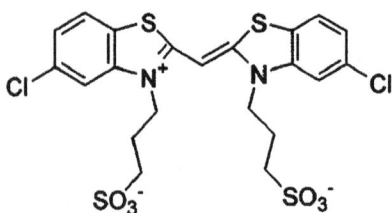

Figure 5.3.7: The structure of a 3,3'-disulfopropyl-5,5'-dichlorothiacyanine (TC) anion.

The methods mentioned above, mainly spectrophotometry and fluorescence, were used to elucidate the changes of optical properties of dye–NPs assemblies. Besides, the application of these techniques enables to elucidate mechanisms of TC dye–NP surface interaction, including evaluating the number of TC dye molecules adsorbed

Table 5.3.1: Some specific characteristics of AuNPs and AgNPs and their interaction with TC dye.

Metal NPs	Average particle size (nm)	Shape	Surface covering	Surface interaction with target molecules
Au	6	sphere	borate-capped	TC J-aggregation
Au	10	sphere	borate-capped	TC adsorption
Au	9	sphere	citrate-capped	TC adsorption
Au	17	sphere	citrate-capped	TC adsorption
Au	30	sphere	citrate-capped	TC adsorption
Au	8 × 40	rods	CTAB	TC adsorption
Ag	6	sphere	borate-capped	TC J-aggregation
Ag	10	sphere	citrate-capped	TC J-aggregation
Ag	40	prisms	polyvinil pirolidon	TC adsorption
Ag	8 × 40	rods	CTAB	TC J-aggregation

on NP surfaces, stability constants of the dye–NPs composite, the surface charge of the assembly, and kinetics of the J-aggregation. The AuNPs and AgNPs are usually synthesized by the reduction of $AuCl_4^-$ and $AgNO_3$ with sodium citrate or borohydride and have various shapes, sizes, and surface capping. Their properties and modes of interaction with NPs are presented in Table 5.3.1. It is worthy of note that TC can be adsorbed on the surface of all selected NPs, but only some specific NPs induce J-aggregation on their surface.

UV-Vis spectrophotometry

In our studies, the formation of TC J-aggregates on the surface of the noble metal NPs (Au and Ag) was followed spectrophotometrically [23–26, 49–51]. The appearance of a characteristic dip in TC absorption spectra is typical for AuNP-mediated J-aggregation [23]. On the contrary, a new peak in the case of AgNPs appears, also at 481 nm (Fig. 5.3.8) [25, 26]. These spectral changes, characteristic of this process, occur due to a coupling of the J-aggregate dye exciton to the metal NPs' plasmon. Figure 5.3.8 also illustrates the absorption spectra of the pure dye and the increase of the intensity of the J-aggregate characteristic peak as a function of the AgNP concentration. In the NP solution, where particles are not agglomerated, absorption spectra can quantify the NP concentration, calculated using the Beer–Lambert law.

In addition, the stability and intensity of the absorption peak of J-aggregates at 481 nm were strongly dependent on TC, NP, and KCl concentrations. J-aggregation also depends on time as well as on the NP type. The spectral dip appeared only for AuNPs with 6 nm particle diameter [23], obtained by borohydride reduction of Au(III) salt. In contrast, J-aggregation occurred on the surface of AgNPs of various sizes and surface capping.

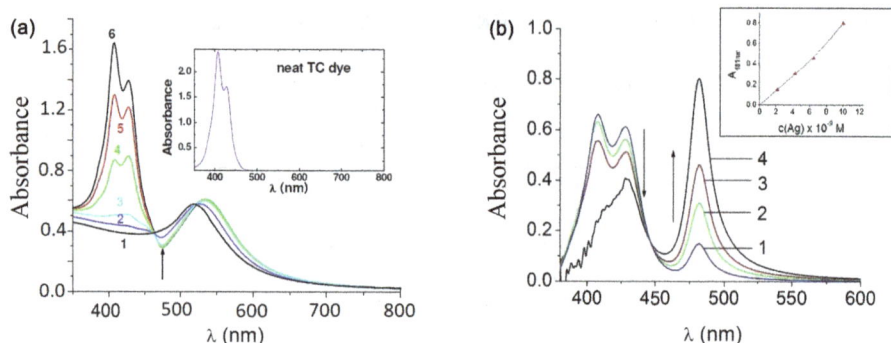

Figure 5.3.8: TC J-aggregation in the presence of 2.5×10^{-8} M AuNPs (a) and 1×10^{-5} M TC (Inset a); TC concentrations (a): 0 (1) to 1.67×10^{-5} (6) M [23]. AgNP concentration (b): 2.17×10^{-9} M (1), 4.3×10^{-9} M (2), 6.5×10^{-9} M (3), 1.0×10^{-8} M (4). Inset (b): absorbance at 481 nm versus AgNP concentration. Reprinted with permission from Ref. [51].

Fluorescence spectroscopy

The sharp emission band of J-aggregates is red-shifted relative to the monomer absorption band (Fig. 5.3.9) [52]. The fluorescence can be excited resonantly via the strong and narrow absorption J band or nonresonantly via the relatively weak and broad absorption bands in the blue region of the J band. The excitations in such systems can relax by incoherent energy transfer from some exciting species to the emitting excitonic states of a particular segment of the J-aggregate. The other way of relaxation is intersegment exciton energy conversion.

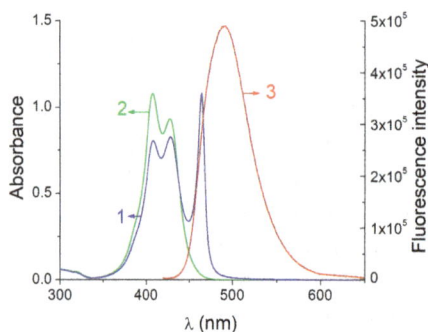

Figure 5.3.9: The absorption (1, 2) and fluorescence (3) spectra of TC dye in an aqueous solution. 1, TC J-aggregate; 2, TC dimer.

The combination of fluorescence microscopy and optical spectroscopy contributed to the development of single-molecule spectroscopy (SMS). This method studies fluorescence light coming from a single molecule (or other nano-objects) [53, 54] and the underlying dynamics that are not observable in experiments on bulk materials. A fluorescence "blinking" indicating the fluctuations of fluorescence intensity of a single molecule or single NP under continuous excitation can be explained by the NP temporarily residing in a long-lived nonfluorescent "dark" state from time to time. This "collective blinking effect" is related to energy exchange between chromophores [6–9]. In the excited states, the J-aggregate excitons are delocalized over many dye molecules depending on disorder and temperature [20–23, 28]. The SMS approach indicated that such systems could work as efficient light-harvesting antennas. Besides, it could be useful to study the structure of the exciton bands at low temperatures by monitoring the excitation spectra of individual aggregates [24].

SMS is among the most powerful methods to reveal the individual exciton levels characteristic for the commonly observed properties of molecules, aggregates, and various nanosystems [31, 34–39]. For example, some reported excitonic properties of tubular aggregates suggest that the ensemble averages due to a single aggregate consisting of tens of thousands of dye molecules and having only a quasi-1D (tubular) geometry.

Raman spectroscopy

Raman spectroscopy is a widely used method to analyze the scattering of photons by molecules or molecular aggregates [55, 56]. The formation of J-aggregates could occur on the surface of NPs, and the plasmonic coupling among NPs can be responsible for electrical field enhancement, which determines the optical and spectroscopic properties.

Figure 5.3.10 shows a schematic presentation of SERS. The blue dots represent molecules attached to metal NPs (orange balls) or can be in the gap between NPs [57]. As an example, an electron micrograph of AuNPs is also presented.

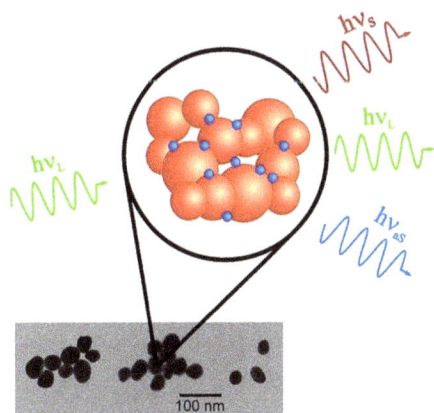

Figure 5.3.10: Schematic presentation of SERS and an electron micrograph of AuNPs. Reprinted with permission from Ref. [58].

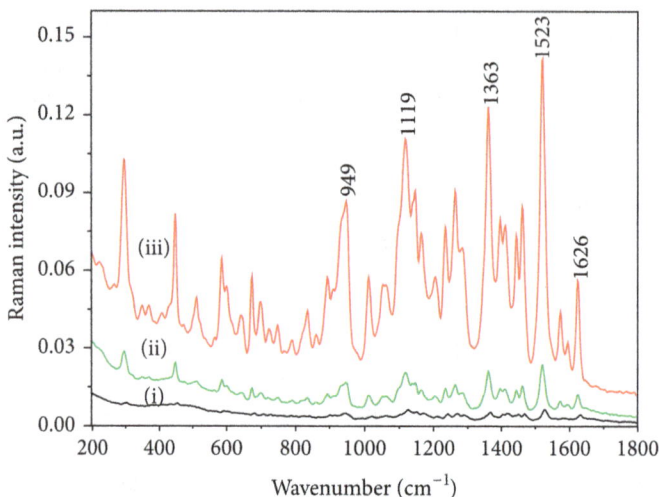

Figure 5.3.11: Raman (i) and SERS (ii, iii) spectra of 1×10^{-4} M indocyanine-type dye IR-820 in solution (i) and in the presence of 0.1×10^{-6} M (ii) and 1×10^{-6} M (iii) AuNPs [60].

SERS is useful for understanding the dye–NPs interaction and detecting dye-containing NP systems [15, 58, 59]. Both LSPR absorption and SERS use plasmonic coupling in the process of $\pi - \pi$ interaction of a dye and NP assembly.

Figure 5.3.11 represents SERS spectra of indocyanine-type IR-820 dye molecules adsorbed on AuNPs. The Raman bands are barely detectable in the absence of the AuNPs, and the SERS effect is evident in their presence.

The Raman spectrum of 5×10^{-5} M TC dye is presented in Figure 5.3.12a. High laser power was needed to obtain this spectrum. However, 5×10^{-6} M TC dye solution did not exhibit Raman peaks even at a laser power of ~200 µW/µm^2, but the addition of AgNPs into the solution induced SERS (Fig. 5.3.12b) [57].

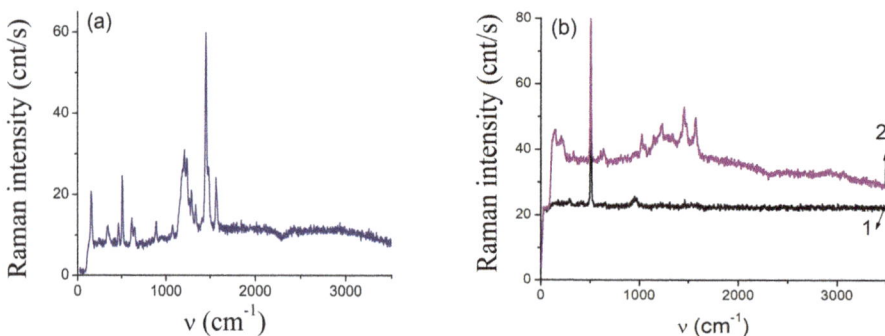

Figure 5.3.12: Raman spectra of 5×10^{-5} M TC (a) and 5×10^{-6} M TC (b) in the presence of AgNPs (line 2) and for TC without NPs (line 1).

Fluorescence microscopy

The aggregation of cyanine dyes under various conditions can be studied using multiple microscopic techniques [61–63]. For example, fluorescence microscopy can evaluate the dynamic processes during aggregation. This is well illustrated in Figure 5.3.13, representing the micrographs of TCC at various concentrations.

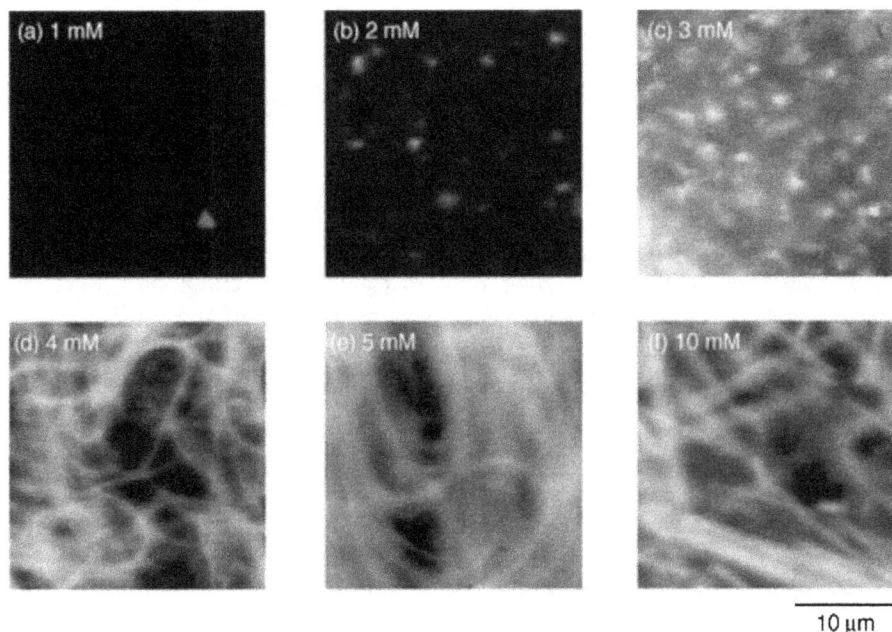

Figure 5.3.13: Concentration-dependent fluorescence micrographs of TCC J-aggregates in aqueous solution. Reprinted with permission from Ref. [28].

The transformation from dot-like emissive fragments in Brownian motion at 1 mM TCC concentration was followed by the further drastic change of the J-aggregates' morphology with a concentration increase up to 10 mM. For TCC concentrations of 1.0 or 2.0 mM there is no indication of J-aggregates in the solution. On the contrary, 3.0 mM TCC enables the observation of dot-like emissive fragments in Brownian motion. The increasing dye concentration induces the drastic transformation of the J-aggregates' morphology. TCC concentrations between 4.0 and 5.0 mM induce a mesoscopic fiber morphology with a fiber length of several tens of micrometers and a submicron width. This fibrous structure probably comes from anisotropic interactions between TCC molecules in the aggregate. The further increase of TCC concentration makes the morphology of the fibers straighter.

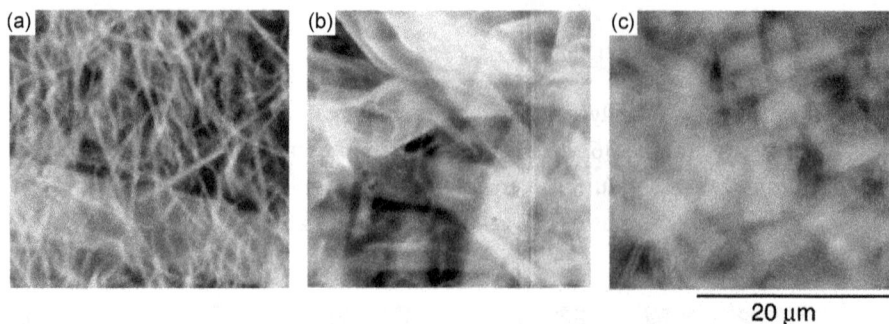

Figure 5.3.14: The concentration dependence of fluorescence micrographs of TC J-aggregates in aqueous solutions. TC concentration: (a) 1 mM, (b) 3 mM, (c) 7 mM. Reprinted with permission from Ref. [64].

Figure 5.3.14 represents concentration-dependent fluorescence micrographs of TC dye, indicating a change of its morphology.

The variation of TC concentration from 1 to 7 mM induced a morphology change from string shapes of about several tens of micrometers length and submicrometer width. The further concentration increase generated a morphology change to sheets that have a rectangular shape, with a length of the short axis of 4–9 µm and the long axis being longer than 20 µm.

Birefringence property of aggregates

Birefringence is a polarized-light microscopy method that uses optical anisotropy, with isotropic and anisotropic regions, which differ in their refractive index. Using this method, J- and H-aggregates of cyanine dyes can be detected by placing samples of various concentrations between a cross-coupled analyzer and detector, both of them being polarizers. For example, the polarized light micrograms of TCC H- and J-aggregates are presented in Figure 5.3.15. The results show that the colors of aggregates change from blue (H-aggregates in the concentration range from 1 to 2 mM TCC) to red (J-aggregates at above 5 mM TCC). With the increase in TCC concentration, both H- and J-aggregates are in equilibrium. The change of color is the consequence of light polarization. The microscope image shows the recombination of two components of light that travel in the same direction and produce an intrinsic interference of color.

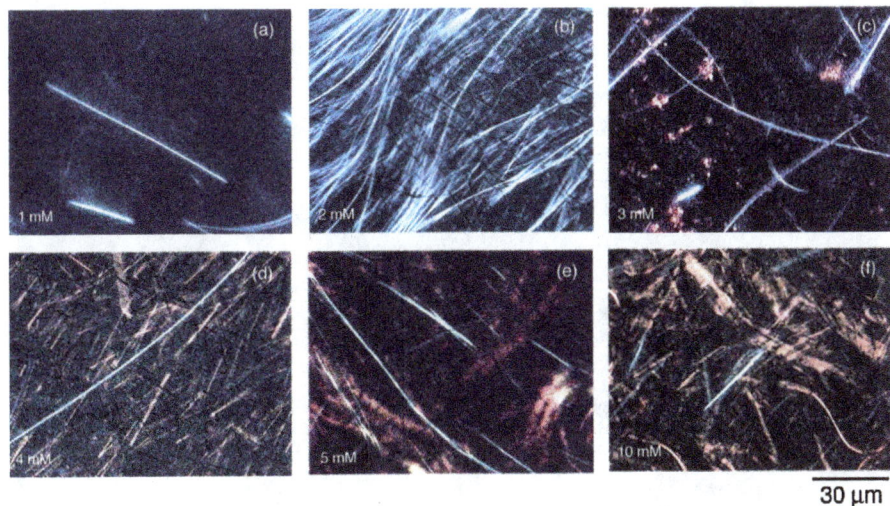

Figure 5.3.15: Polarized light micrographs of TCC in dependence on concentrations from 1 to 10 mM. Blue color, H-aggregates; red color, J-aggregates. Reprinted with permission from Ref. [28].

TEM characterization of dye-coated NPs

TEM belongs to the most important nanomicroscopy techniques for directly imaging nanomaterials. This technique enables to obtain quantitative measures of particle shape, size distribution, and morphology. For example, Figure 5.3.16 represents AuNPs and AgNPs of various shapes, sizes, and surface capping.

Figure 5.3.16: TEM images of borate-capped spherical AuNPs (a), triangular Ag nanoplates (b), and CTAB-capped Ag nanorods (c).

The aggregation of NPs in the presence of TC dye can also be identified on TEM micrographs. These findings are in accordance with the UV-Vis spectrum of the dye–NP assembly showing a slight red shift of the LSPR position (Fig. 5.3.17) [23]. This behavior is attributed to the J-aggregation of TC molecules due to the dye adsorption on NP surfaces. Moreover, the dye molecules are involved in the interlinking of metal NPs.

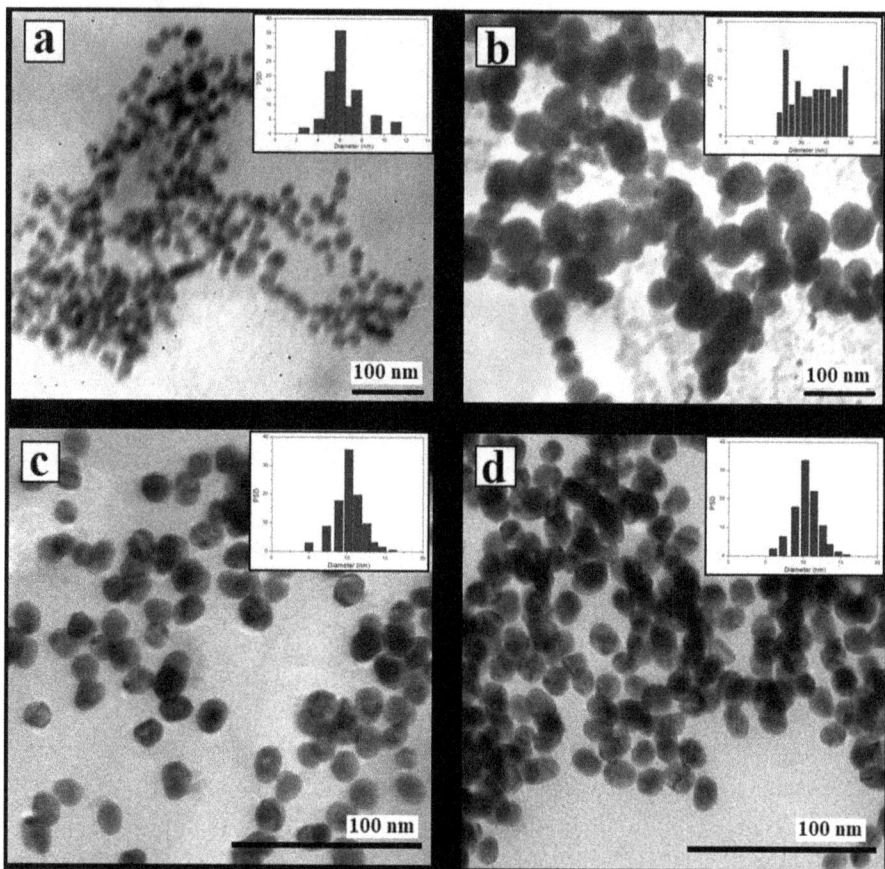

Figure 5.3.17: TEM micrographs of (a, c) bare AuNPs and (b, d) TC-coated AuNPs of various sizes. Inset: particle size distribution (PSD). Reprinted with permission from Ref. [49].

DLS and electrophoretic analysis

DLS is a noninvasive technique for measuring the particles' size and distribution in emulsions or molecules, which have been dispersed or dissolved in a liquid. The method is based on the study of the Brownian motion of particles or molecules in suspension, which induces the scattering of laser light with different intensities. Analysis of the intensity fluctuations yields the particle size distribution (PSD) using the Stokes–Einstein relationship [65]. The average NP diameter (d_{av}) obtained by DLS measurements represents the hydrodynamic diameter of the particle with hydration shell. However, the DLS technique parameters, such as the particle size, zeta potential, electrophoretic mobility, and conductivity, provide the key parameters for quantitative evaluation and understanding of particle stability in suspensions.

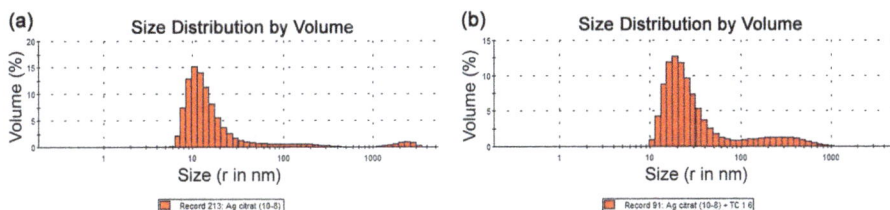

Figure 5.3.18: DLS diagrams for light scattering depending on the volume and diameter of 6-nm AgNPs in the absence (a) and presence (b) of 1×10^{-5} M TC.

Table 5.3.2: The parameters of TC–AgNP assembly obtained by DLS and TEM measurements at 25 °C [26].

	AgNPs	TC	AgNPs + TC
Concentration (M)	2.17×10^{-9}	1×10^{-5}	mixture
d_{av} (nm) (DLS)	25.5 ± 1.58	–	48.97 ± 3.6
d_{av} (nm) (TEM)	10 ± 1	–	10 ± 1
Zeta potential (mV)	-23.3 ± 2.0	–	-38.3 ± 3.1
Conductivity ($\mu S\ cm^{-1}$)	0.082 ± 0.001	0.036 ± 0.001	0.113 ± 0.002
mobility ($\mu m\ cm/Vs$)	-1.82 ± 0.15	-0.682 ± 0.404	-3.005 ± 0.244

d_{av}, the average particle diameter; TC concentration, 1×10^{-5} M.

Typical DLS spectra, which show the distribution of the intensity of scattered light as a function of the PSD, are presented in Figure 5.3.18.

The obtained d_{av} values of colloid dispersions containing AgNPs before and after the addition of TC dye, together with their zeta potential, mobility, and conductivity, are given in Table 5.3.2. The d_{av} values are usually higher than those obtained by TEM measurements since they include the added solvent and stabilizer moving with the particles. Moreover, the presence of TC in the colloid dispersion also increased the particles' diameter. Practically, the adsorbed dye molecules are included as an additional shell at the AgNP surface.

The zeta potential measurements of TC–AgNP assemblies indicated that the negative charge of the AgNPs increased in the presence of TC dye. This increase of the absolute value of negative surface charge due to the surface capping indicates the higher AgNP–TC suspension stability and the electrophoretic mobility. These features are the consequences of stronger repulsion between negatively charged NPs and suggest that the AgNP suspension in the presence and absence of TC should remain stable for a long time. Moreover, there also follows the long-time stability of absorption spectra of suspensions. Besides, the replacement of citrate and borate capping ions with TC molecules on AgNP surfaces significantly increased the conductivity of the colloid solution.

Atomic force microscopy analysis

AFM analysis allows 3D characterization of NPs with subnanometer resolution. AFM imaging mainly provides the aggregate's size, shape, and thickness. Specialized software algorithms identify NPs in the image, which then are counted. The studied material's height, volume, surface area, and perimeter may be calculated and displayed from these data.

Figure 5.3.19 represents AFM images of AgNPs in the absence and presence of TC dye in three dimensions. NPs were attached to a flat substrate, and the heights above this substrate were measured. The topography images (Fig. 5.3.19a) indicate the AgNPs are nearly spherical. The diameter of the bare and TC-coated NPs was determined from the AgNPs' height (z-direction), at a distance where the tip is repelled or attracted by the forces due to the interaction with the surface. It could be pointed out that the average diameters of NPs obtained from the AFM images (approximately 20 nm for bare and 50 nm for TC-coated NPs) are larger than those obtained by TEM, but agree very well with DLS measurements (Table 5.3.2). Furthermore, the broad distribution and increased size of borate/citrate bare and TC-capped AgNPs is observed. This result could be ascribed to the partial agglomeration (aggregation) of TC dye-coated AgNPs. The effect is also visible in the TEM micrographs [26].

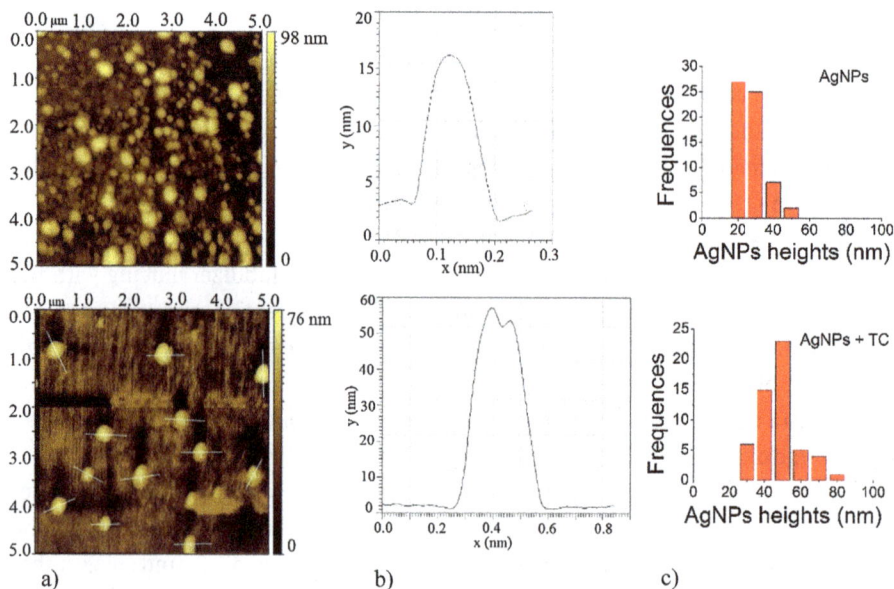

Figure 5.3.19: AFM images of bare (top) and TC-coated (bottom) AgNPs. (a) Topography. (b) Profile. (c) Particle size distribution. Experimental conditions: 2.17×10^{-9} M AgNPs suspension; 1.6×10^{-5} M TC; 1×10^{-3} M KCl. Reprinted with permission after Ref. [26].

Theoretical studies of J-aggregation of cyanine dyes

Some recent publications are devoted to the theoretical studies of the interaction between metallic NPs coated with different cyanine dyes [57, 66–69]. The light absorption and scattering cross-sections of dye-coated Ag, Au, Cu, or Al NPs were the bases to construct a theoretical model capable of explaining and quantitatively describing the absorption spectra of dye J-aggregates in more detail. The theoretical calculations followed the experimental results, which confirmed that the extinction spectra peaks of the NPs depended on their geometric parameters and the optical constants of the core and shell. This is well illustrated in Figure 5.3.20, which shows the dependence of the calculated absorption cross-section on the thickness of the J-aggregate layer.

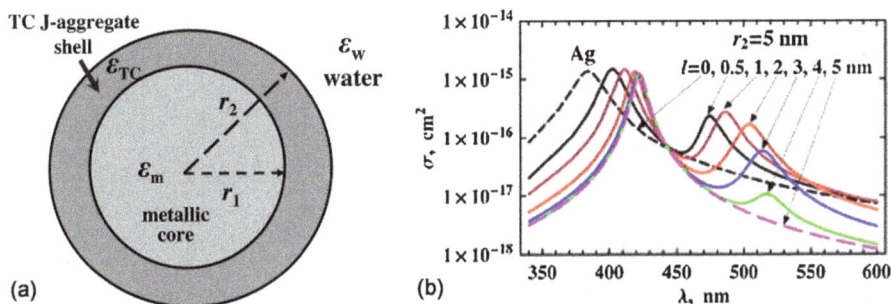

Figure 5.3.20: Schematic presentation of a composite consisting of the NP core and TC J-aggregate shell (a) and calculated absorption cross-section of the Ag/TC J-aggregate for $r_2 = 5$ nm and variation of the thickness ($l = r_2 - r_1$) of the J-aggregate shell (b). The dashed black curve denotes the absorption cross-section of bare 5-nm AgNPs. The dashed magenta curve shows the absorption cross-section of a single-component 5-nm NP of TC dye. Reprinted with permission from Ref. [68].

The characteristic J-aggregate peaks differ in position or height because of different effects in plasmon–exciton coupling in such systems, as well as the outer radius of the particle. Based on the obtained results, an approach to controlling the optical properties of the metal–organic NPs can be proposed. The ratio variation between the inner and outer radii of the concentric spheres influences positions and heights of the peak in the extinction spectrum. As the particle size increases, the changes of extinction spectra of the particles depend on the competition between the absorption and scattering contributions. Also, the polarizability of the hybrid system depends on the permittivity of the core and shell and the electromagnetic core–shell coupling [68].

A simple set of DFT calculations can provide some deeper insight into the positioning of TC dye on the surface of NPs. For example, the obtained results for DFT optimization of the TC dye anion geometry adsorbed on AgNPs revealed a rather planar geometry (Fig. 5.3.21.), which agrees with previously obtained results [70].

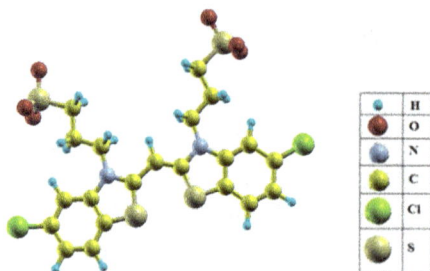

Figure 5.3.21: DFT-optimized geometry of TC dye anion visualized with XcrysDen Software [26, 71].

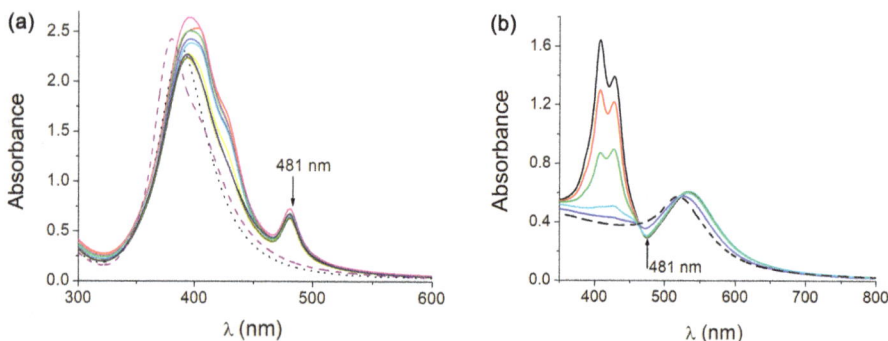

Figure 5.3.22: Differences in extinction spectra of TC J-aggregation on the surface of 6-nm AgNPs (a) and AuNPs (b).

In Figure 5.3.22, the absorption spectra of TC dye in a AgNPs or AuNPs surrounding are compared. Under similar chemical conditions, significant differences are observed in the shape of the exciton spectrum. Namely, a J band appears at 481 nm, having the peak for AgNPs or dip for AuNPs [23, 25, 26, 50]. In the peak-type absorption, the J band and plasmon band independently appear in the absorption spectra of the composite, since dip-type absorption is associated with an increase in the spectral overlap between the J band and the surface plasmon band. Dip-type absorption indicates strong coupling, while the peak type absorption suggests weak coupling [72].

Besides, a slight bathochromic shift (5–10 nm) in the position of the plasmon band was observed. These changes in extinction spectra point out the significant role of the metal boundary in the adsorption process of TC dye. Moreover, it confirms that the dye is bound to the particle nonspecifically. The electrostatic interaction with partially negative citrate or borate capping groups has the main role in dye aggregation.

To provide a semiquantitative picture of the TC dye interaction with the AgNPs, the interaction between different parts of the dye molecule anion with AgNPs was modeled considering the AgNPs' surface as a single Ag atom [26]. The energies of binding between atoms from different functional groups of the TC dye and Ag atoms are pre-

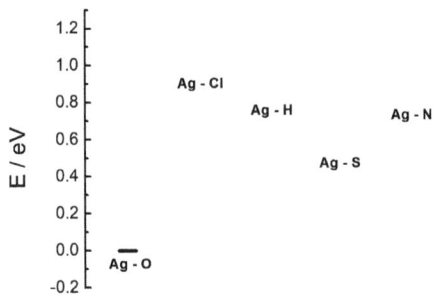

Figure 5.3.23: Relative energies of Ag interactions with TC dye functional groups. Reprinted with permission from Ref. [26].

sented in Figure 5.3.23. Here, as a reference point, i. e., the energetically most stable configuration, the sulfonate oxygen bond with Ag (Ag-O) is taken.

The conclusion could be that partially negative oxygen from the SO_3^- group from the TC dye exhibits the strongest interaction with the Ag atom. Moreover, the interaction between AgNPs via an oxygen atom from SO_3^- groups is favorable from the thermodynamical point of view. The ab initio study of oxygen adsorption on semiconducting surfaces provided similar results [73]. Besides, this type of adsorption mode enables a variety of tilted geometries of adsorbed dye. This is also in good agreement with the results of spectrophotometric and fluorescence studies and also indicates a slanted orientation of the dye on the NP surface.

However, partially negative SO_3^- groups of TC dye must overcome the electrostatic barrier imposed by negatively charged citrate anions tightly bound to the NP surface. A more precise surface model, consisting of an 18-atom Ag cluster, has been established to understand this problem better. Table 5.3.3 illustrates the energies of adsorption processes of pristine citrate and TC anions. The two most probable types of adsorbed borate species, BO_3^{3-} and $B(OH)_4^-$, are included in these calculations since the adsorption of borate anions is also possible on the NP surface.

Table 5.3.3: Adsorption energies of available anionic species on an 18-atom Ag cluster [26].

Species	TC dye	Citrate	BO_3^{3-}	$B(OH)_4^-$
E_{ads} (eV)	−2.48	−7.94	−3.43	−1.74

The results presented in Table 5.3.3 indicate that all anions expected at the investigated pH are bound to Ag significantly stronger than TC dye, which forms J-aggregates on the AgNPs only in the presence of significant excess of monovalent and divalent metal cations [26]. The influence of K^+ ions on the system was modeled by calculating the energy for adding one or two K^+ ions to the citrate-capped AgNPs. The results are

Table 5.3.4: Energies of possible K^+ ions reactions with citrate and TC dye on the AgNPs surface. Six water molecules were taken into account in the K^+ hydration sphere.

	Reaction type	Reaction	Energy (eV)
1	dehydration	$K(H_2O)_6^+ \rightarrow K^+ + 6H_2O$	+1.54
2	dehydration	$2K(H_2O)_6^+ \rightarrow 2K^+ + 12H_2O$	+3.08
3	addition	$K + citrate^* \rightarrow K^*-citrate^*$	−2.71
4	addition	$2K^+ + citrate^* \rightarrow K_2^*-citrate^*$	−2.81
5	addition	$3K^+ + citrate^* \rightarrow K_3^*-citrate^*$	−1.48
6	addition	$K^+ + TC^*$	+6.44
7	exchange	$K-citrate^* + TC \rightarrow TC^* + K-citrate$	+0.87

* Adsorbed species.

presented in Table 5.3.4, together with the K^+ dehydration energy, which must also be considered.

The adsorption energy was then calculated as the sum of the points for reactions 1, 3, and 7 for replacement of binding of one K^+ cation and for reactions 2, 4, and 7 for binding two K^+ cations with the citrate anion capping the AgNP surface. The interaction between K^+ ions and the adsorbed citrate inevitably weakens citrate adsorption on the AgNP surface through its partial neutralization, thus favoring the replacement of citrate by TC dye. On the contrary, the interaction between adsorbed TC dye and K^+ ions is not expected to impact TC dye adsorption on the AgNP surface since it is not energetically favorable.

The results obtained using DFT calculations can partially explain the experimentally obtained increase in zeta potential (Table 5.3.2). The adsorbed citrate species are replaced by TC dye molecules, which preferably bind another negative TC dye molecule rather than positive K^+ ions. The interaction between adsorbed TC dye and K^+ ions is not energetically favorable and has no impact on the TC dye adsorption on the AgNP surface.

J-aggregation kinetics

It has been found that J-aggregation kinetics, i. e., the dependence of J-aggregation formation on time, exhibits a sigmoid-type curve [30, 74] or a nonsigmoid type curve [75, 76]. The kinetics is usually followed by applying a stopped-flow technique, following the absorption spectra or fluorescence immediately after the fast mixing of dye and colloid suspension solutions. A simple model of a stopped-flow device is presented in Figure 5.3.24. This accessory enables the rapid mixing of two or more thermostated solutions and excellently fits any spectrophotometer or fluorimeter with a 1 cm path length cuvette compartment. It consists of two or more syringes for solutions connected with the small volume (e. g., 25 µl) cuvette. Here, the dead time, i. e., the time needed for all solutions to come from the syringe and be mixed in the cuvette, is below 20 ms.

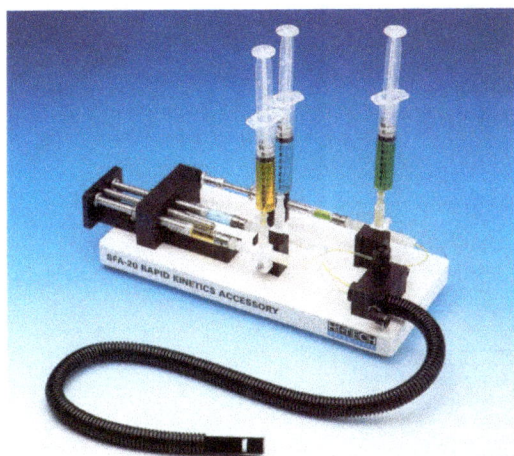

Figure 5.3.24: Stopped-flow accessory (adapted with permission from [77], copyright: HORIBA).

The typical kinetic curves for forming J-aggregates could not be successfully fitted to standard first-order, standard second-order, or coupled first-order equations because of their sigmoid or nonsigmoid shape. Sigmoid kinetic curves are characteristic of the reversible autocatalytic aggregation of dyes [30, 76, 78]. On the other hand, a time-dependent rate constant is typical for the nonsigmoid type kinetic curves having a simple exponential dependence [76, 79].

A typical kinetic curve representing the TC J-aggregate formation over time mediated by AgNPs is presented in Figure 5.3.25. The concentration of J-aggregates can be calculated from the curves by knowing the molar extinction coefficient of J-aggregates.

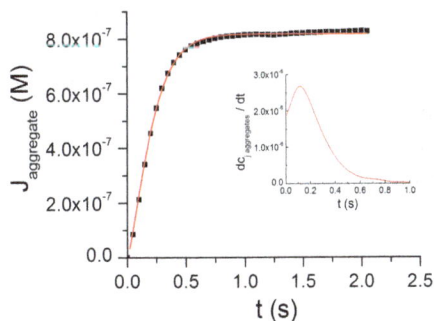

Figure 5.3.25: Typical kinetic curve for forming J-aggregates of 1.4×10^{-5} M TC in the presence of 1×10^{-8} M Ag and 1×10^{-3} M KCl at 25 °C. Points represent the experimental data; the solid line is the fit using equation (5.3.1). Inset: dependence of reaction rate on time. Reprinted with permission from Ref. [25].

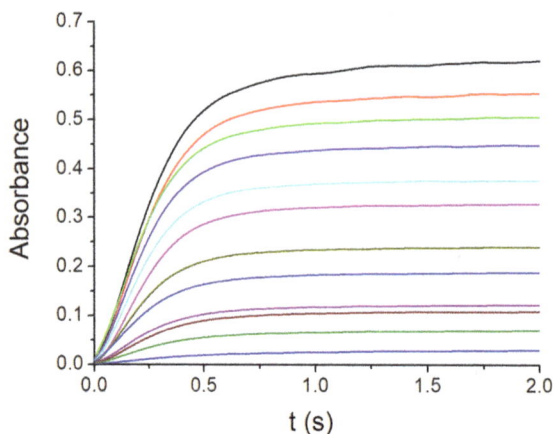

Figure 5.3.26: Kinetic curves of the growth of the J-aggregate concentration in dispersions containing 1×10^{-5} M TC dye and AgNPs in the concentration range from 0.25×10^{-8} M (bottom) to 5×10^{-8} M (top). Reprinted with permission from the Supplementary Information of Ref. [25].

Figure 5.3.26 shows kinetic curves of TC J-aggregation on the AgNPs' surface in dependence of AgNP concentration; the TC dye concentration was constant. An inflection point also characterizes the kinetic curves, as found in the J-aggregation studies of some cyanine dyes in the presence of gelatin [75] or metal ions [74].

A stretched exponential function represented by equation (5.3.1) [75, 76, 78, 80] can be successfully applied to describe J-aggregate formation in systems containing the independent relaxing species, which change exponentially over time and interact with each other [78, 80]:

$$C_J = C_J^o + (C_J^{oo} - C_J^o)(1 - \exp(-(k_{app}t)^n)). \tag{5.3.1}$$

In equation (5.3.1), C_J^o and C_J^{oo} represent the concentrations of J-aggregates formed immediately after mixing and after J-aggregation, respectively, and n represents the stretch exponential parameter, i. e., the degree of the sigmoid character of the kinetic curve. Here, k_{app} is a specific apparent relaxation rate constant of J-aggregate formation. As shown in [25], n usually ranges from 0.90 to 1.11, indicating a low degree of cooperation in aggregate building [78]. The parameter k_{app} represents the "average" relaxation rate. The first derivative of the kinetic curve dC_J/dt versus time has a bell-shaped form with the maximum value at the inflection point (Fig. 5.3.25, inset). This value represents the top rate of J-aggregate formation.

Binding mechanism

The k_{app}, maximal reaction rate dC_J/dt_{max}, and n values presented in dependence on dye concentration lead to saturation graphs. The shape of the graphs leads to the conclusion that a series of second- and first-order reactions must be taken into account. This finding indicates that a few reaction steps in J-aggregation on NP surfaces can be considered [81], according to the following relation:

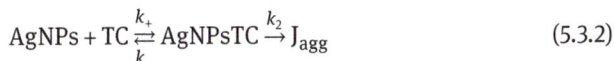

$$AgNPs + TC \underset{k_-}{\overset{k_+}{\rightleftharpoons}} AgNPsTC \overset{k_2}{\rightarrow} J_{agg} \qquad (5.3.2)$$

Here, the second-order reaction step represents the adsorption of TC on the NP surface. However, the first-order step represents the formation of J-aggregates from the TC molecules adsorbed on the NP surfaces [23, 24, 50]. Increasing the TC concentration induces the complete occupation of the AgNPs' surface. Consequently, the further addition of TC does not affect the rate of J-aggregate formation. Therefore, the conclusion can be made that the rate-determining step in the whole process is a first-order reaction of the formation of J-aggregates.

In further elucidation of the mechanism of dyes binding on the NPs' surface, the particles are usually considered macromolecules with several binding sites and the dye as a ligand. The saturation binding curves can be easily constructed from the spectrophotometric data. They represent the dependence of J-aggregate concentration on the equilibrium concentration of dye in the presence of NPs. For example, Figure 5.3.27 illustrates the dependence of J-aggregate formation on equilibrium TC concentration in the presence of various AgNP concentrations [25].

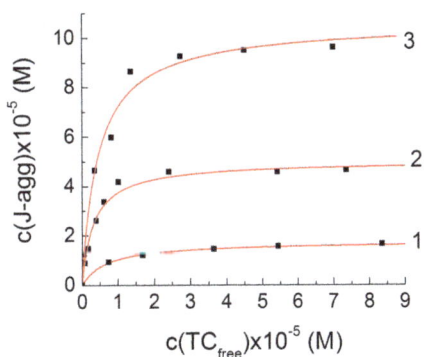

Figure 5.3.27: Influence of AgNPs on J-aggregate formation in dependence of equilibrium TC concentration: 1×10^{-8} M (1); 2.5×10^{-8} M (2), 5×10^{-8} M (3) AgNPs. Reprinted with permission from Ref. [25].

The reaction mechanism of J-aggregation can be elucidated using some approximations which describe biological macromolecules and ligands [82] binding. Some of these methods are Scatchard analysis [83, 84], as a method of linearizing data from saturation binding experiments, and Hill analysis, which enables determining bind-

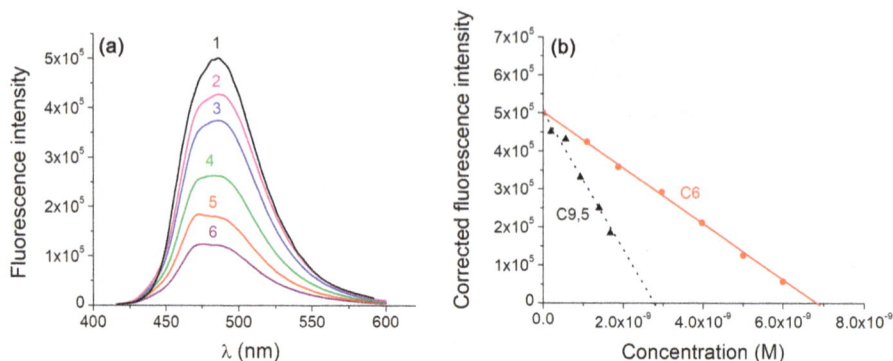

Figure 5.3.28: (a) Change of 1×10^{-6} M TC (1) fluorescence spectra at ~485 nm after the addition of 10-nm AuNPs at concentrations from 0.19×10^{-9} M (2) to 1.68×10^{-9} M (6). (b) Dependence of fluorescence intensity on the 6- and 10-nm AuNPs concentration. Adapted from [49].

ing constants and the number of dye molecules on the NPs' surface. These methods can be successfully applied to the results presented in Figure 5.3.27 [25].

Thiacyanine dyes also have characteristic fluorescence spectra, and the addition of the NPs to their solution usually exerts concentration-dependent fluorescence-quenching properties. This behavior is illustrated in Figure 5.3.28, representing the TC fluorescence intensity change in dependence of borate-capped AuNP concentration.

The Stern–Volmer relationship applied on the results presented in Figure 5.3.28b takes into account both static and dynamic (collision) fluorescence quenching [85] and has the general form

$$F_0/F = 1 + K_{SV}[Q]. \tag{5.3.3}$$

Here, F_0 and F represent the TC fluorescence intensities in the absence and presence of the AgNPs or AuNPs, respectively, and $[Q]$ is the quencher concentration. $K_{SV} = K_S + K_D$ is the Stern–Volmer quenching constant represented as the sum of K_S and K_D, i. e., the static and dynamic quenching constants. An example of a typical Stern–Volmer plot obtained for TC in dependence on the AgNP concentration is presented in Figure 5.3.29.

The results presented in Figure 5.3.29 indicate that the linearity of the Stern–Volmer plot implies that only one type of quenching occurred. This type of NP behavior is found by studies of the AgNP-induced quenching of fluorescence of some other fluorescent dyes [25, 86, 87]. The Stern–Volmer relation analysis suggests that NPs quench TC fluorescence with a high Stern–Volmer constant (K_{SV}) close to 10^8 M^{-1}. The very high value of the quenching constant confirms that the dye molecules are adsorbed on the AgNP surfaces [86] and suggests a strong association between NPs and TC dye [87].

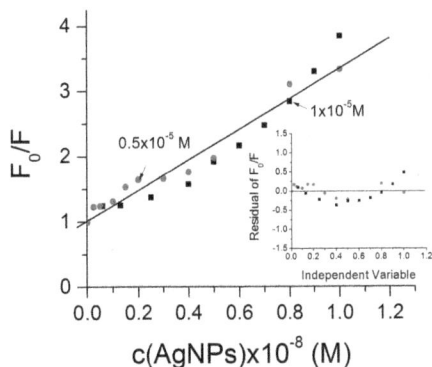

Figure 5.3.29: Dependence of F_0/F versus concentration of AgNPs (Stern–Volmer plot) for 5 × 10^{-6} M TC (circle) and 1×10^{-5} M TC (squares). Inset: residual of F_0/F. Reprinted with permission from Ref. [25].

5.3.5 Some challenges and solutions

Over the last few decades, a wide variety of materials consisting of NPs has drawn the attention of many researchers. Although nanometer-sized particles have been known for a long time, full attention has been paid to their deeper study and applications in recent years. Special attention has been paid to the synthesis of NPs with desired properties. NPs find applications in different areas such as nonlinear optics, physics, chemistry, biosensing, and medicine. Consequently, attention is also focused on the synthesis of hybrid nanosystems, i. e., inorganic–organic composites, which are interesting for fundamental research and their possible practical application.

The synthesis of novel hybrid nanostructures requires their detailed characterization, which can be achieved by spectroscopy and microscopy methods, such as UV-Vis, fluorescence, TEM, DLS, AFM, SNOM, etc. With the development of nanospectroscopy and microscopy, new approaches such as SERS and tip-enhanced Raman spectroscopy (TERS) are developed and applied to evaluate their properties and control them. The synthesis of novel hybrid nanostructures aims to obtain systems with better properties than individual components, inorganic or organic parts. Therefore, various methods for their synthesis with different inorganic–organic compositions and applications were described [88, 89]. In general, these hybrid nanostructures can find applications in other areas. Still, the question is whether it is possible to control the synthesis of hybrid nanostructures for a targeted application or find a specific application for already synthesized hybrid nanostructures. Also, it is important to understand the fundamental interaction between the inorganic and organic parts of these systems. From that point of view, nanospectroscopy can give clearer insight into the properties of hybrid nanostructures and their possible application in specific areas.

Especially, hybrid nanostructures consisting of a noble metal core (Ag or Au) covered with J-aggregates of cyanine dye molecules are still a challenge for experimental and theoretical studies [23–26, 49–51, 57, 66, 67, 69]. Considering that nanotechnology is an area that develops rapidly and assumes increasing importance, it is vital to establish whether NPs pose a risk to human health and the environment [90]. This finding includes their detailed characterization and analysis of their transport through the human body and the mechanism of interaction at the subcellular level [91, 92]. Their toxicity depends on the surface capping, shape, and size associated with the quantity (mass) of incorporated substances [40]. Even the smallest amount (expressed in mass) of the NPs can cause high toxicity due to their large active surface area because of their small size.

Various cyanine dyes have been used to label proteins, nucleic acids, and other biomolecules to apply fluorescence techniques for imaging in biochemical analysis. NPs synthesized by controlled modification of their surface were tested for their application in diagnosing and treating tumor cells [93–95]. Indocyanine green is a photosensitive dye approved by the Food and Drug Administration (FDA) for use in clinical trials; it is used in the diagnostics and therapy of laser radiation in the near-infrared region [96]. Applying SERS using AuNPs and AgNPs, indocyanine green can be detected in living cells [59]. Fluorescent cyanine dyes were also tested in vivo to obtain clear tumor images [97]. From this point of view, the study of cyanine dye–NP assembly is still a challenge because dramatic changes in their optical and fluorescence properties are not properly characterized. However, the elucidation of the formation mechanism and optical properties of dye–NP composites is the starting point for designing assemblies with the desired characteristics.

Several artificial nanosystems with unique optical responses have been developed for application in organic photovoltaics. They are used for light harvesting, where absorption and emission features are altered through morphological engineering [31, 98]. In addition, they control the properties such as the size and shape of nanosystems for the desired application [34]. However, the most important modification in creating light-driven molecular chemical machines is controlling the energy funneling direction.

5.3.6 Summary and impact

The polymethine group of synthetic compounds (thiacyanine dyes) has applications in various spectroscopy detection techniques in many fields of science and technology. Besides increasing the sensitivity of photographic emulsions, these dyes can also label any kind of biomolecules, such as proteins, antibodies, peptides, and nucleic acid probes. They undergo J- and H-aggregation, i. e., the spontaneous self-organization of the dye molecules to form a sandwich-type arrangement. H-aggregation (plane-to-plane stacking) induces a blue-shifted absorption band in the absorption spectrum. On the contrary, a head-to-tail arrangement (end-to-end stacking) forms J-aggregates. Their main characteristic is a red-shifted absorption concerning the monomer absorption.

Metal NPs, with their unique optical properties dependent on the size and surface capping, support the self-organization of thiacyanine dyes. The thiacyanine–NP assemblies are especially interesting and important because of their application in nanoelectronics, medical diagnostics, drug delivery, chemical sensing, and catalysis. Two types of aggregates, i. e., H- and J-aggregates, can be selectively formed under the proper experimental conditions (choice of solvents, dye concentration, presence of metal ions, acidity of suspension). The aggregates' properties significantly differ in material properties and the resultant photovoltaic performance. The J-aggregation is an effective approach to tune the optical/electrical properties in organic optoelectronic devices. In dye/NP bilayer heterojunction photovoltaic cells, a significant contribution of J-aggregation to the photocurrent and a deeper ionization potential could be observed. It is most likely due to the intermolecular charge transfer interaction, giving larger open-circuit voltages than those of H-aggregation-based devices.

For the characterization of dye self-organization, many nanospectroscopy techniques are widely used. These are UV-Vis, Raman and fluorescence spectroscopy, TEM, AFM, FTIR, DLS, or DFT calculations, which also offer insights into the mechanism and rate of the process. The interaction mechanism can be elucidated by applying the methods considering NPs as macromolecules and dyes as ligands. The binding energy between NPs and atoms from the dye structure can be calculated using the theoretical model. These results enable us to predict the orientation of the dye on the NPs' surface.

The mechanism of TC–NP interactions can be successfully understood by examining the changes in the absorption and fluorescence spectra. The quenching of the dyes' fluorescence is quantitatively related to the surface coverage of the dyes on the nanocrystal surfaces. Furthermore, kinetic measurements provide important information for assessing a two-step process involving fast adsorption of the dye on the NPs' surface combined with a slower process – the growth of J-aggregates on the initial TC layer. Finally, knowing the mechanism of self-organization processes (J- and H-aggregation) enables scientists to compose dye–NP assemblies with the desired properties for applications in various science and technology fields, especially in the photographic industry and as sensors in medicine and the environment.

Bibliography

[1]   Weitz DA, Garoff S. The enhancement of Raman scattering, resonance Raman scattering, and fluorescence from molecules adsorbed on a rough silver surface. J Chem Phys. 1983;78:5324–38.
[2]   Johansson P, Xu H, Käll M. Surface-enhanced Raman scattering and fluorescence near metal nanoparticles. Phys Rev B. 2005;72:035427.

[3]   Zhang J, Lakowicz JR. Metal-enhanced fluorescence of an organic fluorophore using gold particles. Opt Express. 2007;15:2598–606.

[4]   Fort E, Grésillon S. Surface enhanced fluorescence. J Phys D, Appl Phys. 2007;41:013001.

[5]   Henglein A, Meisel D. Spectrophotometric observations of the adsorption of organosulfur compounds on colloidal silver nanoparticles. J Phys Chem B. 1998;102:8364–6.

[6]   Williams CG. Trans R Soc Edinb. 1856;21:377–401.

[7]   Konig W. J Prakt Chem. 1906;73:100–8.

[8]   König W. Über die Konstitution der Pinacyanole, ein Beitrag zur Chemie der Chinocyanine. Ber Dtsch Chem Ges (A and B Series). 1922;55:3293–313.

[9]   Tang TB, Yamamoto H, Imaeda K, Inokuchi H, Seki K, Okazaki M, Tani T. Electronic structure of thiacyanine dyes in the solid state. J Phys Chem. 1989;93:3970–3.

[10]   Kirstein S, Daehne S. J-aggregates of amphiphilic cyanine dyes: self-organization of artificial light harvesting complexes. Int J Photoenergy. 2006;2006:21.

[11]   Shelkovnikov VV, Plekhanov AI. Optical and resonant non-linear optical properties of J-aggregates of pseudoisocyanine derivatives in thin solid films, Chapter 16. In: Macro to Nano Spectroscopy. IntechOpen; 2012.

[12]   Prokhorov VV, Pozin SI, Lypenko DA, Perelygina OM, Mal'tsev EI, Vannikov AV. AFM height measurements of molecular layers of a carbocyanine dye. WJNSE. 2011;1:67–72.

[13]   Huang X, Jain PK, El-Sayed IH, El-Sayed MA. Gold nanoparticles: interesting optical properties and recent applications in cancer diagnostics and therapy. Nanomedicine. 2007;2:681–93.

[14]   Shang L, Qin C, Wang T, Wang M, Wang L, Dong S. Fluorescent conjugated polymer-stabilized gold nanoparticles for sensitive and selective detection of cysteine. J Phys Chem C. 2007;111:13414–7.

[15]   Griffin J, Singh AK, Senapati D, Rhodes P, Mitchell K, Robinson B, Yu E, Ray PC. Size- and distance-dependent nanoparticle surface-energy transfer (NSET) method for selective sensing of hepatitis C virus RNA. Chem Eur J. 2009;15:342–51.

[16]   Li J, Guo L, Zhang L, Yu C, Yu L, Jiang P, Wei C, Quin F, Shi J. Donor -π- acceptor structure between Ag nanoparticles and azobenzene chromophore and its enhanced third-order optical non-linearity. Dalton Trans. 2009;0:823–31.

[17]   Behera GB, Behera PK, Mishra BK. Cyanine dyes: self aggregation and behaviour in surfactants: a review. J Surf Sci Tech. 2007;23:1–31.

[18]   Mishra A, Behera RK, Behera PK, Mishra BK, Behera GB. Cyanines during the 1990s: a review. Chem Rev. 2000;100:1973–2012.

[19]   Chibisov AK, Slavnova TD, Görner H. Self-assembly of polymethine dye molecules in solutions: kinetic aspects of aggregation. Nanotechnol Russ. 2008;3:19–34.

[20]   West W, Pearce S. The dimeric state of cyanine dyes. J Phys Chem. 1965;69:1894–903.

[21]   Struganova I. Dynamics of formation of 1, 1'-diethyl-2, 2'-cyanine iodide J-aggregates in solution. J Phys Chem A. 2000;104:9670–4.

[22]   Struganova IA, Lim H, Morgan SA. Influence of inorganic salts and bases on the J-band in the absorption spectra of water solutions of 1, 1'-diethyl-2, 2'-cyanine iodide. J Phys Chem A. 2003;107:2650–6.

[23]   Vujačić A, Vasić V, Dramićanin M, Sovilj SP, Bibić N, Hranisavljević J, Wiederrecht GP. Kinetics of J-aggregate formation on the surface of Au nanoparticle colloids. J Phys Chem C. 2012;116:4655–61.

[24]   Vujačić A, Vasić V, Dramićanin M, Sovilj SP, Bibić N, Milonjić S, Vodnik V. Fluorescence quenching of 5, 5'-disulfopropyl-3, 3'-dichlorothiacyanine dye adsorbed on gold nanoparticles. J Phys Chem C. 2013;117:6567–77.

[25]   Laban B, Vodnik V, Dramićanin M, Novaković M, Bibić N, Sovilj SP, Vasić VM. Mechanism and kinetics of J-aggregation of thiacyanine dye in the presence of silver nanoparticles. J Phys Chem C. 2014;118:23393–401.

[26] Laban B, Zeković I, Vasić Aničijević D, Marković M, Vodnik V, Luce M, Cricenti A, Dramićanin M, Vasić V. Mechanism of 3, 3′-disulfopropyl-5, 5′-dichlorothiacyanine anion interaction with citrate-capped silver nanoparticles: adsorption and J-aggregation. J Phys Chem C. 2016;120:18066–74.

[27] Robinson BH, Loffler A, Schwarz G. Thermodynamic behaviour of acridine orange in solution. Model system for studying stacking and charge-effects on self-aggregation. J Chem Soc Faraday Trans. 1973;1(69):56–69.

[28] Yao H, Domoto K, Isohashi T, Kimura K. In situ detection of birefringent mesoscopic H and J aggregates of thiacarbocyanine dye in solution. Langmuir. 2005;21:1067–73.

[29] Frenkel J. On the transformation of light into heat in solids. I. Phys Rev. 1931;37:17–44.

[30] Chibisov AK, Görner H, Slavnova TD. Kinetics of salt-induced J-aggregation of an anionic thiacarbocyanine dye in aqueous solution. Chem Phys Lett. 2004;390:240–5.

[31] Merdasa A, Jiménez ÁJ, Camacho R, Meyer M, Würthner F, Scheblykin IG. Single Lévy states–disorder induced energy funnels in molecular aggregates. Nano Lett. 2014;14:6774–81.

[32] Kelly KL, Coronado E, Zhao LL, Schatz GC. The optical properties of metal nanoparticles: the influence of size, shape, and dielectric environment. J Phys Chem B. 2003;107:668–77.

[33] https://nanocomposix.com/pages/gold-nanoparticles-optical-properties; https://nanocomposix.com/pages/silver-nanoparticles-optical-properties. Accessed 2017, copyright 2022.

[34] Vuković VV, Nedeljković JM. Surface modification of nanometer-scale silver particles by imidazole. Langmuir. 1993;9:980–3.

[35] Vodnik VV, Nedeljković JM. Adsorption of boron containing molecules on silver nanoparticles. J Serb Chem Soc. 1998;63:995–1000.

[36] Vodnik VV, Nedeljković JM. Influence of negative charge on the optical properties of a silver sol. J Serb Chem Soc. 2000;65:195–200.

[37] Vodnik VV, Bozanić DK, Bibić N, Saponjić ZV, Nedeljković JM. Optical properties of shaped silver nanoparticles. J Nanosci Nanotechnol. 2008;8:3511–5.

[38] Ilić V, Šaponjić Z, Vodnik V, Potkonjak B, Jovančić P, Nedeljković J, Radetić M. The influence of silver content on antimicrobial activity and color of cotton fabrics functionalized with Ag nanoparticles. Carbohydr Polym. 2009;78:564–9.

[39] Vodnik VV, Vuković JV, Nedeljković JM. Synthesis and characterization of silver-poly(methylmethacrylate) nanocomposites. Colloid Polym Sci. 2009;287:847–51.

[40] Vujačić A, Vodnik V, Joksić G, Petrović S, Leskovac A, Nastasijević B, Vasić V. Particle size and concentration dependent cytotoxicity of citrate capped gold nanoparticles. Digest J Nanomater Biostruct. 2011;6:1367–76.

[41] Vukoje I, Lazić V, Vodnik V, Mitrić M, Jokić B, Ahrenkiel SP, Nedeljković JM, Radetić M. The influence of triangular silver nanoplates on antimicrobial activity and color of cotton fabrics pretreated with chitosan. J Mater Sci. 2014;49:4453–60.

[42] Bogdanović U, Vodnik VV, Ahrenkiel SP, Stoiljković M, Ćirić-Marjanović G, Nedeljković JM. Interfacial synthesis and characterization of gold/polyaniline nanocomposites. Synth Met. 2014;195:122–31.

[43] Pajović JD, Dojčilović R, Božanić DK, Kaščáková S, Réfrégiers M, Dimitrijević-Branković S, Vodnik VV, Milosavljević AR, Piscopiello E, Luyt AS, Djoković V. Tryptophan-functionalized gold nanoparticles for deep UV imaging of microbial cells. Colloids Surf B, Biointerfaces. 2015;135:742–50.

[44] Bogdanović U, Pašti I, Ćirić-Marjanović G, Mitrić M, Ahrenkiel SP, Vodnik V. Interfacial synthesis of gold–polyaniline nanocomposite and its electrocatalytic application. ACS Appl Mater Interfaces. 2015;7:28393–403.

[45] Atwater HA, Polman A. Plasmonics for improved photovoltaic devices. Nat Mater. 2010;9:205–13.

[46] Sepúlveda B, Angelomé PC, Lechuga LM, Liz-Marzán, LM. LSPR-based nanobiosensors. Nano Today. 2009;4:244–51.
[47] Liao H, Nehl CL, Hafner JH. Biomedical applications of plasmon resonant metal nanoparticles. Nanomedicine. 2006;1:201–8.
[48] Zeman EJ, Schatz GC. An accurate electromagnetic theory study of surface enhancement factors for silver, gold, copper, lithium, sodium, aluminum, gallium, indium, zinc, and cadmium. J Phys Chem. 1987;91:634–43.
[49] Vujačić A, Vodnik V, Sovilj SP, Dramićanin M, Bibić N, Milonjića S, Vasić V. Adsorption and fluorescence quenching of 5, 5'-disulfopropyl-3, 3'-dichlorothiacyanine dye on gold nanoparticles. New J Chem. 2013;37:743–51.
[50] Laban BB, Vodnik V, Vujačić A, Sovilj SP, Jokić AB, Vasić V. Spectroscopic and fluorescence properties of silver-dye composite nanoparticles. Russ J Phys Chem A. 2013;87:2219–24.
[51] Laban BB, Vodnik V, Vasić V. Spectrophotometric observations of thiacyanine dye J-aggregation on citrate capped silver nanoparticles. Nanospectroscopy. 2015;1:54–60.
[52] Fidder H, Terpstra J, Wiersma DA. Dynamics of Frenkel excitons in disordered molecular aggregates. J Chem Phys. 1991;94:6895–907.
[53] Moerner WE, Kador L. Optical detection and spectroscopy of single molecules in a solid. Phys Rev Lett. 1989;62:2535–8.
[54] Orrit M, Bernard J. Single pentacene molecules detected by fluorescence excitation in a p-terphenyl crystal. Phys Rev Lett. 1990;65:2716–9.
[55] Schrader B. Infrared and Raman Spectroscopy: Methods and Applications. John Wiley & Sons; 2008.
[56] Chalmers JM, Griffiths PR. Handbook of Vibrational Spectroscopy. Chichester, UK: John Wiley & Sons; 2002.
[57] Ralević U, Isić G, Vasić Anićijevic DV, Laban B, Bogdanović U, Lazović VM, Vodnik V, Gajić R. Nanospectroscopy of thiacyanine dye molecules adsorbed on silver nanoparticle clusters. Appl Surf Sci. 2018;434:540–8.
[58] Kneipp K, Kneipp H, Kneipp J. Surface-enhanced Raman scattering in local optical fields of silver and gold nanoaggregates from single-molecule Raman spectroscopy to ultrasensitive probing in live cells. Acc Chem Res. 2006;39:443–50.
[59] Kneipp J, Kneipp H, Rice WL, Kneipp K. Optical probes for biological applications based on surface-enhanced Raman scattering from indocyanine green on gold nanoparticles. Anal Chem. 2005;77:2381–5.
[60] Neves TBV, Andrade GFS. SERS characterization of the indocyanine-type dye IR-820 on gold and silver nanoparticles in the near infrared. J Spectroscopy. 2015;2015:9.
[61] Yao H, Sugiyama S, Kawabata R, Ikeda Hi, Matsuoka O, Yamamoto S, Kitamura N. Spectroscopic and AFM studies on the structures of pseudoisocyanine J aggregates at a mica/water interface. J Phys Chem B. 1999;103:4452–6.
[62] Ono SS, Yao H, Matsuoka O, Kawabata R, Kitamura N, Yamamoto S. Anisotropic growth of J aggregates of pseudoisocyanine dye at a mica/solution interface revealed by AFM and polarization absorption measurements. J Phys Chem B. 1999;103:6909–12.
[63] Yao H, Morita Y, Kimura K. Mesodomain separation in amalgamated J aggregate formation of cyanine dyes at a mica/solution interface. Surf Sci. 2003;546:97–106.
[64] Yao H, Isohashi T, Kimura K. Large birefringence of single J aggregate nanosheets of thiacyanine dye in solution. Chem Phys Lett. 2004;396:316–22.
[65] Edward JT. Molecular volumes and the Stokes–Einstein equation. J Chem Educ. 1970;47:261.
[66] Lebedev V, Medvedev A. Absorption and scattering of light by hybrid metal/J-aggregate nanoparticles: plasmon–exciton coupling and size effects. J Russ Laser Res. 2013;34:303–22.
[67] Lebedev VS, Medvedev AS. Optical properties of three-layer metal-organic nanoparticles with a molecular J-aggregate shell. Quantum Electron. 2013;43:1065–77.

[68] Lebedev VS, Vitukhnovsky AG, Yoshida A, Kometani N, Yonezawa Y. Absorption properties of the composite silver/dye nanoparticles in colloidal solutions. Colloids Surf A, Physicochem Eng Asp. 2008;326:204–9.

[69] Vasić Aničijević D, Nikolić VM, Marčeta Kaninski MP, Pašti IA. Structure, chemisorption properties and electrocatalysis by Pd3Au over layers on tungsten carbide – a DFT study. Int J Hydrog Energy. 2015;40:6085–96.

[70] Valleau S, Saikin SK, Yung MH, Aspuru Guzik A. Exciton transport in thin-film cyanine dye J-aggregates. J Chem Phys. 2012;137:034109.

[71] Kokalj A. XCrySDen—a new program for displaying crystalline structures and electron densities. J Mol Graph Model. 1999;17:176–9.

[72] Kometani N, Tsubonishi M, Fujita T, Asami K, Yonezawa Y. Preparation and optical absorption spectra of dye-coated Au, Ag, and Au/Ag colloidal nanoparticles in aqueous solutions and in alternate assemblies. Langmuir. 2001;17:578–80.

[73] Vasić Aničijević D, Perović I, Maslovara S, Brković S, Žugić D, Laušević Z, Marčeta Kaninski M. Ab initio study of graphene interaction with O_2, O and O^-. Maced J Chem Chem Eng. 2016;35:271–4.

[74] Slavnova TD, Chibisov AK, Görner H. Kinetics of salt-induced J-aggregation of cyanine dyes. J Phys Chem A. 2005;109:4758–65.

[75] Görner H, Chibisov AK, Slavnova TD. Kinetics of J-aggregation of cyanine dyes in the presence of gelatin. J Phys Chem B. 2006;110:3917–23.

[76] Pasternack RF, Fleming C, Herring S, Collings PJ, dePaula J, DeCastro G, Gibbs EJ. Aggregation kinetics of extended porphyrin and cyanine dye assemblies. Biophys J. 2000;79:550–60.

[77] https://static.horiba.com/fileadmin/Horiba/Products/Scientific/Molecular_and_Microanalysis/Fluorescence_Accessories/Stopped_Flow/stopped_flow_accessory_sfa-20.pdf. Accessed 2017, copyright 2022.

[78] Kodaka M. Requirements for generating sigmoidal time–course aggregation in nucleation-dependent polymerization model. Biophys Chem. 2004;107:243–53.

[79] Leyvraz F. Rate equation approach to aggregation phenomena. In: Stanley HE, Ostrowsky N, editors. On Growth and Form. Springer Netherlands; 1986. p. 136–44.

[80] Johnston DC. Stretched exponential relaxation arising from a continuous sum of exponential decays. Phys Rev B. 2006;74:184430.

[81] Schmid R, Sapunov VN. Non-formal Kinetics: In Search for Chemical Reaction Pathways. D-6940 Weinheim: Verlag Chemie GmbH; 1982.

[82] Voet D, Voet JG. Biochemistry. New York: John Wiley & Sons; 1995.

[83] Scatchard G. The attractions of proteins for small molecules and ions. Ann NY Acad Sci. 1949;51:660–72.

[84] Munson P. Ligand binding data analysis: theoretical and practical aspects. In: Cattabeni F, Nicosia S, editors. Principles and Methods in Receptor Binding. Springer US; 1984. p. 1–12.

[85] Lakowicz JR. Principles of Fluorescence Spectroscopy. Springer; 2007.

[86] Umadevi M, Vanelle P, Terme T, Rajkumar BJ, Ramakrishnan V. Fluorescence quenching of 1, 4-dihydroxy-2, 3-dimethyl-9, 10-anthraquinone by silver nanoparticles: size effect. J Fluoresc. 2009;19:3–10.

[87] El-Sayed YS, Gaber M. Excited state interaction of laser dyes and silver nanoparticles in different media. Adv Nanopart. 2012;1:54–60.

[88] Sanchez C, Julián B, Belleville P, Popall M. Applications of hybrid organic-inorganic nanocomposites. J Mater Chem. 2005;15:3559–92.

[89] Sanchez C, Soler-Illia, GJ de AA, Ribot F, Lalot T, Mayer CR, Cabuil V. Designed hybrid organic–inorganic nanocomposites from functional nanobuilding blocks. Chem Mater. 2001;13:3061–83.

400 —— B. Laban et al.

[90] Ostiguy C, Lapointe G, Trottier M, Ménard L, Cloutier Y, Boutin M, Antoun M, Normand C. Health Effects of Nanoparticles. Citeseer; 2006.
[91] Ravindran A, Chandran P, Khan SS. Biofunctionalized silver nanoparticles: advances and prospects. Colloids Surf B, Biointerfaces. 2013;105:342–52.
[92] Banerjee V, Das KP. Interaction of silver nanoparticles with proteins: a characteristic protein concentration dependent profile of SPR signal. Colloids Surf B, Biointerfaces. 2013;111:71–9.
[93] He H, Xie C, Ren J. Nonbleaching fluorescence of gold nanoparticles and its applications in cancer cell imaging. Anal Chem. 2008;80:5951–7.
[94] Deng ZJ, Morton SW, Ben-Akiva E, Dreaden EC, Shopsowitz KE, Hammond PT. Layer-by-layer nanoparticles for systemic codelivery of an anticancer drug and siRNA for potential triple-negative breast cancer treatment. ACS Nano. 2013;7:9571–84.
[95] Guo X, Wu Z, Li W, Wang Z, Li Q, Kong F, Zhang H, Zhu X, Du YP, Jin Y, Du Y, You J. Appropriate size of magnetic nanoparticles for various bioapplications in cancer diagnostics and therapy. ACS Appl Mater Interfaces. 2016;8:3092–106.
[96] Yu J, Yaseen MA, Anvari B, Wong MS. Synthesis of near-infrared-absorbing nanoparticle-assembled capsules. Chem Mater. 2007;19:1277–84.
[97] Xin J, Zhang X, Liang J, Xia L, Yin J, Nie Y, Wu K, Tian J. In vivo gastric cancer targeting and imaging using novel symmetric cyanine dye-conjugated GX1 peptide probes. Bioconjug Chem. 2013;24:1134–43.
[98] Camacho R, Tubasum S, Southall J, Cogdell RJ, Sforazzini G, Anderson HL, Pullerits T, Scheblykina G. Fluorescence polarization measures energy funneling in single light-harvesting antennas—LH2 vs conjugated polymers. Sci Rep. 2015;5:15080.

Gagik Shmavonyan, Dmitry Cheshev, Andrey Averkiev,
Tuan-Hoang Tran, and Evgeniya Sheremet

5.4 Nanospectroscopy of graphene and two-dimensional atomic materials and hybrid structures

5.4.1 Key messages

- Graphene and other 2D materials are becoming more and more popular thanks to their unique properties in comparison to bulk materials. Graphene has found several commercial applications and along with other 2D materials has enormous potential. The properties of these materials are intensively investigated, in particular by optical spectroscopy methods.
- High-resolution spectroscopic and microscopic techniques allow for the identification and characterization of graphene and 2D atomic materials and their hybrid structures synthesized by conventional (e. g., micromechanical exfoliation, liquid phase exfoliation, chemical exfoliation, chemical vapor deposition (CVD) on metal surfaces, epitaxial growth on electrically insulating surfaces) and non-conventional (e. g., substrates rubbing) methods.
- Raman spectroscopy allows the identification of bonding in 2D materials. Based on a Raman spectrum, the layer thickness of graphene and other 2D materials can be identified easily, fast, and nondestructively. Besides, the presence of defects, strain, and doping of materials can be evaluated based on their Raman spectra.
- Photoluminescence (PL) spectroscopy is one of the widespread optical spectroscopy techniques that allows the characterization of several key properties of materials, for instance, the electronic structure of the materials, their quality, and their purity.
- Tip-enhanced Raman scattering (TERS) takes advantage of a scanning probe microscopy (SPM)-Raman platform combined with surface plasmon resonance effects localized at the tip apex to provide accurate information about the sample with nanometer spatial resolution.

5.4.2 Pre-knowledge

Most conventional materials we deal with in everyday life are 3D. We are now starting to understand how the properties of bulk materials change if one moves to the nanoscale. Layered bulk materials

Author contribution: G. Shmavonyan is the lead author who wrote the major part of the chapter, while D. Cheshev, A. Averkiev, T. H. Tran, and E. Sheremet added the part devoted to optical spectroscopy methods as well as edited the other parts of the chapter.

https://doi.org/10.1515/9783110442908-018

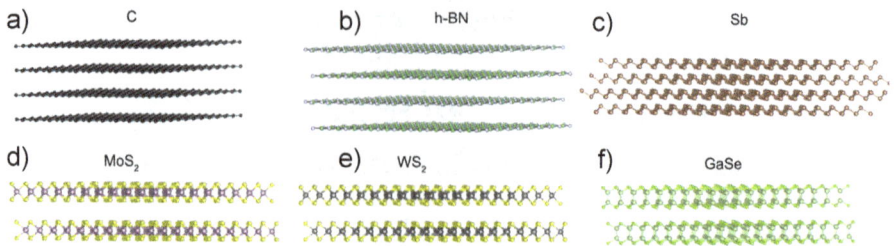

Figure 5.4.1: Crystalline structures of (a) graphene, (b) hexagonal boron nitride, (c) antimony, (d) molybdenum disulfide, (e) tungsten disulfide, and (f) gallium selenide.

exhibit strong in-plane covalent or ionic bonding along two dimensions and weak out-of-plane van der Waals (or hydrogen) bonding. The weak nature of van der Waals bonding (40–70 meV) and surface tension (60–90 mJ/m^2) allow for exfoliating layered bulk materials into 2D atomic materials [1, 2]. 2D materials are divided in three classes: (a) layered van der Waals solids, (b) layered ionic solids, and (c) surface-assisted nonlayered structures (i. e., silicene) [3]. The most studied 2D material is graphene due to its amazing properties. Beyond graphene, there is a wide spectrum of 2D materials, which include 2D layers and layered structures, whose total thicknesses vary from an atomic layer to a few nanometers [4]. The members in the 2D layered materials family are the following: (a) the graphene family (graphene, h-BN, and fluorographene or graphene fluoride [2D carbon sheet of sp^3 hybridized carbons with each carbon atom bound to one fluorine, $(CF)_n$]) [5], BCN compounds (compounds of boron, carbon, and nitrogen atoms), and graphene oxide (compound of carbon, oxygen, and hydrogen atoms), (b) 2D chalcogenides (MoS_2, $MoTe_2$, WS_2, WTe_2, $ZrSe_2$, NBS_2, GaSe, GaTe, InSe, etc.), and (c) 2D oxides (MnO_2, V_2O_5, MoO_3, WO_3, TiO_2, TaO_3, RuO_2, perovskite type $LaNb_2O_7$, etc.). Crystalline structures of some 2D materials are shown in Figure 5.4.1.

A host of 2D layered materials are the transition metal dichalcogenides (TMDCs), transition metal oxides, and nitrides [6]. TMDCs have been known in their bulk form for decades. They are a large class of 2D layered materials with the formula MX_2, where X is a chalcogen (S, Se, Te) and M is a transition metal element from group IV (Ti, Zr, Hf, etc.), V (V, Nb, Ta), or VI (Mo, W, etc.) [7]. The wide range of compositions of 2D layered materials spans the periodic table [6]. Depending on the combination of the atoms of the transition metal and the chalcogen, a variety of TMDCs with properties ranging from semiconducting, metallic, semimetallic, or ferromagnetic to superconducting properties can be obtained [7]. For example, monolayer h-BN and fluorographene are insulators, MoX_2 and WX_2 (i. e., MoS_2, $MoSe_2$, WS_2, WSe_2) are semiconductors with direct band-gaps, and NbX_2 and TaX_2 (i. e., $NbSe_2$, $TaSe_2$) are metals [8]. The band-gaps of monolayer InSe, $MoSe_2$, WSe_2, MoS_2, phosphorene, WS_2, and h-BN at room temperature are equal to ~1.25 eV, ~1.5 eV, ~1.7 eV, ~1.8 eV, ~2.0 eV, ~2.1 eV, and ~5.9 eV, respectively.

There are many 2D atomic materials that go beyond TMDCs, including monochalcogenides (GaSe, GeSe, SiS, etc.) and monoelemental 2D semiconductors (silicene, phosphorene, germanene, etc.) [9]. Silicene is a 2D allotrope of silicon with a hexagonal honeycomb structure (similar to that of graphene) consisting of silicon atomic layers [10]. Contrary to graphene, it is not flat and has a periodically buckled structure. Due to the latter, silicene has a tunable band-gap when applying an external electric field. Phosphorene is a single atomic layer of bulk phosphorus. Germanene is a single atomic layer of germanium, the electronic properties of which are unusual, like those of graphene and silicene. It has no band-gap, but one can be opened by attaching, e. g., a hydrogen atom to each germanium atom.

This chapter will focus on a few 2D materials most studied by optical spectroscopy techniques. It is helpful for the reader to be familiar with the fundamental characteristics of semiconductors (Volume 1, Chapter 1.4), electronic spectroscopy (Volume 1, Chapter 2.2), optical (PL) and vibrational spectroscopy (especially Raman, TERS) (Volume 1, Chapters 2.3 and 3.1), and the basics of materials for nanotechnology (Volume 2, Chapter 6). Regarding materials and characterization methods at the nanoscale, detailed information on quantum dots and nanocrystals is given in Volume 2, Chapter 6.1, whereas scanning-probe microscopy techniques are discussed in Volume 2, Chapter 4.5.

5.4.3 Importance of the application

The most well-researched and prominent representative of the 2D materials is graphene. This is how an atomically thin sheet or single layer of carbon atoms was named by German chemist Hanns-Peter Boehm in 1962. The term "graphene" was combined from the word "graphite" and the suffix "-ene," which was taken from the end of the word "ethylene." Theoretical research into graphene started in the 1940s and continued for the next decades, boosted from the 1980s by the discoveries of fullerenes (graphene curled up into balls) and carbon nanotubes (graphene folded into a cylinder) [11]. Until 2004, it was theoretically believed that single-layer graphene (SLG) could not exist due to thermodynamic instability when separated under ambient conditions. However, once graphene was isolated by Andre Geim and Konstantin Novoselov working at the University of Manchester, UK, it became clear that it was actually possible. The reason graphene can be obtained is due to very strong carbon-to-carbon bonds in it, which prevent thermal fluctuations from destabilizing it. The researchers applied Scotch tape to peel off layers of highly oriented pyrolytic graphite (HOPG) and obtain multilayer, few-layer, and monolayer graphene flakes. After producing them they successfully identified a single layer of carbon atoms or monolayer graphene flakes from multilayer ones. They published their research in the journal *Science* in 2004 and received the Nobel Prize in physics for the discovery of graphene in 2010 [3, 12]. When first discovered, graphene was an oddity, but now it has shown many record-winning properties, such as the thinnest, strongest, and lightest material, excellent electrical and thermal conductivity, and optical transparency. The extraordinary properties of graphene make it an ideal test bed to probe fundamental problems in physics, as well as lending itself to a wide range of applications in electronics, photonics, energy, sensors, bioapplications, etc. [13, 14].

As outlined in the Pre-knowledge section above, the discovery of graphene gave rise to a new class of atomic materials, known as "2D materials." 2D materials include layers of carbon (graphene), boron (borophene), h-BN ("white graphene"), germanium (germanene), silicon (silicene), phosphorus (phosphorene), tin (stanene), molybdenum disulfide (molybdenite), etc. Most of them could be obtained from layered bulk (3D) materials. Though the latter have been studied for more than 150 years, one only recently began to realize their potential for applications in advanced technologies and consumer products such as flexible thin-film field-effect transistors (FETs), photodetectors, sensors, batteries, supercapacitors, etc. A large number of 2D layered materials, such as TMDCs, are semiconductors with direct and indirect band-gaps that are tunable by the choice of material or external factors. The characteristics of 2D materials are very different from those of their 3D counterparts [15] and rapidly result in new advancements and applications. Compared to bulk materials, such as silicon, 2D materials exhibit novel properties: (a) their surfaces are naturally passivated without any dangling bonds, which are important for the stacking of different 2D structures, (b) strong light–matter interaction takes place, and, moreover, (c) they can cover a wide electromagnetic spectrum due to their varied electronic properties. Their physical properties are strongly dependent on the number of layers [16]. For example, by decreasing the number of the MoS_2 layers, i. e., from few (two to four) layers to a monolayer, there is a transition from semiconducting to metallic properties or from an indirect band-gap of the bulk layered material to a direct band-gap of the monolayer [17]. Thus, identifying the number of layers and determining their properties is a key issue in studying 2D materials. Besides, other factors, such as strain, doping, defects, impurities, etc., can also drastically change the properties of 2D materials. As will be discussed later, the synthesis method strongly affects these properties, and spectroscopic techniques such as Raman and PL spectroscopy can play a key role. Based on Raman and PL spectra, the materials and number of layers can be fast and easily identified. The effects of strain, doping, and defects can also be evaluated. However, due to the diffraction limit of light, the resolution of Raman and PL is on the order of 1 μm. Therefore, nanospectroscopy approaches are promising in terms of nanocharacterization.

New materials can change the world, like the Iron Age came to replace the Bronze Age, then came concrete, stainless steel, silicon and plastics, and now 2D materials are entering the modern era. 2D materials will bring us to the flatland [18].

5.4.4 State-of-the-art

5.4.4.1 Graphene and 2D atomic materials, their hybrid structures and physical properties

The current chapter focuses on graphene and 2D atomic materials, hybrid structures, and their properties, synthesis, identification, characterization, and application. This chapter is structured as follows. In the first part, we will look at what graphene and 2D atomic materials are and how they were discovered. Then we will consider the types and structures of 2D atomic materials, and the main physical, electrical, and optical properties associated with them. Further we will explore various conventional and nonconventional synthesis methods for obtaining mono- and few-layer (MFL) graphene and 2D atomic materials. The main characterization techniques and identification methods of 2D atomic materials with the focus on Raman spectroscopy, PL, and TERS will then be discussed in the next section. This will be followed by a review of the 2D hybrid atomic heterostructures. Challenges such as contamination, relia-bility, reproducibility, and scalability for fabrication of large-area 2D materials and heterostructure devices will be discussed. We shall conclude the chapter by nanoengi-neering issues of 2D hybrid structures before turning to some important applications in 2D hybrid devices. At the end of the chapter flexible electronics and its prospects will be dealt with.

The most widely known 2D material is graphene. Graphene is the mother element of several carbon allotropes, including graphite, carbon nanotubes, and fullerenes. Graphite, or pencil lead, has been known as a mineral for ~500 years. Graphite has a layered 3D structure, consisting of graphene layers (very weakly held together by van der Waals forces) stacked parallel to each other in a 3D, crystalline, and long-range order. The carbon atoms of each layer of graphite are arranged in a hexagonal lattice with an interval of 0.142 nm, and the distance between atomic planes is 0.345 nm [19]. Graphene is about five orders of magnitude thinner than printing paper and about three million times thinner than 1 mm thick graphite lead in pencil. A graphene sheet is also extremely light at 0.77 mg/m^2, which was calculated from the 0.052 nm^2 area of a hexagonal graphene cell. For comparison purposes, paper is ~100.000 times hea-vier than a single sheet of graphene. A monolayer of graphene has a high theoretical specific surface area of $2620 \text{ m}^2 \text{ g}^{-1}$ [20].

Graphene has a remarkable energy band structure thanks to its hexagonal crys-tal structure (Fig. 5.4.2a) [5, 12]. Each carbon atom in a SLG sheet has three in-plane covalent bonds (σ bonds) and one orbital bond perpendicular to the plane (π bond). This structure makes it mechanically strong and flexible. Graphene is therefore used in bullet-proof body armor, fabric, and suits. The π bonds hybridize and form the π and π^* bands (Fig. 5.4.2) that contribute to the remarkable conductivity of graphene. The first Brillouin zone of graphene forms a hexagon with high symmetry points at the center of the first Brillouin zone and two inequivalent points K and K' in the corners (Fig. 5.4.2b) [21, 22]. The π band corresponds to the valence band and the π^* band is the conduction band [22]. In contrast to almost all solids except for the Dirac-type ones, as the valence and conduction bands in graphene touch in the discrete K, K' points, the bonding–antibonding gap closes at the corners of the Brillouin zone, so there is no band-gap (Fig. 5.4.2b). Thus, monolayer graphene is a semimetal or a zero-band-gap semiconductor [5]. The density-of-states is zero at the Fermi level, which is crossed by electronic bands near the six corners of the Brillouin zone (Fig. 5.4.2b) [5, 21]. The charge carriers (electrons and holes) in graphene are called Dirac fermions, and the six corners of the Brillouin zone are called Dirac points (Fig. 5.4.2b) [5]. The electronic dispersion for graphene at the six corners of the 2D hexagonal Brillouin zone is linear [23]. Due to this linear or conical dispersion near the Dirac points, the charge carriers of graphene have zero effective mass and behave as relativistic particles described by the Dirac equation [24]. Many other outstanding electronic properties of graphene are also the result of linear dispersion or the bonding and antibonding of π orbitals [25, 26].

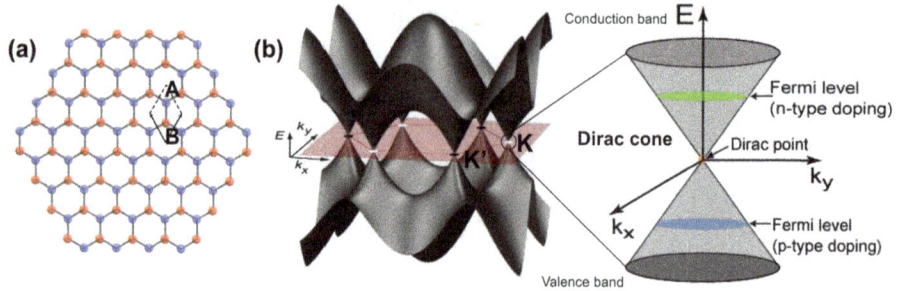

Figure 5.4.2: (a) Hexagonal honeycomb lattice of graphene with two atoms (A and B) per unit cell [27]. (b) Electronic band structure of graphene in the first Brillouin zone and a zoom into one energy band close to the Dirac point. Reprinted (adapted) with permission from Ref. [28]. Copyright 2010 American Chemical Society.

What makes graphene outstanding is that it has the highest known electrical and thermal conductivity. SLG exceeds diamond in electronic mobility ($2.5 \times 10^5 \ cm^2 \cdot V^{-1} \cdot s^{-1}$) [27, 29] and thermal conductivity ($\sim$5000 $Wm^{-1}K^{-1}$, 25 times higher than that of silicon) at room temperature [30]. Graphene also has the highest current density at room temperature (a million times higher than that of copper), it has the highest electronic mobility (200,000 $cm^2 \cdot V^{-1} \cdot s^{-1}$, 100 times more than in silicon), and it conducts electricity in the limit of no electrons (200 times faster than in silicon). However, the limiting factors of the electronic mobility of graphene are the quality of the graphene sheet and the substrate material used. For example, the electronic mobility is limited to 40,000 $cm^2 \cdot V^{-1} \cdot s^{-1}$ in case of monolayer graphene on SiO_2/Si substrate. As electrons in graphene have a longer mean free path (in the order of 65 μm) than in any other material, its charge carriers are able to travel micrometer distances without scattering and operate at ambient temperatures with virtually no resistance. This is similar to superconductivity, but at room temperature. When the mean free path is longer than the dimensions of the material, one can get ballistic transport. The latter is the electron transport in a medium having negligible electrical resistivity caused by scattering. So, graphene exhibits ballistic conduction, the highest electrical conductivity, and the lowest resistivity ($10^{-6} \Omega \cdot cm$, lower than silver).

One of the amazing properties of graphene is its inherent mechanical strength, which is due to its continuous pattern and the strong bonds between the carbon atoms, as each carbon atom in graphene is bound to three other carbon atoms instead of four as in diamond [5]. Due to the mechanical strength of carbon bonds, graphene is the strongest material discovered so far with a Young's modulus of ~1.1 TPa [31] and a high intrinsic strength of 130 GPa (compared to 0.4 GPa for A36 structural steel) [32], which approaches the predicted maximum theoretical value [5]. These strong bonds between the carbon atoms of graphene are also mechanically very flexible. Therefore, it is very elastic and able to regain its initial size after strain [33]. As graphene is bendable and stretchable, it can be relatively easily twisted, pulled, and curved to a certain extent without breaking. One can stretch a graphene sheet to an additional 25 % of its original length without breaking it. Besides, the graphene monolayer is impermeable, with such closely knit carbon atoms that they can work like a fine atomic net, not allowing other material atoms (i. e., helium, hydrogen, etc.) to get through. It can thus act as a very effective surface passivation layer or protective coating for different materials; e. g., monolayer graphene can protect an underlying metal from corrosion, reactions, etc.

Also, it should be noted that graphene is almost completely optically transparent (it transmits ~97.7 % of white light, compared to ~80–90 % for a window glass). The absorption of ~2.3 % of incident visible light is sufficient for one to see a monolayer of graphene by the naked eye on a piece of white paper. This makes it a great candidate for replacing traditional electrodes by transparent and flexible graphene-based electrodes. The developed organic light-emitting diode (OLED) displays [34], which are thinner, bendable, and more energy-efficient than conventional displays, can be applied in e-book readers, smart cards, e-posters, etc. They reflect light like paper, which allows reading more comfortably and provides a wider viewing angle than most light-emitting displays. Inexpensive graphene touch screens or displays can enable cheap, flexible, and rollable portable electronic devices, which will likely appear in the near future [35]. The same properties make it promising for flexible touch screens and electronic skin or artificial retina applications [36].

Furthermore, graphene could enable ultrafast uploads and replace batteries with quickly and long-lasting light-weight rechargeable batteries or supercapacitors due to its large surface area [37]. Medical applications in contact lenses may enable IR vision, smart plasters may reduce the risk of bacterial infection, and neural interfaces may create a direct connection of machines with the human mind [38]. Filtering and absorbing properties can be employed in the desalination of water, treatment of radioactive waste, and cleaning of oil spills by absorbing oil [39].

The most-studied noncarbon 2D material is molybdenum disulfide (MoS_2), which is also obtained from the bulk structure that appears as a mineral molybdenite. Individual MoS_2 layers have a direct band-gap with an enhanced energy of ~1.89 eV. A single monolayer of MoS_2 can absorb 10 % of the incident light with the energies above the band-gap [40]. Moreover, the MoS_2 monolayer has a thermal conductivity of about 35 $Wm^{-1}K^{-1}$, which is about 100 times lower than that of graphene. These distinguished properties give MoS_2 scientific and industrial importance for example for the creation of electronic devices such as transistors, sensors, solar cells, etc. [41]. MoS_2 monolayers are flexible enough, and thin-film FETs retain their electronic properties when bent [42]. They have a stiffness comparable to that of steel and a higher tensile strength than flexible plastics, making them particularly suitable for flexible electronics.

The compound tungsten diselenide (WSe_2) has a hexagonal crystal structure similar to that of MoS_2. Each tungsten atom is covalently bound to six selenium ligands in a trigonal-prismatic coordination sphere, and each selenium is bound to three tungsten atoms in a pyramidal geometry [43]. Bulk WSe_2 is a semiconductor with an indirect band-gap of ~1.3 eV, while monolayer WSe_2 has a direct band-gap. The layers are stacked on top of each other by van der Waals interactions and can be exfoliated into thin 2D layers. Recent studies have shown that the WSe_2 monolayer nanostructure has a smaller gap (1.7 eV) and exhibits excellent mobility (250 $cm^2V^{-1}s^{-1}$), and a high on/off current ratio (10^8) at room temperature [44], making it very interesting for electronic and optoelectronic devices [17]. For instance, exfoliated monolayer WSe_2 can be used to create a high-performance p-type FET [45].

Hexagonal boron nitride (h-BN), also known as "white graphene," has a structure based on sp^2 covalent bonds similar to graphene. The h-BN monolayers have a layered structure and lattice structure similar to graphene, with a lattice divergence of only about 1.8 %. 2D h-BN has no absorption in the visible range, but in the ultraviolet (UV) region with good PL. h-BN and graphene differ in electrical conductivity. h-BN is a semiconductor with a direct band-gap of ~5.9 eV and a very high breakdown voltage (>0.4 V/nm). h-BN is widely used as an insulator for the production of ultrahigh-permeability 2D heterostructures consisting of various types of 2D semiconductors (e. g., WSe_2, MoS_2, etc.). h-BN is also a promising material for building high-resolution optical microscopes. Polaritons formed on the surface of a crystal constructed from 99 % pure boron isotope make it possible to lower the diffraction limit many times and achieve even units of nanometer resolutions [46].

Gallium selenide (GaSe) is a layered semiconductor in which each individual layer consists of covalently bound stacks of Se ions at the top and bottom and two levels of Ga ions in the middle in the Se-Ga-Ga-Se sequence, which are held together by van der Waals forces. This makes it possible to exfoliate the structure mechanically or by liquid means. The fabricated 2D GaSe flakes have a tunable indirect band-gap of ~2.1 eV. For the monolayer, the experimental mobility value is ~0.6 $cm^2V^{-1}s^{-1}$, consistent with the transport properties of FETs. 2D GaSe monolayers exhibit nonlinear layer-dependent optical properties. GaSe layers could be used for the development of electronic and optoelectronic devices to realize functional FET and photodetector applications [47].

5.4.4.2 Synthesis of 2D atomic materials

Currently there are numerous techniques for obtaining graphene and 2D atomic materials of various dimensions, shapes, and quality. There are two main ways for their mass production: large-scale exfoliation (obtaining graphene from graphite) and large-scale growth (CVD, etc.), which can be classified into two main categories [48], i. e., top-down (e. g., micromechanical [12, 49, 50], liquid phase [51, 52], chemical [53], electrochemical [54], microwave-assisted [55], exfoliation methods) and bottom-up (e. g., CVD on metal surfaces [56, 57], epitaxial growth on electrically insulating surfaces, such as SiC [58]) processes [59]. First, we will focus on the main conventional

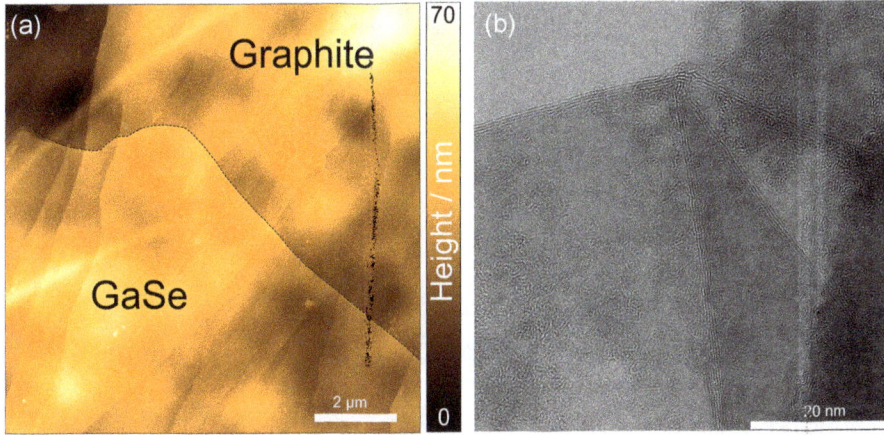

Figure 5.4.3: (a) AFM topography of a GaSe nanoflake on a graphite surface and (b) high-resolution transmission electron microscopy (HR-TEM) image of MFL graphene sheets obtained by the substrates rubbing method. Reprinted (adapted) with permission from [48].

exfoliation and growth methods for obtaining 2D materials and then discuss some nonconventional ones.

Micromechanical exfoliation (Scotch tape method) was used to obtain the first 2D atomic material, graphene, from layered bulk material, graphite, by separation with a Scotch tape. The essence of the method is slicing or multiple exfoliating of layers from the layered bulk material by repeatedly rubbing a piece of graphite against another surface on the Scotch tape until atomically thin flakes are obtained (Fig. 5.4.3b). After that the exfoliated flakes are transferred to a silicon wafer covered with a dielectric SiO_2 layer with a thickness of 300 nm, which allows visualizing graphene flakes through optical interference [12]. Micromechanical exfoliation has been applied to obtain 2D materials such as GaSe (Fig. 5.4.3a), graphene (Fig. 5.4.4a), h-BN, and TMDCs. Though micromechanical exfoliation allows producing high-quality monolayers, it is a nonscalable and time consuming process with extremely low output and does not allow controlling the flake thickness (atomic layer number) and size (few tens of microns); therefore, its application is limited.

To overcome these problems, the CVD method is used to epitaxially grow large-area and uniform-thickness MFL graphene (Fig. 5.4.4b), MoS_2, GaSe, h-BN, TMDCs, etc., on various substrates, such as SiO_2 or sapphire. As the 2D atomic materials are formed by growing (depositing) layers from gas (vapor), this method is called CVD. The CVD process is based on the decomposition of hydrocarbons on catalytic or metallic surfaces, such as copper, at temperatures above 800 °C [60]. Though CVD synthesis is more scalable than mechanical and chemical exfoliation techniques, there are a few drawbacks. Firstly, they have a higher concentration of structural defects, such as grain boundaries, and it is impossible to avoid the grains. Secondly, it is difficult to control the exact number of atomic layers across a substrate. Thirdly, after the CVD

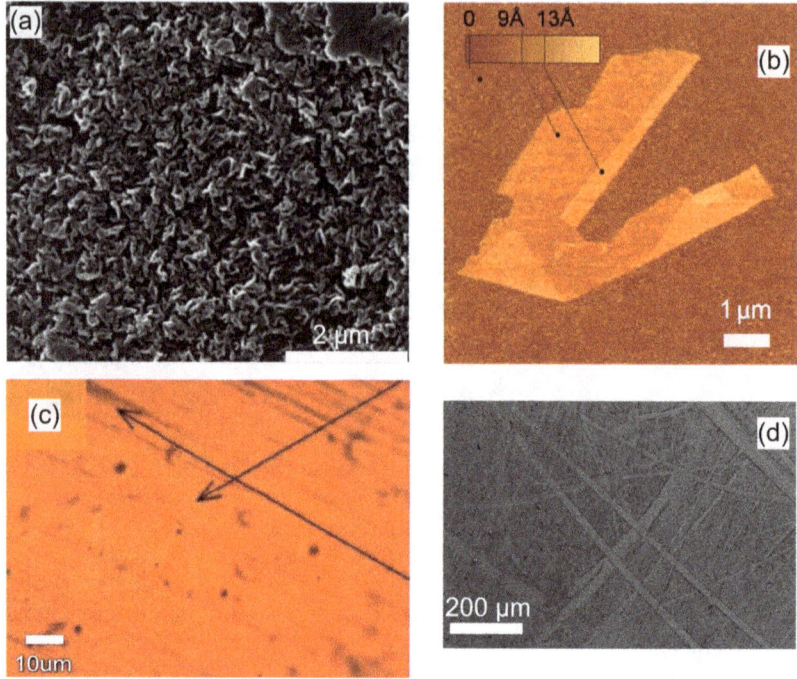

Figure 5.4.4: As-prepared graphene films. (a) AFM image of mechanically exfoliated graphene layers. Reprinted (adapted) with permission from Ref. [78]. (b) Scanning electron microscopy (SEM) image of graphene synthesized by CVD on MgO substrate. Reprinted (adapted) with permission from Ref. [77]. (c) Optical visualization of graphene growth by the CVD method on copper. Reprinted (adapted) with permission from Ref. [79]. (d) SEM image of MFL graphene sheet consisting of nanostripes of quantum dots on silicon substrates obtained by the substrates rubbing method at 1000 rubbing cycles. Reprinted (adapted) with permission from Ref. [48].

epitaxial growth it is necessary to transfer the 2D atomic layer from the catalytic surface to the proper substrate, which is always accompanied by defects. Therefore, until now the reproducible CVD growth of high-quality 2D atomic materials for mass production is still challenging. Another CVD method for producing monolayers is metalorganic CVD (MOCVD), which consists of inserting carbon-based organic gas, such as methane, into a closed container with a metallic film at the bottom and then controlling the temperature and pressure until a monolayer is formed on it. MOCVD growth is realized by chemical reaction and not physical deposition (as in molecular beam epitaxy (MBE)). The possibility to grow graphene by MBE has also been demonstrated [61, 62]. However, the question whether MBE will routinely be used for growing 2D monolayers remains open.

Currently, the simplest, cheapest, and fastest methods for obtaining graphene and other 2D materials are based on the use of graphite-based and raw 2D materials [48] such as mechanical thinning [63], shear exfoliation in liquids [64], hydrothermal exfoliation [65], wet ball milling [66], roll-to-roll [67], and arc-discharge [68] methods. This list of the methods can be continued as currently various methods have been proposed to produce 2D materials. Anyway, most of these methods are not industrially scalable ones and cannot be applied for mass production of high-quality and large-size 2D atomic materials. Their up-scaling is still not straightforward due to the difficulty of the setup, the relatively low yield to meet the potential market demand, and usage of different toxic chemicals, solutions, and sophisticated devices [48].

The quality of the synthesized 2D material layers plays a crucial role for the fabrication of 2D devices as the presence of the defects, impurities, grain boundaries, multiple domains, structural disorders, and wrinkles [48] formed in them during growth or post-growth (exfoliation) or the transfer of 2D layers from one substrate (the one it was grown on) to another has an adverse effect on their electronic and optical properties. Depending on the transfer approach, the methods of obtaining 2D materials can be classified as "wet" (i. e., polymethyl methacrylate-assisted transfer (PMMA)) [69] and "dry" (i. e., polydimethylsiloxane (PDMS) stamps) [70]. The problems with wet and dry transfer methods are that (a) polymer or metal residues are resulting on the surfaces of 2D layers and (b) the transfer can increase damage or folding of the 2D atomic layers. For mass production of high quality, preparing large-size and low-cost 2D materials directly on any substrate via cheap, simple, fast, transfer-free, ecologically clean, and universal technologies is important [71]. Industrially scalable nonconventional methods for obtaining high-quality, large-size, and low-cost 2D atomic materials, hybrid structures, and devices directly on any substrate are developed [48, 71, 72].

The author G. Shmavonyan with coworkers proposed the substrates rubbing method [48, 71–74] for obtaining 2D materials (graphene (Fig. 5.4.3b), MoS$_2$, GaSe, h-BN, etc.), their hybrid structures, dispersion, and powder, which consists of putting pristine bulk (graphite, h-BN, MoS$_2$, GaSe, etc.) layered material (i. e., crystalline powder) between two solid-state atomically flat substrates or stepped (terraced) surfaces and rubbing the substrates against each other mechanically in any direction, so that the crystalline powder uniformly spreads between them and covers the surfaces of the two substrates with it. It leads to the exfoliation of 2D layers upon multiple rubbings [48, 71–76]. The resulting morphology can be seen in Figure 5.4.4c and d.

The 2D atomic material nanostripes investigated by the author G. Shmavonyan, sheets of nanostripes, and hybrid structures have a unique structure (nanostripes consisting of quantum dots and 2D atomic layers consisting of nanostripes), shape, and size (their length can reach hundreds of meters), advantages which allow to obtain 2D materials of high quality (low-defect and nonoxidized) with clean interfaces (due to a rapid and transfer-free process), achievable from many types of layered material on many solid-state substrates.

After synthesis, the visualization and characterization by spectroscopic methods of 2D atomic materials is important for different applications. Thus, in the following we discuss the application of optical methods for 2D material analysis to evaluate the quality of 2D materials.

5.4.4.3 Identification and characterization techniques of 2D atomic materials

In this section we will consider the prospects and challenges connected with the identification and characterization of 2D atomic materials. There are different characterization techniques for the identification of 2D materials, such as optical, AFM, high-resolution transmission electron microscopy (HR-TEM), scanning tunneling microscopy (STM), Raman spectroscopy, TERS, PL spectroscopy, angle-resolved photoemission (ARPES), and UV-Vis spectroscopy. For more information on these techniques, see also Volume 1, Chapters 2.2, 2.3, and 3.1–3.3 and Volume 2, Chapters 4.5, 6.1, and 6.5. The existing analytical approaches via advanced microscopy tools only exploit the information about the optical contrast of the 2D materials [80, 81]. Meanwhile, the chemical identification of 2D materials and their more detailed characterization require spectroscopy aided by complementary methods such as AFM, which are typically more time consuming. So, there is a growing interest in automating these processes. An artificial intelligence-based approach [82] speeds up the preparation, identification, and characterization of 2D materials, accelerates the discovery of new ones, and predicts properties of materials.

Optical microscopy is an efficient tool for visualizing or identifying MFL flakes from multilayer ones, i. e., graphene from graphite. After obtaining MFL flakes by micromechanical exfoliation, their identification is difficult, but it is an important step for their characterization and application. Under an optical microscope many thick and tiny flakes with different shapes and colors can be observed. Some optical images of as-prepared graphite and graphene flakes obtained on Si/SiO$_2$ substrate after micromechanical exfoliation are shown in Figure 5.4.5. After the micromechanical exfoliation it is very difficult to find graphene flakes among millions of thicker graphitic ones (Fig. 5.4.5a). For distinguishing monolayers from few-layer or multilayer graphene (bulk graphite) it is necessary to carefully scan the sample surface by an optical microscope, which is a time-consuming process. In case of a SiO$_2$ layer thickness of 300 nm for interference imaging, the ultrathin flakes have sufficient contrast to differentiate graphene flakes among thicker ones, and thus become visible with an optical microscope. MFL graphene flakes are almost transparent and crystalline shapes, which have darker regions with respect to the substrate (Fig. 5.4.5b). The color of much thicker flakes (more than five layers) does not follow this trend and can change from blue to gray. The SiO$_2$/Si substrate is the most common one used to visualize MFL. The color of the SiO$_2$ dielectric layer on the Si wafer depends on an interference effect from reflection of the two SiO$_2$ layer surfaces. MFL flakes on the dielectric layer modify the interference and create a color contrast between the flake and the substrate. For optimal contrast, the thickness of the dielectric layer needs to be within 5 nm of the ideal value [83]. Visualizing monolayers on other substrates without the need for interference can be done as well. It is possible to directly visualize monolayers (i. e., MoS$_2$, WS$_2$) with direct band-gaps by fluorescence microscopy. However, their fluorescence

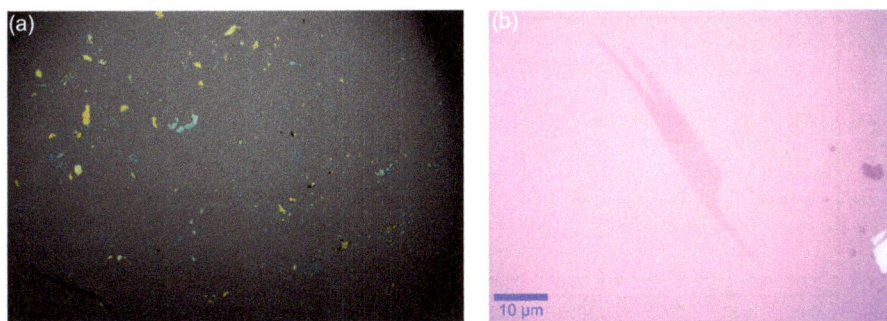

Figure 5.4.5: Optical images with different scales (10× (a) and 100× (b)) of as-prepared MFL graphene and graphite (a) and monolayer graphene (b) on SiO_2/Si substrate obtained after micromechanical exfoliation [84].

reduces as their band-gaps become indirect by increasing the number of the atomic layers.

Raman spectroscopy

After identification of graphene or other 2D material flakes by an optical microscope, they can be precisely distinguished by Raman spectroscopy (cf. Volume 1, Chapter 2.3). The latter is a nondestructive, fast, and powerful characterization technique for characterizing structural properties of 2D materials, such as the number of atomic layers (distinguishing MFL (less than 5 layers)), defect densities, or strain. Raman spectroscopy is an effective method to investigate graphite and its analog. When laser light interacts with a lattice, it induces in-plane or out-of-plane vibrations with single (A, B), double (E), and triple (T) degeneration in the whole lattice, which results in resonance Raman spectra. These vibrations can be longitudinal or transversal with low (acoustic) or high (optical) energy. Let us consider the Raman spectra of graphite and graphene. There are two pronounced peaks in the Raman spectra of high-quality graphene and graphite (Fig. 5.4.6a) [85]. The first peak at a wavenumber of ~1580 cm^{-1} (G peak) corresponds to the in-plane bond-stretching optical vibration of sp^2-hybridized carbon atoms [86]. This vibration in graphene has the E_{2g} symmetry and the G peak is due to the E_{2g} phonon at the Brillouin zone center of graphene [87]. The E_{2g} mode is a doubly degenerate Raman active optical vibration, where the carbon atoms move in the graphene plane [88]. The second peak at a wavenumber of ~2700 cm^{-1} (2D or D$'$ peak) is due to second-order Raman scattering by in-plane transverse optical phonons near the boundary of the Brillouin zone of graphene [85]. The third peak (D peak) at a wavenumber of ~1360 cm^{-1} is due to the breathing modes of sp^2 atoms and requires a defect for its activation (Fig. 5.4.6b) [89], which makes this method particularly suitable for the confirmation of the material quality. In addition

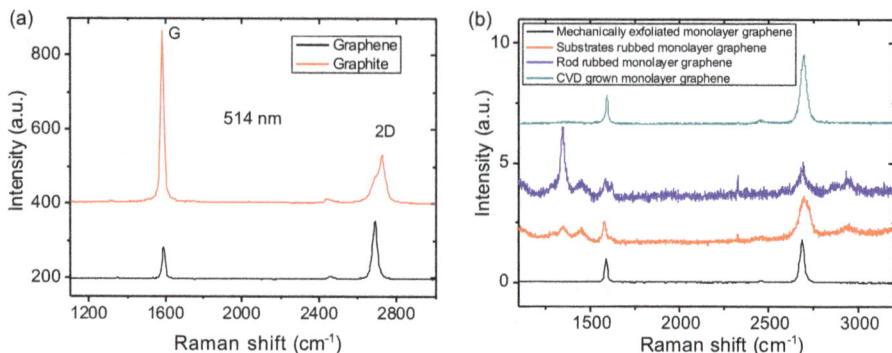

Figure 5.4.6: (a) Raman spectra of monolayer graphene (black) and graphite (red) obtained by micromechanical exfoliation on SiO_2/Si substrate at an excitation wavelength of 514.5 nm [84]. (b) Raman spectra of micromechanically exfoliated (black), substrates-rubbed (red), rod-rubbed (blue), and CVD-grown (gray) monolayer graphene on SiO_2/Si substrates at an excitation wavelength of 514.5 nm. Reprinted (adapted) with permission from Ref. [48].

to the observed intense Raman D, G, and D$'$ peaks, other, less intense D$'$ (1620 cm^{-1}), D+D$''$ (2450 cm^{-1}), and D+D$'$ (2950 cm^{-1}) peaks are observed [90, 91].

Besides, Raman spectroscopy allows observing new peculiarities of 2D materials. For example, in experiments by Shmavonyan and coworkers, a new peak (with its intensity depending on the number of rubbings) at a wavenumber of 1450 cm^{-1} is observed (Fig. 5.4.6) [48].

The Raman signal of monolayer graphene is easily detectable (Fig. 5.4.6). The obvious difference between the Raman bands of monolayer graphene and graphite is in the second-order D peak, the 2D band. The 2D band of monolayer graphene has a single sharp and symmetric peak, while the same band for graphite consists of four peaks [92]. The line shape of the Raman 2D band of MFL and multilayer graphene is strongly dependent on the number of atomic layers (1 to 4), stacking order, and twist angle between graphene layers. By choosing an appropriate excitation energy, the number and stacking order of atomic layers can be determined. Few-layered graphene has two types of stacking order: Bernal (ABA) and rhombohedral (ABC). Bilayer graphene has a much broader and up-shifted (bathochromic shift) 2D band relative to monolayer graphene [92]. The G band intensity increases almost linearly with an increase in the number of layers [93], while the 2D peak profoundly changes its shape, making it great for the identification of the number of layers up to five. The intensity of the D peak depends on the structural defects. If graphene or graphite is of good quality, no Raman D band is observed, which corresponds to the absence of the graphene structural defects (Fig. 5.4.6a). Thus, the most striking differences of the Raman features in monolayer graphene compared to graphite are as follows: (a) the 2D peak is single and sharp; (b) the FWHM of the 2D peak is ~30 cm^{-1}; (c) the ratio of the intensities of the 2D and

G peaks is ≥ 2; (d) the G peak position of graphene has a slightly lower wavenumber than for graphite (Fig. 5.4.6a) [3].

A comparison of the Raman spectra of graphene obtained by micromechanical exfoliation, rod rubbing, substrates rubbing, and CVD methods is shown in Figure 5.4.6b.

Besides, Raman spectroscopy also gives important information about doping, edge type, strain and stress, disorder, oxidation, hydrogenation, chemical functionalization, electrical mobility, thermal conductivity, electron–phonon and electron–electron interactions, magnetic field, and interlayer coupling [94]. For instance, the G peak is lowest when the Fermi level of graphene is at the Dirac point, and when the electron or hole concentration increases, the G peak has a blue shift. The G band width reaches a maximum when there is no doping in graphene and decreases for the increase of n- or p-doping in graphene [95].

Despite the fact that the Raman signal from graphene is strong and comparable to that of bulk graphite, investigating properties of graphene, graphene nanoribbons, or single carbon nanotubes at the nanoscale is not always easy due to the optical diffraction limit of laser beams. TERS is a rising star to investigate properties of materials at the nanoscale. For example, Gadelha and colleagues used TERS to identify local strain and doping in in-plane homojunctions in graphene devices, built as van der Waals heterostructures where half of the graphene sits on talc and the rest sits on SiO_2 [96]. The high spatial resolution of TERS enables the observation of the charge depletion region at the interface between different substrates [97]. Recent experiments show possible TERS resolutions of 1 nm or even below to (sub)molecule resolution [98].

Besides graphene, there are still a large number of semiconducting TMDCs, such as MoS_2, GaSe, InSe, $MoSe_2$, WS_2, WSe_2, etc. For MoS_2, the most-studied TMDC material, two main peaks, $E_{2g}^1 \sim 382\,cm^{-1}$ and $A_{1g} \sim 406\,cm^{-1}$, are used to identify the number of layers. E_{2g}^1 and A_{1g} correspond to in-plane and out-of-plane vibrations, respectively. With increasing flake thickness, the E_{2g}^1 peak exhibits a red shift, but the A_{1g} peak shows a blue shift [27]. The blue shift in the A_{1g} peak can be explained by the van der Waals interaction between layers that suppresses the atom vibration, resulting in stiffening of the vibration. On the contrary, the E_{2g}^1 peak's red shift suggests that the stacking-induced structure changes or long-range Coulombic interlayer interactions in multilayer MoS_2 affect the change of atomic vibration [99] stronger than the increase in van der Waals force [100]. Similar phenomena for E_{2g}^1 and A_{1g} modes also have been observed for other 2D materials such as $MoSe_2$, WSe_2, WS_2. Even though E_{2g}^1 and A_{1g} are good indicators for flake thickness, this method is limited to approximately four layers, and the shift is small and can also be affected by strain, doping, and substrate material. Another way is using the interlayer shearing and breathing modes, which are shown in Figure 5.4.7. These modes are sensitive to layer thickness, and most of the few-layer 2D materials are predicted to have these modes [101], so interlayer shearing and breathing modes can be used to identify the number of layers [16]. The disadvantage of this method is that the interlayer shearing and breathing

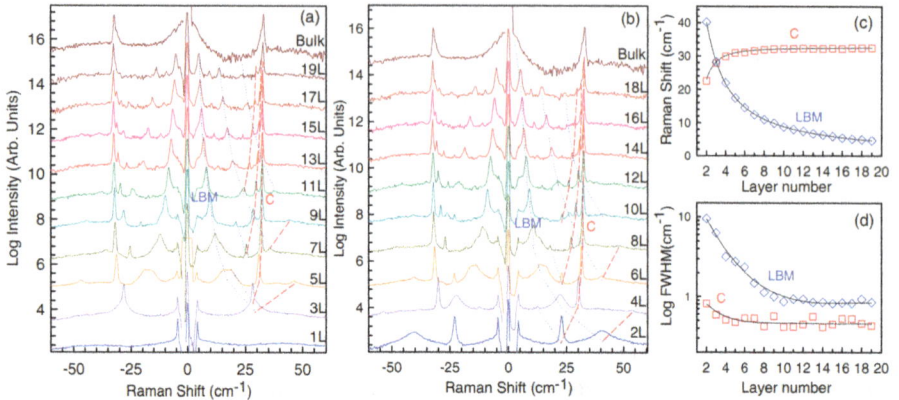

Figure 5.4.7: (a) Stokes and anti-Stokes Raman spectra of odd N layer (ONL)-MoS$_2$ in the low-frequency range. (b) Stokes and anti-Stokes Raman spectra of even N layer (ENL)-MoS$_2$. The spectrum of bulk MoS$_2$ is also included in (a) and (b). Dashed and dotted lines in (a) and (b) are guides to the eye. (c) Position of typical shear (C) and layer breathing (LB) modes as a function of N. (d) FWHM of C and LBM as a function of N. Solid lines in (c) and (d) are guides to the eye. Reprinted (adapted) with permission from Ref. [102].

modes appear at low Raman shifts, so it requires an additional triple stage spectrometer or a single-stage spectrometer with recently developed Bragg notch filters to record these modes [101]. For MoS$_2$, the number of layers can be determined by this method up to 14 layers.

The disorder in graphene can be quantified by the I_D/I_G peak intensity ratio in graphene spectra. Several groups have reported the effect of strain in 2D materials such as graphene, MoS$_2$, or GaSe. In the case of GaSe, the forbidden Raman mode located at 250 cm^{-1} is observed on the strained region created by a step on the HOPG substrate. This peak is not observed on plane and bulk GaSe [103]. Uniaxial strain is reported to cause red shift and splitting of the E_{2g}^1 mode into two subbands E_{2g}^{+1} and E_{2g}^{-1} in MoS$_2$, WS$_2$, WSe$_2$, and MoTe$_2$ [104, 105]. The A_{1g} Raman mode of MoS$_2$ is sensitive to doping. The A_{1g} peak becomes weaker and broader with an increase of n- or p-doping concentration. In contrast, the E_{2g}^1 peak is insensitive to doping, so the A_{1g} phonon vibration is typically used to investigate doping in MoS$_2$ [106]. In the case of WS$_2$, the out-of-plane A_{1g} mode of p-doped or n-doped WS$_2$ shows the red shift and blue shift, respectively, in comparison with undoped WS$_2$ [107, 108].

Photoluminescence

PL spectroscopy is one of the most widespread techniques in 2D material science along with Raman spectroscopy and the other advanced methods listed in this chapter. PL spectroscopy is based on a fluorescence effect, when an observed material illuminates under irradiation. In general, PL spectroscopy allows to investigate electronic properties, a material's structure, and a presence of impurities with submicron spatial resolution.

For instance, [109] investigated the lifetime of PL and fluorescence on CVD-grown MoS_2. In the case of PL, the measurement was performed with a 532 nm continuous laser on a $75 \times 75\ \mu m^2$ range with 375×375 points that results in a high-quality image, shown in Figure 5.4.8. In Figure 5.4.8a the authors demonstrate a PL map of the region of interest where MoS_2 flakes are grown. Dark regions here are the results of nonphotoactive areas, which are cracks or grain boundaries. Figure 5.4.8b shows a PL map of MoS_2 flakes that do not reveal grain boundaries or stress-related cracks. Hence, PL mapping is a useful method for investigating the material's defects.

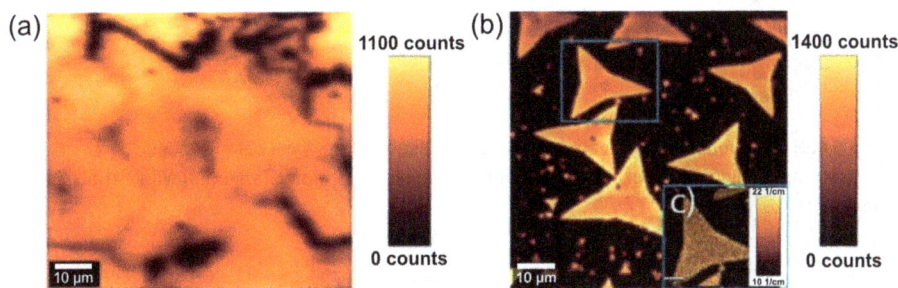

Figure 5.4.8: (a) Total PL map and (b) PL map of clearly seen MoS_2 flakes on the substrate. Reprinted (adapted) with permission from Ref. [109].

Another clear experiment aimed at revealing defects was made by Xu et al. [110]. They studied the PL behavior of high-quality MoS_2 single layers. In this work, a MoS_2 monolayer was measured by PL spectroscopy under different temperature and air/vacuum conditions. Figure 5.4.9a demonstrates a common PL map of a flake on a sapphire substrate. Obviously, the PL map contains several regions with lower PL intensity along the crystal edges. The authors suggest that these fluctuations in PL intensity may be related to the varying material quantity at the edges or contaminants from air (such as dust). But, in general, the PL image shows the material with a homogeneous structure, which is also proven by Raman imaging on this flake in the inset. The authors also discussed a relationship between PL intensity and ambient conditions, such as the presence of air. Decreasing PL intensity is connected with interaction of oxygen or nitrogen from air and sulfur vacancies in MoS_2 and leads to recombination of photogenerated charge carriers. Particularly, the lowering of PL intensity with increasing temperature signals a higher number of nonradiative recombinations. In addition, the peak positions of the PL express a hypsochromic shift while temperature decreases, depicted in Figure 5.4.9c. This blue shift is related to the widening of the band-gap, caused by lattice shrinkage and lower thermal relaxation ability.

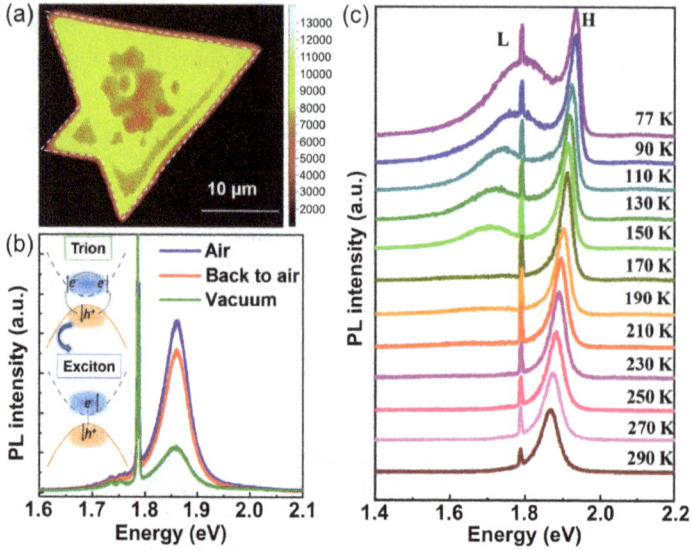

Figure 5.4.9: (a) PL map of high-quality MoS$_2$. (b) PL spectra of the flake being exposed by air/vacuum. (c) PL spectra of MoS$_2$ flake under various temperature. Reprinted (adapted) with permission from Ref. [110].

TERS

One powerful method for the investigation of 2D materials properties is TERS. In TERS, the Raman spectrum is amplified when a gold- or silver-coated tip approaches the illuminated sample surface. Using the AFM cantilever as an amplifier, the physical and chemical properties of the 2D materials could be measured in the nanorange with an integrated Raman spectrometer while obtaining topographic data by a coupled AFM system.

In 2013, Su et al. [111] performed a TERS investigation of SLG flake edges obtained by the Scotch tape method from HOPG. The far- and near-field spectra shown in Figure 5.4.10a were recorded in the middle of the SLG to avoid artifacts caused by intensity changes near the edge. The disappearance of peak D in Figure 5.4.10a indicates the absence of defects in the measured spots. Figure 5.4.10b shows an AFM lateral force image acquired simultaneously with a TERS image. Mechanically exfoliated graphene flakes have been found to have sharp and linear as well as jagged and indeterminate edges. A 2D TERS image of the same area is shown in Figure 5.4.10c.

The authors discovered that the graphene D peak intensity (Fig. 5.4.10d) could be used to accurately determine the SLG edge location, and the phase-breaking length (path length of conduction electrons while maintaining phase coherence) can be precisely measured by using TERS. The phase-breaking length was measured to be

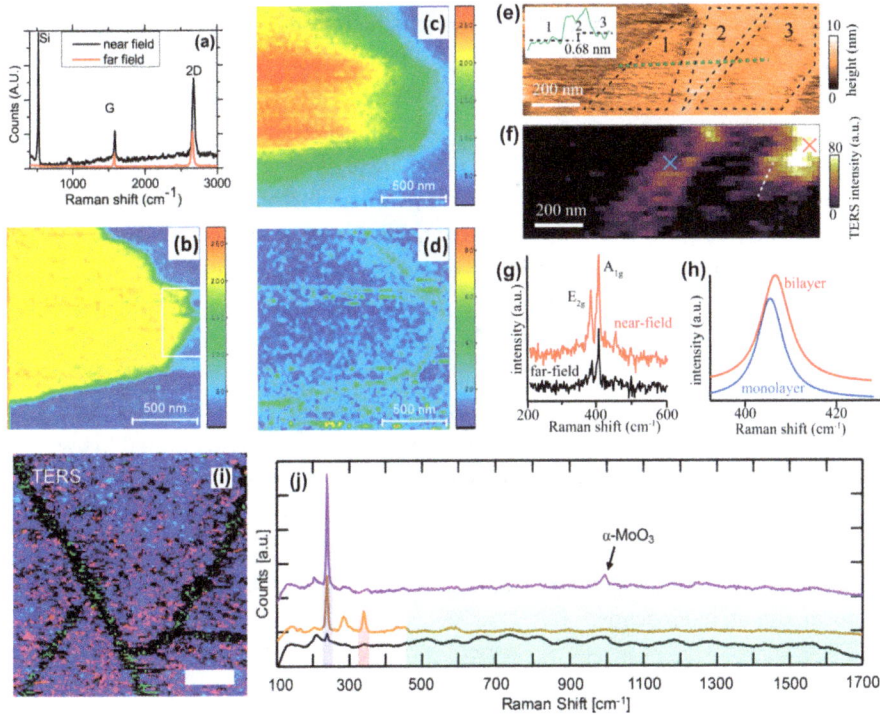

Figure 5.4.10: (a) Tip-enhanced and confocal Raman spectra of a monolayer graphene flake. (b) AFM lateral force image of monolayer graphene recorded during TERS mapping. (c) TERS image formed using the 2D peak height. (d) TERS image formed using the D peak height. Image size: 1.5 μm × 1.5 μm, 60 × 60 pixels. Reprinted (adapted) with permission from Ref. [111]. (e) AFM image of the region of interest. Topographic line profile along the green dotted line is shown in the inset. (f) TERS intensity map of the A_{1g} mode for the same area as that in (e). (g) Near-field and far-field Raman spectra of the MoS_2 layer taken at the same position. (h) Lorentzian fitted TERS spectra of the A_{1g} mode. Reprinted (adapted) with permission from Ref. [112]. (i) Combined TERS maps of $MoSe_2$ polycrystalline monolayer. (j) Averaged TERS spectra showing resonant response (orange), nonresonant response (purple), and signal from the grain boundaries regions (black, showing significantly decreased intensity of Raman signal of $MoSe_2$ over the grain boundaries). Reprinted (adapted) with permission from Ref. [113].

4.2 nm, which is much smaller in comparison with far-field Raman results with a value of 30–50 nm.

Kato et al. [112] investigated atomically thin MoS_2 layers on a glass substrate by TERS coupled with AFM. It was revealed that the inhomogeneous features of the thin MoS_2 layers originate from mechanical and chemical preparation methods.

Figure 5.4.10 demonstrates AFM (e) and corresponding TERS (f) images as well as Raman (g) and Lorentz-fitted spectra constructed using the intensity of the A_{1g} mode (h). In this work, TERS intensity denotes the near-field Raman intensity, which is obtained after subtracting the far-field intensity. Spectral changes of the near-field Raman signal locate the presence of nanometer defects and residual substances with a high spatial resolution of ~20 nm, which would not be visible without TERS using only far-field Raman spectroscopy, since its nanometric volume is much smaller than the diffraction-limited focus spot.

Another interesting work is related to TERS measurements of poly- and monocrystalline MoSe$_2$ monolayer films synthesized by CVD and transferred to gold substrates [113]. Here the authors claim that based on TERS results the concentration of the charge carriers in the grain boundaries of MoSe$_2$ flakes is increased in comparison with the interior of the as-grown monolayer crystals. Also, a decreased capacitance and TERS signal are observed at these boundaries after transferring to the gold substrates.

High-resolution resonance TERS measurements of MoSe$_2$ are performed as shown in Figure 5.4.10i and j. The appearance of the peak at 995 cm^{-1} is attributed to the presence of α-MoO$_3$. The authors conclude that nanoscale inclusions of MoO$_3$ in the MoSe$_2$ matrix are responsible for the abovementioned observations. The high spatial resolution of TERS allowed to see differences in nanoscale domains in layered MoSe$_2$ and to reveal significant inhomogeneity where uniform spectral distribution is observed by confocal Raman microscopy.

5.4.4.4 2D hybrid atomic heterostructures and devices

This section focuses on 2D hybrid atomic heterostructures and devices consisting of various 2D atomic layers, their synthesis, and their nanoengineered properties.

After the isolation of 2D atomic materials it became possible to form 2D hybrid heterostructures based on these for application in 2D devices. As semiconductor heterostructures have a great influence on our daily life, the Nobel Prize in physics was awarded to Zhores Alferov and Herbert Kroemer in 2000 for the discovery of semiconductor heterostructures [2, 115]. Traditional heterostructures usually consist of semiconductors only, while 2D hybrid heterostructures include different 2D atomic layers, such as metallic graphene, insulating h-BN, and semiconducting TMDCs. As an atomic layer can be considered to consist of the surface only, the single atomic layers in the hybrid heterostructure behave as independent entities. The simplest 2D heterostructure consists of double atomic layers [116] such as stacked graphene and h-BN (graphene/h-BN), MoS$_2$/graphene, and WSe$_2$/h-BN, while a trilayer heterostructure consists of triple atomic layers such as stacked semiconductor TMDC, dielectric h-BN, metallic graphene (MoS$_2$/h-BN/graphene), WSe$_2$/MoSe$_2$/graphene, etc.

Due to a large number of 2D materials, different atomic heterostructures with electronic properties ranging from insulators, semiconductors, and metals to superconductors can be prepared, enabling a greater number of combinations than traditional heterostructures. As nanolayers are self-passivated [117], i. e., their surfaces are free of chemically active dangling bonds, the formation of hybrid 2D heterostructures with high-quality heterointerfaces is possible. Atomic layers in the heterostructures interact via weak van der Waals forces, which allow for stacking dissimilar layers with very different lattice constants. For example, even though the large lattice mismatch between graphene and WSe_2 is 23 %, the formed WSe_2/graphene nanoheterostructure shows an atomically sharp interface and nearly perfect crystallographic orientation. The complexity of the 2D heterostructures can be tailored, and different heterostructure devices with desired band alignment can be fabricated [114].

2D layers can reassemble into multilayers or monolayer stacks and form vertically or horizontally stacked heterostructures. Vertically layer-by-layer stacked monolayers are called vertical 2D heterostructures, and monolayers stitched together in a plane are called lateral 2D heterostructures. These 2D heterostructures are also called van der Waals heterostructures as the van der Waals interaction serves as a "weak glue" in the stack.

The vertical 2D heterostructures have different atomic layers that are vertically stacked with various architectures and functionalities such as stacks of graphene (G) on h-BN (G/h-BN) [118], h-BN on graphene (h-BN/G), graphene on TMDC (G/TMDC), TMDC on TMDC (TMDC/TMDC), etc. The lateral 2D heterostructure is an atomically thin structure which consists of different 2D atomic layers arranged in one plane and allows the formation of clean interfaces. For example, monolayers of graphene and h-BN can be assembled and form lateral G/h-BN heterostructures with clean interfaces due to the same crystal structure. The major appeal of the lateral heterostructures for electronic applications certainly lies in their atomically thin nature that offers a wider world of new nanomaterials with various functionalities and opens up new opportunities for 2D atomic devices [119]. They have been demonstrated for use in flexible, transparent, low-power electronics and optoelectronics, such as lateral atomic p-n diodes [120], lateral Schottky diodes [121], lateral p–n junction photodetectors [122], and lateral heterostructure FETs [123], as well as providing a high potential for future high-density atomically thin integrated circuits [124]. 2D hybrid atomic heterostructures with different architectures, such as stacks of various 2D atomic layers with different order and composition, can be formed by both mechanical stacking and direct epitaxial growth methods which are considered below.

The mechanical stacking method allows forming stacks of 2D atomic layers and their hybrid heterostructures. For that purpose, the mechanically or chemically exfoliated atomic layers are mechanically transferred and after manipulation layer-by-layer stacked on the appropriate substrate. During manipulation of the atomic layers the transferred second atomic layer is precisely stacked (aligned) on the first one by a micromanipulator under an optical microscope. As a result, two different atomic layers are attached closely by van der Waals forces. After further continuation of the transfer-stacking process a stack, and finally a multilayer 2D hybrid heterostructure, is formed.

The direct epitaxial growth of 2D hybrid heterostructures is an alternative to the mechanical stacking method. The CVD method (see Section "Synthesis of 2D atomic materials" above) allows direct epitaxial growth of various vertically (out-of-plane) and laterally (in-plane) stacked 2D hybrid heterostructures [125]. CVD-grown 2D stacks can be formed in two ways: (a) sequential multistep epitaxial growth of atomic layers on the substrate and (b) growth of atomic layers separately on the substrate, followed by layer-by-layer transfer and stacking. Vapor-phase growth of 2D heterostructures has much in common with 2D atomic materials growth and is an effective way to obtain 2D materials. Compared to the growth of graphene and h-BN, the vapor-phase growth of TMDCs is less controlled and has poor repeatability. Van der Waals epitaxy refers to the epitaxial growth of layered materials on clean surfaces without dangling bonds, even if there is a large lattice mismatch between the two atomic layers [83]. Now we are in the initial stages of forming and characterizing van der Waals heterostructures. CVD, direct growth, and van der Waals epitaxy methods have already been used for growing different atomic heterostructures [126].

During the fabrication of 2D hybrid heterostructures by mechanical stacking and CVD methods, the layer-by-layer mechanical transfer and stacking causes cracks and contamination to the 2D layers and their interfaces. Nowadays the existing wet [69] and dry [70] transfer methods for forming 2D layers (e. g., graphene, MoS_2) and heterostructures (e. g., graphene/MoS_2 stacks) can lead to uncontrollable atomic layer and heterointerface contamination as adsorbates. Hydrocarbons get absorbed or trapped between the atomic layers. Thus, they may not result in high-quality 2D layers with clean interfaces and 2D hybrid heterostructures with improved characteristics. Therefore, the development of alternative and nonconventional methods for obtaining 2D hybrid heterostructures is important.

The rod [75, 76] and substrate [48, 71, 72] rubbing methods, which are discussed in the Section "Synthesis of 2D atomic materials," may offer an alternative to the transfer-stacking and CVD growth methods. Mechanical methods allow for obtaining lateral and vertical 2D hybrid heterostructures with clean and sharp interfaces on many substrates, with no need for transfer after the formation [49, 127].

Herbert Kroemer, the Nobel Prize winner in Physics in 2000, stated that "the interface is a device" [128]. More than a decade later, Andre Geim and Irina Grigorieva experimentally confirmed that statement, which has opened up new opportunities at the interface of bulk materials. As a result, 2D heterostructures with the desired energy band alignment can be designed by combining various metallic, semiconducting, or insulating 2D atomic layers in the lateral or vertical geometries. This allows forming different atomic heterojunctions with novel functionalities at the nanoscale. The interface of the heterojunction between two atomic layers can be tuned by selecting the 2D materials, applying electric fields, elastic strain, etc. Moreover, the physical properties of atomic heterostructures can be tailored at the nanoscale by the heterogeneous nanoengineering of the interlayer stacking distance, order and orientation, material and number of constituent atomic layers, band structure of the junction interfaces, etc. For example, changing from AA (AA-stacking implies that corresponding atoms in different layers have the same lateral coordinates) to random stacking, the band structure changes from direct to indirect [129]. As a result, heterostructures with clean atomically sharp interfaces will have unprecedented tunability.

The reproducible formation of high-quality heterointerfaces between 2D layers and the control over their location, number, orientation, and stacking order of atomic layers still remains challenging. Even a single atom out of place can strongly affect the properties of the atomic layers. Thus, the quality of the heterolayers and heterointerfaces can be controlled by tailoring the disorder structure at the atomic scale, such as substitutional defects, grain boundaries, anisotropy, or stacking misalignment.

2D materials have tunable band-gaps, which are strongly dependent on the number of the atomic layers in the stack, and can be altered to produce a range of gap energies. The change in the band structure with the change of the number of atomic layers is due to quantum confinement and the resulting change in the hybridization [130]. Novel tunable heterostructure devices can be fabricated by nanoengineering their band-gap. This way, new exciting opportunities and advantages (extremely high surface area, flexibility, etc.) open up for designing and fabricating next-generation devices with improved characteristics. Recently, Xue et al. demonstrated large-scale production of a vertical MoS_2/WS_2 heterojunction in multiple layers by two-step sulfurization of patterned WO_3 and Mo sheets using a thermal pressure reduction process. Based on the grown heterojunctions, arrays of photodetectors were fabricated and a photoresponse was realized on both rigid and flexible substrates [131]. In another work by Xu et al., the authors show the development of a hybrid 2D graphene-MoS_2 phototransistor with high responsiveness due to the combination of the advantages of strong MoS_2 light absorption and high mobility of graphene carriers. The main mechanism is explained by the photoexcited hole transfer from MoS_2 to the graphene layer when exposed to light. The photoexcited holes generated from the MoS_2 flakes are transported into the graphene under the action of the increasing electric field; meanwhile, the electrons are trapped in the MoS_2 flakes, creating an additional p-type photoexcitation effect for the graphene located below them [132]. Xiong and coauthors demonstrated a multifunctional nonvolatile logic-in-memory application based on novel 2D heterostructures: black phosphorus/rhenium disulfide (BP/ReS_2) [133]. The van der Waals heterostructure and device characterization are presented in Figure 5.4.11. The synaptic weight change is one of the characteristics of artificial synaptic devices, which are memristor-based devices mimicking biological synapses, and they are used in neuromorphic computing systems that process information in a parallel, energy-efficient way and store information in an analog, nonvolatile form. A record-high synaptic weight change, i. e., a change in the strength in the connection between two nodes, over 10^4 % has been shown in the devices. Neural network simulations for handwritten digits recognition based on the heterostructure devices were successfully implemented with a recognition accuracy of around 90 %.

The spectroscopic analysis of heterostructure features such as strain and defect concentration is analogous to that of monomaterial layers. However, the number of

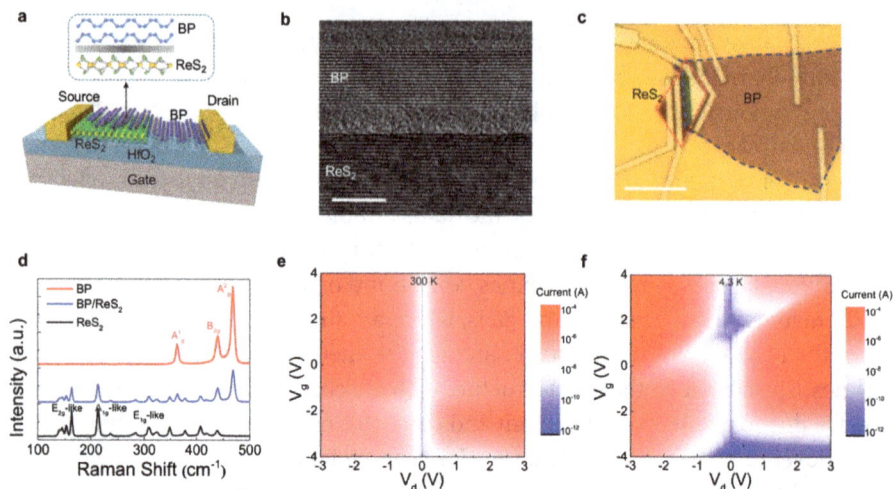

Figure 5.4.11: Van der Waals heterostructure and device characterization. (a) Schematic illustration of a BP/ReS$_2$ heterostructure device and schematic lattice cross-section of the junction interface. The drain electrode is deposited on the BP layer and the source on the ReS$_2$ layer. (b) Cross-section HR-TEM image of the BP/ReS$_2$ interface. A PO$_x$ layer of about 4 nm thick can be observed. The scale bar is 10 nm. (c) Device morphology in an optical microscope. Scale bar: 10 μm. (d) Raman spectra of BP, overlapping BP/ReS$_2$, and ReS$_2$ regions, respectively. Current of the BP/ReS$_2$ junction with different V_d and V_g bias at (e) 300 K and (f) 4.3 K. Reprinted (adapted) with permission from Ref. [133].

layers may prove harder to be identified: the spectroscopic methods may need to be recalibrated due to electronic structure changes in heterostructures that define the Raman and PL sensitivity to the number of layers.

5.4.4.5 Applications of 2D atomic materials and devices

Thanks to their unique properties, 2D atomic materials have the potential to revolutionize many technologies, in areas such as materials and composites, energy, health, space travel, etc., in the same way that graphene promises to do. The various applications of 2D materials mentioned throughout the chapter so far mostly appear as proofs-of-principle in research papers and include (a) wearables (flexible electronics, smart textiles, etc.), (b) mobile and data communications (computer circuits, data storage, ultrafast transistors, optical modulators, etc.), (c) displays (touch screens, conductive inks, transparent electrodes, coatings, light sources, etc.), (d) energy (solar cells, advanced batteries, supercapacitors, fuel cells, hydrogen cells, etc.), (e) transport (corrosion-protective coatings, ultrastrong composites, etc.), (f) environment (desalination membranes, monitors, biofuel, etc.), and (g) food and health (water filters, biosensors, drug delivery nanocarriers, food safety monitors, DNA sequencers, etc.). Still, graphene is the one that currently does have commercial applications, where however in many cases graphene flakes are merely admixed into other materials, e. g., (a) wearables (graphene-enhanced mechanical watches [Richard Mille, France], bracelets with curved graphene film touch sensor [Wuxi Graphene Films, China], graphene-enhanced shoes [inov-8, UK], cycling shoes [Catlike, Spain], graphene-enhanced sportswear with graphene-assisted heat management [Directa Plus, UK, Colmar, Italy], graphene-based smart textiles [Versarien plc, UK, GrapheneUP, Czech Republic, Grafren AB, Sweden], graphene-based fabrics, wearable sensors [GrapheneUP, Czech Republic], graphene-enhanced clothing [Graphene-X, Hong Kong], graphene-coated bulletproof jackets [Vollebak, UK]), (b) mobile and data communications (graphene-based smartphones with flexible touch screen [Galapad Settler, Chongqing Graphene Technology Company, China], bendable graphene smartphones [Moxi Group, China], smartphones with graphene film as a heat sink for cooling [Huawei], graphene integrated circuits [IBM], graphene-based flash memory [University of California, USA]), (c) displays (graphene screens with touch screen panel [Huayuan Display, China; Wuxi Graphene Films, China; Chongqing Graphene Technology, China], updatable foldable electronic newspapers based on graphene-based electronic paper [e-paper] with graphene-based flexible displays [Guangzhou OED Technologies, China], the largest newspaper-size flexible e-paper [LG Display, Korea], e-paper-based touch screens [Sony], graphene e-paper displays [Plastic Logic, Germany]), which have potential for application in next-generation e-readers/books, cell phones, music player displays, smart card displays, wearables (clothing, watches, military), consumer electronics, etc., (d) energy storage (graphene solar cells [IDTechEx, UK], graphene-enhanced lithium [Nanotech Energy, USA] and aluminum-ion batteries [Graphene Manufacturing Group, Australia], graphene anodes and cathodes for lithium-ion batteries [Nanotech Energy, USA], graphene supercapacitors [First Graphene Limited, Australia], graphene power banks [Real Graphene, USA]), (e) transport (graphene-coated airplanes and airplane wings [Haydale Composite Solutions, UK], graphene-coated cars [Briggs Automotive, UK], graphene car wheel arches [Torque, Singapore] and wheels [Vittoria, Italy], anticorrosion and antierosion graphene-enhanced coated cargo ships [Talga Technologies Ltd., Australia], 3D-printed graphene-infused arch structures over railways [AECOM, USA], graphene-enhanced coated satellites with regulated temperature [from too hot to cold] [SmartIR Ltd., UK], graphene-based rockets [Orbex, UK]), (f) environment (graphene membranes [Perforene, Maryland], graphene-ceramic coatings for cars [Adam's Polishes, USA], graphene-based anticorrosive and moisture penetration blocking paint and coatings [Applied Graphene Materials, UK], graphene-enhanced rubber, ink, and coatings [Vorbeck Materials Corporation, USA]), (g) food and health (graphene biosensors [Cardea, USA], functionalized graphene ink for biomedical sensors [Haydale, UK], graphene face masks [Medicevo Corporation, USA]), (h) electronic components and devices (graphene-enhanced earphones [ZOLO, India], headphones [Ora Graphene Audio, Canada], and earbuds [MediaDevil, UK], solid-state drives with graphene heat spreaders [TeamGroup, Taiwan], graphene-enhanced curved light bulbs and lamps [Graphene Lighting, UK], graphene FETs [BGT Materials, UK], graphene FET arrays [Graphenea, Spain], all-graphene optical communication links [modulator and detector] [Graphene Flagship], graphene-based photodetectors [Emberion, Finland], graphene sensors [Biolin Sientific, China]), and (i) different products (cement [Versarien, UK], tennis racquets and skis [HEAD, Austria], bicycles [Dassi, France], helmets [Momodesign, Italy], tires [Vittoria, Italy], graphene-enhanced fishing rods [G-Rods, USA], graphene bows [Win & Win Company, Korea], etc.).

At the beginning let us consider graphene layers as a coating and membrane. Graphene is composed of sp^2 hybridized carbon atoms with high electron density in the aromatic rings. The internuclear pore size considering the van der Waals radius of carbon is 0.064 nm [134]. The latter is smaller than the van der Waals radius of small atoms and molecules like helium (0.14 nm) and hydrogen (0.157 nm). As a result, pristine graphene can be considered as the thinnest coating or ultimate impermeable membrane for gases and liquids, which due to its high mechanical strength can withstand even when a 1 to 5 atm pressure difference is imposed across its atomic thickness at room temperature [5]. An impermeable membrane can become permeable after modifying it, i. e., opening artificial nanopores in it by top-down (e. g., hole drilling by high temperature oxidation, electron beam, UV irradiation, plasma bombardment) or bottom-up (e. g., self-assembly) methods [135]. The main function of a permeable membrane is to effectively separate molecules and ions from a mixture. The membranes can have two main structures: monolayer and stacked multilayer. Graphene membranes can be used for dialysis, filtration, and separation of gases and solvents [133], such as separating organic solvents from water, removing water from gas, removing carbon dioxide in the atmosphere, etc.

By combining flexibility and stretchability with its impermeability to all gases and liquids a graphene layer can be considered as a barrier layer or coating. Some of the existing hazardous anticorrosion materials (hexavalent chromium, cobalt, cadmium, etc.) of metal coatings have recently been fully restricted because of their carcinogenic risk and biocidal properties or classified as toxic [27]. When replacing them with chemically passive graphene layers, the latter can serve as a protective coating due to its good chemical resistance, impermeability to gases, adsorption capacity, antibacterial properties, mechanical strength, lubricity, and thermal stability [27]. MFL graphene acts as a protective coating against (a) water and oxygen, reactive gases, liquids, salts, acids, and microbes or (b) adsorption of toxicants, chemical contaminants, and hazardous gases (i. e., CO, NH_3) from the environment. For example, graphene coatings can stop the transfer of water and oxygen, thus keeping food, perishable goods, or pharmaceutical packaging fresh for longer. The methods for preparation of graphene protective coatings include CVD, high-temperature pyrolysis of organic molecules, rapid thermal annealing, electrophoretic deposition, powder spray (electrostatic powder coating, plasma spray coating), solution spray, dip coating, spin coating, drop casting, vacuum filtration, and brushing [27]. However, there is no unique coating technique that is effective for all applications [27]. For example, CVD-grown graphene is not suitable for corrosion resistance applications for long periods due to a high numbers of grain boundaries, folds, wrinkles, and point defects on the coated surface [27].

The most widespread graphene applications employ composites. Composite materials are formed by combining two or more materials with different properties to make a resulting material with unique characteristics. To obtain graphene-based nanocomposites (mixture of graphene with other materials, including metal nanoparticles, polymers, plastics, small molecules, etc.) graphene should be incorporated and then homogeneously distributed into them [83]. The resulting composite typically shows higher mechanical strength. Additionally, graphene–polymer composites are flexible and good electrical conductors, while ceramic–graphene composites, i. e., SiC-, Si_3N_4-, Al_2O_3-, BN-graphene, etc., enhance electrical properties, thermal conductivity, and refractory, mechanical, antifriction, anticorrosive, and biocompatible properties. Graphene-based composites can be applied in satellites, in the sun's environment as probes, in Li-ion batteries, sensors, supercapacitors, fuel cells, and solar cells, for photocatalysis and corrosion protection, in medical implants, in aerospace materials, etc. However, only some of the mentioned applications have started to see commercial realization, such as electrode materials for Li-ion batteries (anodes and cathodes) [136].

Some further graphene-based electronic devices, such as batteries, supercapacitors, solar cells, touch screen displays, bendable smartphones, sensors, light-emitting diodes (LEDs), and transistors are just beginning to appear. Let us consider a few of them.

The 2D nature of graphene becomes relevant for the enhancement of charge storage by using graphene batteries and supercapacitors, and for energy generation by using graphene-based solar cells, which will be discussed below. Batteries have long been the Achilles heel in energy storage as they are heavy, have long charging times (minutes), and lose charge capacity over time. As graphene is the thinnest material with the highest surface area-to-volume ratio, it potentially can store tremendous amounts of energy, reduce battery charging times (to seconds), prolong lifetime, and lower weight and waste. The first graphene-based batteries were demonstrated in 2014 [37, 137]. Next-generation batteries will move away from electrochemical cells (for example, lithium-ion) towards supercapacitors. A supercapacitor or electric double-layer capacitor stores energy in an electric field instead of in a controlled chemical reaction like batteries. Graphene supercapacitors are flexible, light, durable, consistent across a wider temperature range, and environmentally friendly (without using lithium). They provide high power, while using much less energy than conventional devices. As graphene supercapacitors are light, they can be charged in seconds and store more energy in a smaller space, and they can reduce the weight of smart phones, cars, airplanes, etc. Nowadays graphene supercapacitors are much more expensive than graphene batteries, and by decreasing their cost the batteries could be replaced by supercapacitors, revolutionizing this field.

The integration of 2D materials into optoelectronic devices, such as solar cells, is promising for energy generation, as they are hundreds of thousands of times thinner and lighter than silicon ones and absorb a significant amount of photons per atomic layer. For example, monolayer graphene has a broadband light absorption of 2.3 % below the energy of 3 eV that linearly scales with the number of atomic layers [60]. The main challenge for the durability of the standard solar cells is the exposure to different weather conditions due to which they become less efficient with time (though their lifetime is ~40 years) and as a result they have to be replaced. The combined properties of graphene and other materials could improve the efficiency of solar cells and favor the fabrication of flexible and rollable solar cells. The latter can be wrapped to fit any product, such as building roofs, automobile bodies, furniture, clothing, etc. Due to its transparency, mono- and bilayer graphene can be used as both an optically active medium and transparent conductive electrode or coating material in graphene-based solar cells. Graphene also can improve the performance of perovskite solar cells, the stability of which is still low due to their degradation in air and humidity upon continued exposure to sunlight and heat [138].

Many electronic devices rely on costly platinum- or indium tin oxide (ITO)-based conventional electrodes, while most portable devices with touch screen displays rely on ITO coatings. Although both ITO electrodes and coatings are popular, they have various drawbacks: they are brittle, not fully transparent (85 % of visible light, compared to 97.7 % for graphene), chemically unstable, and costly, they have a high refractive index, and indium has a limited availability. For example, nowadays plasma TVs and portable devices, such as smartphones and tablets, have thick, inflexible (ITO is layered on glass to protect them), and expensive (the most expensive component of the portable devices) ITO displays or screens. Furthermore, since graphene is mechanically strong, transparent, ultrathin, highly conductive, and flexible, it can be an alternative to conventional ITO electrodes and coatings.

Another futuristic application is smart windows. For example, at the flick of a switch (a) smart glass windows of buildings could block part or all sunlight, only heat, or both, helping control the temperature of buildings, (b) smart glass windows of cars could temporarily turn into a transparent display (for example, a car windscreen can light up to show you an urgent message), (c) transparent transistors on top of a contact lens could continuously monitor eye function (i. e., ocular pressure), etc.

2D materials are very sensitive to the environment and are ideal materials for sensors due to a large surface area-to-volume ratio and unique optical properties. Ultrasensitive 2D material sensors could detect dangerous compounds helping to protect the environment, reduce food wastage, prevent illnesses, protect crops, react to chemical warfare agents and explosives, etc.

Due to their outstanding properties, graphene and 2D materials are promising for biomedical applications, such as targeted drug and gene delivery, improved cancer treatment, biosensing, bioimaging, tissue or cell engineering, DNA sequencing, and "smart" implants. Currently the challenges are biodistribution, biocompatibility, and toxicity of 2D materials, which are either poorly studied or need to be tackled by functionalization. As functionalization improves the selectivity of graphene, graphene sensors can be used for biosensing, e. g., to detect biomolecules including glucose, cholesterol, hemoglobin, and DNA. They may be employed in food analysis, water testing, drug development, forensic analysis, medical diagnosis, environmental field monitoring, industrial process control, manufacturing of pharmaceuticals, etc. [139]. Bodily worn biosensors can collect data about the physical and chemical properties of the body and monitor day-to-day activity. Heat, humidity, salt, or pressure biosensors can be used in bed sheets to monitor a patient. Wearables act as extra accessories to the clothing and can be worn on the wrists, faces, ears, and feet. Wearables include smart glasses, smart watches, wristbands, lenses, smart fabrics, etc. For example, a graphene-based smart and wearable thermometer can continuously monitor the body temperature and be read out by a smartphone. Combined from the words "biology" and "electronic," bionic devices refer to devices that improve an organ or tissue, replacing organs or other body parts by mechanical ones. Bionic implants differ from simple prostheses by mimicking the original function very closely, or even surpassing it [140]. Bionic devices can be incorporated in living tissues or connected directly to neurons. Recent research demonstrates that CVD graphene material causes less inflammation, determined by the number of microglial tags, compared to biocompatible polymers. Therefore, graphene could be used in electrodes for retinal prostheses or even in any electronic devices for the registration and stimulation of the central nervous system. Graphene-based retina implants could allow blind people to see in the near future [141]. Graphene surfaces are more biocompatible compared to some traditional materials due to their flexibility and chemical durability, while toxicity mechanisms need further investigation since they strongly depend on the sheet size and functionalization. Thus, graphene-based medical devices may improve healthcare.

Silicon transistors have consistently become smaller and more powerful over the last few decades, following Moore's law. Subsequent reductions in the transistor scale will soon approach limits due to statistical and quantum effects and difficulty with heat dissipation [17]. Due to its outstanding properties graphene can be applied as an alternative to silicon. Graphene may allow microprocessors to produce very little heat and run at speeds several times faster than existing microprocessors [142]. Flexible ultrathin and ultrafast computers are still waiting to emerge. While graphene shows promise for transistors, due to the absence of a band-gap, graphene has a major drawback: it cannot switch the flow of electricity "off" like silicon. This means the current in graphene will flow constantly and it cannot serve as a transistor on its own. This main problem must be resolved before graphene can be employed as a transistor. There exist multiple approaches for opening and engineering the band-gap in graphene such as nanostructuring (formation of graphene nanoribbons, quantum dots, etc.), chemical doping and functionalization, applying a high electric field, changing the number of the atomic layers, and introducing buffer atomic layers between two atomic layers. However, these modifications in graphene structures (a) cause a change in the electronic structure and result in the degradation of the electron transport properties of graphene and (b) add complexity and diminish the high mobility of graphene. For example, after functionalization of graphene, the sp^2 structure can be converted to sp^3 by converting metallic graphene to an insulator. Anyway, after these modifications the graphene transistor can be switched "on" and "off," but the speed of the electrons is slowed down somewhat. Semiconducting 2D materials are particularly useful here. The first graphene [143] and MoS$_2$ [144] FETs, as well as graphene integrated circuits [145] with graphene FETs were demonstrated. Due to downscaling of the length of the device or conducting channel in graphene or MoS$_2$ FETs short-channel effects become significant [146], which is an important challenge for the electronic industry. With the extremely small size of such structures, nanospectroscopic measurements allow to study the channel length scaling, clarify whether the reduced thickness is beneficial or detrimental to the contacts, and finally offer new insights and solve nowadays existing major problems, such as self-heating, hot carrier effects, interfacial effects between 2D layers, and usage of various architectures and geometries of the channel, which affect the electrical performance of the FETs. Combinations of graphene (as a channel material) and TMDCs (as photosensitive material) allow the fabrication of phototransistors. Recently only one atom thick and 10 atoms wide graphene transistors have already been created. The smaller the transistor size, the better they perform within circuits [147]. Still, graphene technology is a long way from the reliability and reproducibility of the established silicon-based technology, and will require significant development.

5.4.5 Some challenges and solutions

New advancements in the synthesis and characterization of 2D atomic materials and heterostructures, as well as their application in next-generation advanced technologies, are intensely pursued. Future flexible electronics technology will evolve from rigid devices to bendable, rollable, and foldable ones [31]. These devices are expected to be advantageous compared to rigid ones due to their better durability, lighter weight, lower material requirements, and improved comfort [31]. Multiple nanofabrication possibilities allow producing a great variety of new generation 2D hybrid multilayer heterostructures and flexible devices [48]. Nevertheless, until now we have seen only few commercial devices based on 2D materials, mostly with graphene. The reason for this is that there are still a lot of technological challenges toward 2D material devices and applications. Each specific application has its own requirements from synthesis to final products, for instance, electronics, photonics, optoelectronics, twistronics, or solar cells require 2D material thin films with high crystal quality: thickness, defect concentration, grain size, etc.

Synthesis of 2D materials is the first critical step to build devices based on 2D materials. Despite many achievements in this area, there are still a lot of questions about scalability, quality, controlling the defects, and fundamental understanding of synthetic processes. Thickness control is critical for high-performance electronics. Nowadays, CVD, liquid exfoliation, and wet chemical synthesis are the most popular methods to synthesize 2D materials on a larger scale. Wafer-scale polycrystalline monolayer and multilayer graphene films have been successfully synthesized. Still, some major challenges in 2D material device fabrication remain: growing 2D materials (in large volumes or over large areas with a consistent quality), making Ohmic contacts (to minimize parasitic effects) and passivation layers, controlling the etching (to achieve sophisticated device structures), and improving the quality of the atomic layers and heterointerfaces of the heterostructure devices. Besides, now most research is focused on inorganic 2D materials (with graphene as a pure allotrope of carbon being considered as inorganic, too), while later on organic ones also should be considered. In order to achieve such results, it is also necessary to evolve existing spectroscopic methods and develop new ones for a more detailed investigation and characterization of 2D materials on the basis of which new devices will be built. These methods are capable of delivering information not only about the defects, electron–phonon coupling, number of layers, and contamination, but also about lattice orientation and edge types. Most significantly, nanoscale methods such as TERS and nano-IR can provide this information at the scale of the device channel and grain edges. Opportunities are plenty, but there will be many challenges ahead, in particular since some of the details of these methods are still poorly understood. Investigating the fundamentals of spectroscopic methods in application to 2D materials and improving spatial, temporal, and spectral resolution and sensitivity will be necessary as well as engineering new nanosystems for gaining basic knowledge of light–matter interaction and energy flow at the nanoscale. These small sample volumes also require extreme sensitivity. Finally, these approaches are experimentally complex, and the methods need to be developed further to make them user-friendly. 2D materials in particular are ideal targets for fundamental TERS studies since TERS overcomes two main drawbacks for conventional Raman spectroscopy, which are a low signal response and a limited spatial resolution. This in turn allows us to obtain much more detailed results for the investigation and characterization of 2D material properties.

5.4.6 Summary and impact

The twentieth century was the age of plastics, while the twenty-first century may become the age of graphene [11]. Maybe one day we will talk about "graphenes" as we now speak of "plastics." In the twentieth century, our life completely changed by replacing older materials, such as metal and wood, with plastics. New possibilities might lie ahead if graphene and other 2D materials lead us to ultralight, ultrathin, strong, optically transparent, electrically conducting, and semiconducting materials. To date, they show great promise due to the variety of properties and high-quality interface formation. Graphene and 2D materials may likely soon be present all around us in composites, sensors, batteries, electronics, and optoelectronics, if the technology can be optimized to a sufficient degree. Optical spectroscopies, and in particular nanospectroscopy, play an integral role in 2D material development and characterization, as well as in the novel device analysis. For example, Raman spectroscopy allows us to quickly, easily, and precisely characterize the obtained samples in terms of layer thickness, doping, impurities, defects, etc. In addition, TERS results demonstrate nanoscale detection of structural defects in 2D materials and offer detailed information on inhomogeneities as well as the local chemical composition of the material through the presence, possible shifts or splittings, and linewidths of the observed Raman peaks. PL spectroscopy allows confirming or excluding the presence of impurities or adsorbents in 2D materials and studying their electronic structure. As a result, the characterization data give us more details on the success or limitations of different synthesis methods, which we can use to improve the synthesis methods and device performance. Lots of commercial products based on graphene have already been developed and are available to purchase, while other devices based on the discussed 2D materials have great potential to be introduced in the near future, so we might see them on the market soon.

Bibliography

[1] Nicolosi V, Chhowalla M, Kanatzidis MG, Strano MS, Coleman JN. Liquid exfoliation of layered materials. Science. 2013;340:1226419.

[2] Madkour LH. Carbon nanomaterials and two-dimensional transition metal dichalcogenides (2D TMDCs). Nanoelectron Mater. 2019;165–245.

[3] Celasco E, Chaika AN, Stauber T, Zhang M, Ozkan C, Ozkan C, Ozkan U, Palys B, Harun SW, editors. Handbook of Graphene Set, I-VIII. Scrivener Publishing LLC, Wiley; 2019. p. 375–500.

[4] Di Bartolomeo A. Emerging 2D materials and their Van Der Waals Heterostructures. Nanomater (Basel). 2020:10.

[5] Miró P, Audiffred M, Heine T. An atlas of two-dimensional materials. Chem Soc Rev. 2014;43(18):6537.

[6] Kaul AB. Two-dimensional layered materials: structure, properties, and prospects for device applications. J Mater Res. 2014;29(3):348–61.

[7] Chacko L, Swetha AK, Anjana R, Jayaraj MK, Aneesh PM. Wasp-waisted magnetism in hydrothermally grown MoS_2 nanoflakes. Mater Res Express. 2016;3(11):116102.

[8] Vogel EM, Robinson JA. Two-dimensional layered transition-metal dichalcogenides for versatile properties and applications. Mater Res Soc Bull. 2015;40:558–63.

[9]   Gablech I et al. Monoelemental 2D materials-based field effect transistors for sensing and biosensing: phosphorene, antimonene, arsenene, silicene, and germanene go beyond graphene. Trends Anal Chem. 2018;105:251–62.

[10]   Vogt P, De Padova P, Quaresima C, Avila J, Frantzeskakis E, Asensio MC, Resta A, Ealet B, Lay GL. Silicene: compelling experimental evidence for graphenelike two-dimensional silicon. Phys Rev Lett. 2012;108(15):155501.

[11]   Chatterjee B. Wonder material of 21st Century, Part 1: Development of graphene – a unique material, 1–8, 2021.

[12]   Novoselov KS, Geim AK, Morozov SV, Jiang D, Zhang Y, Dubonos SV et al. Electric field effect in atomically thin carbon films. Science. 2004;306:666–9.

[13]   Novoselov KS, Fal'ko VI, Colombo L, Gellert PR, Schwab MG, Kim K. A roadmap for graphene. Nature. 2012;192–200.

[14]   Geim AK. Graphene: status and prospects. Science. 2009;324(5934):1530–4.

[15]   Roman RE, Pugno NM, Cranford SW. Mechanical characterization of 2D nanomaterials and composites. In: Silvestre N, editor. Advanced Computational Nanomechanics. 2016. p. 201–42.

[16]   Liang L, Zhang J, Sumpter BG, Tan Q, Tan P-H, Meunier V. Low-frequency shear and layer-breathing modes in Raman scattering of two-dimensional materials. ACS Nano. 2017;11(12):11777–802.

[17]   Wang QH, Kalantar-Zadeh K, Kis A, Coleman JN, Strano MS. Electronics and optoelectronics of two-dimensional transition metal dichalcogenides. Nat Nanotechnol. 2012;7:699–712.

[18]   Novoselov KS. Nobel lecture: graphene: materials in the flatland. Rev Mod Phys. 2011;83:837–49.

[19]   Soldano C, Mahmood A, Dujardin E. Production, properties and potential of graphene. Carbon. 2010;48:2127–50.

[20]   Peigney A, Laurent C, Flahaut E, Bacsa RR, Rousset A. Specific surface area of carbon nanotubes and bundles of carbon nanotubes. Carbon. 2001;39:507–14.

[21]   Castro Neto AH, Guinea F, Peres NMR, Novoselov KS, Geim AK. The electronic properties of graphene. Rev Mod Phys. 2009;81:109–62.

[22]   Hulman M. Raman spectroscopy of graphene. In: Graphene: Properties, Preparation, Characterisation and Applications, Woodhead Publishing Series in Electronic and Optical Materials. 2021. p. 381–411.

[23]   Electronic properties of graphene probed at the nanoscale (Chapter 14). In: Mikhailov S, editor. Physics and Applications of Graphene – Experiments. 2011. p. 353–78. https://doi.org/10.5772/590

[24]   Gupta A, Sakthivel T, Seal S. Recent development in 2D materials beyond graphene. Prog Mater Sci. 2015;73:44–126.

[25]   Wallace PR. Erratum: the band theory of graphite. Phys Rev. 1947;71(622):258.

[26]   Dubois SM-M, Zanolli Z, Declerck X, Charlier J-C. Electronic properties and quantum transport in graphene-based nanostructures. Eur Phys J B. 2009;72:1–24.

[27]   Nine MJ, Cole MA, Tran DNH, Losic D. Graphene: a multipurpose material for protective coatings. J Mater Chem A Mater Energy Sustain. 2015;3:12580–602.

[28]   Avouris P. Graphene: electronic and photonic properties and devices. Nano Lett. 2010;10:4285–94.

[29]   Mayorov AS, Gorbachev RV, Morozov SV, Britnell L, Jalil R, Ponomarenko LA et al. Micrometer-scale ballistic transport in encapsulated graphene at room temperature. Nano Lett. 2011:2396–9.

[30]   Balandin AA, Ghosh S, Bao W, Calizo I, Teweldebrhan D, Miao F et al. Superior thermal conductivity of single-layer graphene. Nano Lett. 2008;8:902–7.

[31]   Kim SJ, Choi K, Lee B, Kim Y, Hong BH. Materials for flexible, stretchable electronics: graphene and 2D materials. Annu Rev Mater Res. 2015;45(1):63–84.

[32]   Fonash S, Van de Voorde M. Engineering, Medicine and Science at the Nano-Scale. 2018. 296p.

[33]   Radadiya TM. A properties of graphene. Eur J Mater Sci. 2015;2(1):6–18.

[34]   Park I-J, Kim TI, Yoon T, Kang S, Cho H, Cho NS et al. Flexible and transparent graphene electrode architecture with selective defect decoration for organic light-emitting diodes. Adv Funct Mater. 2018;28:1704435.

[35]   Wang J, Liang M, Fang Y, Qiu T, Zhang J, Zhi L. Rod-coating: towards large-area fabrication of uniform reduced graphene oxide films for flexible touch screens. Adv Mater. 2012;24:2874–8.

[36]   Xia K, Wu W, Zhu M, Shen X, Yin Z, Wang H et al. CVD growth of perovskite/graphene films for high-performance flexible image sensor. Sci Bull (Beijing). 2020;65:343–9.

[37]   Kim H, Park K-Y, Hong J, Kang K. All-graphene-battery: bridging the gap between supercapacitors and lithium ion batteries. Sci Rep. 2014;4:5278.

[38]   Kostarelos K, Vincent M, Hebert C, Garrido JA. Graphene in the design and engineering of next-generation neural interfaces. Adv Mater. 2017;29.

[39]   Hong J-Y, Sohn E-H, Park S, Park HS. Highly-efficient and recyclable oil absorbing performance of functionalized graphene aerogel. Chem Eng J. 2015;269:229–35.

[40]   Piper JR, Fan S. Broadband absorption enhancement in solar cells with an atomically thin active layer. ACS Photonics. 2016;3(4):571–7.

[41]   Mao J, Wang Y, Zheng Z, Deng D. The rise of two-dimensional MoS_2 for catalysis. Front Phys. 2018;13.

[42]   Tsai M-Y, Tarasov A, Hesabi ZR, Taghinejad H, Campbell PM, Joiner CA et al. Flexible MoS_2 field-effect transistors for gate-tunable piezoresistive strain sensors. ACS Appl Mater Interfaces. 2015;7:12850–5.

[43]   Kloprogge JT, Ponce CP, Loomis TA. The Periodic Table: Nature's Building Blocks: An Introduction to the Naturally Occurring Elements, Their Origins and Their Uses. Elsevier; 2021. p. 633–883.

[44]   Hao G, Kou L, Lu D, Peng J, Li J, Tang C, Zhong J. Electrostatic properties of two-dimensional WSe2 nanostructures. J Appl Phys. 2016;119(3):035301.

[45]   Fang H, Chuang S, Chang TC, Takei K, Takahashi T, Javey A. High-performance single layered WSe_2 p-FETs with chemically doped contacts. Nano Lett. 2012;12:3788–92.

[46]   Wang J, Ma F, Liang W, Wang R, Sun M. Optical, photonic and optoelectronic properties of graphene, h-BN and their hybrid materials. Nanophotonics. 2017;6:943–76.

[47]   Ko PJ, Abderrahmane A, Takamura T, Kim N-H, Sandhu A. Thickness dependence on the optoelectronic properties of multilayered GaSe based photodetector. Nanotechnology. 2016;27:325202.

[48]   Shmavonyan GSh, Vázquez-Vázquez C, López-Quintela MA. Single-step rubbing method for mass production of large-size and defect-free 2D materials. Transl Mater Res. 2017;4:025001.

[49]   Novoselov KS, Jiang D, Schedin F, Booth TJ, Khotkevich VV, Morozov SV et al. Two-dimensional atomic crystals. Proc Natl Acad Sci USA. 2005;102:10451–3.

[50]   Jayasena B, Subbiah S. A novel mechanical cleavage method for synthesizing few-layer graphenes. Nanoscale Res Lett. 2011;6:95.

[51]   Lotya M, King PJ, Khan U, De S, Coleman JN. High-concentration, surfactant-stabilized graphene dispersions. ACS Nano. 2010;4:3155–62.

[52]   Hernandez Y, Nicolosi V, Lotya M, Blighe FM, Sun Z, De S et al. High-yield production of graphene by liquid-phase exfoliation of graphite. Nat Nanotechnol. 2008;3:563–8.

[53]   Park S, Ruoff RS. Chemical methods for the production of graphenes. Nat Nanotechnol. 2009;4:217–24.

[54] Hofmann M, Chiang W-Y, Nguyễn TD, Hsieh Y-P. Controlling the properties of graphene produced by electrochemical exfoliation. Nanotechnology. 2015;26:335607.

[55] Zhu Y, Murali S, Stoller MD, Velamakanni A, Piner RD, Ruoff RS. Microwave assisted exfoliation and reduction of graphite oxide for ultracapacitors. Carbon. 2010;48:2118–22.

[56] Li X, Cai W, An J, Kim S, Nah J, Yang D et al. Large-area synthesis of high-quality and uniform graphene films on copper foils. Science. 2009;324:1312–4.

[57] Chae SJ, Güneş F, Kim KK, Kim ES, Han GH, Kim SM et al. Synthesis of large-area graphene layers on poly-nickel substrate by chemical vapor deposition: wrinkle formation. Adv Mater. 2009;21:2328–33.

[58] Emtsev KV, Bostwick A, Horn K, Jobst J, Kellogg GL, Ley L et al. Towards wafer-size graphene layers by atmospheric pressure graphitization of silicon carbide. Nat Mater. 2009;8:203–7.

[59] Keith E, Whitener JR, Sheehan PE. Graphene synthesis. Diam Relat Mater. 2014;46:25–34.

[60] Schwierz F, editor. Two-Dimensional Electronics – Prospects and Challenges. Printed Edition of the Special Issue Published in Electronics. MDPI AG; 2016. p. 1–250.

[61] Moreau E, Ferrer FJ, Vignaud D, Godey S, Wallart X. Graphene growth by molecular beam epitaxy using a solid carbon source. Phys Status Solidi. 2010;207:300–3.

[62] Hernández-Rodríguez I, García JM, Martín-Gago JA, de Andrés PL, Méndez J. Graphene growth on Pt(111) and Au(111) using a MBE carbon solid-source. Diam Relat Mater. 2015;57:58–62.

[63] Janowska I, Vigneron F, Bégin D, Ersen O, Bernhardt P, Romero T et al. Mechanical thinning to make few-layer graphene from pencil lead. Carbon. 2012;50:3106–10.

[64] Paton KR, Varrla E, Backes C, Smith RJ, Khan U, O'Neill A et al. Scalable production of large quantities of defect-free few-layer graphene by shear exfoliation in liquids. Nat Mater. 2014;13:624–30.

[65] Liu X, Zheng M, Xiao K, Xiao Y, He C, Dong H et al. Simple, green and high-yield production of single- or few-layer graphene by hydrothermal exfoliation of graphite. Nanoscale. 2014;6:4598–603.

[66] Zhao W, Fang M, Wu F, Wu H, Wang L, Chen G. Preparation of graphene by exfoliation of graphite using wet ball milling. J Mater Chem. 2010;20:5817–9.

[67] Bae S, Kim H, Lee Y, Xu X, Park J-S, Zheng Y et al. Roll-to-roll production of 30-inch graphene films for transparent electrodes. Nat Nanotechnol. 2010;5:574–8.

[68] Subrahmanyam KS, Panchakarla LS, Govindaraj A, Rao CNR. Simple method of preparing graphene flakes by an arc-discharge method. J Phys Chem C Nanomater Interfaces. 2009;113:4257–9.

[69] Haigh SJ, Gholinia A, Jalil R, Romani S, Britnell L, Elias DC et al. Cross-sectional imaging of individual layers and buried interfaces of graphene-based heterostructures and superlattices. Nat Mater. 2012;11:764–7.

[70] Pan D, Zhang J, Li Z, Wu M. Hydrothermal route for cutting graphene sheets Into blue-luminescent graphene quantum dots. Adv Mater. 2010;22:734–8.

[71] López-Quintela MA, Shmavonyan GSh, Vázquez Vázquez C. Method for producing sheets for graphene, Patent No: ES2575711 B2, November 3, 2016. EP3246286 A1.

[72] López-Quintela MA, Shmavonyan GSh, Vázquez Vázquez C. Method for producing sheets of graphene, USPTO patent US 10,968,104 B2, April 6, 2021.

[73] Shmavonyan G. Unique nanostripes and sheets of 2D atomic materials obtained by substrates rubbing technology. In: The 8th Annual International Conference on Chemistry, July 20–23, 2020, Athens, Greece.

[74] Shmavonyan G, Vázquez-Vázquez C, López-Quintela MA. Technology for mass production of 2D atomic materials. In: EuroSciCon Conference on Graphene and Carbon Nanotechnology, Nanoscience and Graphene Nanotechnology, November 25–27, 2019, Tokyo, Japan.

[75] Shmavonyan GSh, Mailian A, Mailian R. Anisotropy of carrier mobility in multilayered graphite structures obtained by rubbing. In: The 10th International Conference on Nanosciences and Nanotechnologies, July 9–12, 2013, Thessaloniki, Greece.

[76] Shmavonyan GSh, Mailian AR, Mailyan MR, López-Quintela MA. Obtaining graphene on paper from pencil drawn lines. European Conference on the Synthesis, Characterization and Applications of Graphene, February 18–21, 2014, Lanzarote, Canary Island, Spain.

[77] Wang X, You H, Liu F, Li M, Wan L, Li S et al. Large-scale synthesis of few-layered graphene using CVD. Chem Vap Depos. 2009;15:53–6.

[78] Geim AK, Novoselov KS. The rise of graphene. Nat Mater. 2007183–91.

[79] Bhuyan MSA, Uddin MN, Islam MM, Bipasha FA, Hossain SS. Synthesis of graphene. Int Nano Lett. 2016;6:65–83.

[80] Zhao Q, Puebla S, Zhang W, Wang T, Frisenda R, Castellanos-Gomez A. Thickness identification of thin InSe by optical microscopy methods. Adv Photon Res. 2020;1:2000025.

[81] Wang YY et al. Thickness identification of two-dimensional materials by optical imaging. Nanotechnology. 2012;23:495713.

[82] Han B. Deep-learning-enabled fast optical identification and characterization of 2D materials. Adv Mater. 2020;32(29):2000953.

[83] Butler SZ, Hollen SM, Cao L, Cui Y et al. Progress, challenges, and opportunities in two-dimensional materials beyond graphene. ACS Nano. 2013;7(4):2898–926.

[84] Shmavonyan GS, Sevoyan GG, Aroutiounian VM. Enlarging the surface area of monolayer graphene synthesized by mechanical exfoliation. Armen J Phys. 2013;6:1–6.

[85] Koh YK. Heat transport by phonons in crystalline materials and nanostructures. Ph.D. Thesis. University of Illinois at Urbana-Champaign; 2010.

[86] Basko DM. New J Phys. 2009;11:095011.

[87] Hasan T et al. Nanotube and graphene polymer composites for photonics and optoelectronics. In: Hayden O, Nielsch K, editors. Molecular- and Nano-Tubes. Boston, MA: Springer; 2011.

[88] Javed H, Pani S, Antony J, Sakthivel M, Drillet J-F. Synthesis of mesoporous carbon spheres via a soft-template route for catalyst supports in PEMFC cathodes. Soft Matter. 2021;17:7743–54.

[89] Zhang Y-L, Guo L, Xia H, Chen Q-D, Feng J, Sun H-B. Photoreduction of graphene oxides: methods, properties, and applications. Adv Opt Mater. 2014;2(1):10–28.

[90] Peña-Álvarez M, del Corro E, Baonza VG, Taravillo M. Probing the stress effect on the electronic structure of graphite by resonant Raman spectroscopy. J Phys Chem C. 2014;118(43):25132–40.

[91] Beyssac O, Lazzeri M. Application of Raman spectroscopy to the study of graphitic carbons in the Earth sciences. EMU Notes Mineral. 2012;12(1):415–54.

[92] Kuila T, Bose S, Mishra AK, Khanra P, Kim NH, Lee JH. Chemical functionalization of graphene and its applications. Prog Mater Sci. 2012;57(7):1061–105.

[93] Kuila T, Sheng Y, Murmu NC. Graphene/conjugated polymer nanocomposites for optoelectronic and biological applications. In: Fundamentals of Conjugated Polymer Blends, Copolymers and Composites: Synthesis, Properties and Applications. 2015. p. 229–79.

[94] Shvets V. Polymer masks for nanostructuring of graphene. PhD Thesis. DTU Nanotech; 2015.

[95] Bruna M, Ott AK, Ijäs M, Yoon D, Sassi U, Ferrari AC. Doping dependence of the Raman spectrum of defected graphene. ACS Nano. 2014;8:7432–41.

[96] Gadelha AC, Vasconcelos TL, Cançado LG, Jorio A. Nano-optical imaging of in-plane homojunctions in graphene and MoS_2 van der Waals heterostructures on Talc and SiO_2. J Phys Chem Lett. 2021:7625–31.

[97] Malard L, Lafetá L, Cunha R, Nadas R, Gadelha A, Cancado G, Jorio A. Studying 2D materials with advanced Raman spectroscopy: CARS, SRS and TERS. Phys Chem Chem Phys. 2021;23:23428–44.

[98] Lee J, Crampton KT, Tallarida N, Apkarian VA. Visualizing vibrational normal modes of a single molecule with atomically confined light. Nature. 2019;568:78–82.

[99] Li H, Zhang Q, Yap CCR, Tay BK, Edwin THT, Olivier A, Baillargeat D. From bulk to monolayer MoS2: evolution of Raman scattering. Adv Funct Mater. 2012;22(7):1385–90.

[100] Lee C, Yan H, Brus LE, Heinz TF, Hone J, Ryu S. Anomalous lattice vibrations of single- and few-layer MoS$_2$. ACS Nano. 2010;4:2695–700.

[101] Lee J-U, Kim M, Cheong H. Raman spectroscopic studies on two-dimensional materials. Appl Microsc. 2015;45(3):126–30.

[102] Zhang X, Han WP, Wu JB, Milana S, Lu Y, Li QQ et al. Raman spectroscopy of shear and layer breathing modes in multilayer MoS$_2$. Phys Rev B, Condens Matter Mater Phys. 2013;87.

[103] Rodriguez RD, Müller S, Sheremet E, Zahn DRT, Villabona A, Lopez-Rivera SA et al. Selective Raman modes and strong photoluminescence of gallium selenide flakes on sp2 carbon. J Vac Sci Technol, B Nanotechnol Microelectron: Materials, Processing, Measurement, Phenomena. 2014:04E106.

[104] Wu Q, Tagani MB, Zhang L, Wang J, Xia Y et al. Electronic tuning in WSe$_2$/Au via van der Waals interface twisting and intercalation. ACS Nano. 2022;16(4):6541–51.

[105] Iqbal MW, Shahzad K, Akbar R, Hussain G. A review on Raman finger prints of doping and strain effect in TMDCs. Microelectron Eng. 2020:111152.

[106] Chakraborty B, Bera A, Muthu DVS, Bhowmick S, Waghmare UV, Sood AK. Symmetry-dependent phonon renormalization in monolayer MoS$_2$ transistor. Phys Rev B. 2012.

[107] Khalil HMW, Khan MF, Eom J, Noh H. Highly stable and tunable chemical doping of multilayer WS$_2$ field effect transistor: reduction in contact resistance. ACS Appl Mater Interfaces. 2015;7:23589–96.

[108] Peimyoo N, Yang W, Shang J, Shen X, Wang Y, Yu T. Chemically driven tunable light emission of charged and neutral excitons in monolayer WS$_2$. ACS Nano. 2014;8:11320–9.

[109] Özden A, Şar H, Yeltik A, Madenoğlu B, Sevik C, Ay F et al. CVD grown 2D MoS$_2$ layers: a photoluminescence and fluorescence lifetime imaging study. Phys Status Solidi RRL. 2016;10:792–6.

[110] Xu L, Zhao L, Wang Y, Zou M, Zhang Q, Cao A. Analysis of photoluminescence behavior of high-quality single-layer MoS$_2$. Nano Res. 2019;12:1619–24.

[111] Su W, Roy D. Visualizing graphene edges using tip-enhanced Raman spectroscopy. J Vac Sci Technol, B Nanotechnol Microelectron. 2013;31:041808.

[112] Kato R, Umakoshi T, Sam RT, Verma P. Probing nanoscale defects and wrinkles in MoS$_2$ by tip-enhanced Raman spectroscopic imaging. Appl Phys Lett. 2019;114:073105.

[113] Smithe KKH, Krayev AV, Bailey CS, Lee HR, Yalon E, Aslan ÖB et al. Nanoscale heterogeneities in monolayer MoSe$_2$ revealed by correlated scanning probe microscopy and tip-enhanced Raman spectroscopy. ACS Appl Nano Mater. 2018;1:572–9.

[114] Jit S, Das S, editors. 2D Nanoscale Heterostructured Materials – Synthesis, Properties, and Applications. Elsevier; 2020. https://www.elsevier.com/books/2d-nanoscale-heterostructured-materials/jit/978-0-12-817678-8.

[115] The Nobel Prize in Physics 2000. NobelPrize.org. Nobel Prize Outreach AB 2022. Sun. 17 Jul 2022.

[116] Zhang T, Fu L. Controllable chemical vapor deposition growth of two-dimensional heterostructures. Chem. 2018;4(4):671–89.

[117] Pant A, Mutlu Z, Wickramaratne D, Ca H, Lake RK, Ozkan CS, Tongay S. Fundamentals of lateral and vertical heterojunctions of atomically thin materials. Nanoscale. 2016;8:3870–87.

[118] Gao T, Song X, Du H et al. Temperature-triggered chemical switching growth of in-plane and vertically stacked graphene-boron nitride heterostructures. Nat Commun. 2015;6:6835.

[119] Novoselov KS, Mishchenko A, Carvalho A, Castro Neto AH. 2D materials and van der Waals heterostructures. Science. 2016;353:aac9439.

[120] Pospischil A, Furchi MM, Mueller T. Solar-energy conversion and light emission in an atomic monolayer p–n diode. Nat Nanotechnol. 2014;9:257–61.

[121] Fontana M, Deppe T, Boyd AK, Rinzan M, Liu AY, Paranjape M et al. Electron-hole transport and photovoltaic effect in gated MoS2 Schottky junctions. Sci Rep. 2013;3:1–6.
[122] Xu Z-Q, Zhang Y, Wang Z, Shen Y, Huang W, Xia X et al. Atomically thin lateral p–n junction photodetector with large effective detection area. 2D Mater. 2016041001.
[123] Moon JS, Seo H-C, Stratan F, Antcliffe M, Schmitz A, Ross RS et al. Lateral graphene heterostructure field-effect transistor. IEEE Electron Device Lett. 2013;34:1190–2.
[124] Wang H, Yu L, Lee Y-H, Shi Y, Hsu A, Chin ML et al. Integrated circuits based on bilayer MoS$_2$ transistors. Nano Lett. 2012;12:4674–80.
[125] Geim AK, Grigorieva IV. Van der Waals heterostructures. Nature. 2013;499:419–25.
[126] Kai N, Xiaocha W, Wenbo M. Spin-dependent electronic structure and magnetic anisotropy of two-dimensional SnO/Fe$_4$N heterostructures. J Phys Chem C. 2019. https://doi.org/acs.jpcc.9b06896.
[127] Yuan L, Ge J, Peng X, Zhang Q, Wu Z, Jian Y et al. A reliable way of mechanical exfoliation of large scale two dimensional materials with high quality. AIP Adv. 2016;6:125201.
[128] Kroemer H. Quasi-electric and quasi-magnetic fields in nonuniform semiconductors. RCA Rev. 1957;18:332–42.
[129] Hu X, Kou L, Sun L. Stacking orders induced direct band gap in bilayer MoSe$_2$-WSe$_2$ lateral heterostructures. Sci Rep. 2016;6:31122.
[130] Chen Y, Fan Z, Zhang Z, Niu W, Li C, Yang N et al. Two-dimensional metal nanomaterials: synthesis, properties, and applications. Chem Rev. 2018;118:6409–55.
[131] Xue Y, Zhang Y, Liu Y, Liu H, Song J, Sophia J et al. Scalable production of a few-layer MoS$_2$/WS$_2$ vertical heterojunction array and its application for photodetectors. ACS Nano. 2016;10:573–80.
[132] Xu H, Wu J, Feng Q, Mao N, Wang C, Zhang J. High responsivity and gate tunable graphene-MoS$_2$ hybrid phototransistor. Small. 2014;10:2300–6.
[133] Xiong XX, Kang J, Hu Q, Gu C, Gao T et al. Reconfigurable logic-in-memory and multilingual artificial synapses based on 2D heterostructures. Adv Funct Mater. 20201909645.
[134] Berry V. Impermeability of graphene and its applications. Carbon. 2013;62:1–10.
[135] Bernardo P, Drioli E. Membrane technology. In: Drioli E, Giorno L, Fontananova E, Editors-in-Chief. Comprehensive Membrane Science and Engineering. 2nd ed. Elsevier Science; 2017. p. 164–88.
[136] Cai X, Lai L, Shen Z, Lin J. Graphene and graphene-based composites as Li-ion battery electrode materials and their application in full cells. J Mater Chem A Mater Energy Sustain. 2017;5:15423–46.
[137] Son IH, Park JH, Park S, Park K, Han S, Shin J et al. Graphene balls for lithium rechargeable batteries with fast charging and high volumetric energy densities. Nat Commun. 2017;8:1561.
[138] Acik M, Darling SB. Graphene in Perovskite solar cells: device design, characterization and implementation. J Mater Chem A Mater Energy Sustain. 2016;4(17):6185–235.
[139] Justino CIL, Gomes AR, Freitas AC, Duarte AC, Rocha-Santos TAP. Graphene based sensors and biosensors. Trends Analyt Chem. 2017;91:53–66.
[140] Warnke PH. Orthopaedic tissue engineering. In: Sivananthan S, Sherry E, Warnke P, Miller M, editors. Mercer's Textbook of Orthopaedics and Trauma. Tenth edition. London: CRC Press; 2012.
[141] Nguyen D, Valet M, Dégardin J, Boucherit L, Illa X, de la Cruz J et al. Novel graphene electrode for retinal implants: an biocompatibility study. Front Neurosci. 2021;15:615256.
[142] Kim K, Choi J-Y, Kim T, Cho S-H, Chung H-J. A role for graphene in silicon-based semiconductor devices. Nature. 2011;479:338–44.
[143] Lin Y-M, Jenkins KA, Valdes-Garcia A, Small JP, Farmer DB, Avouris P. Operation of graphene transistors at gigahertz frequencies. Nano Lett. 2009;9:422–6.

[144] Nourbakhsh A, Zubair A, Sajjad RN, Tavakkoli KGA, Chen W, Fang S et al. MoS$_2$ field-effect transistor with sub-10 nm channel length. Nano Lett. 2016;16:7798–806.

[145] Lin Y-M, Valdes-Garcia A, Han S-J, Farmer DB, Meric I, Sun Y et al. Wafer-scale graphene integrated circuit. Science. 2011;332:1294–7.

[146] Liu H, Neal AT, Ye PD. Channel length scaling of MoS$_2$ MOSFETs. ACS Nano. 2012;6:8563–9.

[147] Chahardeh JB. A review on graphene transistors. Int J Adv Res Comput Commun Eng. 2012;1.

Index

https://doi.org/10.1515/9783110442908-019

www.ingramcontent.com/pod-product-compliance
Lightning Source LLC
Chambersburg PA
CBHW061926190326
41458CB00009B/2664